KU-301-958

633.85

| SCOTTISH CROP RESEARCH |
| INSTITUTE LIBRARY |
| 2 1 FEB 1996 |
| Accession No. AKTJ |
| Class No. 633.85 K |

CANCELLED

Brassica Oilseeds

Production and Utilization

Brassica Oilseeds

Production and Utilization

Edited by

D.S. Kimber

Formerly of the National Institute of Agricultural Botany, Cambridge, UK

and

D.I. McGregor

*Agriculture and Agri-Food
Canada Saskatoon
Research Centre, Canada*

CAB INTERNATIONAL

CAB INTERNATIONAL
Wallingford
Oxon OX10 8DE
UK

Tel: +44 (0)1491 832111
Telex: 847964 (COMAGG G)
E-mail: cabi@cabi.org
Fax: +44 (0)1491 833508

©CAB INTERNATIONAL 1995. All rights reserved. No part of this publication may be reproduced in any form or by any means, electronically, mechanically, by photocopying, recording or otherwise, without the prior permission of the copyright owners.

A catalogue record for this book is available from the British Library.

ISBN 0 85198 960 8

Typeset in $10\frac{1}{2}$/12pt Baskerville by Colset Pte Ltd, Singapore
Printed and bound in the UK at the University Press, Cambridge

Contents

Contributors   vii

Preface   ix

**1   The Species and Their Origin, Cultivation and World
     Production**   1
     D.S. Kimber and D.I. McGregor

PART I: THE FIELD CROP   9

  2   Physiology: Crop Development, Growth and Yield   11
      N.J. Mendham and P.A. Salisbury

  3   Agronomy   65
      A. Pouzet

  4   Weeds and Their Control   93
      J. Orson

  5   Diseases   111
      S.R. Rimmer and L. Buchwaldt

  6   Insect Pests   141
      B. Ekbom

  7   Plant Breeding   153
      G.C. Buzza

  8   Biotechnology   177
      D.J. Murphy and R. Mithen

v

9   **Environmental Impact of Rapeseed Production**   195
    R. Marquard and K.C. Walker

PART II: PROCESSING AND UTILIZATION   215

10 **Seed Chemistry**   217
    B. Uppström

11 **Seed Analysis**   243
    J.K. Daun

12 **Processing the Seed and Oil**   267
    R.A. Carr

13 **Oil Properties of Importance in Human Nutrition**   291
    B.E. McDonald

14 **Meal and By-product Utilization in Animal Nutrition**   301
    J.M. Bell

15 **Industrial Utilization of Long-chain Fatty Acids and Their
    Derivatives**   339
    N.O.V. Sonntag

16 **Utilization of Oil as a Biodiesel Fuel**   353
    W. Körbitz

17 **The Mustard Species: Condiment and Food Ingredient Use
    and Potential as Oilseed Crops**   373
    J.S. Hemingway

Index   385

Contributors

J.M. BELL
Department of Animal and Poultry
  Science,
University of Saskatchewan,
Saskatoon, Saskatchewan,
Canada S7N 0W0.

L. BUCHWALDT
Department of Plant Science,
University of Manitoba,
Winnipeg, Manitoba,
Canada R3T 2N2.

G.C. BUZZA
Pacific Seeds Pty Ltd,
268 Anzac Avenue,
PO Box 337, Toowoomba,
Queensland 4350, Australia.

R.A. CARR
POS Pilot Plant Corp.,
118 Veterinary Road,
Saskatoon, Saskatchewan,
Canada S7N 2R4.

J.K. DAUN
Canadian Grain Commission,
Grain Research Laboratory,
1404-303 Main Street,

Winnipeg, Manitoba,
Canada R3C 3G8.

B. EKBOM
Department of Entomology,
Swedish University of Agricultural
  Sciences,
PO Box 7044,
S-750-07 Uppsala, Sweden,

J.S. HEMINGWAY
18 Postwick Lane,
Brundall, Norwich,
Norfolk NR13 5LR, UK.

D.S. KIMBER
44 Church Street,
Haslingfield,
Cambridge CB3 7JE, UK.

W. KÖRBITZ
Graben 14/3,
PO Box 97,
Vienna A1014, Austria.

B.E. MCDONALD
Department of Food and Nutrition,
University of Manitoba,
Winnipeg,
Canada R3T 2N2.

D.I. McGregor
Agriculture and Agri-Food Canada
Saskatoon Research Centre,
107 Science Place,
Saskatoon, Saskatchewan,
Canada S7N 0X2.

R. Marquard
Institut für Pflanzenbau und
    Pflanzenzüchtung,
Ludwigstrasse 23,
D-6300 Giessen, Germany.

N.J. Mendham
Department of Agricultural Science,
University of Tasmania,
GPO Box 252C, Hobart,
Tasmania 7001, Australia.

R. Mithen
John Innes Centre,
Colney,
Norwich NR4 7UH, UK.

D.J. Murphy
John Innes Centre,
Colney,
Norwich NR4 7UH, UK.

J. Orson
ADAS Cambridge,
Brooklands Avenue,
Cambridge CB2 2BL, UK.

A. Pouzet
CETIOM,
174 avenue Victor Hugo,
75116 Paris, France.

S.R. Rimmer
Department of Plant Science,
University of Manitoba,
Winnipeg, Manitoba,
Canada R3T 2N2.

P.A. Salisbury
Victoria Institute of Dryland
    Agriculture,
Private Bag 260, Horsham,
Victoria 3400, Australia.

N.O.V. Sonntag
306 Shadow Wood Trail,
Ovilla,
Texas 75154, USA.

B. Uppström
Svalöf Weibull AB,
S-268-00, Svalöf,
Sweden.

K.C. Walker
Scottish College of Agriculture,
581 King Street,
Aberdeen AB9 1UD, UK.

Preface

Over the past couple of decades *Brassica* oilseed production has increased to become one of the most important world sources of vegetable oil. Improvements in the quality of rapeseed oil and meal have culminated in recognition of the oil as a nutritionally superior edible oil, and the meal as an important source of protein for animal feeds. Amenability of the *Brassica* genus to biotechnology holds promise for further increases in yield and quality and for *Brassica* oilseed crops to become an important source of chemicals including industrial enzymes and pharmaceuticals.

Brassica oilseeds have been subjected to extensive scientific research and development over the last half century. This book presents a comprehensive review of results that have led to advancements in their production, processing and utilization.

It begins with the origin and inter-relationships of the species, which are now being intercrossed with each other and with related genera to increase germplasm diversity. The first section on production opens with a chapter on physiology, which considers how crops grow, develop and interact with the environment. Then agronomy addresses the range of climates, soils and economic conditions under which *Brassica* oilseeds are produced. This is followed by chapters on the weeds, diseases and insect pests which limit production. Strategies and methods of germplasm improvement are examined in the chapters on plant breeding and biotechnology. The environmental impact of *Brassica* oilseed production rounds off the section on production.

Since *Brassica* oilseed crops are grown primarily for their seed, the second section on processing and utilization opens with consideration of the chemical composition of the seed and methods for analysis of the constituents. Advances in the technology of seed processing to produce oil and meal of superior quality are presented followed by consideration of oil quality for human nutrition and meal quality for animal nutrition. In addition to edible uses the potential for utilization of the oil in the developing biodiesel market and of long-chain fatty acids and their derivatives by the oleochemical industry is reviewed. The book concludes with consideration of the mustards which, although primarily grown for condiment purposes, are important sources of oil in the Indian subcontinent, the former USSR and China. Because of their superior drought resistance mustards also hold promise as oilseed crops for semiarid regions of the world.

The editors are indebted to the contributors not only for their submissions but also for their help and advice in preparing the book. We would also like to record our appreciation for assistance from the following: in Canada – K. Kirkland, Agriculture and Agri-Food Canada, Scott Research Centre; and A.G. Thomas, Agriculture and Agri-Food Canada, Saskatoon Research Centre; and in the UK – M. Askew and P. Bowerman, ADAS, Cambridge; J. Bowman, Nickerson Seeds, Lincoln; W. Hollands, BEOCO, Liverpool; G. Hughes, PBIC, Cambridge; J. Law, A. Morgan and J. Thomas, NIAB, Cambridge; J. Ward, King's Lynn; and V. Wheelock, University of Bradford.

1 The Species and Their Origin, Cultivation and World Production

D.S. Kimber[1] and D.I. McGregor[2]

[1]*Formerly of the National Institute of Agricultural Botany, Cambridge;* [2]*Agriculture and Agri-Food Canada Saskatoon Research Centre*

THE OILSEED SPECIES

The ability of *Brassica* seeds to germinate and the plants to thrive at low temperatures have made *Brassica* oilseeds one of the few edible oil crops that can be cultivated in the temperate agricultural zones of the world, at high elevations and, as winter crops, under relatively cool growing conditions. The small spherical seed normally contains over 40% oil, while the meal residue after oil extraction contains 36 to 44% protein. Five related species are cultivated worldwide as a source of vegetable oil.

Brassica napus L., AC genomes; n = 19

Brassica napus is the rape or rapeseed commonly grown in Europe and in Canada, where it was formerly known as Argentine rape after the origin of its introduction. It is swede rape, or the non-bulbing form of the true swede or rutabaga. Both spring and winter cultivars are grown as sources of vegetable oil, the winter cultivars usually being the more productive where conditions are favourable. Most *Brassica* oilseed cultivation in Europe and China is of

winter cultivars. These are also grown in eastern Canada and in recent years in parts of the United States, notably the northwest. However, cultivation gives way to spring cultivars as latitudes and altitudes rise and the chances of survival of winter crops are reduced. All crops in western Canada are essentially of spring cultivars.

The seed is dark in colour and no natural yellow-seeded forms are known. However, development of yellow-seeded cultivars is an objective of some plant breeding programmes. Yellow seed colour is thought to be associated with lower tannin content and with a thinner seed coat, offering potential for proportionally more oil and protein in the seed and less fibre in the meal.

Brassica rapa L., A genome; n = 10

Brassica rapa, formerly *Brassica campestris* L., is known as turnip rape and, in Canada, as Polish rape after its country of origin. It is the non-bulbing form of the true turnip. Both spring and winter cultivars are grown as a source of seed. The most cold-hardy cultivars of the *Brassica* oilseeds belong to this species, which has a relatively high growth rate under low temperatures. About half the rapeseed crop grown in western Canada is *B. rapa* because of its earlier maturity at more northerly latitudes. Both brown- and yellow-seeded *B. rapa* cultivars are grown in Canada.

Three ecotypes of *B. rapa* are grown on the Indian subcontinent, yellow sarson, brown sarson and toria. Brown sarson is the prevalent crop, while toria is confined to the foothills of the Himalayas and yellow sarson to areas in certain states (Prakash, 1980).

Brassica juncea L. Czern. and Coss., AB genomes; n = 18

Brassica juncea is distinguished by the colour of its seed. The brown-seeded cultivars are known as brown mustard, while the yellow-seeded cultivars are referred to as yellow or Oriental mustard. It is well adapted to drier conditions and matures relatively quickly. *Brassica juncea* is grown widely as an oilseed crop in the north of the Indian subcontinent and in various parts of China, where variations of the cultivated forms abound. It has been mainly grown for condiment purposes in western Canada but has considerable potential as an oilseed crop (Woods *et al.*, 1991), notably in Australia.

Brassica carinata Braun, BC genomes; n = 17

Brassica carinata is relatively slow growing. Cultivation is limited to the Ethiopian plateau and adjacent areas in east Africa. The seed is large and predominantly dark, although there are yellow-seeded forms and *B. carinata* is being investigated as a source of large yellow seeds.

Sinapis alba L., D genome; n = 12

Sinapis alba, often referred to as (*Brassica hirta* Moench), is known as white mustard in Europe and as yellow mustard in North America. It is grown extensively in these countries as a source of condiment. The seed is large and of a light yellow colour and is being investigated by plant breeders as a source of these traits.

GENOMIC RELATIONSHIP OF THE SPECIES

The botanical relationship between the *Brassica* oilseed species was established as a result of taxonomic studies carried out in the 1930s (Morinaga, 1934; U, 1935) (Fig. 1.1) although recent studies using biotechnology indicate that some revision of these species and related genera may be necessary (Song and Osborn, 1991; Warwick and Black, 1993). The three species with higher chromosome numbers, *B. napus*, *B. juncea* and *B. carinata*, are amphidiploids derived from the diploid species, *Brassica nigra* (L.) Koch, *B. rapa* and *Brassica oleracea* L. These relationships have been confirmed by the artificial synthesis of *B. napus* (U, 1935), *B. rapa* and *B. oleracea* (Downey *et al.*, 1975; Olsson and Ellerstrom, 1980).

Chromosome analysis has further suggested an early evolution from a common progenitor species with a basic chromosome number of $n = 6$, and that the diploid *Brassica* species, with $n = 8$, 9 and 10, resulted from secondary balanced polyploidy (Röbbelen, 1960). Although *S. alba* is more distantly related, when grouped with diploid *Brassica* and more closely related *Cruciferae* species, a continuous series of haploid chromosome numbers from $n = 7$ to 12 is evident (Downey and Röbbelen, 1989). Some interspecific crosses of *S. alba* with related *Brassica* species have been reported (Downey and Röbbelen, 1989).

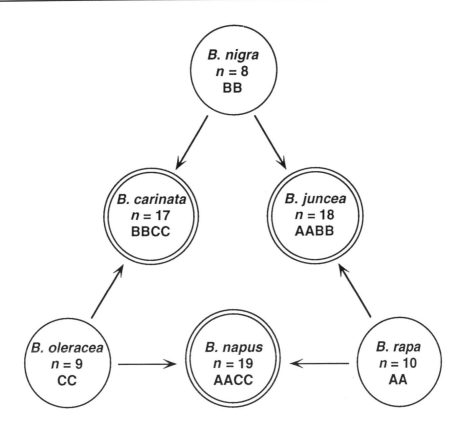

Fig. 1.1. Genomic relationship of the *Brassica* species.

CENTRES OF ORIGIN

Although the origin of the different species is not entirely clear, *B. napus* must be derived from a cross between *B. rapa* and *B. oleracea*. Since the latter was originally confined to the Mediterranean region, it is believed that *B. napus* must have originated in southern Europe from where it was introduced into Asia in the early 18th century (Downey and Röbbelen, 1989).

Brassica rapa is believed to be the oldest species and to have had the widest distribution. It could be found at least 2000 years ago over an area extending from the west of Europe to the east of China and Korea, and from Norway to the north of the Sahara and India (Hedge, 1976). Burkhill (Burkhill, 1930 cited in Prakash, 1980)

proposed that *B. rapa* originated somewhere in Europe as a biennial form and that the annual form evolved later. Central Asia, Afghanistan or India may have been another centre of origin (Sinskaia, 1928; Vavilov, 1949).

Brassica juncea is generally thought to have originated in the Middle East where the *B. rapa* and *B. nigra* species overlapped in the wild, but central Asia and China have been suggested as sites of primary origin (Prakash, 1980 and references therein). However, Hemingway (1976) considers that *B. juncea* may also have arisen by independent hybridization at secondary centres in India, China and the Caucasus as *B. nigra* was widely used as a commercial spice from early times.

Brassica carinata is believed to have

arisen in north-eastern Africa where the parent species, *B. nigra* and *B. oleracea*, overlapped in the wild (Downey and Röbbelen, 1989). Distribution is still limited to Ethiopia and neighbouring countries (Downey and Röbbelen, 1989).

The centre of origin of *S. alba* is thought to be the eastern Mediterranean region. Low-growing, branched, short-season wild forms with brown or black seeds occur around most of the Mediterranean shores (Hemingway, 1976).

CULTIVATION

Brassica oilseeds are closely related to the green vegetables and condiment mustards, which were esteemed by early man for their pungency, acerbic taste and reputed medicinal properties (Nieuwhof, 1969; Crisp, 1976). Domestication probably occurred at different times and places as the value of locally adapted forms was recognized. Initially, it seems that the vegetative parts of *Brassica* plants were used for medicinal purposes and as a vegetable. Use of the seed as a source of oil probably came later.

It is likely that *Brassica* was among the earliest domesticated plants since vegetable forms were apparently in common use in the Neolithic age (Downey and Röbbelen, 1989). Mustards are referred to in Indian Sanskrit writings of about 1500 BC (Prakash, 1980) and in Chinese writings of 1122 to 247 BC (Li, 1980). The use of mustard as a condiment and for medicinal purposes was noted by Pythagoras (about 520 BC) and Hippocrates (about 400 BC) and Pliny (AD 23 to 79) (Fenwick *et al.*, 1983 and references therein).

Cultivation of *Brassica* crops is believed to have spread across Europe in the early Middle Ages. They gradually became better known, perhaps initially for their medicinal properties, through cultivation in monastic gardens, by armies, ambulant merchants and seafarers. Later they were certainly grown as a vegetable for human consumption. In the latter half of the 17th century, for example, *Brassica* crops were grown in 'kailyards' (presumably in the form of kale) in Scotland and eaten with barley and oatmeal (Smout, 1969). It was the only vegetable in the diet for European peasants prior to the introduction of the potato in the 18th century and, as such, undoubtedly reduced the incidence of vitamin-related disorders. In the 18th century the English explorer James Cook discovered the curative effects of pickled cabbage on scurvy, which was a vital factor in the success of his explorations as it promoted the well-being of his sailors.

There is uncertainty as to when *Brassica* was first used as a source of oil. Toxopeus (1979) suggested that a small form of red cole (presumably *B. oleracea*) was perhaps first used in Europe as a source of oil. Appelqvist (1972) suggested that *B. napus* may first have been grown as a source of oil in Holland in the 17th century from where the practice spread to other parts of Europe. Linnaeus suggested in the mid-18th century that mustard or turnip would produce as good a source of oil (Appelqvist, 1972).

Studies by Schröder-Lembke (1989) provide evidence from church tithe records dating from 1421 indicating that rapeseed was certainly grown in Holland as a crop but much earlier than Appelqvist had thought. The earliest comprehensive account of rapeseed cultivation in Europe dates from 1570, when Heresbach referred to winter rapeseed being grown in the Rhineland area of Germany as a lamp oil, to provide a cheaper alternative to olive oil, and as a cooking fat for 'poor men's kitchens' (Heresbach, 1570).

A horse-driven oil mill of 1830 vintage still stands in the open-air museum near Arnhem, Holland. Appelqvist (1972) noted with interest, from a similar description by Linnaeus of early oil milling in Sweden, that the cake remaining from pressing linseed was fed to horses, while the rapeseed cake was used as a fuel for cooking.

Rapeseed oil was primarily used for making soap and for illumination in Europe during the latter part of the Middle Ages (Appelqvist, 1972). It was widely used as a lamp oil on the railways in the mid-19th century and is still used in sanctuary lamps in some churches today as the oil is favoured for its slow-burning and relatively odourless properties. But the heyday ended with the availability of cheap mineral oil in the latter half of the 19th century. With the advent of the steam engine, *Brassica* oil with a naturally high erucic acid content became the lubricating oil of choice for water-washed parts and remained so until after the Second World War. It was first grown in Canada in 1943 to meet this demand (National Research Council of Canada, 1992).

Despite wide acceptance for centuries as a cooking oil in Asia, the Indian subcontinent and among poorer people in Europe, the crop has only been used extensively in the western world as an edible oil since the Second World War. This became possible due to improved agronomic methods in the field and processing techniques in the factory. Subsequent progress in breeding for quality of both oil and meal ensures that use as an edible oil now greatly exceeds all other uses, although industrial uses are many and are likely to become more significant.

The status of the oil has improved in recent years with the discovery that it has beneficial nutritional properties. Its value for industrial purposes and as a fuel is enhanced by the perceived benign effect on the environment. Although the meal is still used as a fertilizer in some Asian countries, the value of the crop has been further enhanced by its utilization as a protein-rich animal feed. These nutritional and industrial uses are considered in detail in later chapters.

The different uses of rapeseed have

Table 1.1. Five-year 1989/93 average annual *Brassica* oilseed production by country.

	Production (tonnes $\times 10^{-3}$)
Canada	3,758
China	6,506
European Union (12)	6,161
India	4,878
Poland	1,158
Other countries	3,565
Total	26,026

Source: *Oil World*, Hamburg.

given rise to specialized cultivars and it is important to differentiate between the three principal types of seed quality: high erucic acid (HEAR), low erucic acid (LEAR) and canola. The term canola refers to seed or seed products with less than 2% erucic acid in the oil and less than 30 μmol g^{-1} meal of aliphatic glucosinolates (Canola Council of Canada, 1990).

WORLD PRODUCTION

Man has valued the *Brassica* species for many centuries but no more so than in the latter half of the 20th century. The recent increase in cultivation has been largely for vegetable oil production and has involved shifts in the traditional production areas. Whereas in the late 1940s Asia produced 70% of the world's *Brassica* oilseeds (Bunting, 1986), by the 1990s approximately 40% was produced by Canada and Europe (Howard, 1993) (Table 1.1). Expanding world population and improving standard of living should promote continuing production increase for edible oil use. Development of speciality cultivars for niche industrial markets should also add to overall cultivation.

Changes in world production of the six principal vegetable oils are shown in Table 1.2 (Howard, 1993). Annual growth of

Table 1.2. Five-year average annual world production (tonnes $\times$ 10^{-3}) of oilseeds and palm oil.

	Year							
	57/58 61/62	62/63 66/67	67/68 71/72	72/73 76/77	77/78 81/82	82/83 86/87	87/88 91/92	92/93 96/97[1]
Soyabean	25,400	30,520	42,669	58,311	82,356	93,000	111,159	126,820
Cotton seed	17,846	20,613	21,712	23,500	25,171	29,331	32,563	34,830
Groundnut	9,413	10,806	11,926	11,535	12,479	13,696	15,473	16,833
Sunflower	5,874	7,872	9,924	10,537	13,957	17,795	21,917	24,150
Rapeseed	3,599	4,126	6,178	7,563	10,527	16,986	23,152	26,126
Palm oil	1,262	1,358	1,725	2,763	4,482	6,745	10,101	12,234

Source: *Oil World*, Hamburg.
[1] Forecast.

Table 1.3. Five-year 1988/92 average annual rapeseed seed and oil exports and imports by country.

Exports	Production (tonnes $\times$ 10^{-3})	Imports	Production (tonnes $\times$ 10^{-3})
Seed			
Canada	2030	European Union	2166
European Union	1972	Japan	1716
Other countries	452	Other countries	568
Total	4454		4450
Oil			
Canada	378	China	114
European Union	1185	European Union	468
Other countries	257	India	215
		USA	152
		Other countries	861
Total	1820		1810

Source: *Oil World*, Hamburg.

rapeseed exceeded that of soyabean, cotton seed, sunflower and groundnut over the last decade, rapeseed moving from fifth to third in world production. In India, Poland and the European Union (EU) the growth of rapeseed production has been even more pronounced. This growth has occurred despite, or perhaps because of, the fact that rapeseed oil on the world market normally has traded at a discount compared with other oilseeds (Howard, 1993).

Most rapeseed-producing countries utilize the crop internally. Of those that export Canada and the EU share about 90% of world trade almost equally between them (Table 1.3). However, the EU imports almost half the rapeseed on the world market, while Japan imports 40%. Some 65% of the rapeseed oil marketed internationally is refined in the EU, while 20% comes from Canada. The main importers are China, the EU, India and the United States.

REFERENCES

Appelqvist, L.-Å. (1972) Historical background. In: Appelqvist, L.-Å. and Ohlson, R. (eds) *Rapeseed: Cultivation, Composition, Processing and Utilization.* Elsevier, Amsterdam, pp. 1–8.

Bunting, E.S. (1986) Oilseed rape in perspective. In: Scarisbrick, D.H. and Daniels, R.W. (eds) *Oilseed Rape.* Collins, London, pp. 10–31.

Canola Council of Canada (1990) *Canola Oil and Meal: Standards and Regulations.* Canola Council of Canada, Winnipeg, Canada, 4 pp.

Crisp, P. (1976) Trends in the breeding and cultivation of cruciferous crops. In: Vaughan, J.G., MacLeod, A.J. and Jones, A.J. (eds) *The Biology and Chemistry of the Cruciferae.* Academic Press, New York, pp. 69–118.

Downey, R.K. and Röbbelen, G. (1989) *Brassica* species. In: Röbbelen, G., Downey, R.K. and Ashri, A. (eds) *Oil Crops of the World.* McGraw-Hill, New York, pp. 339–362.

Downey, R.K., Stringam, G.R., McGregor, D.I. and Stefansson, B.R. (1975) Breeding rapeseed and mustard crops. In: Harapiak, J.T. (ed.) *Oilseed and Pulse Crops in Western Canada.* Western Cooperative Fertilizers, Calgary, Alberta, Canada, pp. 157–183.

Fenwick, G.R., Heaney, R.K. and Mullin, W.J. (1983) Glucosinolates and their breakdown products in food and food products. *CRC Critical Reviews in Food Sciences and Nutrition* 18, 123–201.

Hedge, I.C. (1976) A systematic and geographical survey of the world Cruciferae. In: Vaughan, J.G., MacLeod, A.J. and Jones, B.M.G. (eds) *The Biology and Chemistry of the Cruciferae.* Academic Press, New York, pp. 1–45.

Hemingway, J.S. (1976) Mustards: *Brassica* spp. and *Sinapis alba* (Cruciferae). In: Simmons, N.W. (ed.) *Evolution of Crop Plants.* Longman, London, pp. 56–59.

Heresbach, K. (1570) *Rei rustica libri quator.* Cologne (translated by G. Markam, 1631, London).

Howard, B. (1993) *Oils and Oilseeds to 1996.* Economist Intelligence Unit, Commercial Outlook Series.

Li, C.W. (1980) Classification and evolution of mustard crops (*Brassica juncea*) in China. *Cruciferae Newsletter* 5, 33–36.

Morinaga, T. (1934) Interspecific hybridization in *Brassica.* VI. The cytology of F₁ hybrids of *B. juncea* and *B. nigra. Cytologia* 6, 62–67.

National Research Council of Canada (1992) *From Rapeseed to Canola: the Billion Dollar Success Story.* National Research Council of Canada, Publication 33537, Saskatoon, Canada, 79 pp.

Nieuwhof, M. (1969) *Cole Crops, Botany, Cultivation and Utilization.* Leonard Hill, London, 353 pp.

Olsson, G. and Ellerstrom, S. (1980) Polyploidy breeding in Europe. In: Tsunoda, S., Hinata, K. and Gómez-Campo, C. (eds) *Brassica Crops and Wild Allies.* Japan Scientific Press, Tokyo, Japan, pp. 167–190.

Prakash, S. (1980) Cruciferous oilseeds in India. In: Tsunoda, S., Hinata, K. and Gómez-Campo, C. (eds) *Brassica Crops and Wild Allies.* Japan Scientific Press, Tokyo, Japan, pp. 151–163.

Röbbelen, G. (1960) Beitrage zur Analyse des Brassica-Genoms. *Chromosoma* 11, 205–228.

Schröder-Lembke, G. (1989) *Die Entwicklung des Raps- und Rübsenanbaus in der deutschen Landwirtschaft.* Verlag Th. Mann, Gelsenkirchen-Buer, 35 pp.

Sinskaia, E.N. (1928) Geno-systematical investigations of cultivated *Brassica. Bulletin of Applied Botany and Plant Breeding* 17, 1–166.

Smout, T.C. (1969) *A History of the Scottish People 1560§1830.* Collins Clear-Type Press, Glasgow, 576 pp.

Song, K. and Osborn, T.C. (1991) Origins of *Brassica napus*: new evidence based on nuclear and cytoplasmic DNAs. In: McGregor, D.I. (ed.) *Proceedings of the Eighth International Rapeseed Congress*, Saskatoon, Canada. Organizing Committee, Saskatoon, pp. 324–327.

Toxopeus, H. (1979) The domestication of *Brassica* crops in Europe. Evidence from the herbal books of the 16th and 17th centuries. *Proceedings of the Eucarpia Cruciferae Conference*, Wageningen, The Netherlands. Organizing Committee, Wageningen, pp. 47–52.

U, N. (1935) Genome analysis in *Brassica* with special reference to the experimental formation of *B. napus* and peculiar mode of fertilization. *Japanese Journal of Botany* 7, 389–452.

Vavilov, N.I. (1949) Origin, variation, immunity and breeding of cultivated plants. *Chronica Botanica* 13, 15–39.

Warwick, S.I. and Black, L.D. (1993) Molecular relationships in subtribe Brassicinae (Cruciferae, tribe Brassiceae). *Canadian Journal of Botany* 71, 906–918.

Woods, D.L., Capcara, J.J. and Downey, R.K. (1991) The potential of mustard (*Brassica juncea* L. Coss) as an edible crop on the Canadian prairies. *Canadian Journal of Plant Science* 71, 195–198.

Part I

The Field Crop

---

2 Physiology: Crop Development, Growth and Yield

N.J. Mendham[1] and P.A. Salisbury[2]
[1]*Department of Agricultural Science, University of Tasmania;*
[2]*Victorian Institute of Dryland Agriculture, Australia*

INTRODUCTION

This chapter will examine how the rape-seed crop develops and grows, and how these two aspects of the crop life cycle inter-act with each other and environmental factors to produce seeds, oil and protein. An understanding of this is necessary to make progress in breeding, i.e. altering the genotype, and for management of the crop, either to increase yield and quality, or to reduce costs of production.

The main environments

Physiological studies have been carried out in most of the main rapeseed-growing areas of the world. Examples to bring out the main contrasts are listed below.

1. *Western Canada.* Cold winters generally limit the crop to spring sowing, and a short growing season requires early maturity to avoid killing frosts in autumn. Crops flower in midsummer, and ripen under increasingly cool conditions.
2. *Western Europe.* Milder winters generally allow crops to be sown in late summer–autumn and harvested in mid-summer. A long growing season (up to a year) allows a high yield potential. Crops ripen usually during the warmest part of the year but summer rainfall can make harvesting difficult.
3. *Southern Australia.* An even milder winter than in Europe is followed by a hot, dry summer. Crops are usually sown in late autumn, although winter or spring sowings are possible in some areas. Crops are then flowering in spring and ripening under increasing daylength, temperatures and water stress in early summer.
4. *India.* Rapeseed is grown as a dry-season (winter) crop in subtropical areas, mainly on stored soil moisture from the previous wet season. Crops thus flower and ripen under increasing water stress so drought resistance and efficiency of water use are key factors.

The main genotypes

The four oilseed *Brassica* species are listed below, with an indication of the ecological niche which each occupies (Bunting, 1986). 'Winter' and 'spring' cultivars are used of the first two species, the winter ones requiring a substantial period of cold

11

(vernalization) to flower without delay. This mechanism will be discussed later, but it should be noted that many spring cultivars also show some response to vernalization.

1. *Brassica napus.* The highest yield potential, at least in favourable environments, is generally shown by this species, probably due to its allopolyploid origin. Winter types are generally grown in Europe, and spring types in other areas. Where crops are grown over mild winters, as in Australia and Japan, cultivars with a vernalization response are often used but this may not be essential. Much of the published work on crop physiology has been carried out on *B. napus* and, unless otherwise specified, work reviewed here will pertain to that species.

2. *Brassica rapa.* While the synonym *B. campestris* is still widely used *B. rapa* is preferred. The winter type of this species tends to be grown in areas where winters are marginally too cold for *B. napus*, as growing points are usually better protected from frost (Torssell, 1959). The spring type of *B. rapa* is normally earlier flowering and maturing than *B. napus*, hence it tends to be grown in shorter-season areas. In Canada the proportion of each species grown is about equal, but *B. rapa* predominates at the northern margins. The crop was grown in Australia in areas with short seasons due to heat and drought, but has been replaced by higher-yielding, early-maturing *B. napus* cultivars, and potentially by *Brassica juncea*.

3. *Brassica juncea* (Indian mustard) is the major oilseed on that subcontinent. Interest has been generated in Canada (Love *et al.*, 1991) and Australia (Oram and Kirk, 1992) recently since low erucic acid and low glucosinolate types are becoming available. It has been shown to be more heat and drought tolerant than the other species, with a range of contributing characters. For example, the better germination on dry-

ing soils appears to be due (Oram and Kirk, 1992) to the higher concentration of mucilage in the testa of *B. juncea* seeds than in *B. napus*, allowing water attraction and retention. In Australia the crop is likely to extend oilseed production into short-season winter-growing areas of southern Australia (Salisbury *et al.*, 1991), allowing earlier sowing after the first rains in autumn, and giving higher yield potential during rising temperatures and water stress in spring. In Canada the crop shows promise for the southern low-rainfall prairie areas (Woods, 1992) where currently only cereals are grown. There it shows higher yield and earlier maturity than *B. napus* or *B. rapa* whereas it matures later in the cooler northern regions such as Peace River in Alberta.

4. *Brassica carinata* (Ethiopian mustard) is only widely grown in Ethiopia, and to a limited degree in India, but interest has been shown in its drought resistance characteristics (Kumar *et al.*, 1984), plus resistance to pod shattering and disease (Alonso *et al.*, 1991). In India, Malik (1990) showed that this species performed better under saline conditions and after late sowing. Its tropical origin may allow it to tolerate high temperatures during the reproductive period better than the other species. While of poor agronomic type it has yielded well in Victoria, Australia (P.A. Salisbury, unpublished data). Poor quality of both oil and meal remains a major problem.

There is enormous variation in all the species discussed above, and the scope for interspecific crosses and other genetic transfers provides the potential for breeding and selection to make the best use of physiological and other characteristics wherever they are found in the *Brassica* gene pool.

Table 2.1. Overview of development and growth of rapeseed. Codes are HB (Harper and Berkenkamp, 1975) and SM (Sylvester-Bradley and Makepeace, 1984) (see text), and refer to the mainstem.

Development	Code HB	Code SM	Growth
Sowing	0	0.0	
Emergence	1	0.8	Expansion of cotyledons, growth of taproot
Leaf production (as rosette)	1.1–1.2[1] 2.1 ↓ 2.8	1.00 ↓ 1.20	Establishment of root system and expansion of leaves, interception of solar radiation, photosynthesis, increased leaf dry weight
Inflorescence initiation (vernalization and photoperiod responses)	–	–	
Stem elongation (photoperiod responses)	–	2.00 ↓ 2.20	Stem dry weight increases, stem photosynthesis commences, reserves laid down
Flower bud development (ovule numbers determined)	3.1 ↓ 3.3	3.0 ↓ 3.9	
Flowering (pollination, seed set)	4.1 ↓ 4.4	4.1 ↓ 4.9	Leaf area and root extension close to maximum. Flowers shade leaves, young pods
Pod development (pod and seed abortion, final numbers determined)	– –	5.1 ↓ 5.9	Pod and stem photosynthesis replaces declining leaf area as leaves senesce. Pod walls reach maximum size and seed growth commences
Seed development (formation of embryo, storage cells)	5.1 ↓ 5.4	6.1 ↓ 6.9	Seed growth with assimilate from leaves, stems, pods. Oil and protein synthesis and storage

[1] First two true leaves not counted in original HB scale–added as 1.1 and 1.2 by Morrison *et al.* (1989).

OVERVIEW OF DEVELOPMENT, GROWTH AND YIELD

It is important to distinguish between 'development', which is the progress of a crop through the stages of its life cycle, and 'growth', which is the increase in size of organs, and the accumulation of dry matter, firstly as sugars, then as structural and storage materials in leaves, stems and fruits. Table 2.1 attempts to make this distinction clear.

The stages of development are often needed to be quantified and more precisely defined. The use of numerical keys is useful in research, for repeatability and communication, and for commercial production, for example for timing of management operations. The first key to be widely used was that of Harper and Berkenkamp (1975) in Canada, which they unfortunately termed a 'growth stage key'. This is briefly compared in Table 2.1 with a more elaborate key developed in the United Kingdom by Sylvester-Bradley and Makepeace (1984). Details of the latter are contained in Table 2.2. Those authors review other keys which have been used, and set out the general principles needed for establishment of a usable key. The life cycle needs to be divided into no more than ten principal stages, each subdivided as required into secondary stages for more refined use. The principal stages are not mutually exclusive, for example plants can be elongating stems at the same time as flower buds are developing. The numerical codes are mainly for recording and processing, and should not be used in verbal or written communication without the relevant definition. Drawings of the stages are given by the authors. In this review, written descriptions rather than codes will normally be used.

Development includes much more than these visually identified stages, for example inflorescence initiation just before stem elongation begins, and the determination of numbers of seeds per pod in the 2 to 3 weeks after flowering. These will be addressed in detail later.

The interaction between development and growth at each stage builds up the potential, and then the actual yield of the crop. Each stage and process is under a greater or lesser degree of genetic control, and is affected in varying ways by environmental factors such as temperature.

This review will firstly examine development of the crop and its environmental control. Growth of roots, leaves, branches, pods and seeds will then be discussed within the framework of development, including the sequence of yield components and the quality changes that occur during seed growth. Applications of the physiological approach to agronomic prob-

lems will then be studied separately, including time and rate of sowing, nutrition, water supply, drought and cold tolerance and the use of growth regulators. Finally, a summary of morphological characters of physiological value is presented as a possible crop ideotype.

DEVELOPMENT

The basic effect of temperature

Rapeseed, being 'cold-blooded' like all plants, has a basic response to temperature which should be kept in mind while other factors are being examined. This was brought out clearly by Morrison et al. (1989) for the spring cultivar Westar grown in generally temperature-limited conditions in Canada. In the long days of the growing season there, any daylength response is likely to be saturated, and vernalization responses either not present in the cultivars used or overridden by the effect of the long days. When grown in controlled environment cabinets with 16-hour days but a range of temperatures, the rate of development to maturity was shown to be a linear function of the log of the mean temperature (Fig. 2.1). The rate of development was expressed as per cent of the total time to maturity per day. When the fitted relationship was extrapolated to the x axis a base temperature of $4.8°C$ was indicated, close to the figure of $5°C$ used for some other temperate crops.

Using the base of $5°C$ (or other base if appropriate), growing degree-days (GDD) or day-degrees ($°C$ d) can be used as a measurement scale for crop development, usually called 'thermal time'. Morrison et al. (1989) used

$$\text{GDD} = \sum_{S_1}^{S_2} (T_m - b_0)$$

where T_m is the mean daily temperature, b_0 is the base temperature, and S_1 and S_2

Table 2.2. Definitions and codes for stages of development in oilseed rape (*B. napus*).

Definition	Code
Germination and emergence	
Dry seed	0.0
Imbibed seed	0.2
Radicle emerged	0.4
Hypocotyl extending	0.6
Cotyledons emerged	0.8
Leaf production (lost leaves are counted by their scars)	
Both cotyledons unfolded and green	1.00
First true leaf exposed	1.01
Second true leaf exposed	1.02
Third true leaf exposed	1.03
Fourth true leaf exposed	1.04
Fifth true leaf exposed	1.05
.	.
.	.
.	.
Tenth true leaf exposed	1.10
.	.
.	.
.	.
Twentieth true leaf exposed	1.20
Stem extension	
No internodes detectable ('rosette')	2.00
One internode detectable	2.01
Two internodes detectable	2.02
Three internodes detectable	2.03
Four internodes detectable	2.04
Five internodes detectable	2.05
.	.
.	.
.	.
Ten internodes detectable	2.10
.	.
.	.
.	.
Twenty internodes detectable	2.20

Note: The following descriptions should normally be applied to the raceme on the mainstem. Otherwise, the raceme position should be stated.

Flower bud development	
Only leaf buds present	3.0
Flower buds present but enclosed by leaves	3.1
Flower buds visible from above ('green bud')	3.3
Flower buds raised above leaves	3.5
First flower stalks extending	3.6
First flower buds yellow ('yellow bud')	3.7
More than half flower buds on raceme yellow	3.9

Table 2.2. (*continued*)

Definition	Code
Flowering	
First flowers opened	4.1
20% all buds on raceme flowering or flowered	4.2
30% all buds on raceme flowering or flowered	4.3
40% all buds on raceme flowering or flowered	4.4
50% all buds on raceme flowering or flowered	4.5
60% all buds on raceme flowering or flowered	4.6
70% all buds on raceme flowering or flowered	4.7
80% all buds on raceme flowering or flowered	4.8
All viable buds on raceme finished flowering	4.9
Pod development	
Lowest pods more than 2 cm long	5.1
20% potential pods on raceme more than 2 cm long	5.2
30% potential pods on raceme more than 2 cm long	5.3
40% potential pods on raceme more than 2 cm long	5.4
50% potential pods on raceme more than 2 cm long	5.5
60% potential pods on raceme more than 2 cm long	5.6
70% potential pods on raceme more than 2 cm long	5.7
80% potential pods on raceme more than 2 cm long	5.8
All potential pods on raceme more than 2 cm long	5.9

Note: The following descriptions should normally be applied to the lowest third of the raceme on the mainstem. Otherwise, position on the raceme should be stated.

Seed development	
Seeds present	6.1
Most seeds translucent but full size	6.2
Most seeds green	6.3
Most seeds green–brown mottled	6.4
Most seeds brown	6.5
Most seeds dark brown	6.6
Most seeds black but soft	6.7
Most seeds black and hard	6.8
All seeds black and hard	6.9

Source: Sylvester-Bradley and Makepeace (1984).

are two development stages. Table 2.3 shows how the development of crops at ten field sites as recorded on the Harper and Berkenkamp scale can be measured in days, GDD or predicted per cent development to maturity, the latter essentially also measuring thermal time. The coefficients of variation of the two thermal time measures were similar, and less than the calendar time measure at each stage except from sowing to emergence. Emergence time would depend mainly on soil tem-perature and water status, neither of which may be closely related to air temperature. The coefficients of variation decreased as the crop developed because cumulative measures were used, i.e. time to each stage also includes the previous ones.

Other workers have used differing base temperatures. Hodgson (1978b), using Canadian spring cultivars but sown in autumn in Australia, estimated base tem-peratures for *B. rapa* to vary between 3 and 7°C for different stages, and for *B. napus*

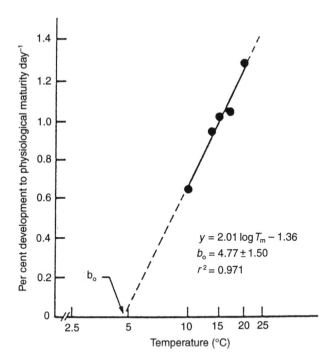

Fig. 2.1. The effect of temperature ($\log_{10}$ scale) on per cent development to physiological maturity per day for cultivar Westar in Canada. Source: Morrison *et al.* (1989).

to vary between 0 and 6°C. Similar results were obtained using either the intercept method (as in Fig. 2.1, except that a linear temperature scale was used), or using a range of base temperatures and selecting the one with the lowest coefficient of variation. In most cases using the latter method there was a range of 2 to 4°C over which the coefficient was at a similarly low level, i.e. the base is difficult to determine with precision. It is to some degree an artefact of the methods used, as in most cases development will still proceed slowly above 0°C. Leterme (1988a) in France used 0°C as a base for development and 5°C as a base for growth. In Hodgson's experiments, where crops were grown over winter, a daylength response may have raised the apparent base temperature, for example for *B. rapa* a base of 7°C from sowing to bud appearance is probably too high.

Germination and establishment

Soil temperature is the main factor promoting germination once seeds have imbibed water. Kondra *et al.* (1983) in laboratory studies showed that *B. rapa* had significantly lower germination percentages than *B. napus* at 7°C or below, or at 25°C and presumably above. *B. napus* cultivars recorded at least 90% germination from 2 to 25°C, whereas at 2°C and 3°C only 34 and 67% respectively of *B. rapa* seeds germinated. Germination time varied with temperature, from 11 to 14 days at 2°C to 1 day at 21 to 25°C. The response at low temperature is critical in Canada, where sowing needs to be as early as possible to take full advantage of a short season. In areas where autumn sowing is normal, soil moisture will generally be at least as important as temperature.

Table 2.3. Mean ($\bar{x}$) and coefficient of variation (cv) for calendar days, growing degree-days (GDD) and predicted per cent development to physiological maturity (% DPM) for cv. Westar in Canada. HB indicates Harper and Berkenkamp (1975) stage of development.

Sowing to	HB	Days		GDD		% DPM	
		$\bar{x}$	cv	$\bar{x}$	cv	$\bar{x}$	cv
Emergence	1.0	8.7	(15.7)[1]	98	32.2	9.3	22.6
Early vegetative	2.1	21.4	15.9	246	11.3	22.7	(11.0)
Late vegetative	2.4	32.9	11.9	373	(10.4)	34.5	10.5
Bolting	3.1	38.7	7.7	449	9.9	40.6	(6.4)
Flowering	4.1	46.9	8.5	576	6.7	51.4	(5.6)
Pod fill	5.1	65.9	4.9	860	5.2	74.8	(4.3)
Maturity	5.3	86.6	5.2	1157	(2.2)	100.0	2.3

Source: Morrison *et al.* (1989).
[1] Parentheses denote model with the lowest cv at each stage.

Leterme (1988a) reported that 130 to 140°C d (above zero) are required for emergence (as distinct from germination) or around 9 days at 15°C, but that a range of other factors will affect rate and final percentage emergence. Seed that was fully mature (black) on the mother plant was observed to have faster radicle emergence than immature (red) seed. Larger seed, for example from mainstem or upper branches, produced larger cotyledons and hence more vigorous seedlings. This was also noted by Major (1977) who graded out three sizes of seeds and showed greater seedling vigour from large seeds, but no effect on final population or yield. Mendham *et al.* (1981b) also showed an early benefit to growth from large seed, which only affected later crop performance when plant size at flowering was limited by late sowing.

Barber *et al.* (1991) showed substantial differences in performance of commercial certified seed lots from different sources, which could be at least as large as cultivar differences, and hence care should be used in comparisons, and the use of high-quality seed encouraged.

Leaf initiation and appearance

The growing point or apex of the plant produces leaf initials in a helical arrangement, or phyllotaxy of about 130° between leaves. Leaf initiation rates (plastochron) are faster than appearance rates (phyllochron), so leaf primordia accumulate around the apex. Smith and Scarisbrick (1990) (Table 2.4) showed that, for autumn-sown crops of *B. napus* cultivar Bienvenu, the number of days and day-degrees varied between sowings for both leaf initiation and appearance, the latter corresponding with 'exposed' leaves (Sylvester-Bradley and Makepeace, 1984), where the majority of the surface and petiole is visible but not fully expanded. Leterme (1988a) reported plastochron (initiation) times of around 20 to 60°C d for *B. napus* cultivar Jet Neuf, depending on nitrogen supply and plant population, comparable figures to those in Table 2.4. Winter cultivars in Europe will produce from around 12 leaves on the mainstem after late sowing to 30 after early sowing, before initiation of inflorescences takes place. In Canada, however, spring cultivars grown by Morrison and McVetty (1991) only produced about seven leaves,

Table 2.4. Development of winter rapeseed cv. Bienvenu up to inflorescence initiation in the United Kingdom. Thermal times as °C d above zero were calculated from date of sowing.

| Date of sowing | Date of emergence | Date of initiation | Mean no. of leaves | | Time to initiation | | Mean interval | | | |
| | | | | | | | For leaf appearance | | For leaf initiation | |
			Expanded	Initiated	Days	°C d	Days	°C d	Days	°C d
24 Aug 1983	7 Sep	10 Nov	8.9	22.8	78	1030	8.8	116	3.4	45
31 Aug 1984	13 Sep	7 Nov	10.4	22.8	65	868	6.3	83	2.9	38
4 Sep 1985	12 Sep	7 Nov	10.3	27.7	64	793	6.2	77	2.3	29

Source: Smith and Scarisbrick (1990).

at 4 days per leaf appearance, or 46 GDD (above 5°C, or about 66°C d above zero). These values were calculated from emergence, rather than sowing, in Smith and Scarisbrick's work so data are comparable.

Inflorescence initiation

This stage was described briefly by Mendham and Scott (1975), and in greater detail with diagrams and photographs by Tittonel et al. (1982), Daniels et al. (1986) and Smith and Scarisbrick (1990). Figure 2.2 from Tittonel et al. summarizes the main features. Each leaf primordium has the potential to develop a further meristem in its axil. Once any requirements for vernalization or photoperiod have been met (discussed in the next section), the first sign of flower initiation will be a swelling in the axil of one of the leaf primordia at the apex. Successive primordia then develop from the axillary rather than leaf meristems, becoming the flowers of the mainstem or raceme. Once the mainstem flower buds are forming, axillary buds lower down will then begin to develop sequentially into the primary branches in a basipetal direction. At the mainstem apex, floral development continues acropetally as in Fig. 2.2, with the sepals on each bud becoming obvious, and then peduncles. Smith and Scarisbrick (1990) describe development of sepals, stamens, petals and gynoecium, the latter

three occurring at about the time that buds become visible without dissection (green bud stage, 3.3 of Sylvester-Bradley and Makepeace). During the green bud stage, meiosis occurs in pollen mother cells and microspores are released from the tetrads. Ovules begin to differentiate in the ovary, and flower development is completed during the yellow bud stage.

Factors affecting timing of inflorescence initiation and flowering

The two key points in development are inflorescence initiation and flowering. The number of leaf initials produced before initiation has a major influence on leaf area and hence photosynthetic potential of the crop, which is realized when all the mainstem leaves expand by flowering time. At flowering the yield potential is set, as a balance between vegetative growth and the potential number of flowers, pods and seeds. Hence the timing of the key points is crucial to the match between genotype and environment. Successful adaptation to an environment means that the unfavourable risk factors such as frost at flowering and drought between then and maturity are minimized, and the favourable factors such as optimum radiation, temperature and moisture conditions for pod and seed growth are maximized.

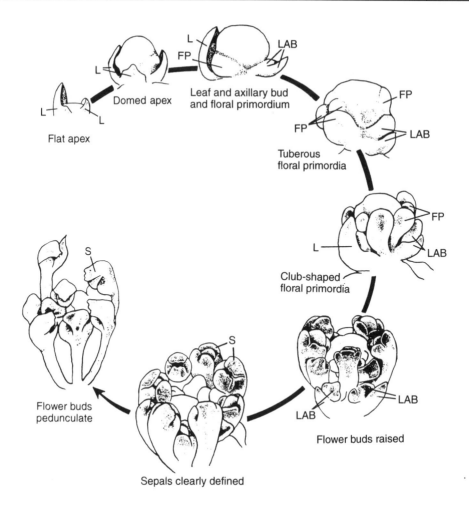

Fig. 2.2. Changes to the apex of rapeseed during inflorescence initiation. L = leaf, LAB = leaf and axillary bud, FP = floral primordium, S = sepal. Source: Tittonel *et al.* (1982).

The main controls over initiation and flowering are as follows:

1. A minimum number of leaf initials before initiation can take place, which varies with genotype.
2. The basic temperature response or plastochron, in °C d per leaf.
3. Vernalization responses, mainly operating before initiation.
4. Daylength responses, operating before both initiation and flowering.

The presence of a vernalization or day length response in a genotype means that, if not fully satisfied, initiation or flowering will be delayed past the minimum number of leaf initials. This will have the effect of delaying flowering until either the winter is past, or days are lengthening, or both. Later flowering may also mean a larger vegetative structure, more pods and a higher yield potential.

Minimum leaf number
Evidence for a minimum leaf number for each genotype comes from both controlled environment and field studies. European winter types when sown late in autumn can

Table 2.5. Number of nodes to initiation, number of days to flower and mean number of days per node (P) of three cultivars subjected to a range of temperature (temp), vernalization (vern) and photoperiod (PP) treatments.

Temp (°C)	PP (h)	Vern (weeks)	Target (Canada)			Bronowski (Europe)			Isuzu (Japan)		
			Nodes	Days	P	Nodes	Days	P	Nodes	Days	P
15	12	0	12	124	10.3	20	168	8.4	17	134	7.9
		4	11	110	10.0	14	119	8.5	12	81	6.7
		8	10	87	8.7	11	93	8.4	10	71	7.1
	24	0	7	54	7.7	9	74	8.2	13	105	8.1
		4	6	51	8.5	7	62	8.9	8	52	6.5
		8	6	50	8.3	6	53	8.8	8	50	6.2
25	12	0	15	76	5.1	*	*		*	*	
		4	11	58	5.3	36	165	4.6	38	163	4.3
		8	11	56	5.1	24	117	4.9	11	64	5.8
	24	0	10	41	4.1	37	132	3.6	*	*	
		4	10	41	4.1	24	85	3.5	10	44	4.4
		8	10	39	3.9	14	52	3.7	11	43	3.9

Source: Thurling and Vijendra Das (1977).
*Indicates no initiation by 160 days.

be fully vernalized as small plants, and then initiate once the minimum number of about 12 leaves is reached (Mendham *et al.*, 1981a). This may not be until spring on small plants delayed by a cold winter. Canadian spring cultivars may have a minimum of seven leaves, not normally exceeded greatly under field conditions there, with a minimum of six leaves in some cultivars (Thurling and Vijendra Das, 1977). Indian or Chinese cultivars may also have a minimum of about six leaves if responses are satisfied.

An example of responses

A wide range of genotypes has been used in physiological studies and breeding programmes in Australia. Table 2.5 shows an example, extracted from Thurling and Vijendra Das (1977). Imbibed seeds were chilled (vernalized) at 3°C for between 0 and 8 weeks, and then subjected to the combination of two temperatures and two photoperiods shown, in controlled environments. Data for 2 and 6 weeks' vernalization were included in the original paper,

which also divided days to flowering into two phases, sowing to beginning of stem elongation, and from there until flowering.

Vernalization as imbibed seeds was effective on all three *B. napus* cultivars. The Canadian *B. napus* cultivar Target showed a small response to vernalization in short days but almost none in long days (continuous light). The European *B. napus* cultivar Bronowski, normally regarded as a spring cultivar, responded to vernalization under all combinations of temperature and photoperiod, with a larger response at the higher temperature. The Japanese *B. napus* cultivar Isuzu responded in a similar way to Bronowski to vernalization, except in high temperatures and long days where there was a large response to four weeks' vernalization but no extra response to six or eight weeks. Bronowski responded to the 8-week treatment in all environments.

The enormous variation in number of nodes to initiation was the main factor causing variation in time to flowering. The number of days per node, approximating the phyllochron, was around six to ten at low temperature and four to five at high temperature (approximately 90 to 150°C d in each case), with some differences between cultivars. The number of nodes to flowering is often a better guide to 'physiological age' of a crop than calendar age, as temperatures have such a large effect on rate of leaf appearance.

All cultivars responded to photoperiod, with the minimum node number in each case being recorded under long days at the lower temperature. The minimum number of days to flowering was in some cases under long days at high temperature, but this was at a greater node number. Photoperiod and vernalization in most cases had an additive effect, although in Target under long days there was little extra effect of vernalization at either temperature.

Thurling and Vijendra Das (1979a) showed that the vernalization response is governed by least four genes in *B. napus*. Target, the early cultivar in Table 2.5, was dominant for all four, with two in the recessive form in Bronowski and the other two recessive in Isuzu, presumably reflecting separate evolution of European and Asian material. The genes have differing effects and thus in various combinations can give a full spectrum of responses. Plants which flowered earlier than Target were recovered, probably implying that extra genes for earliness could be found in one of the later-flowering parents.

The range of responses and interactions

A wide range of *Brassica* germplasm was studied by Myers *et al.* (1982) and Rao and Raymer (1991). In both cases plants were established first for about 2 weeks before vernalizing for up to 12 weeks. European winter cultivars, probably with all four of the main vernalization genes in the recessive form, required vernalization before flowering, with in most cases a considerable further promotion of flowering by long days. In most other cultivars, either vernalization or long days were necessary for prompt flowering, with only a few lines such as some very early Indian *B. rapa* showing little response to either.

Most cultivars respond to vernalization as imbibed seeds as well as established plants. Most plants of the winter cultivar Primor (Mendham *et al.*, 1984 and unpublished), when vernalized as seed for four or eight weeks flowered within 160 days when transferred to a glasshouse at 15 to 25°C in natural winter–spring photoperiod, and within 120 days under continuous light. At these high temperatures, however, some plants reverted to producing leaves, and hence were 'devernalized'. This was also seen on the late-flowering Australian cultivar Wesbell (Mendham *et al.*, 1990) where in a semi-controlled environment it behaved like earlier-flowering cultivars under most conditions of vernalization, daylength and temperature, except under high temperatures and long days. Then the effect of seed vernalization was lost. A slow rate of leaf appearance in this cultivar appeared to contribute to this effect.

Winter cultivars sown in the field in spring show a similar effect, in that they may flower eventually, but usually in an irregular and incomplete fashion (Mendham and Scott, 1975; Mendham *et al.*, 1990). This reflects either segregation for vernalization response, or differences in plant size during marginal temperatures for vernalization. Older plants require less chilling than young plants, as shown by Tittonel (1988) for cultivar Jet Neuf where plants 2 weeks old before vernalizing required 9–10 weeks, but 4-week-old plants only required 5–6 weeks at 5°C, and slightly longer in each case at 12°C. Temperatures above 17°C were not effective.

Temperatures of around 3 to 7°C appear to be most effective for vernalization. It is an active process requiring energy, and hence may be less effective at lower temperatures. Hodgson (1978a) found that, at 1°C, seed vernalization was not effective on winter cultivars, and established plants required 15 to 19 weeks of chilling before they would flower. It is therefore difficult to build vernalization into a model of plant development, as it is generally working in reverse to the basic temperature response. The number of weeks at vernalizing temperatures is therefore more appropriate than a reverse thermal time measure, e.g. day-degrees below a base, since, although there may be an optimum temperature for vernalization, there is a wide range over which it is effective.

When comparing 'spring' cultivars, Salisbury and Green (1991) showed that, in general, European cultivars were most responsive to both vernalization and photoperiod. Australian cultivars were intermediate, and Canadian cultivars least responsive. Time to flower in all cultivars was significantly delayed by low light intensity, with differences between cultivars in the size of the delay. Higher temperatures generally promoted flowering, but in the case of the French cultivar Drakkar, 27/20°C delayed flowering compared to 22/17°C, similar to the effect on Wesbell noted above.

Interactions with the stem elongation phase

The duration of the stem elongation phase from initiation to flowering may be of more significance to final yield than the length of the pre-initiation phase. Thurling and Vijendra Das (1979b) showed that, for cultivar Target, the lengths of these phases could be manipulated by photoperiod, and that a longer stem elongation phase was associated with production of more pods on a larger inflorescence.

A range of other cultivars showed an association between the length of this phase and yield. Thurling and Kaveeta (1992a) studied the factors affecting the lengths of these phases (Fig. 2.3). *Brassica rapa* had earlier been shown to flower too early

in Western Australia, before sufficient leaf area and hence yield potential had been accumulated. An experimental later-flowering population, Chinoli C42, was compared with an early-flowering *B. napus* line RU2, with Japanese and Chinese parentage, and a standard *B. napus* cultivar Wesbrook. Vernalization or long photoperiod reduced the length of the vegetative phase in all cases. Long days reduced the phase from initiation to stem extension to just a few days in RU2, and also reduced the duration of stem extension. Vernalization also surprisingly reduced the length of the stem elongation phase in the *B. napus* lines grown in natural daylengths. Vernalization is normally considered to operate only on the first phase, to initiation.

Crosses between the three lines and subsequent F_2 or backcross generations showed that genes for flowering time could be successfully transferred between the two species, and lengths of the different phases thereby adjusted. Significant variation was found for rate of leaf appearance as well as node number.

The responses under field conditions

Most of the work reviewed above was conducted under at least partly controlled conditions. In field plantings it is very difficult to distinguish between the environmental responses. Mendham *et al.* (1990) sowed three cultivars with maturity varying from early to late over a range of times from autumn to spring (May to October) in Tasmania, Australia. Figure 2.4a shows that the rate of development from sowing to flowering (as 1/d, inverse of number of days) was closely correlated with mean daylength over this phase. Temperatures, however, are also changing over this time, and thermal development rate or 1/°C d should be more appropriate as there will still be the basic temperature response as well as any others such as photoperiod. When the same data are plotted on a

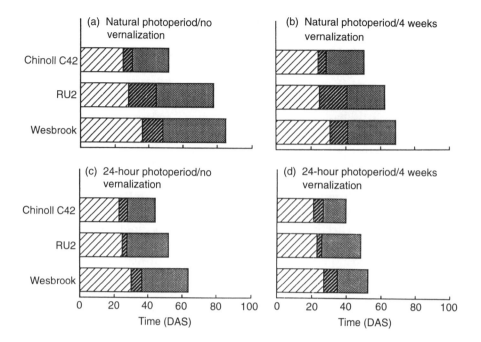

Fig. 2.3. Effects of photoperiod and vernalization on pre-anthesis development in selected *B. rapa* (Chinoli) and *B. napus* populations. Three phases in sequence: vegetative (▨), post-initiation (▨) and stem elongation (▨). DAS = days after sowing. Source: Thurling and Kaveeta (1992a).

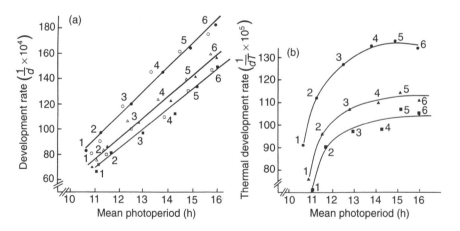

Fig. 2.4. (a) Relationship between mean photoperiod and development rate (inverse of number of days, d) for six sowings at two sites in Tasmania of three cultivars: RU1, Marnoo and Wesbell (upper, middle and lower fitted lines respectively). (b) Relationship between mean photoperiod and thermal development rate (d = number of days, T = mean temperature) for six sowings at one site in Tasmania of three cultivars, as in (a) above. Source: Mendham *et al.* (1990).

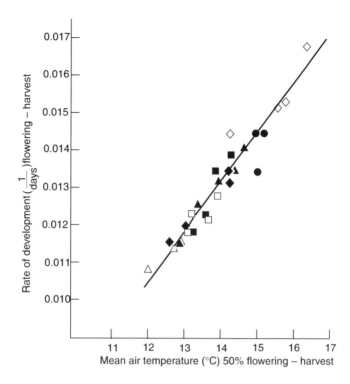

Fig. 2.5. The relationship between mean air temperature and the rate of development (inverse of number of days) from 50% flowering to final harvest of crops sown on a range of dates in seven seasons in the United Kingdom. Source: Mendham *et al.* (1981a).

thermal development rate scale against photoperiod (Fig. 2.4b), strong curvature is apparent. The first sowing (May) in autumn developed slowly until the vernalization response was fulfilled. At long photoperiods (fifth and sixth sowings), temperatures were also high, but this meant that vernalization responses were unsatisfied, hence development was little faster than at shorter daylengths. The use of a higher base temperature, e.g. 5°C, does not reduce the curvature shown in Fig. 2.4b.

Development from flowering to harvest

The details of pod and seed set will be discussed later. At this stage it should be observed that temperature is the main factor controlling development here, as in earlier phases, but without the extra responses. As an example, a range of sowings of crops of a single cultivar (Victor) over seven seasons in the United Kingdom (Mendham *et al.*, 1981a) (Fig. 2.5) developed at a rate proportional to temperature from flowering to physiological maturity (windrowing). When the fitted line is extrapolated, a base temperature of 4.2°C is indicated. Crops thus took 715°C d above 4.2°C to maturity, compared with cultivar Westar in Canada (Morrison *et al.*, 1989) which took 581°C d above 5°C.

Over the range of temperatures shown in Fig. 2.5, a direct plot of days to maturity against mean temperature gave nearly as

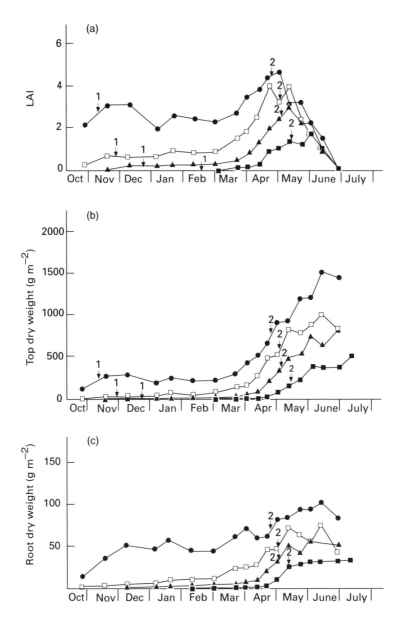

Fig. 2.6. Changes with time in (a) leaf area index LAI; (b) top dry weight; and (c) root dry weight of crops sown in the United Kingdom on 25 August (●), 13 September (□), 24 September (▲) and 13 October (■). Numbers above arrows: 1 = inflorescence initiation, 2 = 50% of plants flowering. Source: Mendham *et al.* (1981a).

good a linear relationship ($y = 185 - 7.8x$, $r^2 = 0.94$), indicating that each degree rise in temperature gave nearly 8 days' earlier maturity. This relationship would be unlikely to hold outside a range of 12 to 16°C, as the equation indicates that crops would flower and ripen on the same day at 24°C, and that crops would ripen in 185 days at 0°C, both being impossible. The inverse relationship is much more generally useful, but there must still be an upper limit. When Morrison et al. (1989) grew plants at 27/17°C and 30/20°C (day/night) in controlled environments flowers were sterile and produced no seeds.

CROP GROWTH TO FLOWERING

The general pattern of growth of a European winter rapeseed crop is shown in Fig. 2.6, from Mendham et al. (1981a), as a basis for discussion. Rapid growth after establishment in autumn, at least on early-sown crops, is followed by very slow growth or even dormancy in winter. Leaf area index (LAI) may even decline, as large older leaves are replaced by new smaller leaves developing under low temperatures, hence restricting cell size and expansion rate. Plants are largely in the rosette stage, until rapid stem extension begins in spring. Early sowings may, however, extend the first few internodes in autumn (Leterme, 1988a), particularly in high population density crops or in those with a large leaf area. Inflorescence initiation occurs on the large plants of early (August) sowings by November, but may be delayed on the very small plants of late sowings until late winter or spring. Flowering takes place about half-way through the main spring growth period.

Root growth and reserves

Rapid root growth after establishment consists of taproot extension vertically, growth of the secondary roots laterally, and then laying down of reserves, principally in the taproot. Most studies of roots include only those reasonably accessible to digging, i.e. to about 30 cm as in Fig. 2.6, or where root activity is inferred from water extraction patterns. Few workers have taken the trouble and expense to measure roots, either from extracted cores or in situ. Kjellstrom (1991) in Sweden took core samples to 1 metre and showed that, while 85% of root dry matter was found in the top 23 cm, root length and surface area were more homogeneously distributed over the profile. Maximum dry matter (200 g m^{-2}), length (2 km m^{-2}) and surface area (5 m^2 m^{-2}) were all reached late in the flowering phase.

Almond et al. (1984) dug pits to 1.5 m and measured root distribution on a grid to show the effects of cultivation method. Root numbers were distributed 27 to 52% in the top 10 cm, 14 to 26% from 10 to 20 cm, down to 1% or less below 1 m, again being close to their maximum at the end of flowering. This may be why the earlier-flowering B. rapa lacks an extensive root system (Richards and Thurling, 1978b) as well as being limited by pre-flowering above ground biomass. Almond et al. (1984) showed that if direct drilling takes place on a compacted soil then rooting may be mainly at the surface. On less compacted soils, the advantages of moisture conservation with direct drilling gave a well-developed and well-distributed root system with depth. Crops extracted water to 40 cm early in the season and to 110 cm by maturity, as measured by neutron probe readings, and the differences were related to measurements of compaction and root distribution.

The root system (mainly taproot) acts as an assimilate and nutrient reservoir over the autumn–winter period (Quilleré and Triboi-Blondel, 1988a), with a maximum of non-structural carbohydrate and nitrogen stored at the end of the rosette stage in early spring. Root reserves appear to be mainly used in regrowth of leaves in spring.

The leaf canopy

Leaf production on the mainstem, and later as bracts on branches, produces the pattern of leaf area expansion and decline as shown in Fig. 2.6. An LAI of up to 3 may be produced before winter on early sowings, which corresponds to interception of over 80% of incoming solar radiation (see below), whereas late sowings may go through the winter as very small plants with leaf and ground cover almost non-existent.

The pattern of individual leaf appearance and senescence of a typical autumn sowing of cultivar Jet Neuf (sown 7 September) in France is shown in Fig. 2.7, from Triboi-Blondel (1988), on both calendar and day-degree scales. Rapid production in autumn was followed by slow production in winter at low temperatures, although the phyllochron remained at about 75°C d up to leaf 16. Later leaves corresponding with stem extension followed more rapidly at only about 20°C d per leaf. The area of individual leaves increased from about 24 cm^2 on the first true leaf to 95 cm^2 on leaf 5, and then declined to around 40 to 50 cm^2 for most of the remaining leaves (9 to 21). The duration of life of each leaf ranged from 440°C d for leaf 1 to about 850°C d for leaves 26 to 33.

Controlled environment studies of leaf expansion and duration in the Canadian cultivar Westar by Morrison et al. (1992) showed that expansion rates of individual leaves were linearly related to temperature over the range 10–25°C. Final area per leaf increased to a maximum at leaf 4 or 5, as in the French work discussed above, but, since crops flowered at about eight leaves, all except the first two were still present at flowering.

There are substantial differences in leaf growth and expansion rates between individual genotypes, especially at low temperatures. While European winter types may be virtually dormant, with new leaves just replacing old, Asian lines of particularly B. rapa may grow at a steady rate. Rao and Mendham (1991a) showed that an experimental 'Chinoli' population produced up to double the leaf area of B. napus or normal B. rapa cultivars by late winter in Tasmania. The Chinoli was produced by crossing oilseed B. rapa with Chinese mustard, also B. rapa (Thurling and Kaveeta, 1992a), in an attempt to introduce greater vigour. Mendham et al. (1984) also showed substantially faster winter growth and leaf expansion from Chinese- or Japanese-derived cultivars of B. napus compared with a Canadian cultivar, particularly under dry as well as cool conditions. Japanese breeders in the 1930s used populations derived from B. napus/B. rapa crosses in selecting for early flowering and vigorous winter growth (Thurling and Kaveeta, 1992a).

Higher LAIs by flowering can be achieved by either faster leaf expansion rates or a longer period between sowing and flowering. The latter can be achieved by earlier sowing, or the use of a later-flowering cultivar. In a long growing season this may be an advantage (Mendham and Scott, 1975). The use of a winter cultivar Primor by Mendham et al. (1984) in Tasmania led to slow growth over winter, and flowering several weeks after adapted Australian cultivars, but at a much higher LAI. Crops were, however, already running out of water then, so high yields on the local cultivars were followed by very low yields on Primor.

Leaf area, interception of radiation and growth analysis

Crops are only likely to be limited by lack of leaf area if peak index is less than about 4, as this is enough to intercept about 90% of solar radiation (Mendham et al., 1981a).

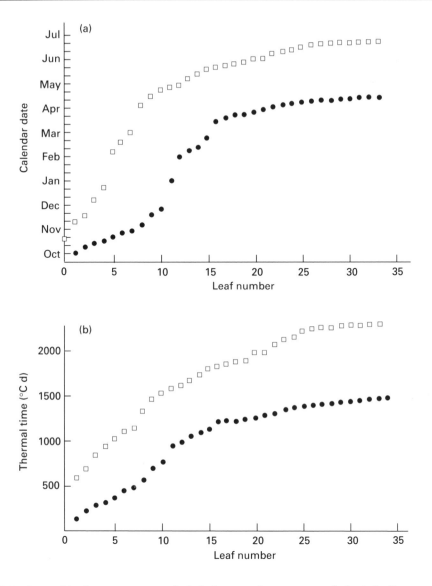

Fig. 2.7. Rate of leaf appearance and abcission on winter rapeseed plants in France: (a) with calendar time , and (b) with thermal time (°C d). ● indicates emergence and □ indicates abcission or loss, with distance between indicating duration. Source: Triboi-Blondel (1988).

Very high yields can be produced on crops with an LAI of 4 at flowering and more may be wasteful of resources. Extended leaf area duration may be of value to build up reserves before flowering, but the photosynthetic role of leaves is mainly lost after flowering.

The characteristics of the leaf canopy change with time. In the rosette stage the large leaves have an intermediate angle with the stem, and bend to a horizontal orientation, giving an extinction coefficient k of 0.6 or greater, in the commonly used Beer's Law application of

$$I_L/I_0 = e^{-kL}$$

where I_0 is incident radiation, I_L is amount of radiation transmitted through a layer of leaves of area index L and k is the extinction coefficient.

During stem elongation the smaller leaves or bracts subtending branches on the mainstem are more upright, with a k of 0.4–0.6. A value of 0.6 is satisfactory for most whole-crop measurements relating leaf area to transmission or interception of solar radiation, and corresponds to interception of 45, 70 and 84% of total solar radiation at LAI values of 1, 2 and 3, respectively.

Solar radiation intercepted by the crop is converted to dry matter via photosynthesis. The efficiency of crops in spring in the United Kingdom was shown to be about 1.2 g MJ^{-1} total solar radiation (Mendham et al., 1981a), equivalent to 2.4 g MJ^{-1} photosynthetically active radiation.

Similar values were reported in France by Gelfi et al. (1988). Crops pre-flowering in Tasmania (Mendham et al., 1990) grew at about 1.5 g MJ^{-1} (total solar) for early sowings growing under cool, moist conditions, down to about 0.7 g MJ^{-1} for late spring sowings growing under high temperatures, water stress and increasing pest damage. At low temperatures in the United Kingdom before active spring growth began, the efficiency of growth was only 0.4 g MJ^{-1} (Mendham et al., 1981a).

Some workers have used conventional growth analysis to compare genotypes or treatments. Clarke and Simpson (1978a) showed that crop growth rates were related to LAI up to 3, with little further increase above that, corresponding to near-full interception of radiation. Net assimilation rates decreased with time as the leaf canopy built up and lower leaves were shaded to a greater degree.

Thurling (1974a) showed that the mean net assimilation rate of B. rapa was greater than that of B. napus in later sowings under lower temperatures but moderate radiation levels. Thurling and Kaveeta (1992b) were able to demonstrate considerable variation in relative growth rate and net assimilation rate between Chinoli, Wesbrook and crosses between them. A high net assimilation rate in one of the crosses (INT 88) meant that, even though it flowered earlier at a lower LAI than in Wesbrook, it still produced a similar amount of dry matter by flowering.

GROWTH AND DEVELOPMENT: FLOWERING TO MATURITY

This section will be an integrated study of the processes between flowering and maturity. Firstly, yield components will be introduced with examples to show how they can vary. Then branching, flowering, seed set, pod and seed abortion or growth, and synthesis of oil and protein will be examined, in relation to supply of assimilates and nutrients, and interaction with environmental factors.

Yield components

These are often measured at harvest to try to determine which is the most important in discriminating between genotypes or treatments. Examples are given in Table 2.6. In a short season in Western Australia characterized by hot, dry finishing conditions (Richards and Thurling, 1978b), when five cultivars of each species were sown in early winter, B. napus produced a very similar yield to B. rapa, even though much later flowering. Less than half the number of pods were produced, but each retained more seeds of greater mean weight.

In a cooler, longer-season environment in Tasmania (Rao and Mendham, 1991a), B. napus produced about ten times the yield of that in Western Australia, and about

Table 2.6. Examples of yields and components from a range of crops (for standard errors see original papers).

	No. of pods $(m^{-2} \times 10^{-3})$	Seeds (per pod)	Mean seed wt (mg)	Seed yield $(g\ m^{-2})$	Oil content (%)
Western Australia					
(Richards and Thurling, 1978b)					
B. napus	1.10	13.1	3.38	48	–
B. rapa	2.49	9.8	1.94	46	–
Tasmania					
(Rao and Mendham, 1991a)					
B. napus					
Marnoo	8.40	14.4	4.07	492	50.4
RU1	7.28	13.7	4.32	428	49.3
B. rapa					
Jumbuck	8.40	7.8	2.79	183	49.8
Chinoli	7.53	9.8	3.58	256	48.7
United Kingdom					
(Mendham et al., 1981a)					
B. napus					
Early sown	12.2	5.7	5.72	358	39.8
Late sown	3.9	7.1	4.45	123	38.2
Late sown	4.3	20.1	4.53	454	–

–, Not measured.

double the yield of B. rapa at the same site. The major difference between environments was that many more pods were produced on well-grown plants in Tasmania, although numbers of seeds per pod were very similar to those recorded in Western Australia. Better finishing conditions (moderate temperatures, adequate water supply) allowed seeds to fill out to a larger size, with high oil content. The major difference between species was that B. napus retained more seeds per pod, of increased size. Numbers of pods were very similar. Within B. napus, the early-flowering RU1 appeared to produce fewer pods and seeds per pod than Marnoo, with greater mean seed weight, although differences were not significant. In B. rapa the faster-growing but earlier-maturing Chinoli was able to retain more seeds, each of greater weight than in Jumbuck, offsetting a reduced number of pods to give a higher yield.

In the United Kingdom (Mendham et al., 1981a), a typical well-grown early autumn sowing produced a very large number of pods. Very heavy seed losses in the dense canopy of pods produced were offset to some degree by large seeds, with a moderate yield resulting. Contrasting results were achieved with late sowings in different years. A very low yield resulted in one year from few pods on poorly grown plants, which also retained few seeds. A well-grown late sowing in a different season produced more than three times the yield by retaining nearly three times as many seeds per pod, other components being similar to those in the poor-yielding late sowing.

The sequence of development of the yield components and their timing in relation to internal crop factors and interaction with the environment (mainly temperature, radiation and water supply) are

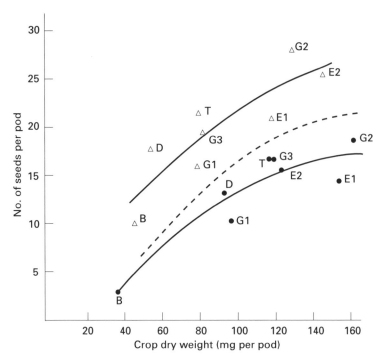

Fig. 2.8. The relationship between crop dry weight at flowering (expressed in mg per pod) and number of seeds per pod in cultivars Marnoo (upper line, △) and Midas (lower line, ●) in Tasmania (Mendham *et al.*, 1984). Letters next to symbols indicate different sites. Broken line indicates relationship for United Kingdom data given by Mendham *et al.* (1981a).

the keys to understanding how crop yields vary, and will be addressed in later sections. This then enables either genotype or management factors to be changed to improve yield and profitability in the range of situations found in practice.

Just measuring yield components at harvest without taking into account crop growth and the timing of formation of the components can be misleading and confusing, as there is usually some compensation between them. A restriction on flower production followed by favourable subsequent conditions, for example (Mendham *et al.*, 1984), gave a crop with few pods (5000 m^{-2}) but with very high numbers of seeds per pod (28), whereas at the same site with irrigation the same cultivar produced 8000 pods m^{-2}, only retained

20 seeds per pod, but produced a higher yield.

Thurling (1974b) showed how components could be treated statistically to isolate each from the previous one in the time sequence of formation, hence removing the effect of compensation between them. In *B. napus* the simple correlation of number of seeds per pod with yield was not significant, but, if the effect of number of pods was held constant, number of seeds per pod could be shown to have a marked effect on yield.

While there will often be an inverse relationship between number of pods and seeds per pod, there will also be a relationship between numbers of both and the size of the crop, or ability to carry them. Mendham *et al.* (1984) (Fig. 2.8) showed how

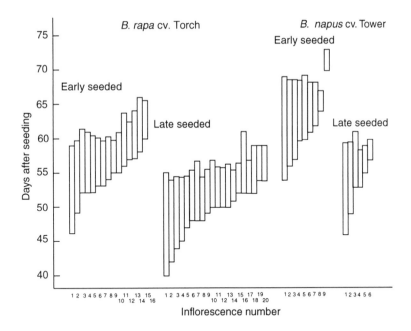

Fig. 2.9. Number of days to flower and duration of flowering of individual inflorescences (mainstem and branches) on plants of early- and late-seeded *B. rapa* cultivar Torch (left) and *B. napus* cultivar Tower (right), in Canada. Source: McGregor (1981).

numbers of seeds per pod increased with crop dry weight at flowering, expressed in mg per pod, which is also in effect an inverse measure of pod number. The two cultivars differed substantially in their ability to retain seeds to final harvest, but both produced similar curvilinear relationships. The Australian cultivar Marnoo retained more seeds at any given crop dry weight per pod, and the maximum was close to the maximum ovule number of about 30. The Canadian cultivar Midas appeared unable to produce more than about 18 seeds per pod even on well-grown crops with few pods, and may be a result of selection for large seeds and high oil content. The broken line on Fig. 2.8 represents data from crops of Victor in the United Kingdom (Mendham *et al.*, 1981a), and is intermediate between the above two cultivars.

Flowering and branching pattern

Tayo and Morgan (1975) describe the pattern of development of plants grown in pots, which is generally similar to the field situation (Mendham and Scott, 1975; McGregor, 1981). Flowering commences on the mainstem, which becomes the terminal inflorescence or raceme, and proceeds acropetally, i.e. from the base towards the tip of the raceme. Axillary buds, i.e. in the axils of leaves on the mainstem, develop into primary branches in sequence downwards, or basipetally. Tayo and Morgan (1975) recorded first flowering on the first, second, third, fourth and fifth branches 3, 4, 6, 7 and 8 days later than the beginning of flowering on the terminal raceme. McGregor (1981) (Fig. 2.9) in Canada identified a similar pattern for *B. napus*, with eight and five branches on early and late sowings respectively, most finishing flowering within a few days of each other. *B. rapa* showed a very different pattern, with much more profuse branching, particularly on the late sowing, in contrast to *B. napus*, and with delayed cessation of flowering on the lower branches.

Early-sown crops of *B. napus* in Europe (Mendham *et al.*, 1981b) at low density (less than 30 plants m^{-2}) may produce more than 12 primary branches, corresponding to the large number of nodes on the main-stem, whereas late- or spring-sown crops or those at higher density may produce between two and seven primary branches. Secondary branches can also develop in the axils of bracts on the primary branches, but in most crops do not produce many pods with seeds.

Pollination and seed set

The biology of the flowers of *B. napus* and the pollination process was studied by Eisikowitch (1981). While self-fertile, pollination was usually poor in the still air of a glasshouse. The sticky, entomophilous pollen was not transferred efficiently by wind, but, when the anthers were touched by an insect or other object, small clouds of pollen grains were observed to burst out. All insects visiting the flowers to collect nectar or pollen were seen to transfer pollen grains to the stigma, effecting either self- or cross-pollination.

Under field conditions (Williams *et al.*, 1987), wind improved the transfer of pollen within or between flowers, probably by shaking and contacting the floral parts. Plants in plots caged with bees had their flowers pollinated faster, shed petals sooner, finished flowering earlier and were shorter than plants caged without bees, but there was little difference in yield. Numbers of seeds per pod were usually higher just after flowering in the plants caged with bees, but in some years more seeds were subsequently lost. Bees will thus assist the maximum number of ovules to be ferti-

lized, but survival to harvest depends on other factors such as supply of assimilate and water.

The availability of beehives close to the crop should therefore be at least of potential benefit on crops able to support large numbers of seeds. In hybrid seed production (Hogarth, 1993), where the female line is male sterile, bees or other insects are essential for effective pollen transfer from the male line. Numbers of seeds per pod in some female lines declined with distance from the pollen source, but in some cases, where plants were well grown, yields were not greatly reduced as seed size was considerably increased to compensate.

Brassica rapa generally benefits from cross-pollination (Williams, 1978), with self-incompatibility present to varying degrees. Both cvs Span and Torch set some seed when self-pollinated by hand, but poor seed set in the absence of a pollinator resulted in extended flowering.

Weather at flowering will thus influence pollination in several ways. Bee activity is less on cold, dull or windy days and a spell of such weather certainly causes poor pollination in hybrid seed crops. In normal open-pollinated *B. napus* crops there is a greater proportion of self-pollination under such conditions. Becker *et al.* (1992) showed that the outcrossing rate in the Swedish cultivar Topas varied from 12 to 47% over five locations, depending on environmental conditions.

Frost at flowering (Mendham and Scott, 1975) may be seen later as a section of stem with aborted buds (very small peduncles remaining), flowers or pods (larger peduncles with the remains of petals or small pods attached).

In the same way as in *B. napus*, the other two amphidiploid species are self-fertile. *Brassica juncea* showed average interplant outcrossing values of 18.7% compared to 21.8% for *B. napus* (Rakow and Woods, 1987). Although there are no published studies of outcrossing rate in

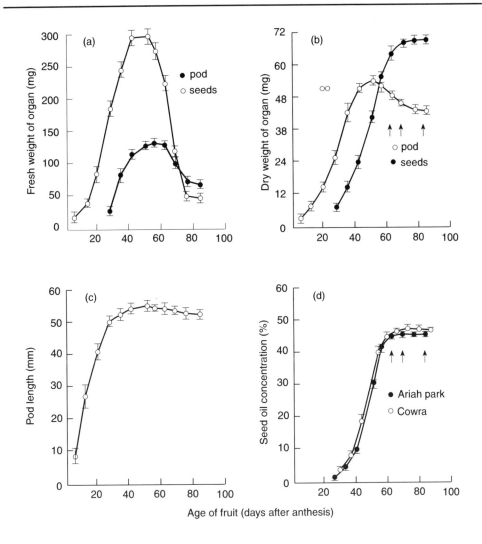

Fig. 2.10. Changes with time in pod walls and seeds of cultivar Barossa grown at two sites in New South Wales, of (a) fresh weight, (b) dry weight, (c) pod length, and (d) oil concentration, with (a) to (c) as a mean of sites. Source: Hocking and Mason (1993).

B. carinata, Salisbury (1991) observed that it has a very similar mode of pollination to the other two species, where the flower is open and the stigma receptive for some time before pollen is released.

Pattern of growth of pods and seeds

An example of growth of pods and seeds under field conditions in Australia is shown in Fig. 2.10, from Hocking and Mason (1993). Flowers numbered 4 to 6 from the base of the mainstem were tagged on a large number of plants and then sampled regularly until maturity. All pods were therefore of similar age and position on the plants, and were some of the first formed. Pods commenced rapid growth in length and then weight within a few days of anthesis, whereas rapid seed growth was delayed for about 20 days. Pods were therefore nearly full length before rapid seed growth started, and seeds had only attained 35% of their final dry weight when pod walls reached full size and weight. The pod walls gained no more dry weight after dehydration commenced, at 50 days from anthesis, whereas seeds increased dry matter by 42% during their period of dehydration, from 50 to 75 days after anthesis. Oil content increased in a similar way to seed dry weight, reaching a maximum percentage at about 60 days, but the total oil content increased further with dry matter accumulation. The mature fruits (pods) contained around 23 seeds, or 61% of total fruit dry weight.

When rapid pod growth is commencing, stem and branch dry weight accumulation and extension are close to their maximum (Mendham et al., 1981a), and leaf area is already declining rapidly (Fig. 2.6). There is thus a clear sequence of growth in which leaves are followed by stems, pod walls and finally seeds.

Determination of the number of pods

The first of the components to be determined, number of pods m^{-2}, overlaps to a considerable degree with the second, number of seeds per pod, and similar factors affect both.

Tayo and Morgan (1975) (Fig. 2.11) recorded flowering and pod production on a daily basis on plants of a spring cultivar grown in a glasshouse, with results generally similar to those recorded in less detail in the field. The number of days were recorded from the start of flowering on the mainstem, which carried 25% of the total flowers on the plant, and 38% of the final number of pods. The success rate of flowers which formed pods decreased from 68% on the mainstem to 22% on the fifth branch, with only 45% over the whole plant. Very few of the many flowers which opened after the 18th day on any branch carried pods through to harvest. While the period of effective flowering is longer in the field under lower temperatures, the lost flowers, in addition to being a wasted resource, are a deleterious factor as the flower layer may reflect or absorb up to 60% of incoming solar radiation (Mendham et al., 1981a), shading leaves and earlier-formed pods.

While position on the plant is a major determinant of the likely success of a flower in forming a pod, environmental factors will also operate. Leterme (1988b) showed that there was a relationship between the amount of solar radiation intercepted (by the whole canopy) per flower and its likelihood of success. With less than 20 kJ per flower during its flowering period the success rate was roughly proportional to radiation, although the more extreme losses (down to 20% success) were produced by shading treatments rather than in normal field situations. With more than 20 kJ per flower about 70 to 80% success was recorded regardless of radiation level. Water or nutrient stress will also either curtail flowering or limit its success rate and a direct relationship was shown between the date of last nitrogen application and success rate of flowers.

Crops in field situations may also lose half or more of potential pod sites (Fig. 2.12, from Mendham et al., 1981a). The early sowing produced an enormous number of flowers (25,000 m^{-2}). A small proportion was lost at and just after flowering, presumably due to poor pollination. The majority were lost as either abscised flowers or young pods, or where all seeds aborted and pods remained small. Fewer than half survived as pods until final harvest, but even on the latest sowing with 6000 potential pods m^{-2}, only slightly more than half survived. Most of the losses occurred in a 3–4-week period after 'full flower', taken as 7 days after 50% of plants had at least one flower open.

This 3–4-week period corresponds

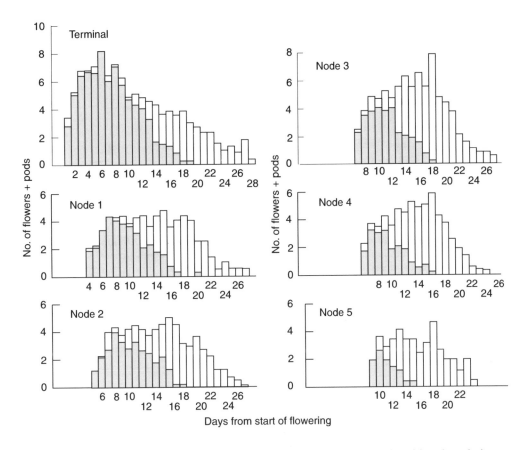

Fig. 2.11. Numbers of flowers which opened on each inflorescence (total bars) at daily intervals after the beginning of flowering of glasshouse-grown plants, and the numbers of these which formed pods which were retained to maturity (shaded bars). Source: Tayo and Morgan (1975).

to the time of maximum growth rate of pod hulls, which would also be a period of maximum demand for assimilate and nutrients.

Determination of number of seeds per pod

Figure 2.13 (from Mendham *et al.*, 1981a) shows the changes in number of seeds per pod as pods grew in the early and late sowings shown in Fig. 2.12. From a consistent ovule number of around 30 per pod at flowering, the numbers of surviving seeds declined over about a 3-week period, being stable in the last 3 to 4 weeks before

maturity. Data given are the averages of pods of different age, although stratification of the canopy into upper, middle and lower sections would have reduced variation within each level to about a week.

In the denser pod canopy of the early sowing, seed losses were heavier and concentrated in lower levels of the canopy, whereas in the late sowing, with less competition between pods, there was little difference in seed abortion between levels within the crop.

The period over which seeds were lost coincided with the main growth of pod hulls, as seen in Fig. 2.10, and was before

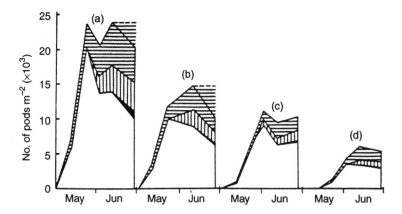

Fig. 2.12. Development of pod canopies in crops sown in the United Kingdom on (a) 25 August, (b) 13 September, (c) 24 September and (d) 13 October. Open (lower) section = pods with seeds, solid section = shattered pods, vertical shading = empty pods and horizontal shading = abscised flowers or pods. Source: Mendham *et al.* (1981a).

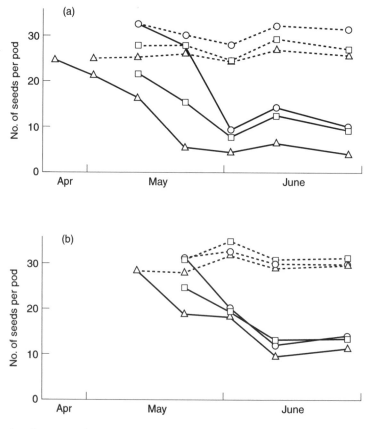

Fig. 2.13. The change with time in potential (dotted line) and surviving (solid line) numbers of seeds per pod in crops sown in the United Kingdom on (a) 25 August and (b) 13 October. Upper, middle and lower sections of the pod canopy denoted by circles, squares and triangles, respectively. Source: Mendham *et al.* (1981a).

the seeds themselves commenced rapid increase in dry weight.

Tayo and Morgan (1979), working with glasshouse-grown plants which had leaves shaded or removed at intervals around flowering time, also concluded that there was a critical 2–3-week period after flowering for seed abortion, the shorter period being due to higher temperatures in the glasshouse. The above work used destructive sampling to obtain averages but was broadly confirmed by *in vivo* observations of developing pods by Pechan and Morgan (1985). They used X-radiation and photography of growing pods of plants in a controlled environment at 20°C, and removed inflorescences other than the mainstem. Pods commenced rapid increase in length between 2 and 8 days after flower opening, and extension continued for 10 days. Some ovules failed to develop in the first few days after anthesis, presumably due to failure of fertilization. Others aborted 4 to 8 days after flowering, most frequently in pods near the apex of the inflorescence and in distal seeds in each pod rather than in proximal ones. Increases in pod width and seed size commenced 2 days after increase in length and continued for about 20 days. The above pattern is similar to that observed in the field, but all times are shorter. The higher temperature would account for most of the difference, but the lack of development of a pod canopy and the resulting competitive stresses might also explain the absence of a prolonged phase of seed abortion.

Leterme (1988b) estimated the phase of increase in pod length, and likely abortion of seeds, as between 200 and 300°C d in France. The 19 to 25 days estimated by Mendham *et al.* (1981a) as the length of this phase in some late sowings took about 300°C d (mean temperatures of 15 to 12°C, respectively), and the 20 days' lag between growth of pods and seeds in Australia (Fig. 2.8) was also at a mean temperature of about 15°C. The 10 days recorded at 20°C

(i.e. 200°C d) by Pechan and Morgan (1985) therefore suggests curtailed pod length growth at high temperatures, and also early abortion. They showed that sections of pod near aborted seeds made little further growth in length or width, so the two processes are interrelated.

Determination of final seed size

Final seed size varies greatly between genotypes (Table 2.6) and also within a cultivar over a range of conditions, for example 3.5 to 5.7 mg in crops of the winter cultivar Victor in the United Kingdom recorded by Mendham *et al.* (1981a). They fitted a linear approximation to the main period of seed growth with time, which when extrapolated back to zero dry weight gave an approximate beginning to the seed growth phase and an end to the phase of seed number determination. For the data of Fig. 2.10, for example, this would give a point about 23 days after flowering. A 'duration of seed growth' can then be determined in days, and for the data of Mendham *et al.* (1981a) ranged from 35 to 55 days. The rate of development, as the inverse of number of days, was proportional to temperature. The rate of growth per seed, as distinct from development, ranged from 0.08 to 0.12 mg day^{-1} and was a function of assimilate and water supply, to be examined in later sections in terms of internal (crop size, leaf and other photosynthetic area, carbohydrate reserves) and external factors (radiation, water supply, temperature).

Yield components within the pod canopy

Vigorous early-sown winter rapeseed crops produce dense canopies of pods, within which there is intense competition, as shown by the losses of pods (Fig. 2.12) and seeds (Fig. 2.13). In these crops, pod and seed survival is usually better (Norton, G., *et al.*, 1991) near the top of the mainstem and upper branches, with better distribution of solar radiation. The winter cultivar Rafal (Daniels *et al.*, 1986) also retained more seeds in the top 15 pods of the mainstem than in the basal 15 pods (about 13 seeds compared to 7), offset to a small degree by lower mean seed weight.

Mendham and Scott (1975) also showed that wastage of pods and seeds was greatest at lower levels of the canopy in early sowings, but that in the lighter pod canopies of late sowings losses may be at least as high near the apex of inflorescences. The hierarchical arrangement of branches and pods would suggest that the lower, earlier-formed pods on the mainstem and first few branches should have a developmental advantage (Daniels *et al.*, 1986) in terms of timing as well as proximity to assimilate from leaves and stems lower down the plants. This has been shown in glasshouse-grown plants (Tayo and Morgan, 1975), and on spring-sown crops in Canada (Clarke, 1979) with fewer pods. There, the number of seeds per pod and their mean weight were higher in pods from the lower half of the mainstem than those on the upper half. The mainstem as a whole carried the most seeds per pod and the heaviest seeds, with a progressive decline in both components on branches 1 to 3, in accordance with their sequence of development. The number of developing seeds per pod at the late flowering stage was just as high (30) on branch 3 as on the mainstem, but greater abortion in the subsequent 4 weeks reduced numbers to 21 instead of 28 on the mainstem. All numbers of seeds were much higher than typically survive in European winter crops.

Sources of assimilate for pod and seed growth

Radiation supply and seed number determination

Before flowering, leaves are almost the only green photosynthetic organs. During and after flowering, however, the major part of the leaf area is rapidly lost on most crops (Fig. 2.6), and green areas on stems and then pods have been shown to take over photosynthesis, to be discussed below. There appears to be a critical time in rapeseed, however, much more so than in other annual crops, when the mass of yellow flowers at the top of the crop typically reflects or absorbs around 60% of incoming total solar radiation (Mendham *et al.*, 1981a). Norton, G., *et al.* (1991), working with dense early autumn-sown crops in the United Kingdom, recorded 70 to 80% of incident photosynthetically active radiation absorbed or reflected by the top 60 cm of the crop, consisting of flowers and young pods, neither of which had significant photosynthetic capacity. The flowers reflected around 50% of total radiation and 20% of photosynthetically active radiation.

Rao *et al.* (1991) (Fig. 2.14) produced similar results for a conventional cultivar with petals in Tasmania, with only 40% of total radiation reaching the leaf canopy in crops at full flower. An experimental line 'Apetalous', related to the conventional cultivar but with virtually no petals, allowed 70–75% of radiation through to the leaf canopy, resulting in greater leaf persistence, better seed survival and growth, and 8 to 48% higher seed yields in different experiments and treatments.

Yates and Steven (1987) further examined the reflection and absorption of radiation by conventional winter rapeseed crops. Photon reflectivity in the visible region of the spectrum increased linearly with per cent flower cover from 0.047 during vegetative growth to 0.195 at full flower (75% cover), due to the properties of the bright yellow petals. Flowering canopies reflected more and absorbed less radiation than vegetative canopies between 500 and 700 nm, but reflected less and absorbed more between 400 and 500 nm. The

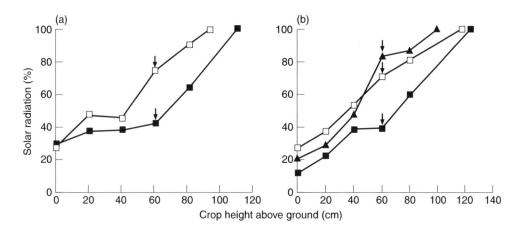

Fig. 2.14. Percentage solar radiation transmitted at different heights within the crop canopies of cultivar Marnoo (■) and an apetalous line (□) in (a) 1984 and (b) 1985 at peak flowering in Tasmania. ▲ denote Marnoo only at 20% flowering. Arrow indicates base of inflorescence. Source: Rao *et al.* (1991).

changed spectral composition as well as the reduced amount of radiation available to leaves underneath the flower canopy therefore contributed to reduced photosynthetic rate and accelerated senescence. It was estimated that crop growth over the flowering period was reduced between 6 and 27% for a range of cultivars which differed in flower cover. The least reduction was in cultivar Perle, which produced the fewest flowers and pods. If sparse flowering is associated with many seeds per pod, this could be an appropriate strategy to increase yield, as was shown for well-grown late sowings by Mendham *et al.* (1981a) (Table 2.6).

At the time of the 'radiation blackout' in normal crops, the earlier-formed pods which should make the major contribution to yield are going through the phase of seed number determination. The excessive flowers shown in Fig. 2.11 contribute to seed abortion and extend it further.

Pods at that stage are largely 'heterotrophic' (Leterme, 1988b), in that they have not established their photosynthetic capacity, and so, at the time that pod walls are growing rapidly, the assimilate supply

from leaves is greatly reduced and hence the abortion of young seeds is even more likely.

Direct relationships were shown by Mendham *et al.* (1981a) (Fig. 2.15) between an estimate of the amount of radiation intercepted by each pod over the phase of seed number determination and the number of seeds per pod finally achieved. Late sowings with relatively few pods were much more responsive than early sowings, with many pods probably at or below the compensation point for photosynthesis, or

Table 2.7. Differences in stomatal frequency (number mm^{-2}) among epidermal tissue types of rapeseed.

Tissue	Species	
	B. rapa	*B. napus*
Adaxial leaf	45	92
Abaxial leaf	105	125
Stem	49	32
Pedicel	56	48
Pod	85	72
Beak	62	75

Source: Major (1975).

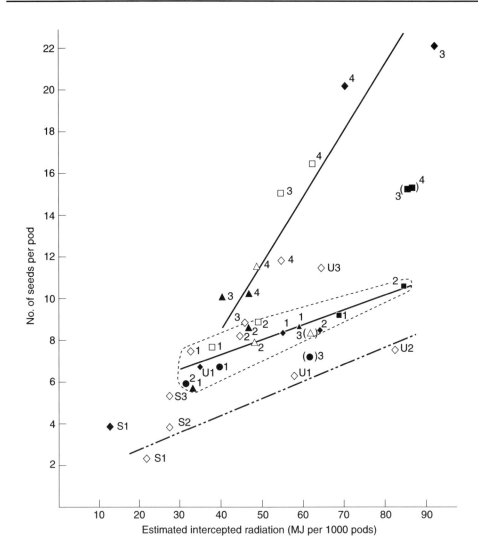

Fig. 2.15. The relationship between the estimated amount of radiation intercepted per thousand pods over the phase of determination of number of seeds, and number of seeds per pod of crops of cultivar Victor sown on a range of dates (numbers next to symbols indicate order of sowing), in seven seasons (different symbols) in the United Kingdom. Regression lines fitted for late sowings (3–4, except those in brackets) and early sowings (1–2). Lower (broken) line indicates shaded (S) and unshaded (U) crops of cultivar Rapol in separate experiments. Source: Mendham et al. (1981a).

Table 2.8. Green area index of different organs of two cultivars. SM indicates approximate stage of development (Sylvester-Bradley and Makepeace, 1984).

Stage	SM	Total		Leaf		Stem		Pod	
		Rafal	Victor	Rafal	Victor	Rafal	Victor	Rafal	Victor
Vegetative	2,00	4.15	4.95	3.45	4.10	0.80	0.85	–	–
Late flowering	4,9	4.25	4.50	2.10	2.80	1.60	1.35	0.40	–
Pod walls fully grown	5,9	3.65	5.10	0.25	1.85	1.70	1.50	1.70	1.85
Physiological maturity	6,6	3.20	4.10	0	0.80	1.60	1.60	1.60	1.70

Source: Norton, G., *et al.* (1991).
–, Not applicable.

at least for active supply from pod wall to seed. Shading of some early-sown crops produced a similar response to either lower incident radiation levels or more pods in unshaded crops.

Contribution of photosynthate from different organs

Major (1975) showed that all green surfaces of the crop contain stomata, essential for photosynthesis in higher plants, but that the frequency varied between species and organ (Table 2.7). The abaxial leaf surfaces had the highest stomatal frequency in both *B. napus* and *B. rapa*, with the adaxial surface next highest in *B. napus* but lowest in *B. rapa*. Pod and beak (style) tissues were intermediate, and stem and pedicel lowest in stomatal frequency. Pods may also be important in refixing respired CO_2 from growing seeds as well as external CO_2 uptake.

The green areas of the different surfaces were measured (Table 2.8) by Norton, G., *et al.* (1991) on a taller (160 cm), older cultivar (Victor) and a more recent, shorter cultivar Rafal (100 cm). Leaf area of both declined sharply during flowering, more so on Rafal, but was largely replaced by stem and then pod area. By the time maximum pod hull weight and area were attained, leaf area had declined to almost nil on Rafal. About one-third of the maximum area persisted on Victor, although

very little radiation penetrated to that level. When the photosynthetic capacity of the different organs was measured by feeding with labelled $^{14}CO_2$, pod and stem tissue were found to be less efficient than leaves on an equivalent area basis, probably related to the lower stomatal frequency. Due to position in the canopy, 95% of the gross photosynthesis of Rafal took place in the pod layer (mainly pods and stems). Measurements were not made on Victor, which did, however, lodge badly and compress the pod layer, resulting in heavy seed abortion and even less effectiveness of the remaining leaf canopy.

Other workers have used similar techniques to estimate activity in different organs and their contribution to overall assimilate supply. In Canada, Major *et al.* (1978) fed $^{14}CO_2$ to different parts of field-grown plants at the early pod-fill stage and determined distribution through the crop, either immediately after feeding or after 48 h (Table 2.9). Material fixed by the lower stem mainly stayed there, but the upper stem exported to pods and particularly seeds. Lower leaves mainly exported basipetally to stem and roots, whereas upper leaves translocated acropetally to stems, pods and seeds. Lower pods translocated strongly to their own growing seeds, whereas upper pods at a young stage (before active seed growth began) largely retained assimilate in pod walls.

Table 2.9. Distribution of $^{14}CO_2$ to different parts of plants as a percentage of total 48 h after treatment of the nominated parts.

Plant part treated	Percentage after 48 h				
	Root	Stem	Leaf	Pod	Seed
Lower stem	3.7	88.9	1.6	2.0	3.9
Upper stem	1.2	38.1	4.7	20.7	35.2
Lower leaf	31.8	31.9	24.4	1.9	1.0
Upper leaf	0.6	19.8	9.6	27.6	38.3
Lower pod	0.5	0	0	47.8	51.5
Upper pod	4.3	0.5	0	84.2	11.1

Source: Major *et al.* (1978).

Chapman *et al.* (1984) showed how uptake and export changed with time, by feeding $^{14}CO_2$ to whole plants in the field and monitoring amounts in different organs 0.5 and 24 hours later (Fig. 2.16). At early flowering, leaves were still the major area of activity (67% of total carbon fixed) with significant amounts fixed by stems (30%) which also imported carbon. By the rapid pod-fill stage, pods in the upper canopy were most active (47%), then stems (40%) which exported strongly to pods. The small remaining area of leaf lower in the profile fixed the remaining 13%.

Keiller and Morgan (1988a) fed $^{14}CO_2$ to individual leaves at 1–4-day intervals on plants in a growth cabinet at 20°C. Between 1 and 13 days after the start of anthesis, the apical 3 to 5 cm of inflorescence above the labelled leaf was a strong sink for assimilate, whereas older flowers and newly formed pods were weak sinks. This corresponds to the time when seed abortion is likely, and when upper flowers are absorbing most of the incoming radiation in field crops. Active seed growth, beginning on the lower pods, then attracted assimilate strongly and appeared to be involved in switching off the assimilate supply to the apex after about 16 days,

and hence cessation of flowering (times were shorter than under lower temperatures in the field, as explained earlier). The lowest branch measured (number 3) was less competitive for assimilate than the mainstem or upper branches. Progressively less carbon was invested in pods of lower branches, and more in the branch structure, which as previously noted has to be longer to raise pods up to the height of the upper branches.

Factors affecting cessation of flowering and seed abortion

The beginning of active seed growth on the first formed pods on the mainstem coincides with the apparently coordinated cessation of flowering in apical regions of all inflorescences on the plant. Keiller and Morgan (1988b) removed the basal 10 to 20 buds, flowers or pods from the mainstem. This resulted in a greater duration of flowering and longer stems, with the largest effect when removal was done at anthesis, lesser effects 6 days later, and no effect after 12 days. Similar results were achieved by application of the plant growth regulator benzyl adenine, with or without gibberellins, with no effect from treatment at 14 days after anthesis or later. The

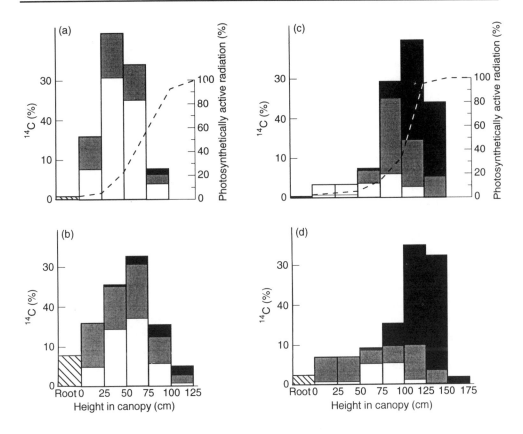

Fig. 2.16. Percentage ^{14}C taken up by different parts of field-grown plants at 0.5 h after feeding (a, c), and taken up or transported by 24 h after feeding (b, d), at early flowering (a, b) and rapid pod fill (c, d). ■, Reproductive parts; ▨, stem; □, leaves; ◫, roots; --- percentage transmission of photosynthetically active radiation (PAR). Source: Chapman *et al.* (1984).

conclusion was that there was a critical change in the apex at about 10 days after first flower opening (under their high temperature conditions) after which the apex was no longer receptive to extended development.

Limitation of assimilate supply to the apex would appear to be the main cause of cessation of flowering, and hence limitation of pod number, as well as the cause of seed abortion in pods further down which are passing through the critical stage. Bouttier and Morgan (1992) endeav-

oured to determine whether the limitation of assimilate supply was the controlling factor, or whether it was a function of loss of sink strength due to changes in phytohormones. Techniques were developed for growing flowers or pods on stem explants *in vitro*. Buds could not be grown on a basal medium of mineral nutrients and sucrose without addition of hormones, whereas open flowers and pods could, indicating that by flowering these organs can synthesize their own phytohormones. Addition of gibberellins and cytokinins did enhance pod elongation and growth to a small extent, suggesting suboptimal production by the pods themselves.

Increased sucrose concentration, however, in the presence of minerals had a large effect on number of seeds retained per pod, and hence also pod length and weight. Without minerals, the addition of sucrose caused virtually all seeds to abort. These results indicate that insufficient assimilate supply rather than hormonal factors is likely to be the major cause of pod and seed abortion in the absence of other stress (e.g. water or mineral nutrients).

Improvement of the environment within the pod canopy

Removal of flowers and pods, as by Keiller and Morgan (1988b) or Tayo and Morgan (1979), changes more than just source/sink relationships. In an attempt to keep the radiation profile of field-grown crops undisturbed, and plant damage to a minimum, Tommey and Evans (1992) removed just the stigma, style and anthers from flowers on a daily basis, leaving petals and sepals intact. When pods were prevented from forming on the mainstem and upper branches, the lower branches produced more flowers and pods, although not enough to compensate for the removal. The reduced assimilate demand from the normally active upper sections was presumed responsible for the extra activity lower on the plant, fed primarily from lower leaves supplying assimilate acropetally. Removal of flower parts on branches below the first had little effect on total yield per plant, and removal from the fifth branch downwards actually increased yield, with better pod survival and growth on untreated branches and the mainstem. Removal of competition at the time when seed numbers were being determined was probably the reason for the improvement, as the radiation interception by flowers was still operating. Restriction

of branching may therefore be beneficial, by either breeding or manipulation of management factors such as plant growth regulators.

The limited number of pods produced by late sowings has been shown to be an advantage in situations where crops were well grown before flowering. These included occasional seasons in the English Midlands where spring began early (Mendham et al., 1981a), or in areas with a mild climate where early spring growth was a normal occurrence, as observed in south Wales by Jenkins and Leitch (1986). There, consistently high yields were achieved from late sowings, which were better than early sowings in two out of three seasons. This was for the same reasons as shown by Mendham et al. (1981a), namely that many more seeds per pod were retained, more than compensating for fewer pods.

Other potential improvements to light relationships within the pod canopy include the apetalous character, as discussed previously, which improves radiation supply to leaves and young pods by allowing more radiation through the canopy of later-opening flowers. The characteristics of the pods themselves could also be improved. Some B. rapa lines have characteristically upright pods, set at an acute angle to the stem. By analogy to the narrow, erect-leafed habit in cereals such as rice, this should allow better distribution of radiation over the pod surfaces. This was shown by Rao and Mendham (1991a) in Chinoli to be of some potential benefit in terms of efficiency of conversion of intercepted radiation to dry matter. However, crops of higher yield potential tended to lodge, negating the effect of pod angle. While lodging could be controlled by growth regulators or plant density, a yield advantage could not be shown in this genotype.

The 'long pod' characteristic found in B. napus (Chay and Thurling, 1989) also has potential. Long-pod lines generally produced more ovules per pod, but, even where fewer ovules were produced, seed survival was better and seed numbers per pod greater. Under the conditions experienced in Western Australia, yields were not significantly greater as numbers of pods per unit area were less, but this may be an advantage in high potential yield situations. There is some concern, however, that the extra length of pod wall is being produced at the time seeds are going through the abortion phases, so the pod walls will require even more assimilate at this critical time. When long and short pod types were compared, the short pod produced more seed per unit length of pod, suggesting a more efficient distribution of assimilate than in the long pod. Clearer evidence of the value of this character is needed before its use could be recommended in breeding.

Contribution of assimilate reserves to yield

As reviewed above, photosynthesis at each stage of development appears to be the main source of assimilate for current growth. Excess photosynthate, particularly over winter when growth is slow, is stored in roots, leaves and stems. Quilleré and Triboi-Blondel (1988b) in France measured stored carbohydrate (starch and ethanol-soluble sugars) as glucose equivalents. In the plant as a whole, reserves built up by early winter were maintained, and then fell during stem extension and flowering. Reserves then increased as pods grew, and fell to almost zero by the end of seed fill. In individual organs, stem reserves were built up during stem elongation, but more than offset by rapid declines in leaf and root reserves. Reserves built up in pod walls were transitory, being moved to seeds when their rapid growth commenced. Reserves built up pre-anthesis were only estimated to contribute about 10% directly to the final harvested yield, but post-

anthesis conditions were generally favourable for assimilate supply from current photosynthate. Bilsborrow and Norton (1984) and Norton, G., et al. (1991) also found little evidence that mobilization of reserves was important for seed growth. In experiments where crops had leaves either removed or shaded for varying times before anthesis, Evans (1984) showed that restriction of the assimilate supply late in the vegetative phase had little effect on yield, and interpreted this as the effect of mobilized reserves from stems contributing to the compensatory growth. Under stress conditions where post-anthesis growth is curtailed, reserves presumably contribute a much larger proportion, although seed yields may still not exceed post-anthesis growth (e.g. late sowings by Mendham et al., 1990).

Seed growth and final size

As noted previously, the duration of seed growth is largely determined by temperature. The rate of growth, however, is determined by the supply of assimilate, nutrients and water. Assimilate supply will depend upon the complex interaction of the range of factors reviewed above. Pre-flowering growth sets up the photosynthetic potential of leaves and stems, with reserves making a significant contribution to final growth in some circumstances. The number of pods and seeds within them which survive the competitive stresses, assimilate shortage and changing nature of the photosynthetic surfaces during flowering determine the size of the sink for photosynthates. The green areas of the pod walls, together with the amount of radiation they intercept, are then the final elements in the supply of assimilate to the seeds they contain.

Mendham et al. (1981a) interpreted the complex supply and demand situation when comparing seed growth rates with mean daily solar radiation receipts during the seed fill period. In a year with heavy losses of seeds but excellent crop growth before flowering, followed by favourable conditions for seed fill, the crop appeared 'sink limited' and individual seeds made rapid growth per day and per unit of radiation received, to a large final size. In years when large numbers of seeds had been retained on crops of moderate size, and particularly with subsequent high temperatures and water stress, rates of seed growth per unit of radiation received were low and interpreted as 'source limited'. Other years including those with high-yielding late sowings appeared to have source and sink roughly balanced, with an intermediate seed growth rate per day and per unit of radiation, and with final seed weight near average for the cultivar.

Seed size varies less than the earlier-formed components, and hence the likely strategy for highest yield is to retain as many seeds per pod as the size of the crop will permit, and then allow seed growth rate and duration to respond to prevailing conditions. This could not be taken to extreme levels where seed size and oil content were lowered to unacceptable levels, but seed retention was shown (Mendham et al., 1984) to be the key to high yield in new Australian cultivars with a substantial proportion of Asian parentage. By comparison, a Canadian cultivar produced a lower yield because it appeared that the strategy of the breeder had been to aim for maximum seed size and oil content. Under the range of conditions experienced, the main Australian cultivar tested produced consistently higher yields, but seeds were always smaller. Oil content was only markedly (2 to 4%) lower when crops ripened under stress conditions (high temperature and/or water stress).

A summary of pod and seed growth

The model of Leterme (1988b) summarizes the factors affecting a pod from flowering to harvest, in three main phases.

1. *Increase in pod length.* Duration 200 to 300°C d. Pods are 'heterotrophic', i.e. relying on imported assimilates. They attain their maximum length, and the number of seeds is largely determined. The main variables responsible are leaf area index at the beginning of the period, number of flowers per unit area, duration of flowering, radiation, and temperature. These factors interact to determine the number of pods and seeds.

2. *Maximum growth rate of pod walls.* Duration about 300°C d. Pod walls attain maximum size and area, but seed growth is limited. Pods are 'autotrophic', fixing most of the carbon required for pod and seed growth at a rate of about 1.3 g MJ^{-1} solar radiation.

3. *Maximum growth rate of seeds.* Duration about 300°C d. Largely governed by the surface area of pods and stems and the amount of radiation received. Most carbon goes into seed growth.

During the second and third phases, growth rates are limited by either the amount of radiation intercepted, or the potential maximum growth rate of pods and seeds. For pod walls this appears to be about 3.4 mg per 100°C d per 1 cm of length, and for seeds about 1.5 mg per 100°C d per seed.

Quality changes during seed development

The duration of seed development is largely determined by temperature, with seed of spring cultivars produced under Canadian growing conditions taking approximately 50–60 days to mature, and seed of winter cultivars produced under European conditions taking 90–100 days. During seed development a number of quality changes occur before the final chemical composition of the mature seed is realized.

Oil content

Oil is the most valuable component of the seed, being utilized for both edible and industrial purposes. It serves primarily as a source of energy and carbon precursors in germinating seed.

Synthesis of storage lipids occurs in the seed and thus oil composition is genetically determined by the embryo. During seed development the rate of oil deposition follows a sigmoid curve (Fig. 2.10). Droplets of storage oil are first evident about 18 days after pollination. They increase in size and number between approximately 20 and 30 days after pollination. Oil content reaches a plateau at physiological maturity with little further change occurring until seed maturity (Fowler and Downey, 1970; Rakow and McGregor, 1975). In a comparison of zygotic and microspore-derived rapeseed embryos Pomeroy and Sparace (1992) observed a similar pattern of oil accumulation. Fatty acids represented only 3.5% of the dry matter of microspore-derived embryos after 10 days in culture but rose steadily to 33% by day 42, after which there was little further increase.

At seed maturity, about 80% of the oil is concentrated in droplets in the cotyledonary cells. Oil content of the hypocotyl and radicle of mature seed is low and the seed coat contains only 7–12% seed oil (Fowler and Downey, 1970; Stringam *et al.*, 1974).

Among environmental factors which regulate oil content, temperature has been found to be one of the most important, with high temperatures reducing oil content (Canvin, 1965). Irrigation can increase oil content (Krogman and Hobbs, 1975), while waterlogging (Cannell and Belford, 1980) and water stress (Mailer and Cornish, 1987) can reduce it. High nitrogen fertility tends to reduce oil content (Krogman and Hobbs, 1975; Holmes and Ainsley, 1979). Oil content can also be reduced when frost prematurely arrests seed development (Daun *et al.*, 1985). This finding is in agreement with observations of Diepenbrock and Giesler (1979) who showed that the percentage of protein in developing seed is established well in advance of the oil percentage. Environmental factors tend to have a reciprocal effect on oil and protein content.

Oil quality

Fatty acids occur in seeds predominantly in the form of triacylglycerols. The properties and uses of *Brassica* oils are determined primarily by their fatty acid composition.

Romero (1991) found that the fatty acid composition of developing seed of low erucic acid *B. rapa* lines changed substantially 15–36 days after pollination. The percentage of oleic acid (18 : 1) increased rapidly while palmitic (16 : 0), stearic (18 : 0) and linolenic (18 : 3) acids declined. Linoleic acid (18 : 2) decreased initially, then stabilized at levels comparable to that found in the mature seed. During the latter part of seed development, from 36 days onwards, no substantial change in fatty acid composition occurred. A similar pattern of fatty acid accumulation was observed for microspore-derived embryos of low erucic acid rapeseed (Pomeroy and Sparace, 1992). High erucic acid *B. rapa* lines initially showed similar changes in fatty acid composition with development (Romero, 1991). However, significant amounts of eicosenoic (20 : 1) and erucic (22 : 1) acids were evident at 24 days after pollination and the amounts increased rapidly over the next 14 to 21 days. The pattern of accumulation was consistent with a stepwise elongation of oleic acid to eicosenoic acid, then to erucic acid.

Different parts of mature seed of high erucic acid rapeseed lines have been observed to have different fatty acid compositions with the hypocotyl and seed coat having relatively higher palmitic and linoleic acids, but lower erucic acid content than the cotyledons (Appelqvist, 1969). Bechyne and Kondra (1970) observed that pod position had a significant effect on fatty acid composition when plants were grown in a controlled environment, with the seed from earlier-formed pods having a lower linolenic and higher erucic acid content than later-formed pods.

The level of the polyunsaturated fatty acids, linoleic and linolenic acids, is strongly influenced by environment during oil deposition and seed maturation. Downey (1983) reported that in low erucic acid lines higher temperatures during oil deposition tended to reduce the level of the polyunsaturated fatty acids, leading to higher oleic acid. Canvin (1965) and Appelqvist (1968) both reported that, in high erucic acid lines, higher temperatures caused a reduction in linoleic and erucic acids and an increase in oleic acid. Daun et al. (1985) observed that frost-damaged seed contains more saturated fatty acids than fully mature seed, presumably due to a premature termination of fatty acid accumulation.

Protein content

The relatively high protein content of rapeseed meal, in combination with a well-balanced amino acid composition, makes rapeseed meal a valuable source of protein in animal diets, especially for non-ruminants.

Finlayson and Christ (1971) reported a rapid nitrogen accumulation in the early stages of seed development. Storage protein begins to accumulate when the embryo commences to grow rapidly to replace the endosperm and fill the fully expanded seed coat. Crouch and Sussex (1981) found that the onset of seed protein accumulation coincided with rapid cell expansion and rapid increase in embryo weight.

Most of the protein in mature seed is found in the cotyledons. King et al. (1977) reported 76% of the protein in the cotyledons, 17% in the rest of the embryo and only 7% in the seed coat.

Glucosinolate content

High levels of glucosinolate hydrolysis products can adversely affect the feeding value of rapeseed meal. Studies have shown that glucosinolates are derived from amino acids and have a common biosynthetic pathway (Fenwick et al., 1983). While pod tissues appear to be the main site of synthesis of glucosinolate accumulated in the seed, some glucosinolate may be synthesized in other parts of the plant and transported to the seed (Lein, 1972; De March et al., 1989). Decrease of glucosinolate content in the pod was less than the accumulated amounts in the seed (De March et al., 1989) indicating either that glucosinolates, or glucosinolate precursors, were supplied by other plant parts, or glucosinolates synthesized in the pod were transferred to the seed prior to accumulation in the pod. Booth et al. (1991) reported a noticeable decline in the glucosinolate content of the vegetative tissue while the seed was filling, consistent with a redistribution of glucosinolates.

Trace amounts of seed glucosinolates were observed as early as 8 days after pollination in spring rapeseed by Kondra and Downey (1969) and De March et al. (1989). They also observed that seed glucosinolate content increased rapidly between 15 and 30 to 35 days after pollination, with accumulation subsequently occurring at a much slower rate.

Fieldsend et al. (1991) reported differences in the pattern of glucosinolate accumulation in winter rapeseed between plot sites. After a lag, seed glucosinolate content increased almost linearly with seed age. At Newcastle, in the north of England, the phase of linear increase in content started 30 days after pollination and lasted for a further 30 days, with glucosinolate content then remaining relatively constant for the final 30 days of seed development. At Rothamsted, 200 miles south of Newcastle, however, the concentration started to increase only after 45 days from pollina-

tion and continued to increase for a further 45 days, almost up to seed maturity.
Booth *et al.* (1990) observed cultivar differences in the timing of the onset of
accumulation. Merrien *et al.* (1991) found
that for the high glucosinolate cultivar,
Darmor, after a lag phase, glucosinolate
accumulation continued up to maturity,
120 days after pollination. For the low
glucosinolate cultivar, Samourai, glucosinolate accumulation stopped 30 days
before physiological maturity. Changes in
the relative content of individual glucosinolates during seed development have
also been reported (Booth *et al.*, 1990;
Merrien *et al.*, 1991). Josefsson (1970)
found that the embryo contained most of
the glucosinolates in the mature seed, while
the seed coat contained only a relatively
small amount.

A significant relationship between pod
and position on the plant and glucosinolate
content of the seeds has been observed by
Kondra and Downey (1970) and Booth *et
al.* (1990) with earlier-formed seed having
higher levels of glucosinolates than later-
formed seeds. Seed glucosinolate content
can vary greatly in rapeseed, depending on
growing conditions and agronomic practices, with factors such as nutrition, sowing time and water regime implicated
(Fenwick *et al.*, 1983). However, attempts
to demonstrate a consistent effect of sowing or maturity time on glucosinolate
content have been largely unsuccessful
(Cooke *et al.*, 1987; Salisbury *et al.*,
1987). Higher seed glucosinolate contents
have been associated with water stress
(Mailer and Cornish, 1987), high temperatures during seed growth, and disease
(Salisbury *et al.*, 1987). Sulphur application can also increase seed glucosinolates,
with the glucosinolate content of high
glucosinolate lines more responsive to
sulphur than that of low glucosinolate lines
(Booth and Walker, 1992). Slightly
reduced glucosinolate content in frost-
damaged seed is consistent with growth

being arrested prematurely (Daun *et al.*,
1985).

Chlorophyll

High seed chlorophyll content at harvest
downgrades rapeseed and mustard, and
increases oil refining costs. It is a problem
particularly associated with environments
where ripening occurs under decreasing
temperatures, and where frost may kill seed
prematurely. Temperature during seed
ripening was found to have an influence
on the rate of seed chlorophyll degradation,
degradation being slower under cooler
temperatures (Ward *et al.*, 1992). Rapid
moisture loss from the seed can fix chlorophyll at a higher content than that obtained
with a slow moisture loss (Ward *et al.*,
1992). Compared to *B. rapa*, *B. napus*
cultivars tend to produce seed with higher
residual chlorophyll content (McGregor,
1991; Ward *et al.*, 1991). Cultivars which
require a longer growing season are more
likely to contain a high residual chlorophyll
content at harvest (Ward *et al.*, 1991).
Premature harvest, or frost damage of
immature plants, can also result in higher
residual chorophyll content (Daun *et al.*,
1985). The average chlorophyll content
of seeds from the later-maturing side branches was 1.5 to 2.0 times as high as seeds
from the mainstem (Ward *et al.*, 1992).

A PHYSIOLOGICAL APPROACH TO AGRONOMIC PROBLEMS

The above review of crop development,
growth and yield discusses physiological
work which in many cases was necessary
to try to understand the agronomy of the
crop. In most instances the agronomist/
crop physiologist takes a whole-crop or
community approach, analysing growth
and development to a sufficient level of
detail to understand variation between
genotypes or treatments. This section will
illustrate this approach.

Sowing time

In most instances yield declines with later sowing, in autumn or spring. Mendham *et al.* (1990) in Tasmania showed how hastened development of later sowings, combined with reduced post-anthesis growth, was the main cause, particularly with later-flowering genotypes. Efficiency of growth post-anthesis declined from about 1 g MJ^{-1} intercepted radiation on early-sown crops to about 0.2 g MJ^{-1} on late-sown crops which suffered from increasing temperature, water stress and pest damage. To some extent experiments comparing different sowing times in small plots can accentuate the effects as it is more difficult to manage later sowings compared with early sowings (ground preparation, weed and pest control, etc.). Late-sown commercial crops in the same environment which are well managed are now achieving better yields than were recorded in the experiments.

In some cases yields can increase with later sowing, as noted earlier for crops in the United Kingdom by Mendham *et al.* (1981a) and Jenkins and Leitch (1986). Late-sown crops with sufficient time before flowering for interception of radiation and growth could turn a disadvantage, namely fewer pods m^{-2}, into an advantage by retaining more seeds per pod in a sparser but more efficient canopy of pods.

Plant population density

Rapeseed typically gives similar yields over a wide range of seeding rates and hence plant density. Ogilvy (1984), for example, showed that, although optimum seed rates for winter crops in the United Kingdom were around 4–8 kg ha^{-1}, the full range tested of 3 to 12 kg ha^{-1} gave yields within 10% of the maximum. Populations in spring of 80–100 plants m^{-2} appeared optimal, with lower populations tending to be patchy and subject to pest damage, and

higher populations more likely to lodge. Plants at higher density were thinner stemmed and carried fewer branches.

While most crop species have a fairly wide optimum population range, the particular characteristics of rapeseed at flowering appear to reinforce this. High densities support a dense cover of flowers and then pods, which quickly shade out leaves, whereas at lower density the fewer flowers may allow leaf area to expand further and persist longer, as shown by Morrison *et al.* (1990) in Canada. Crop growth rates and net assimilation rates were higher during flowering in low seeding rate crops (1.5 and 3 kg ha^{-1}) than at high rates (6 and 12 kg ha^{-1}). For a late-sown crop in the United Kingdom (Mendham *et al.*, 1981b), plants at 28 m^{-2} produced a maximum LAI of 4.5, but at 88 m^{-2} it was only 3.5. Individual leaf area per plant in the low density was four times that of the high density. Early sowings all produced dense flower covers, so even though the low-density plants had double the leaf area of those at high density, the latter still produced a higher LAI.

In another season (Mendham *et al.*, 1981b) severe freezing in early winter killed many plants, but even at 8 m^{-2} worthwhile yields were obtained from plants which grew to a large size and produced an average of 400 pods, many on secondary branches, and where leaves and bracts expanded to a much larger size than normal and persisted until late in the crop life. McGregor (1987) in Canada hand-thinned plant stands to as low as 3.6 m^{-2} to simulate hail damage. Leaf dry matter and presumably area peaked progressively later and persisted longer with lower density, delaying maturity by as much as 16 days. Increased branching and pod production per plant allowed the population to be reduced from 200 to 40 m^{-2} with less than a 20% loss in yield except in a year when dry conditions prevented compensatory growth by the remaining plants. Densities less than 20 m^{-2} gave much larger reductions in yield, but reduction to less than 10 m^{-2} was required before yield dropped below about 50% of the unthinned control.

The apetalous characteristic (Rao et al., 1991), with better radiation distribution through the canopy at flowering, allowed leaves to persist better on all densities so the high density maintained a greater leaf cover throughout (symbols on their Fig. 2 were reversed in error). The conventional cultivar, however, showed the benefit of low density in extra leaf expansion and persistence as noted above. Removing petals may therefore allow the benefit of higher density to be realized in crops with high yield potential.

Water supply and drought tolerance

Irrigation

Rapeseed crops will suffer if water supply is limited, but the effects will vary depending on genotype, stage of development and previous exposure to stress. In the dry environment of southern Alberta (Krogman and Hobbs, 1975), B. rapa responded to irrigation to progressively later stages of development, the extra water promoting growth in the stage when it was applied. The green area of leaves and pods was increased and extended, but the increase in seed yield was proportionately greater still, indicating increased photosynthetic efficiency per unit of green area. Oil content was increased by irrigation until pod ripening, the stage when maximum oil accumulation is taking place. Clarke and Simpson (1978b), in a higher-rainfall environment in Saskatchewan, showed that irrigation of B. napus increased the number of pods by lengthening the flowering period, and also increased the number of seeds per pod, attributed to greater leaf area during flowering. The extra assimilate supply therefore avoided a negative correlation between numbers of pods and seeds per pod.

The most critical time for water supply is during flowering and early pod development (Richards and Thurling, 1978a), when numbers of pods and seeds are being determined. Limited irrigation during that time (Mendham et al., 1984) resulted in more pods m^{-2} when the control plots suffered from water stress early, or more seeds per pod if stress occurred late in the phase. Seed yields were increased from 3 to 5 t ha^{-1} by the application of 50 mm water at the critical time, in a year when subsequent conditions were moderately favourable. A single irrigation in many situations, however, may not be enough to make a substantial difference. Rao and Mendham (1991a) applied one irrigation of 50 mm to cultivar Marnoo during flowering, but just increased number of seeds per pod from 14 to 16 with little change to yield. Three irrigations increased both number of pods (from 6000 to 8000 m^{-2}) and seeds per pod (to 21), and hence yields increased by about 50%. When the apetalous line was compared to Marnoo (Rao et al., 1991), the former grew better and gave higher yields than Marnoo when unirrigated, and showed less response to extra water. The absence of petals allowed leaves to persist longer, which was associated with continued root growth and water extraction from a greater depth (Rao and Mendham, 1991b), as well as better osmotic adjustment (discussed below). Responses to irrigation are therefore likely to depend on characteristics of the genotype as well as previous crop growth, soil water reserves and current environmental conditions.

Drought avoidance and tolerance

Most of the world's rapeseed is grown under rainfed conditions, so performance under water stress is crucial. In Australia, crops often encounter water stress during flowering and pod development. This has restricted the crop to the wetter margins of the cereal zone, as it has in Canada. Richards and Thurling (1978b) showed that the only advantage of *B. rapa* was in earlier flowering, up to 4 weeks before *B. napus*. This, however, meant that most dry matter was produced after anthesis when the crop was in a more sensitive stage. In *B. napus*, later flowering meant that more reserves were available to fill seed in dry conditions, and hence yields were usually higher, even from later sowing. Suggested strategies to improve *B. rapa* included breeding for faster early growth and/or later flowering, and for *B. napus* included selection of early-flowering lines also with rapid growth. These ideas were pursued further by Thurling and Kaveeta (1992a, b), as previously discussed.

B. napus showed better drought tolerance, shown in a range of characteristics including a higher root/shoot ratio, and a greater diversion of post-anthesis dry matter to seed rather than branches and pod walls. There was significant variation between and within species (Richards, 1978), particularly *B. napus*, for a range of drought tolerance characteristics such as proline accumulation, chlorophyll stability and germination rate under stress. Winter cultivars possessed considerable drought tolerance, probably involving the same factors as determine cold tolerance.

Clarke and McCaig (1982) used a range of techniques to try to detect differences between cultivars in their response to water stress. While leaf conductance, temperature and osmotic potential were all useful to discriminate between environments, and hence potentially useful for irrigation scheduling, they did not pick up cultivar differences. Greater tolerance of *B. napus* to stress was shown by higher osmotic potential, lower leaf temperatures and greater uptake of $^{14}CO_2$ during stress conditions. The production of short, tuberized roots by *B. napus* under dry conditions (Potfer *et al.*, 1988) appears to be an adaptive response which can act as a reserve of water-absorbing potential, and may be capable of further selection.

Osmoregulation or osmotic adjustment has emerged in work (Kumar *et al.*, 1984) as a more dynamic measure of response to water stress, capable of better discrimination between genotypes than leaf conductance, osmotic potential, etc. as point-in-time measurements. Water stress induces osmotic potential of leaf cells to be reduced (i.e. solute concentration increased) over time to allow leaf turgor and hence transpiration and growth to be maintained as soil water potential is reduced (i.e. as soil dries out). This is of particular importance in India where crops are grown on stored moisture, where Kumar *et al.* (1984) showed that a *B. carinata* line was able to adjust much more effectively than *B. napus*. This allowed leaves to continue transpiration and remain cooler. More wax bloom on the leaves, stems and pods assisted the cooling process by reflecting more radiation. Growth and yield were almost double that of the *B. napus* line. Kumar *et al.* (1987) extended the study to a wider range of genotypes. Indian *B. juncea* and *B. carinata* lines showed the greatest osmoregulation and the highest yields. Canadian *B. juncea* and a range of *B. napus* lines showed the least osmoregulation and low yields, with *B. rapa* lines intermediate for both. The higher-yielding lines generally used less water in the vegetative phase, allowing more for reproductive growth, termed 'water savers' rather than 'spenders'.

As well as drought tolerance, water-use efficiency is a key indicator of productivity with a limited water supply. It is measured as dry matter produced (total or seed) per unit of water used, including both crop transpiration and soil evaporation. Transpiration efficiency just takes account of water transpired by the crop.

The greater water use efficiency of *B. juncea* has attracted attention in Canada and Australia (Oram and Kirk, 1992).

B. juncea generally produces more dry matter, particularly in stem, pod wall and root system, than *B. napus*, but often a lower seed yield under adequate moisture conditions. Under drought stress conditions (P. Wright, University of New England, personal communication) *B. juncea* had an increased number of seeds per pod, probably due to osmoregulation maintaining turgor during the critical time for seed abortion, and up to 50% greater water-use efficiency.

The use of *B. carinata* and wild *Brassica* relatives with improved water use or drought tolerance characteristics is likely to be more feasible with biotechnological methods to transfer characters to cultivars with higher yield potential. Salisbury (1991) observed good drought tolerance in *B. tournefortii* under such severe drought that all *B. napus* cultivars died before setting seed. A related genus, *Moricandia*, has some intermediate C_3–C_4 species, including *M. arvensis* (McVetty et al., 1989), which showed much higher water-use efficiency than *B. napus* and *B. rapa*.

Water-use efficiency of a rapeseed crop will be affected by management factors as well as genotype. For example, Norton, R.M. (1989) showed that application of nitrogen could double growth, yield and water-use efficiency, the latter increasing from about 3 to 6 kg of seed ha^{-1} mm^{-1}. This increase, however, was due (Norton, R.M., 1993) to the nitrogen having a much larger effect on growth and leaf area than on total water use, which only increased about 15%. The nitrogen thus increased the proportion of water used by the crop and decreased soil evaporation due to more rapid ground cover by the vigorous crop. Changes in water-use efficiency were thus seen as a consequence of changes in either or both of: (i) the proportion of carbon allocated to seed yield; (ii) the proportion of the water supply partitioned to transpiration as compared with soil evaporation.

Transpiration efficiency tended to be constant between treatments.

Nitrogen supply

Many studies have shown the importance of nitrogen nutrition to growth and yield of rapeseed, as reviewed by Almond et al. (1986). While effects on development are usually small, growth is affected through protein synthesis, leaf expansion and growth of all components of the crop. High rates of nitrogen may lower oil content, if not primarily used in previous crop growth. Hocking and Stapper (1993) showed that leaf number as well as area could be increased. Most nitrogen required by the crop was taken up by flowering, and then redistributed to pods and seeds from leaves and stems. Hocking and Mason (1993) showed that pod walls could act as a temporary reservoir for nitrogen supplying up to 25% of the requirement of seeds, and also as storage for other nutrients, particularly phosphate, zinc and magnesium.

Nitrogen applied in the seedbed in autumn in the United Kingdom (Mendham et al., 1981b and others) was shown to be of little value, whereas spring applications (Bilsborrow et al., 1993) boosted growth in the critical period before flowering when yield potential was determined. The effects of nitrogen on growth have been shown to be expressed normally in the components of yield as extra pods m^{-2}, with little effect on later-formed components.

Nitrogen application at the rosette to early stem extension stages has generally been shown to be of more benefit (Bernardi and Banks, 1993) than earlier or later application. The use of petiole nitrogen levels shows promise as an indicator of the need for nitrogen, at least in irrigated crops.

Cold tolerance and winter survival

While the Canadian prairie winters are regarded as too severe for survival of winter rapeseed in most circumstances, Topinka *et al.* (1991) demonstrated that a range of factors affect survival, with genotype differences of relatively minor significance. Sowing into standing cereal stubble modified the microclimate by increasing snow capture. A combination of sowing date and plant density could be used to produce plants of optimum size, with a root neck diameter of 5 to 16 mm and stem elongation not exceeding 20 mm. A middle sowing date (20 August) produced plants which were large enough to trap snow but with growing points still close to the ground. On a day when air temperature was − 32°C, this gave a crown temperature of 'only' − 13°C compared to − 19°C and − 25°C on earlier and later sowings, respectively. Survival was 94%, compared to 28 and 3% for early and late sowings, respectively. Periods of warmer weather during the winter in southern Alberta can reduce snow cover and cold hardening, and hence the survival rate in subsequent severe weather.

In the south-eastern United States, temperatures in winter can also fluctuate widely. M.S.S. Rao and colleagues (personal communication) used laboratory freezing tests to show how tolerance increased with extra exposure to low temperature (acclimatization). Tolerance of winter cultivars increased at a faster rate than in spring cultivars, and reached a greater level. Both spring and winter cultivars failed to survive freezing if previously grown at 25/13°C (day/night), whereas after 21 days exposure to 4/2°C the lowest survival temperature reached − 6°C for winter and − 3°C for spring cultivars.

Species differences were shown by Torssell (1959) in Sweden. While winter *B. rapa* is usually considered hardier, the difference was shown to be due to the apex of plants in the rosette stage being kept close to the ground or even below it, being pulled down by contractile roots. Tissues of *B. napus* were more cold-hardy, and survived freezing better than *B. rapa*. In conditions of a strong radiation frost with a large temperature gradient, *B. rapa* survived better, whereas with an overall freeze due to a mass of cold air, *B. napus* was better adapted. Torssell also noted that crops at lower density survived better due to reduced stem elongation and hence exposure of the apex.

Plant growth regulators

Daniels and Scarisbrick (1986) reviewed work on growth regulators, and their potential to improve crop characteristics such as excessive height, lodging, inefficient canopy structure, uneven maturity and lack of winter hardiness. Scarisbrick *et al.* (1985) reviewed earlier work on the anti-gibberellins, chlormequat and Terpal, which were useful in cereals but gave inconsistent results or small benefits in rapeseed. Paclobutrazol was shown to be a much more potent anti-gibberellin on rapeseed. When applied in autumn, there were benefits leading to a yield improvement in one experiment, via inhibition of pod set on lower branches and enhancement on the mainstem and upper branches. This was associated with redistribution of assimilate (Addo-Quaye *et al.*, 1985) as measured after $^{14}CO_2$ uptake. When applied in spring, just prior to rapid stem growth, there were more substantial effects on reduction of height but also lower yields. When the regulator was applied at an early stage fewer pods were set, or when applied later it resulted in fewer seeds per pod, particularly on the normally more productive mainstem and upper branches.

Palcobutrazol, as well as the dramatic dwarfing effect, induces the production of smaller, thicker, dark green leaves. This modified canopy was shown to give an increased depth of penetration of solar radiation (Addo-Quaye *et al.*, 1985). An improvement in efficiency of conversion of radiation to dry matter was shown by Rao and Mendham (1991a) for treated Chinoli, which in the control had large, light green leaves (1.2 and 1.0 g MJ^{-1}, respectively), but not for the *B. napus* cultivar Marnoo with normal leaves. While both cultivars were shortened and lodging prevented, yields were substantially reduced.

In an area of high fertility and favourable climate in New South Wales, Armstrong and Nicol (1991) concluded that, although responses in yield and its components were inconsistent, application of Paclobutrazol could be justified where crops were likely to grow too tall and bulky, and were likely to lodge. The practical advantages of easier windrowing and harvesting of the shorter, more compact, more uniformly ripening and less likely to shatter pod canopy gave higher machine-harvested yields, offsetting any potential loss in yield from fewer pods or seeds.

The production of shorter, sturdier crops through breeding is likely to be more effective in the long term, and it may be that plant growth regulators are mainly seen as a research tool to give a guide to potential benefits from plant modification by other means, rather than as a permanent part of crop husbandry.

PHYSIOLOGY AND PLANT BREEDING: A CROP IDEOTYPE

Throughout this review a range of characters have been discussed which have potential to improve rapeseed yields and efficiency of production. The incorporation of morphological characters which have been shown to give a physiological advantage into an ideotype, or model crop plant, has been shown to be of value in many crops (Donald and Hamblin, 1983). This is to set a goal for breeders, and also

to test the physiological value of the characters in association with others, as there may be some negative interactions. Some features of a potential ideotype for rapeseed (Mendham *et al.*, 1991; Thurling, 1991) are listed below:

1. Rapid early growth, particularly at low temperature except where winter survival is a problem. Some Japanese and Chinese lines have demonstrated this feature.
2. Flowering as early as possible, after frost risk has passed for autumn sowings, but consistent with LAI first reaching near-full interception of radiation.
3. A moderately short, sturdy stem to reduce lodging and facilitate harvesting of a more compact pod canopy. A range of shorter genotypes are available.
4. Apetalous flowers to improve radiation distribution through the canopy at the critical time for seed abortion.
5. The use of genotypes with greater ability to retain seeds during the phase when abortion is likely, again shown to be a feature of some Japanese material.
6. Moderate numbers of pods only (5000–8000 m^{-2}) to allow a better radiation environment for pod photosynthesis.
7. Possible use of the long-pod characteristic to produce more seeds with fewer pods, although this needs confirmation.
8. The use of more upright pods, again to allow better distribution of radiation.
9. Reduced lower branching and enhanced pod production on mainstem and upper branches. Some genotypes already show this feature to some extent, e.g. the French cultivar Jet Neuf (Mendham *et al.*, 1981a).

Several of these characteristics are controlled by few genes and should be relatively easy to handle in a breeding programme. Thurling (1991) has suggested how some of these ideotype characteristics, plus others relating to disease resistance, product quality and pod

shatter resistance, could be combined. Production of dihaploid lines from crosses
involving the separate characteristics could

be the quickest method, and some progress
along these lines has since been made.

References

Addo-Quaye, A.A., Daniels, R.W. and Scarisbrick, D.H. (1985) The influence of paclobutrazol on
the distribution and utilization of C^{14}-labelled assimilate fixed at anthesis in oil-seed rape (*Brassica
napus* L.). *Journal of Agricultural Science, Cambridge* 105, 365–373.

Almond, J.A., Dawkins, T.C.K., Done, C.J. and Ivins, J.D. (1984) Cultivations for winter oilseed
rape (*Brassica napus* L.). *Aspects of Applied Biology* 6, 67–79.

Almond, J.A., Dawkins, T.C.K. and Askew, M.F. (1986) Aspects of crop husbandry.
In: Scarisbrick, D.H. and Daniels, R.W. (eds) *Oilseed Rape*. Collins, London, pp. 127–175.

Alonso, L.C., Fernández-Serrano, O. and Fernández-Escobar, J. (1991) The outset of a new oilseed
crop: *Brassica carinata* with a low erucic acid content. In: McGregor, D.I. (ed.)
Proceedings of the Eighth International Rapeseed Congress, Saskatoon, Canada. Organizing Committee, Saskatoon, pp. 170–176.

Appelqvist, L.-A. (1968) Lipids in Cruciferae III. Fatty acid composition of diploid and tetraploid
seeds of *Brassica campestris* and *Sinapis alba* grown under two climatic extremes. *Physiologia Plantarum* 21, 615–625.

Appelqvist, L.-A. (1969) Lipids in Cruciferae IV. Fatty acid patterns in single seeds and seed populations of various Cruciferae and in different tissue of *Brassica napus* L. *Hereditas* 61, 9–44.

Armstrong, E.L. and Nicol, H.I. (1991) Reduced lodging in rapeseed with growth regulators. *Australian
Journal of Experimental Agriculture* 31, 245–250.

Barber, S.J., Rakow, G. and Downey, R.K. (1991) Field emergence and yield of certified seed lots
of cvs. Westar and Tobin canola on the Canadian prairie. In: McGregor, D.I. (ed.) *Proceedings
of the Eighth International Rapeseed Congress*, Saskatoon, Canada. Organizing Committee, Saskatoon, pp. 659–664.

Bechyne, M. and Kondra, Z.P. (1970) Effect of seed pod location on the fatty acid composition of
seed oil from rapeseed (*Brassica napus* and *B. campestris*). *Canadian Journal of Plant Science* 50, 151–154.

Becker, H.C., Damgaard, C. and Karlsson, B. (1992) Environmental variation for outcrossing rate
in rapeseed (*Brassica napus*). *Theoretical and Applied Genetics* 84, 303–306.

Bernardi, A.L. and Banks, L.W. (1993) Petiole nitrate nitrogen: is it a good indicator of yield potential in irrigated canola? In: Wratten, N. and Mailer, R.J. (eds) *Ninth Australian Research Assembly
on Brassicas*, Wagga Wagga, New South Wales, pp. 51–56.

Bilsborrow, P.E. and Norton, G. (1984) A consideration of factors affecting the yield of oilseed rape.
Aspects of Applied Biology 6, 91–99.

Bilsborrow, P.E., Evans, E.J. and Zhao, F.J. (1993) The influence of spring nitrogen on yield, yield
components and glucosinolate content of autumn-sown oilseed rape (*Brassica napus*). *Journal of
Agricultural Science, Cambridge* 120, 219–224.

Booth, E.J. and Walker, K.C. (1992) The effect of site and foliar sulfur on oilseed rape: comparison
of sulfur responsive and non-responsive seasons. *Phyton* 32, 9–13.

Booth, E.J., Walker, K.C. and Griffiths, D.W. (1990) Effect of harvest date and pod position on
glucosinolates in oilseed rape (*Brassica napus*). *Journal of the Science of Food and Agriculture* 53, 43–61.

Booth, E.J., Walker, K.C. and Griffiths, D.W. (1991) A time-course study of the effect of sulphur
on glucosinolates in oilseed rape (*Brassica napus*) from the vegetative stage to maturity.
Journal of the Science of Food and Agriculture 56, 479–493.

Bouttier, C. and Morgan, D.G. (1992) Development of oilseed rape buds, flowers and pods *in vitro*.
Journal of Experimental Botany 43, 1089–1096.

Bunting, E.S. (1986) Oilseed rape in perspective. In: Scarisbrick, D.H. and Daniels, R.W. (eds)
Oilseed Rape. Collins, London, pp. 1–31.

Cannell, R.Q. and Belford, R.K. (1980) Effects of waterlogging at different stages of development
on the growth and yield of winter oilseed rape (*Brassica napus* L.). *Journal of the Science of
Food and Agriculture* 31, 963–965.

Canvin, D.T. (1965) The effect of temperature on the oil content and fatty acid composition of the oils from several oilseed crops. *Canadian Journal of Botany* 43, 63–69.

Chapman, J.F., Daniels, R.W. and Scarisbrick, D.H. (1984) Field studies on ^{14}C. Assimilate fixation and movement in oil-seed rape (*B. napus*). *Journal of Agricultural Science, Cambridge* 102, 23–31.

Chay, P. and Thurling, N. (1989) Variation in pod length in spring rape (*Brassica napus*) and its effect on seed yield and components. *Journal of Agricultural Science, Cambridge* 113, 139–147.

Clarke, J.M. (1979) Intra-plant variation in number of seeds per pod and seed weight in *Brassica napus* 'Tower'. *Canadian Journal of Plant Science* 59, 959–962.

Clarke, J.M. and McCaig, T.N. (1982) Leaf diffusive resistance, surface temperature, osmotic potential and $^{14}CO_2$-assimilation capability as indicators of drought intensity in rape. *Canadian Journal of Plant Science* 67, 785–789.

Clarke, J.M. and Simpson, G.M. (1978a) Growth analysis of *Brassica napus* cv. Tower. *Canadian Journal of Plant Science* 58, 587–595.

Clarke, J.M. and Simpson, G.M. (1978b) Influence of irrigation and seeding rates on yield and yield components of *Brassica napus* cv. Tower. *Canadian Journal of Plant Science* 58, 731–737.

Cooke, R.J., Kimber, D.S. and Morgan, A.G. (1987) Changes in the glucosinolate content of oilseed rape varieties. In: *Proceedings of the Seventh International Rapeseed Congress*, Poznan, Poland. The Plant Breeding and Acclimatization Institute, Poznan, pp. 532–535.

Crouch, M.L. and Sussex, I.M. (1981) Development and storage-protein synthesis in *Brassica napus* L. embryos *in vivo* and *in vitro*. *Planta* 153, 64–74.

Daniels, R.W. and Scarisbrick, D.H. (1986) Plant growth regulators for oilseed rape. In: Scarisbrick, D.H. and Daniels, R.W. (eds) *Oilseed Rape*. Collins, London, pp. 176–194.

Daniels, R.W., Scarisbrick, D.H. and Smith, L.J. (1986) Oilseed rape physiology. In: Scarisbrick, D.H. and Daniels, R.W. (eds) *Oilseed Rape*. Collins, London, pp. 83–126.

Daun, J.K., Clear, K.M. and Mills, J.T. (1985) Effect of frost damage on the quality of canola (*B. napus*). *Journal of the American Oil Chemists Society* 62, 715–719.

De March, G., McGregor, D.I. and Seguin-Shwartz, G. (1989) Glucosinolate content of maturing pods and seeds of high and low glucosinolate summer rape. *Canadian Journal of Plant Science* 69, 929–932.

Diepenbrock, W. and Giesler, G. (1979) Compositional changes in developing pods and seeds of oilseed rape (*Brassica napus* L.) as affected by pod position on the plant. *Canadian Journal of Plant Science* 59, 819–830.

Donald, C.M. and Hamblin, J. (1983) The convergent evolution of annual seed crops in agriculture. *Advances in Agronomy* 36, 97–143.

Downey, R.K. (1983) Origin and description of the *Brassica* oilseed crops. In: Kramer, J.K.G., Sauer, F.D. and Pigden, W.J. (eds) *High and Low Erucic Acid Rapeseed Oils: Production, Usage, Chemistry, and Toxicological Evaluation*. Academic Press, Toronto, Canada, pp. 1–20.

Eisikowitch, D. (1981) Some aspects of pollination of oil-seed rape (*Brassica napus* L.). *Journal of Agricultural Science, Cambridge* 96, 321–326.

Evans, E.J. (1984) Pre-anthesis growth and its influence on seed yield in winter oilseed rape. *Aspects of Applied Biology* 6, 81–90.

Fenwick, G.R., Heaney, R.K. and Mullin, W.J. (1983) Glucosinolates and their breakdown products in food and food plants. *CRC Critical Reviews in Food Science and Nutrition* 18, 123–201.

Fieldsend, J.K., Murray, F.E., Bilsborrow, P.E., Milford, G.F.J. and Evans, E.J. (1991) Glucosinolate accumulation during seed development in winter sown oilseed rape (*B. napus*). In: McGregor, D.I. (ed.) *Proceedings of the Eighth International Rapeseed Congress*, Saskatoon, Canada. Organizing Committee, Saskatoon, pp. 689–694.

Finlayson, A.J. and Christ, C.M. (1971) Changes in the nitrogenous components of maturing rapeseed (*Brassica napus*). *Canadian Journal of Botany* 49, 1733–1735.

Fowler, D.B. and Downey, R.K. (1970) Lipid and morphological changes in developing rapeseed, *Brassica napus*. *Canadian Journal of Plant Science* 50, 233–247.

Gelfi, N., Hilaire, A., Rozelle, P., Martinez, E., Bouniols, A. and Blanchet, R. (1988) Comparison de l'assimilation nette de quelques cultures d'hiver: colza, blé, pois, féverole. In: *Colza — Physiologie et Elaboration du Rendement*. CETIOM, Paris, pp. 121–123.

Harper, F.R. and Berkenkamp, B. (1975) Revised growth-stage key for *Brassica campestris* and *B. napus*. *Canadian Journal of Plant Science* 55, 657–658.

Hocking, P.J. and Mason, L. (1993) Accumulation, distribution and redistribution of dry matter and mineral nutrients in fruits of canola (oilseed rape), and the effects of nitrogen fertilizer and windrowing. *Australian Journal of Agricultural Research* 44, 1377–1388.

Hocking, P.J. and Stapper, M. (1993) Effects of sowing time and nitrogen fertilizer rate on the growth, yield and nitrogen accumulation of canola, mustard and wheat. In: Wratten, N. and Mailer, R.J. (eds) *Proceedings Ninth Australian Research Assembly on Brassicas*, Wagga Wagga, New South Wales, pp. 33–46.

Hodgson, A.S. (1978a) Rapeseed adaptation in Northern New South Wales. I. Phenological responses to vernalization, temperature and photoperiod by annual and biennial cultivars of *Brassica campestris* L., *Brassica napus* L. and wheat cv. Timgalen. *Australian Journal of Agricultural Research* 29, 693–710.

Hodgson, A.S. (1978b) Rapeseed adaptation in Northern New South Wales. II. Predicting plant development of *Brassica campestris* L. and *Brassica napus* L. and its implications for planting time, designed to avoid water deficit and frost. *Australian Journal of Agricultural Research* 29, 711–726.

Hogarth, C. (1993) Evaluation of a canola hybrid seed production system. Unpublished BAgrSc(Hons) Thesis, University of Tasmania.

Holmes, M.R.J. and Ainsley, A.M. (1979) Nitrogen top-dressing requirements of winter oilseed rape. *Journal of the Science of Food and Agriculture* 30, 119–128.

Jenkins, P.D. and Leitch, M.H. (1986) Effects of sowing date on the growth and yield of winter oilseed rape (*Brassica napus*). *Journal of Agricultural Science, Cambridge* 105, 405–420.

Josefsson, E. (1970) Content of p-hydroxybenzylglucosinolate in seed meals of *Sinapis alba* as affected by heredity, environment and seed part. *Journal of the Science of Food and Agriculture* 21, 94–97.

Keiller, D.R. and Morgan, D.G. (1988a) Distribution of [14]carbon-labelled assimilates in flowering plants of oilseed rape (*Brassica napus* L.). *Journal of Agricultural Science, Cambridge* 111, 347–355.

Keiller, D.R. and Morgan, D.G. (1988b) Effect of pod removal and plant growth regulators on the growth, development and carbon assimilate distribution in oilseed rape (*Brassica napus* L.). *Journal of Agricultural Science, Cambridge* 111, 357–362.

King, J.R., McNeilly, T. and Thurman, D.A. (1977) Variation in the protein content of single seeds of four varieties of oil seed rape. *Journal of the Science of Food and Agriculture* 28, 1065–1070.

Kjellstrom, C. (1991) Growth and distribution of the root system in *Brassica napus*. In: McGregor, D.I. (ed.) *Proceedings of the Eighth International Rapeseed Congress*, Saskatoon, Canada. Organizing Committee, Saskatoon, pp. 722–726.

Kondra, Z.P. and Downey, R.K. (1969) Glucosinolate content of developing *Brassica napus* and *B. campestris* seed. *Canadian Journal of Plant Science* 49, 623–624.

Kondra, Z.P. and Downey, R.K. (1970) Glucosinolate content of rapeseed (*Brassica napus* L. and *B. campestris* L.) meal as influenced by pod position on the plant. *Crop Science* 10, 54–55.

Kondra, Z.P., Campbell, D.C. and King, J.R. (1983) Temperature effects on germination of rapeseed (*Brassica napus* L. and *B. campestris* L.). *Canadian Journal of Plant Science* 63, 1063–1065.

Krogman, K.K. and Hobbs, E.H. (1975) Yield and morphological response of rape (*Brassica campestris* L. cv. Span) to irrigation and fertilizer treatments. *Canadian Journal of Plant Science* 55, 903–909.

Kumar, A., Singh, P., Singh, D.P., Singh, H. and Sharma, H.C. (1984) Differences in osmoregulation in *Brassica* species. *Annals of Botany* 54, 537–541.

Kumar, A., Singh, D.P. and Singh, P. (1987) Genotypic variation in the responses of *Brassica* species to water deficit. *Journal of Agricultural Science, Cambridge* 109, 615–618.

Lein, K.-A. (1972) Genetische und physiologische Untersuchungen zur Bildung von Glucosinolaten in Rapssamen: Lokalisierung des Haupt-biosyntheseortes durch Pfropfungen. *Zeitschrift für Pflanzenphysiologie* 67, 333–342.

Leterme, P. (1988a) Croissance et développmment du colza d'hiver: les principales étapes. In: *Colza: Physiologie et Elaboration du Rendement*. CETIOM, Paris, pp. 23–33.

Leterme, P. (1988b) Modélisation du fonctionnement du peuplement de colza d'hiver en fin de cycle: élaboration des composantes finales du rendement. In: *Colza: Physiologie et Elaboration du Rendement*. CETIOM, Paris, pp. 124–129.

Love, H.K., Rakow, G., Raney, J.P. and Downey, R.K. (1991) Breeding improvements towards canola quality *Brassica juncea*. In: McGregor, D.I. (ed.) *Proceedings of the Eighth International Rapeseed Congress*, Saskatoon, Canada. Organizing Committee, Saskatoon, pp. 164–169.

McGregor, D.I. (1981) Pattern of flower and pod development in rapeseed. *Canadian Journal of Plant Science* 61, 275–282.

McGregor, D.I. (1987) Effect of plant density on development and yield of rapeseed and its significance to recovery from hail injury. *Canadian Journal of Plant Science* 67, 43–51.

McGregor, D.I. (1991) Influence of environment and genotype on rapeseed/canola seed chlorophyll content. In: McGregor, D.I. (ed.) *Proceedings of the Eighth International Rapeseed Congress*, Saskatoon, Canada. Organizing Committee, Saskatoon, pp. 1743–1748.

McVetty, P.B.E., Austin, R.B. and Morgan, C.L. (1989) A comparison of the growth, photosynthesis, stomatal conductance and water use efficiency of *Moricandia* and *Brassica* species. *Annals of Botany* 64, 87–94.

Mailer, R.J. and Cornish, P.S. (1987) Effects of water stress on glucosinolate and oil concentrations in the seeds of rape (*Brassica napus* L.) and turnip rape (*Brassica rapa* L. var. *silvestris* [Lam.] Briggs). *Australian Journal of Experimental Agriculture* 27, 707–711.

Major, D.J. (1975) Stomatal frequency and distribution in rape. *Canadian Journal of Plant Science* 55, 1077–1078.

Major, D.J. (1977) Influence of seed size on yield and yield components of rape. *Agronomy Journal* 69, 541–543.

Major, D.J., Bole, J.B. and Charnetski, W.A. (1978) Distribution of photosynthates after $^{14}CO_2$ assimilation by stems, leaves and pods of rape plants. *Canadian Journal of Plant Science* 58, 783–787.

Malik, R.S. (1990) Prospects for *Brassica carinata* as an oilseed crop in India. *Experimental Agriculture* 26, 125–129.

Mendham, N.J. and Scott, R.K. (1975) The limiting effect of plant size at inflorescence initiation on subsequent growth and yield of oilseed rape (*Brassica napus*). *Journal of Agricultural Science, Cambridge* 84, 487–502.

Mendham, N.J., Shipway, P.A. and Scott, R.K. (1981a) The effects of delayed sowing and weather on growth, development and yield of winter oil-seed rape (*Brassica napus*). *Journal of Agricultural Science, Cambridge* 96, 389–416.

Mendham, N.J., Shipway, P.A. and Scott, R.K. (1981b) The effects of seed size, autumn nitrogen and plant population density on the response to delayed sowing in winter oil-seed rape (*Brassica napus*). *Journal of Agricultural Science, Cambridge* 96, 417–428.

Mendham, N.J., Russell, J. and Buzza, G.C. (1984) The contribution of seed survival to yield in new Australian cultivars of oil-seed rape (*Brassica napus*). *Journal of Agricultural Science, Cambridge* 103, 303–316.

Mendham, N.J., Russell, J. and Jarosz, N.K. (1990) Response to sowing time of three contrasting Australian cultivars of oilseed rape (*Brassica napus*). *Journal of Agricultural Science, Cambridge* 114, 275–283.

Mendham, N.J., Rao, M.S.S. and Buzza, G.C. (1991) The apetalous flower character as a component of a high yielding ideotype. In: McGregor, D.I. (ed.) *Proceedings of the Eighth International Rapeseed Congress*, Saskatoon, Canada. Organizing Committee, Saskatoon, pp. 596–600.

Merrien, A., Merle, C., Quinsac, A., Ribaillier, D. and Maisonneuve, C. (1991) Accumulation of glucosinolates during the ripening period in the seeds and pod walls of winter oilseed rape. In: McGregor, D.I. (ed.) *Proceedings of the Eighth International Rapeseed Congress*, Saskatoon, Canada. Organizing Committee, Saskatoon, pp. 1720–1726.

Morrison, M.J. and McVetty, P.B.E. (1991) Leaf appearance rate of summer rape. *Canadian Journal of Plant Science* 71, 405–412.

Morrison, M.J., McVetty, P.B.E. and Shaykewich, C.F. (1989) The determination and verification of a baseline temperature for the growth of Westar summer rape. *Canadian Journal of Plant Science* 69, 455–464.

Morrison, M.J., McVetty, P.B.E. and Scarth, R. (1990) Effect of altering plant density on growth characteristics of summer rape. *Canadian Journal of Plant Science* 70, 139–149.

Morrison, M.J., Stewart, D.W. and McVetty, P.B.E. (1992) Maximum area, expansion rate and duration of summer rape leaves. *Canadian Journal of Plant Science* 72, 117–126.

Myers, L.F., Christian, K.R. and Kirchner, R.J. (1982) Flowering responses of 48 lines of oilseed rape (*Brassica* spp.) to vernalization and daylength. *Australian Journal of Agricultural Research* 33, 927–936.

Norton, G., Bilsborrow, P.E. and Shipway, P.A. (1991) Comparative physiology of divergent types of winter rapeseed. In: McGregor, D.I. (ed.) *Proceedings of the Eighth International Rapeseed Congress*, Saskatoon, Canada. Organizing Committee, Saskatoon, pp. 578–582.

Norton, R.M. (1989) Applied nitrogen and water use efficiency of canola. In: Buzza, G.C. (ed.) *Proceedings of the Seventh Workshop of Australian Rapeseed Agronomists and Breeders*, Toowoomba, Queensland, Australia, pp. 107–110.

Norton, R.M. (1993) The effect of applied nitrogen on the growth, yield, water use and quality of rapeseed (*Brassica napus* L.) in the Wimmera. PhD Thesis, Latrobe University, Melbourne, Australia.

Ogilvy, S.E. (1984) The influence of seed rate on population structure and yield of winter oilseed rape. *Aspects of Applied Biology* 6, 59–66.

Oram, R.N. and Kirk, J.T.O. (1992) Breeding Indian mustard for Australian conditions. In: Hutchinson, K.J. and Vickery, P.J. (eds) *Proceedings of the Sixth Australian Agronomy Conference*. Australian Society of Agronomy, Armidale, New South Wales, pp. 467–470.

Pechan, P.A. and Morgan, D.G. (1985) Defoliation and its effects on pod and seed development in oil seed rape (*Brassica napus* L.). *Journal of Experimental Botany* 36, 458–468.

Pomeroy, M.K. and Sparace, S.A. (1992) Lipid biosynthesis in microspore-derived embryos of *Brassica napus* L. In: MacKenzie, S.L. and Taylor, D.C. (eds) *Seed Oils for the Future*. AOCS Press, Illinois, pp. 61–69.

Potfer, J.P., Merrien, A. and Vartanian, N. (1988) Etude *in situ* du système racinaire du colza de printemps en condition de sécheresse. In: *Colza: Physiologie et Elaboration du Rendement*. CETIOM, Paris, pp. 47–53.

Quilleré, I. and Triboi-Blondel, A.M. (1988a) Les mouvements d'assimilats chez le colza d'hiver. I. Mise en réserve racinaire et influence des techniques culturales. In: *Colza: Physiologie et Elaboration du Rendement*. CETIOM, Paris, pp. 54–58.

Quilleré, I. and Triboi-Blondel, A.M. (1988b) Les mouvements d'assimilats chez le colza d'hiver. II. Importance et rôle des reserves carbonées. In: *Colza: Physiologie et Elaboration du Rendement*. CETIOM, Paris, pp. 73–77.

Rakow, G. and McGregor, D.I. (1975) Oil, fatty acid and chlorophyll accumulation in developing seeds of two 'linolenic acid lines' of low erucic acid rapeseed. *Canadian Journal of Plant Science* 55, 197–203.

Rakow, G. and Woods, D.L. (1987) Outcrossing in rape and mustard under Saskatchewan prairie conditions. *Canadian Journal of Plant Science* 67, 147–151.

Rao, M.S.S. and Mendham, N.J. (1991a) Comparison of Chinoli (*Brassica campestris* subsp. *oleifera* × subsp. *chinensis*) and *B. napus* oilseed rape using different growth regulators, plant population densities and irrigation treatments. *Journal of Agricultural Science, Cambridge* 117,177–187.

Rao, M.S.S. and Mendham, N.J. (1991b) Soil–plant–water relations of oilseed rape (*Brassica napus* and *B. campestris*). *Journal of Agricultural Science, Cambridge* 117, 197–205.

Rao, M.S.S. and Raymer, P.L. (1991) Vernalization and photoperiod requirements for the adaptability of rapeseed to the south eastern United States. In: McGregor, D.I. (ed.) *Proceedings of the Eighth International Rapeseed Congress*, Saskatoon, Canada. Organizing Committee, Saskatoon, pp. 738–742.

Rao, M.S.S., Mendham, N.J. and Buzza, G.C. (1991) Effect of the apetalous flower character on radiation distribution in the crop canopy, yield and its components in oilseed rape (*Brassica napus*). *Journal of Agricultural Science, Cambridge* 117, 189–196.

Richards, R.A. (1978) Variation within and between species of rapeseed (*Brassica campestris* and *B. napus*) in response to drought stress. III. Physiological and physicochemical characters. *Australian Journal of Agricultural Research* 29, 491–501.

Richards, R.A. and Thurling, N. (1978a) Variation between and within species of rapeseed (*Brassica campestris* and *B. napus*) in response to drought stress. I. Sensitivity at different stages of development. *Australian Journal of Agricultural Research* 29, 469–477.

Richards, R.A. and Thurling, N. (1978b) Variation between and within species of rapeseed (*Brassica campestris* and *B. napus*) in response to drought stress. II. Growth and development under natural drought stresses. *Australian Journal of Agricultural Research* 29, 479–490.

Romero, J.E. (1991) Fatty acid composition of the triacylglycerol fraction in developing *Brassica rapa* seeds. In: McGregor, D.I. (ed.) *Proceedings of the Eighth International Rapeseed Congress*, Saskatoon, Canada. Organizing Committee, Saskatoon, pp. 1826–1830.

Salisbury, P.A. (1991) Genetic variability in Australian wild crucifers and its potential utilisation in oilseed *Brassica* species. PhD Thesis, La Trobe University.

Salisbury, P.A. and Green, A.G. (1991) Developmental responses in spring canola cultivars. In: McGregor, D.I. (ed.) *Proceedings of the Eighth International Rapeseed Congress*, Saskatoon, Canada. Organizing Committee, Saskatoon, pp. 1769–1774.

Salisbury, P., Sang, J. and Cawood, R. (1987) Genetic and environmental factors influencing glucosinolate content in rapeseed in southern Australia. In: *Proceedings of the Seventh International Rapeseed Congress*, Poznan, Poland. The Plant Breeding and Acclimatization Institute, Poznan, pp. 516–520.

Salisbury, P., Potter, T., Castleman, G., Robson, D., Hyett, J. and Holding, B. (1991) Potential for a wider maturity range in canola and mustard. In: *Proceedings of the Eighth Australian Research Assembly on Brassicas*, Horsham, Victoria, pp. 71–74.

Scarisbrick, D.H., Addo-Quaye, A.A., Daniels, R.W. and Mahamud, S. (1985) The effect of paclobutrazol on plant height and seed yield of oil-seed rape (*Brassica napus* L.). *Journal of Agricultural Science, Cambridge* 105, 605–612.

Smith, L.J. and Scarisbrick, D.H. (1990) Reproductive development in oilseed rape (*Brassica napus* cv. Bienvenu). *Annals of Botany* 65, 205–212.

Stringam, G.R., McGregor, D.I. and Pawlowski, S.H. (1974) Chemical and morphological characteristics associated with seedcoat color in rapeseed. In: *Proceedings of the 4th Internationaler Rapskongress*, Geissen, Germany. GCIRC, Geissen, pp. 99–108.

Sylvester-Bradley, R. and Makepeace, R.J. (1984) A code for stages of development in oilseed rape (*Brassica napus* L.). *Aspects of Applied Biology* 6, 399–419.

Tayo, T.O. and Morgan, D.G. (1975) Quantitative analysis of the growth, development and distribution of flowers and pods in oil-seed rape (*Brassica napus* L.). *Journal of Agricultural Science, Cambridge* 85, 103–110.

Tayo, T.O. and Morgan, D.G. (1979) Factors influencing flower and pod development in oil-seed rape (*Brassica napus* L.). *Journal of Agricultural Science, Cambridge* 92, 363–373.

Thurling, N. (1974a) Morphophysiological determinants of yield in rapeseed (*Brassica campestris* and *Brassica napus*). I. Growth and morphological characters. *Australian Journal of Agricultural Research* 25, 697–710.

Thurling, N. (1974b) Morphophysiological determinants of yield in rapeseed (*Brassica campestris* and *Brassica napus*). II. Yield components. *Australian Journal of Agricultural Research* 25, 711–721.

Thurling, N. (1991) Application of the ideotype concept in breeding for higher yield in the oilseed brassicas. *Field Crops Research* 26, 201–219.

Thurling, N. and Kaveeta, R. (1992a) Yield improvement of oilseed rape (*Brassica napus* L.) in a low rainfall environment. I. Utilization of genes for early flowering in primary and secondary gene pools. *Australian Journal of Agricultural Research* 43, 609–622.

Thurling, N. and Kaveeta, R. (1992b) Yield improvement of oilseed rape (*Brassica napus* L.) in a low rainfall environment. II. Agronomic performance of lines selected on the basis of pre-anthesis development. *Australian Journal of Agricultural Research* 43, 623–633.

Thurling, N. and Vijendra Das, L.D. (1977) Variation in the pre-anthesis development of spring rape (*Brassica napus* L.). *Australian Journal of Agricultural Research* 28, 597–607.

Thurling, N. and Vijendra Das, L.D. (1979a) Genetic control of the pre-anthesis development of spring rape (*Brassica napus* L.). II. Identification of individual genes controlling development pattern. *Australian Journal of Agricultural Research* 30, 261–271.

Thurling, N. and Vijendra Das, L.D. (1979b) The relationship between pre-anthesis development and seed yield of spring rape (*Brassica napus* L.). *Australian Journal of Agricultural Research* 31, 25–36.

Tittonel, E.D. (1988) La phase automnale chez le colza d'hiver. In: *Colza: Physiologie et Elaboration du Rendement*. CETIOM, Paris, pp. 59–67.

Tittonel, E.D., Desplantes, G., Grangeret, I. and Pinochet, X. (1982) Modifications morphologiques d'un bourgeon de colza (*Brassica napus*) au cours de la formation des ébauches florales. *Informations Techniques* CETIOM number 78-I-82.

Tommey, A.M. and Evans, E.J. (1992) Analysis of post-flowering compensatory growth in winter oilseed rape (*Brassica napus*). *Journal of Agricultural Science, Cambridge* 118, 301–308.

Topinka, A.K.C., Downey, R.K. and Rakow, G.F.W. (1991) Effect of agronomic practices on the overwintering of winter canola in southern Alberta. In: McGregor, D.I. (ed.) *Proceedings of the Eighth International Rapeseed Congress*, Saskatoon, Canada. Organizing Committee, Saskatoon, pp. 665–670.

Torssell, B. (1959) *Hardiness and Survival of Winter Rape and Winter Turnip Rape*. Department of Plant Husbandry (Crop Production), Royal School of Agriculture, Sweden, Publication No. 15.

Triboi-Blondel, A.M. (1988) Mise en place et fonctionnement des feuilles de colza d'hiver: relations azote-carbone et sénescence. *Agronomie* 8, 779–786.

Ward, K.A., Scarth, R., McVetty, P.B.E. and Daun, J.K. (1991) Genotypic and environmental effects on seed chlorophyll levels in canola (*Brassica napus*). In: McGregor, D.I. (ed.) *Proceedings of the Eighth International Rapeseed Congress*, Saskatoon, Canada. Organizing Committee, Saskatoon, pp. 1241–1245.

Ward, K., Scarth, R., Daun, J. and McVetty, P.B.E. (1992) Effects of genotype and environment on seed chlorophyll degradation during ripening in four cultivars of oilseed rape (*Brassica napus*). *Canadian Journal of Plant Science* 72, 643–649.

Williams, I.H. (1978) The pollination requirements of swede rape (*Brassica napus* L.) and of turnip rape (*Brassica campestris* L.). *Journal of Agricultural Science, Cambridge* 91, 343–348.

Williams, I.H., Martin, A.P. and White, R.P. (1987) The effect of insect pollination on plant development and seed production in winter oil-seed rape (*Brassica napus* L.). *Journal of Agricultural Science, Cambridge* 109, 135–139.

Woods, D.L. (1992) Comparative performance of mustard and canola in the Peace River region. *Canadian Journal of Plant Science* 72, 829–830.

Yates, D.J. and Steven, M.D. (1987) Reflexion and absorption of solar radiation by flowering canopies of oil-seed rape (*Brassica napus* L.). *Journal of Agricultural Science, Cambridge* 109, 495–502.

3 Agronomy

A. Pouzet
CETIOM, Paris

INTRODUCTION

The growth cycle of rapeseed may be as short as 70 days (*Brassica rapa*) or as long as 380 days for winter cultivars in China (Sun *et al.*, 1991). Oil content of the seed varies from 23 to about 50%, while the glucosinolate content ranges from <10 to >150 μmol g^{-1} of seed, the former level being a feature of modern 'double low' and canola cultivars. The diversity of species and cultivars, together with the range of crop management and cultural practices, is too extensive to summarize in a chapter. The object is to set out the general principles starting with choice of cultivar, before going on to consider soil preparation, sowing, fertilizer and irrigation requirements, and the harvesting and storage of the crop.

Descriptions of agronomic practices in various countries are available: Argentina (Murphy and Pascale, 1991), Canada (Thomas, 1984), China (Sun *et al.*, 1991), Ethiopia (Belayneh and Alemayehu, 1987), Europe (Anon., 1994; Farman *et al.*, 1989), Pakistan (Munir and Rahman Khan, 1987) and the United States (Auld and Mahler, 1991).

CHOICE OF CULTIVAR

Oilseed brassicas are grown on all five continents under a wide range of climate, soil type and economic conditions. Countries the size of the United States or China have a wide range of soils and climate, some suited to growing spring cultivars and some to winter cultivars (Auld and Mahler, 1991; Porter *et al.*, 1991). Different species are adapted to certain climates, as are particular cultivars. Thus cultivar selection is critical to successful crop production. Species, cultivar type and adaptability, seed quality, soil characteristics and intended market are all factors which must be considered when making this selection.

Species

The cultivated *Brassica* species are reviewed in Chapter 1. The main species, *B. napus*, is widely grown, mostly as winter cultivars in Europe and China, and also in the United States and Canada, where spring cultivars predominate. However, half the rapeseed area in Canada is of spring *B. rapa* cultivars and this species also provides the traditional crop in the Indian subcontinent

Table 3.1. Effect of seed lot on yield.

	1988	1989		1990	
	Westar	Westar	Tobin	Westar	Tobin
No. of seed lots	12	12	12	12	9
Mean yield (kg ha^{-1})	1828	1044	967	1128	942
Standard deviation	103	44	46	107	35
Highest yield as % of mean	110	106	108	108	105
Lowest yield as % of mean	86	92	88	70	92

Source: Barber *et al.* (1991a).
Tobin = *B. rapa*; Westar = *B. napus*.

in the form of yellow sarson, brown sarson or toria ecotypes. *Brassica juncea* is another traditional crop in northern India and in China, but is now grown in Australia and is favoured in Canada and Europe for condiment purposes.

Winter or spring cultivar

Choice is usually determined by the length of the growing season. Spring cultivars have a short cycle with no vernalization requirement, whereas winter cultivars vary in vernalization requirement and take longer to mature.

Cultivar adaptability

Due to wide variability in climate and the range of farming systems cultivars are usually best adapted to a region when bred specifically for that region. For example, European spring rapeseed cultivars have usually been later maturing when introduced into Canada than cultivars bred in Canada for western Canadian conditions. Most rapeseed-producing countries issue lists of recommended cultivars with descriptions of performance based on comparative trials. Romagosa and Fox (1993) have described recent genotype/environment interaction studies in which stability of yield as a complement to average yield performance has also been evaluated to

assist in choice of cultivar (Van de Putte and Messéan, 1994).

Seed quality

Barber *et al.* (1991a) observed bigger yield differences between certified seed lots of the same cultivar than between different cultivars (Table 3.1). However, they were unable to establish a reliable method for predicting seed lot performance based on germination or seedling growth under standard or stress conditions (Barber *et al.*, 1991b). Tests carried out in accordance with the procedures defined by the International Seed Testing Association should ensure that chosen seed lots are satisfactory in terms of germination and purity, while seed produced under an officially recognized certification scheme should ensure cultivar authenticity.

Soil characteristics

Brassica oilseeds are tolerant of a wide range of soils in rapeseed-growing countries. Crops can tolerate a range in pH from 5.5 to 8.0 and, ideally, the soil should be well structured, free draining, moisture retentive and deep (Almond *et al.*, 1986). The history of previous cropping for a field may determine the appropriate break for rapeseed in a rotation. Level of fertility will affect the level of lodging resistance required in a cultivar and weed problems may dictate the need for a herbicide-tolerant cultivar.

Intended market

Although most *Brassica* seed production is for the edible oil market, substantial quantities are used as a source of industrial oil. Market potential and the availability of special contracts may prescribe cultivar selection. The edible oil market mainly requires rapeseed with low erucic acid and low glucosinolate content but there is increasing diversity of cultivars with specialized fatty acid profiles for niche markets. New cultivars, developed by both conventional breeding and biotechnology, offer new or alternative sources of raw materials for edible, industrial or pharmaceutical use. The relative potential of these niche markets will need to be taken into consideration in cultivar selection.

SOIL PREPARATION

Brassica seed must be placed into a firm, moist, warm, aerated, well-structured seedbed for rapid germination and seedling growth. Soil preparation aims to provide these favourable conditions. Factors affecting crop establishment include the previous crop, crop residues, soil type, fertility, moisture availability, weeds, pests and diseases. Management for some of these factors may have an adverse effect on others: for example, tillage for good seedbed preparation and weed control breaks down soil structure, reduces organic matter, encourages germination of weeds and volunteer seeds from the previous crop, and can leave the soil susceptible to erosion. Management requires a compromise between these factors to achieve optimum results.

Eliminating previous crop residues

Residues from the preceding crop, often a cereal with 3 to 10 tonnes ha^{-1} of straw, may be unfavourable for rapeseed establishment. Large quantities are not easy to dispose of by cultivations and, when incorporated into the soil, may generate a highly porous soil texture. This inhibits soil–root contact and is not conducive to good plant nutrition. Growers usually try to eliminate residues by harvesting the straw or burning, the latter being a dangerous and increasingly unpopular alternative in some countries due to environmental concerns. Burning was banned in the United Kingdom in 1993, a move which prompted investigations into the establishment of winter rapeseed crops on cereal stubble (Bowerman, 1991). Results indicate that rapeseed can be broadcast into the stubble of a preceding wheat crop immediately after harvest, if weather conditions are favourable.

Effect on subsequent crops

Several advantages follow from including rapeseed in the rotation: yields of the first wheat following rapeseed are invariably improved, grass weeds are controlled and the level of cereal pathogens is reduced. The early harvest is a bonus in spreading the workload on the farm at a busy time of year. Furthermore, little, if any, specialist equipment is required (Almond *et al.*, 1986).

Land management

A summer fallow before drilling winter rapeseed allows weed seeds to germinate. Seedlings can then be destroyed mechanically or with chemicals before sowing the crop. The cultivation technique was often used by farmers before the development of herbicides.

It may not be possible to use the same

technique for spring-sown rapeseed as low temperatures impede weed germination and sowing cannot be delayed without risking later maturity and lower yields. However, depending on climatic conditions, a similar technique may be used immediately after harvesting the preceding crop when weed seed germination can be encouraged by shallow cultivations. Seedlings may then be destroyed by low temperatures during the winter or by cultivations before sowing in the spring.

Other *Cruciferae* species can be a major weed problem in rapeseed as, apart from reducing yield, they can have a serious effect on quality due to high glucosinolate content.

Direct drilling

Direct drilling is becoming increasingly popular with spring rapeseed in western Canada as a means of conserving soil moisture. It may not be possible elsewhere as the equipment is not adapted to wet conditions and, moreover, where autumn ploughing is possible, the winter provides an opportunity for weed control and drainage. Otherwise, shallow cultivation is normally best.

Swedish and French studies showed that direct drilling of winter rapeseed gave similar results to conventional ploughing and drilling, depending on site conditions (Cedell, 1983; Pouzet and Rollier, 1983). Cereal residues form a mulch under dry conditions, which encourages germination after direct drilling. However, ploughing is better in soils with a low organic matter content or with a poor structure. In such soils, root penetration and drainage are poor without deep cultivation and direct drilling is liable to make crops more susceptible to drought and to enhance nutritional deficiencies.

Control of pests and weeds can also be difficult with direct drilling. Slugs (*Deroceras reticulatum* and other species), which are a major pest in Western Europe, are found in crops in the autumn. Burning or ploughing in cereal straw often helps to limit slug attack.

Direct drilling does not appear to differ from ploughing in affecting the subsequent weed flora: the main disadvantage is that herbicides requiring incorporation into the soil cannot be used and these are often the more effective and cheaper. Pre-emergence herbicides can be used to control a wide range of broadleaved weeds and post-emergence herbicides can be used to control monocotyledons, but the cost is high. Nevertheless, direct drilling is often favoured as it is quicker and reduces costs of labour and machinery. Choice between direct drilling and ploughing often depends upon the availability of equipment, nature of the soil, assessment of likely risks of pests and weeds, and the costs involved.

SOWING

Time of sowing and seed rate

The object of seedbed preparation is to encourage strong seedling establishment and good root development with plants spaced uniformly. Timing will be influenced by soil temperature and moisture level and there are substantial distinctions between the autumn- and spring-sown crop. Sometimes it is difficult to prepare soil adequately for the small seeds and it is tempting, but not necessarily rewarding, to compensate for poor preparation by raising the seed rate.

Winter oilseed rape
Many studies have been undertaken on the effect of seed rate, sowing date and interactions between them. These have been reviewed in Chapter 2. Mendham *et al.* (1981) showed that, with winter rapeseed under English conditions, yield is determined by spring growth and translocation

of carbohydrates from vegetative parts of the plant. Spring growth is often, but not always, related to plant size at the time of floral initiation, which generally occurs at the end of the autumn (Mendham and Scott, 1975).

Two periods of plant development in relation to accumulated temperatures (the sum of mean daily temperatures above 0°C) were differentiated in France by Pouzet et al. (1983a). The first is from emergence to a total 400°C, during which growth rate is very slow and corresponds to the time necessary to develop the first leaves and roots. During the second period that follows, the main limiting factor for growth is the amount of photosynthetic active radiation (PAR); dry matter accumulation is then linearly related to intercepted PAR.

Early sowing encourages establishment of strong, vigorous seedlings and plant population has little effect: consequently, seed rate is of less significance in relation to total accumulated dry matter or seed yield. But it may have a negative effect on winter hardiness, which decreases after initiation of stem elongation (Szczygielski and Owczarek, 1987). Competition between plants is low with late sowing and total accumulated dry matter will be related to plant population at emergence. Furthermore, winter hardiness will be poor until plants attain a minimum level of vigour (Jasinska et al., 1987). In situations where air and soil temperatures are higher at the end of summer than in early autumn, the active growth period is longer with earlier sowing. Thus experience in eastern France shows dry matter yields as high as 2000 kg ha^{-1} after sowing in August, compared with only 200 kg for mid-September sowing.

So physiological development is favoured by early sowing, which also has economic advantages in terms of lower seed rates and higher yields. However, very early sowing is not always favourable for winter rapeseed: seedbeds are often comparatively dry so that emergence may be delayed or irregular. Weed control is also more difficult as herbicides, particularly trifluralin, incorporated in the soil before sowing are sensitive to high soil temperatures and radiation. Moreover, early-sown crops are more vulnerable to attack by cabbage stem flea beetle (*Psylliodes chrysocephala*), virus-vector aphids (*Myzus persicae*) and cabbage root fly (*Delia radicum*) than those sown later (Williams et al., 1991). Even in the spring, early-sown crops will be more susceptible to lodging and to disease attack as the canopy is much more developed (Morall et al., 1991).

Time of sowing and seed rate are determined in relation to winter hardiness: this is at an optimum when the stem base is 5–16 mm in diameter and stem elongation less than 30 mm (Topinka et al., 1991). French studies showed that, in order to obtain this growth before winter, a period is required with accumulated temperatures of 600–800°C (above 0°C) from emergence to the first frost with about 60 plants m^{-2}. Analyses of annual climatic statistics can help in determining the optimal date for sowing.

Spring oilseed rape

Clearly the constraints determining sowing date are different with the spring crop. The main one is soil temperature as 6°C is required for germination: seed may rot at lower temperatures which adversely affect emergence. On the other hand, in some situations spring droughts may impede germination. Late germination is likely to delay maturity and impair seed quality, especially chlorophyll content (Daun et al., 1983; Ward et al., 1991). As the growing season is shorter than for winter rapeseed, recommended plant densities are higher to ensure that dry matter accumulation before flowering is sufficient to achieve a high yield.

Plant population studies in the United

States indicated that plant densities of spring *B. napus* in excess of 40 plants m^{-2} produced comparable seed yields (Heikkinen and Auld, 1991). Experiments with *B. rapa* and *Eruca sativa* indicate that the shorter the growth cycle the higher the desirable sowing rate, except when climatic conditions are abnormal: for example, experiments in Alaska showed that the lower seeding rate gave the highest yields when precipitation was abnormally high (Lewis and Knight, 1987).

Plant density can influence seed quality: Daun *et al.* (1983) observed that high-density *Brassica napus* crops had a lower chlorophyll content at harvest and Kjellstrom (1987) made similar observations in *E. sativa* crops. This may be because high-density crops mature earlier and more evenly.

Depth of sowing

Ideally the small seeds should be just covered in the seedbed but under dry conditions it may be necessary to drill a little deeper to ensure contact with moisture. Moisture retention is essential for rapid germination and the seedbed may need a light rolling for compaction after sowing. However, rolling can induce poor emergence under dry conditions when rain causes 'capping' or the formation of a surface crust (Pouzet *et al.*, 1983b). Some soils, usually the heavier ones, are more susceptible to this than others.

Row width

An important component of sowing rate is row width as this may affect weed control, susceptibility to lodging and nitrogen fertilization. Plants in narrow rows (<20 cm) are more competitive and smother weeds, especially in winter crops, but wider rows (>50 cm) facilitate mechanical weeding. Susceptibility to lodging is linked to plant density in the row as, in a com-

parison of wide and narrow rows at the same seed rate, Vullioud (1993) found more lodging with the wide rows.

Good root development is essential to maximize utilization of nitrogen. The dynamics of root growth was studied by Kjellstrom (1991) who found that, in the absence of water stress, average daily root growth was 8 mm and that roots can grow to a depth of 90 cm. In fact, it seems that root growth averages 25 cm for every accumulated 100°C (above 0°C) period from emergence to the onset of stem elongation, and only 10 cm per 100°C thereafter. At the stem elongation stage, the preferential allocation of assimilates is probably changing under hormonal control (de Bouillé *et al.*, 1987). At the time of flowering about 85% of total root dry matter (but only 30% total root length) is in the top 0–23 cm of the soil where mineralization occurs.

There are indications that narrow row spacing is preferable to wider rows when nitrogen fertilizer supply is restricted. However, better seed placement and spacing of seeds with precision sowing can counterbalance the negative effects of wider rows, especially at low plant densities (Merrien and Pouzet, 1989).

Results obtained by Christensen and Drabble (1984) in Canada showed that both *B. rapa* and *B. napus* gave higher yields with narrow rows (7.5 cm compared with 15 or 23 cm), and that seed rate (7 or 14 kg ha^{-1}) had no significant effect on yield. The nitrogen application (100 kg ha^{-1}) was the same for the different row treatments and yields up to 2.5 tonnes ha^{-1} for *B. rapa* and 3.4 tonnes ha^{-1} for *B. napus* were obtained.

Methods of sowing

Broadcasting

Broadcasting the seed may demand a slightly higher seed rate for good results as placement is more hazardous due to variable moisture availability and predators. The seedbed should be harrowed and lightly rolled after seed distribution so as to ensure good contact between seed and soil.

Seed drill

Cereal drills are often used to avoid purchasing specialized machinery but may be difficult to adjust for sowing depth and seed rate. Also, caution is required with tractor speed. Variability with depth and rate of sowing increases at higher speeds, especially in the absence of a fine tilth or with stony soils.

Precision drill

Seed placement with pneumatic drills has recently been developed following experience with sugarbeet, which is now often sown in Western Europe 'to a stand', i.e. requiring no subsequent thinning. They move clods to the sides, leaving fine soil where conditions are ideal for germination and seeds can be placed at constant depth with a regular distance between seeds. Experiments have confirmed the advantages under French conditions (Merrien and Pouzet, 1989), but much depends on the likelihood of suitable weather conditions for sowing as precision drilling is slower than conventional drilling.

Seedbed fertilizer

Fertilizer requirements will be considered in detail below but rapeseeds are very sensitive to the proximity of fertilizers, being similar to linseed but more so than wheat or barley in this respect (Nyborg, 1961; Nyborg and Hennig, 1969). Fertilizer placement in the seedbed resulted in seedling injury, and hence impaired emergence, which increased with lower soil temperatures and moisture content.

FERTILIZER REQUIREMENTS

In using fertilizers efficiently the aim is to ensure sufficient nutrients for crop growth with minimum waste. Requirements in terms of major nutrients and trace elements are outlined below. In spite of many references to fertilizer requirements (Holmes, 1980; Grant and Bailey, 1993), environmental constraints have become increasingly burdensome in recent years, especially with respect to nitrate content of drinking-water in Western Europe.

Nitrogen

Deficiency symptoms

Plants deprived of nitrogen are dwarfed and the foliage is pale green to yellow. According to Thomas (1984), nitrogen in older leaves is redistributed to younger leaves, the older leaves withering. Remaining leaves often show purple discoloration. The canopy remains thin and open, pod numbers are low and yield is much reduced (Holmes, 1980).

Crop requirement

A spring rapeseed crop accumulates 50–60 kg nitrogen for every tonne of seed produced (Geisler and Kullman, 1991; Grant and Bailey, 1993). The equivalent for winter rapeseed is about 70 kg nitrogen. One tonne of harvested seed, with 42% oil and 38% protein in the meal, contains 35 kg nitrogen. So, for high yields, rapeseed needs a lot of nitrogen (150–210 kg nitrogen for 3 tonnes ha^{-1}), particularly since the efficiency of nitrogen take-up from the soil is low. By comparison, winter wheat produced in France needs about 30 kg nitrogen for each tonne of seed produced. Each tonne of seed removes about 22 kg nitrogen (13.75% of protein, whole-seed basis) and 210 kg nitrogen produces 7 tonnes of grain ha^{-1}. It is more difficult to determine the appropriate fertilizer rate for rapeseed than for wheat because of a poorer relationship between nitrogen utilization and seed yield. Crop requirements are provided by the addition of fertilizer and subsequent soil mineralization. Coefficients of usage have been determined using ^{15}N-labelling and with winter rapeseed this coefficient is low, around 50%, i.e. only 50% of the nitrogen from mineral fertilizers is used by the crop (Merrien et al., 1991).

In studying nitrogen transfer from the canopy to the seeds, Geisler and Kullmann (1991) grew spring rape under controlled conditions in a hydroponic system. Three levels of nitrogen concentration in the hydroponic solution were compared: 2, 7 and 12 mmol. The results showed that translocation activity relates to the level of nitrogen nutrition. Nitrogen translocation was observed at the 2 and 7 mmol levels but not at the highest level. Leaves and pod walls contributed most to translocation at the 2 and 7 mmol levels. Nitrogen translocation to the seeds utilized on average the following proportions of nitrogen available in the source: 66% of that available from the leaves, 53% from the pods, 27% from the stem and 17% from the roots.

Economic and environmental interest

Management of nitrogen fertility is complex and requires substantial attention by farmers internationally. However, it is of less economic interest to farmers in Western Europe receiving about US $150 tonne^{-1} seed, when the price for nitrogen as ammonium nitrate is about US ¢50 kg^{-1} nitrogen. As response to nitrogen varies with soil and climate conditions, moderately excessive applications provide the best solution. The cost is negligible compared with the returns likely to be realized when environmental conditions enable the crop to respond to a high nitrogen application.

The economics of Western Europe applications is illustrated by results of a nitrogen experiment carried out at 32 sites in France (Merrien and Pouzet, 1989),

where the optimal nitrogen rate per hectare appeared to be <150 kg ha^{-1} at four sites, 150–180 kg ha^{-1} at ten sites, 180–210 kg ha^{-1} at five sites, 210–240 kg ha^{-1} at eight sites and >240 kg ha^{-1} at five sites. The optimal rate was 198 kg ha^{-1} nitrogen on average but the extra cost of an application at 200 kg ha^{-1} averaged only US $5 ha^{-1}, at 240 kg ha^{-1} only US $7 ha^{-1}, while at 160 kg ha^{-1} a loss of US $21 ha^{-1} would be incurred. The strategy for maximum profits is clear – apply 200–240 kg nitrogen ha^{-1}. But this is environmentally unsatisfactory. An application of 200 kg nitrogen ha^{-1} would lead to an average excess of 42 kg nitrogen at 16 of the 32 sites, and 240 kg would lead to an average excess of 57 kg nitrogen at 27 of the sites.

The effect of this excess on water quality would be serious. An excess of 10 kg nitrogen is equivalent to 42 kg nitrates. Assuming that all excess nitrogen is in the form of nitrate and that drainage over winter amounts to 2×10^{6} l ha^{-1} (200 mm), the average nitrate concentration in drainage water would be 21 mg l^{-1}. It follows that when nitrogen levels are 25 kg ha^{-1} above the optimum rate, the European standard for drinking-water (<50 mg NO_3 l^{-1}) will be exceeded.

Excess nitrogen also has an impact on the production of biofuels as rapeseed oil methyl esters (see Chapter 15). There is controversy about the balance between the energy expended on production compared with that released when utilized as a fuel. As nitrogen fertilizer accounts for one third of total energy input (PROLEA, 1993), high inputs may be difficult to justify. Further study is needed to improve the efficiency of nitrogen utilization in the light of environmental worries and the competitiveness of oilseed markets.

Nitrogen application to winter oilseed rape

Knowledge of nitrogen metabolism and translocation of assimilates to the seed pro-

Table 3.2. Effect of autumn nitrogen application on winter oilseed rape yield in 1982–1983.

Time of sowing	Autumn nitrogen	Yields at experimental site (t ha^{-1})		
		Indre	Meuse	Gers
Early	N0	1.82	2.34	2.79
	N1	2.06	2.75	3.67
	N2	2.11	2.63	3.92
	N3	2.19	2.38	3.63
	N4	2.12	2.36	3.74
Normal	N0	2.42	2.21	2.24
	N1	2.53	2.53	2.39
	N2	2.76	2.71	2.68
	N3	2.74	2.42	2.64
	N4	2.82	2.41	2.67
Late	N0	1.70	2.18	2.32
	N1	1.95	2.12	2.47
	N2	2.01	2.03	2.70
	N3	1.71	2.13	2.83
	N4	2.03	2.03	2.69

Source: Pouzet *et al.* (1984).
N0, no nitrogen;
N1, 50 units nitrogen on seedbed;
N2, 100 units nitrogen on seedbed;
N3, 50 units nitrogen at four-leaf stage;
N4, 100 units nitrogen at four-leaf stage.

vides the basis for good crop management. The dynamics of nitrogen uptake in winter rapeseed can be considered at four growth stages: autumn, beginning of spring, vegetative growth stage, and from end of flowering to maturity. The studies of Geisler and Kullman (1991) on spring rape, described above, showed that excess nitrogen appears to decrease nitrogen efficiency and one may infer that this also applies to the winter crop. This may not be of economic importance but has environmental implications. If the amount of nitrogen in the straw, which will be subsequently ploughed in, can be limited by avoiding excess nitrogen applications, the environmental status of rapeseed would be improved.

Under French conditions nitrogen mobilization in the autumn will average about 50–100 kg ha^{-1}. This rises in the spring and can amount to 250 kg ha^{-1}, being used firstly for development of leaves and then stems. As rapeseed has an indeterminate growth habit, nitrogen mobilization for development of the canopy continues after flowering begins. When flowering ends, nitrogen accumulates mainly in the seeds. There is a balance in the crop between nitrogen lost when leaves drop and that absorbed by roots, and generally speaking, there is no net nitrogen accumulation during this last stage.

Apart from the amount, consideration should also be given to timing of nitrogen application but there does not seem to be a particular growth stage when nitrogen nutrition has a major influence. Palleau and Tittonnel (1991) mention a positive effect of nitrogen applied at the onset of reproductive development, but this has not been confirmed by other workers.

Mineralization of nitrogen in the soil occurs when temperature and moisture conditions are favourable for microbial activity: the need for nitrogen fertilizer will then be low. During vegetative development at the beginning of spring, soil temperatures are too low and soil moisture too high for soil mineralization to supply sufficient nitrogen for the crop. When stem and seed development starts at the end of spring, soil mineralization is active again and the need for nitrogen fertilizer is less. Consequently, the main period for additional nitrogen seems to be at the end of winter so that leaf growth is not restricted.

Several workers have studied autumn nitrogen applications in Western Europe. Pouzet *et al.* (1984) concluded that early applications can increase foliage growth, especially after early sowing. However, these rarely increase yield (Table 3.2) and should perhaps be forbidden for environmental reasons.

Nitrogen applications are difficult to

manage just before or around flowering time, because soils begin to dry out and availability is dependent on rainfall or irrigation. Late applications are not always efficient and farmers must decide how much nitrogen to apply as soon as possible after the end of winter when it is not easy to predict yields or estimate nitrogen availability from soil mineralization.

Yield predictions can be based on previous experience with a particular field, which can be modified according to crop status. For example, poor autumn growth will generally result in lower yields. Nitrogen requirement can then be calculated on the basis that 70 kg will be used for every tonne of seed produced.

Before deciding fertilizer requirement for a particular crop, it will be necessary to consider the amount of nitrogen likely to be provided by mineralization of soil and crop residues. This amount can be divided in two parts. First, the amount of mineralized nitrogen in the soil at the end of the winter (Soper, 1971); this is generally low as nitrogen absorption is high in the autumn (Pouzet, 1985; Merrien and Pouzet, 1989; Reau et al., 1994) and the amount available will only be high if crop growth was restricted before the onset of winter. Second, the amount that will be mineralized during spring, which is very dependent on climatic conditions, and results from nitrogen response experiments are needed to provide a guide.

French experiments show that the total amount of nitrogen from mineralization of soil and crop residues is generally about 50–80 kg ha^{-1} but, in some situations, yields up to 3.5 tonnes ha^{-1} can be obtained without any nitrogen fertilizer (Bilsborrow et al., 1993). In fact, nitrogen absorption in the autumn should be taken into account as this can be high.

The nitrogen balance can be predicted thus (after Reau et al., 1994):

$$70 \times Y = X - N_h + N_a + N_s$$

where 70 is the amount of nitrogen in kilograms required for one tonne of seeds at about the optimal rate, Y is the expected yield (tonnes ha^{-1}), X is the amount of nitrogen from mineral fertilizer, N_h is the amount of mineral nitrogen remaining in the soil after harvest, N_a is the amount of nitrogen absorbed during autumn and N_s is the amount of mineral nitrogen supplied by the soil during spring, including mineral nitrogen present at the end of winter. Losses from denitrification are considered negligible although their impact on the global greenhouse effect should be borne in mind.

The amount of nitrogen fertilizer required (X) can be estimated if Y, N_h and N_s can be evaluated with precision: these may be available from regional agronomy stations. Research is now focused on the nitrogen status of the plant as a complement to evaluation of nitrogen supply from the soil. Two analytical methods are being tested: first, total nitrogen content of the plant and, second, nitrate content in petioles or at the stem base. To allow for dilution effects (Palleau, 1989; Greenwood et al., 1990), total nitrogen content of the plant should be related to dry matter biomass, which limits practicability of this method on farms. Methods based on nitrate content seem very positive but more developmental work is needed.

Nitrogen and spring rapeseed

The general principles for nitrogen on spring rapeseed are the same as for winter rapeseed. Soil mineralization will generally be more efficient and fertilizer rates lower. But dry conditions may limit fertilizer efficiency (Henry and MacDonald, 1978; Rollier and Pouzet, 1983) and the probability of drought should be considered in estimating nitrogen requirement.

Applications at sowing were more effective than top-dressing at bud formation stage on low- to medium-fertility sites in Australia (Sykes and Mailer, 1991). In

Canada, Grant and Bailey (1993) found that autumn application of nitrogen could be advantageous for short-season regions with dry soils. It saves time at sowing and the nitrogen can be used by crops even when the first layer of soil has dried out. However, this can be risky, especially under wet conditions, when nitrogen is subject to denitrification and leaching.

Form of nitrogenous fertilizer

Nitrogen is most immediately available to plants when applied in the form of nitrate, although Malhi *et al.* (1988) showed that rapeseed can utilize the ammoniacal form. Rapid availability is important at the end of winter, when climatic conditions do not favour soil mineralization.

Organic manure is also a source of nitrogen, but is variable in content and, consequently, standard rates of application cannot be determined. Moreover, livestock are usually comparatively rare in the major rapeseed-growing regions so that the practicalities of manuring rapeseed have not been studied. However, there is an interest in growing winter rapeseed with heavy rates of organic manure in autumn in places like Britanny in France. This benefits the growers, who get cheap fertilizer, and the livestock producers, who can use whole rapeseed in their animal feed. There is also an environmental benefit as the organic manure is spread over a larger area rather than being restricted to the livestock farms.

Ploughing in a legume cover crop can partially substitute for nitrogen fertilizer and this technique is used in India (Baddesha and Pasricha, 1991).

Splitting applications

Splitting spring nitrogen applications to winter crops has been studied by several workers, but there is no general conclusion. Compared with one application, splitting has never shown negative results and may have a positive effect in some years. Climatic conditions and soil type influence the results: when conditions are favourable to leaching (light soils, heavy rain), a split application is to be preferred (Rollier, 1970). But here again there is a conflict between what is profitable for the farmer and the environmental consequences. The cost of splitting, taking into account labour and machinery, is normally greater than the yield benefit but splitting could reduce nitrate leaching.

Nitrogen application in relation to other agronomic practices

Although a key factor in determining yield, nitrogen nutrition will interact with other agronomic practices: choice of cultivar, plant density and use of growth regulators or fungicides. As far as is known, direct effects are more important than interactions when other agronomic practices are optimal. From multifactorial experiments (Debouzie *et al.*, 1987; Williams *et al.*, 1991), there is no indication that yield potential can be improved by using high plant densities, high nitrogen fertilizer applications, a growth regulator and two or three fungicides, compared to standard practice, and the economic advantage is generally low.

Gerath and Schweiger (1991) have shown that cultivars may differ in nitrogen uptake and translocation. They classified cultivars into three types: type I – the higher the nitrogen application, the higher the yield; type II – as nitrogen is increased, yield increases at first, then remains stable; type III – as nitrogen is increased, yield increases at first, is stable for a while and then decreases. Type III cultivars show a marked decrease in oil content as nitrogen levels are increased. Barszczak *et al.* (1991a) also found statistical interactions between cultivar, drought, soil acidity and nitrogen rates for plants grown in pots under field conditions.

Effect on quality

As glucosinolates are synthesized from sulphur amino acids, glucosinolate content is affected by both nitrogen and sulphur availability and the nitrogen : sulphur ratio should be taken into account when assessing the effect of nitrogen on glucosinolate content. A plentiful supply of both elements can result in high levels of glucosinolates and, according to Grant and Bailey (1993), the optimal ratio of nitrogen to sulphur is 12 at flowering time. Yet glucosinolates may reflect the physiological status of the plants (Clossais-Besnard and Larher, 1991) or indicate the presence of water stress after flowering (Mailer and Wratten, 1987).

High glucosinolate content has been recorded after restricted nitrogen fertilizer application: the effect of nitrogen on seed glucosinolate content varied from one year to the other and was higher in some years following no nitrogen application than in other years when high rates had been applied (Bilsborrow et al., 1993).

A negative correlation between oil and protein content in the traditional determination of seed quality is well documented. High nitrogen applications reduce oil content and increase protein content. Economic changes (especially the 1993 Common Agriculture Policy) and increased concern about the environment in Western Europe have already led to reduced nitrogen usage and protein content of the meal is decreasing (ONIDOL-CETIOM, 1993). The protein fraction in the meal is replaced by non-dietary fibre when nitrogen rates are reduced (Delorme, 1993).

Phosphates

Deficiency symptoms

The leaves and stems of deficient plants turn dark bluish green before turning purple. Roots become stunted and plants dwarfed. In severe cases there may be marginal necrosis of the leaves and older leaves may wither prematurely (Holmes, 1980).

Crop requirements

The amount of phosphate in plants is not very high (about $60\,kg\,ha^{-1}$ in winter rapeseed) and about 12 kg is removed per tonne of seeds. Oil and protein content is hardly affected by phosphate applications (Grant and Bailey, 1993).

Environmental considerations

As phosphorus is relatively immobile in the soil, the risk of phosphate application increasing levels in ground or surface water is very limited.

Form of phosphatic fertilizer

The solubility of phosphates differs among the various forms and is more available as monoammonium phosphate and polyphosphate than as rock phosphate, which dissolves only under acid conditions (Grant and Bailey, 1993).

Rate of phosphate application

Rates of application should take into account the same parameters as for potassium, mainly availability in the soil (Soper, 1971). Solubility varies with the form of phosphate used in the fertilizer formulation. The efficiency of phosphorus uptake by oilseed rape is better than for other crops (Grant and Bailey, 1993).

Phosphate application and factors influencing plant response

Because of their low mobility, phosphates should be incorporated into the soil before sowing. Placement of phosphatic fertilizer in bands may be more economical than incorporation, but can be toxic to the seeds if placed too close. It is better to place phosphate in a band below or to the side of the seed in the row. Deep banding may be an alternative to side banding when high rates of phosphates are needed (Thomas, 1984).

Factors influencing plant response to phosphates include:

1. soil pH: this greatly affects release into the soil. Maximum availability occurs in soils with a pH range 6.0 to 7.0 (Thomas, 1984).

2. soil texture: the clay content of the soil influences the rate of phosphate diffusion to the roots.

3. soil moisture and temperature: conditions favourable to microbial activity and root growth increase phosphate availability and crop uptake. Excessive moisture and low temperatures will prevent plants utilizing phosphates even in soils with high levels.

Yield response and effect on quality

Response to phosphatic fertilizers is seldom as significant as that to nitrogen fertilizers (Holmes, 1980). It is influenced by plant root development and distribution, the method of phosphate application and the available phosphate level in the soil (Thomas, 1984).

Generally speaking, the oil and protein content of rapeseed is not greatly affected by phosphate fertilizers (Thomas, 1984) and the effect on glucosinolate content of the seed is small (Holmes, 1980).

Potassium

Deficiency symptoms

Leaves of potassium-deficient plants turn dark green and may curl, while the margins and areas between the veins may appear scorched. Young plants are dwarfed and wilt readily. In extreme cases affected leaves die completely but tend to remain attached to the stem (Holmes, 1980).

Crop requirements

Oilseed rape needs large amounts of potassium as more than 200 kg K_2O is mobilized in the plants, although only about 25 kg K_2O is removed per tonne, so that a cereal crop grown after rape will not normally need potassium unless the soil content is low. Maximum absorption occurs during stem elongation, when the daily rate can be up to 15 kg K_2O ha^{-1}.

The amount of fertilizer applied should relate to the soil characteristics, especially content of available potassium (Soper, 1971). As soil analytical methods for potassium availability differ between countries, and even from one laboratory to another, it is virtually impossible to propose a universal reliable standard for potassium application.

Potassium application and factors influencing plant response

As potassium is immobile in the soil, fertilizer should be applied before sowing and incorporated in the upper layer of soil. As with phosphates, placement in a band below and to the side of the seed will be more efficient than broadcasting at low rates. Placement with the seed reduces germination.

There is no evidence that soil pH has a strong influence on availability but soil texture is important as the potassium is adsorbed on clay particles. Potassium is similar to phosphates in the influence that soil moisture and texture have on plant response.

Yield response and effect on quality

Potassium fertilizer has a marked yield response but only on deficient soils. It has no effect on seed quality, other than on highly deficient soils (Grant and Bailey, 1993).

Sulphur

Deficiency symptoms

As sulphur is involved in photosynthesis, deficiency decreases chlorophyll content and leaves turn yellow showing interveinal chlorosis. Growth is stunted and the flowers have smaller petals which are pale yellow.

When symptoms become noticeable rape-seed is severely lacking in sulphur (Thomas, 1984). Pods form slowly, are small and poorly filled with shrivelled seeds.

Crop requirements

As with other *Cruciferae*, there are many sulphur compounds present in the vegetative parts of the plant and the seed. Of these compounds, the glucosinolates are of the greatest significance in oilseed rape, the levels of which have been reduced in double-low and canola cultivars. Sulphur therefore has a direct effect on quality as well as growth and yield.

It is not easy to provide a reliable review of the literature as fertilization units are not the same from one country to another. In some countries (e.g. France) the official unit for fertilizer is SO_3 equivalents. In other countries, the unit is based on sulphur alone (Booth *et al.*, 1991; Donald *et al.*, 1993; Grant and Bailey, 1993). As the $SO_3 : S$ ratio is 2.5, care is needed when interpreting results.

Another difficulty is that there is some SO_4 in rain and the quantity is not negligible in comparison to crop needs. It is difficult to evaluate the amount of sulphur provided by rain when undertaking field experiments. This is especially so as the amount is declining as efforts are made to reduce emissions from cars and industry, and sulphur deficiency is now found in countries where it was unknown ten years ago (Booth *et al.*, 1991; Evans *et al.*, 1991; Marquard *et al.*, 1991; Donald *et al.*, 1993).

Sulphur in the plant

Plants absorb sulphates only from the soil and sulphur nutrition will largely depend on soil mineralization. Like nitrates, sulphates are very susceptible to leaching and, under French conditions, sulphur defi-ciency is probable when rainfall exceeds 300 mm during winter. Sulphur deficiency is rare in spring rape crops as soil mineralization is higher during spring.

The amount of sulphur in a crop is at a maximum at the end of flowering – about 60 kg ha^{-1} for a good winter crop. The optimum dry matter content of whole plants is about 0.6% during stem elongation (Merrien and Pouzet, 1989). Although the risk of sulphur deficiency may be determined, taking into account soil characteristics (KH_2PO_4- or $NaHCO_3$-extractable sulphur content according to Scott (1981)) and weather (mainly the amount of rain during winter), it is impossible to estimate the amount needed precisely: sulphur fertilizers are impure and generally contain other nutritive elements (e.g. nitrogen, potassium, magnesium, phosphate), which will affect the amount required. On average a rate of 30 kg sulphur ha^{-1} has generally been found adequate in France.

Form of sulphur application

It is important to note that mineral micronized, or finely ground elemental, sulphur is inadequate to correct sulphur deficiency. Merrien and Pouzet (1989) found that sulphur in this form increased yield in only one of three experiments, while sulphates were effective in all three. Application of micronized sulphur did not show a positive effect on yield in Scotland but did affect total sulphur concentration in the plant at harvest (Donald *et al.*, 1993).

In the same country, positive effects of micronized sulphur on yield and on seed glucosinolate content were registered on a sulphur-deficient soil (Booth *et al.*, 1991). These results also showed an interaction between nitrogen and sulphur fertilization. Increasing nitrogen on a sulphur-deficient site reduced yield and seed glucosinolate content when no extra sulphur was applied. However, increasing nitrogen on a high-sulphur site raised yields and tended to enhance seed glucosinolate content when no extra sulphur was applied (Table 3.3). These data confirm that nitrogen and sulphur interact on yield and glucosinolate content and that a model based on source–

Table 3.3. Effect of site and nitrogen and sulphur applications on yield and glucosinolate content of rapeseed–cv. Tapidor.

Nitrogen (kg ha^{-1})	Sulphur (kg ha^{-1})	Glucosinolate content (mol g^{-1} seed)		Yield (t ha^{-1})	
		Low sulphur	High sulphur	Low sulphur	High sulphur
150	0	4.0	5.2	1.18	4.80
	8	3.9	5.5	2.23	4.60
	16	4.1	9.4	2.83	4.84
	32	4.9	8.0	3.50	4.61
	64	4.9	8.7	3.40	4.79
250	0	3.9	6.9	0.82	4.52
	8	4.2	6.9	1.69	4.75
	16	3.5	7.0	2.04	5.09
	32	5.2	7.6	3.13	5.02
	64	7.6	10.5	3.50	5.08

Source: Booth *et al.* (1991).

sink relationships should be adequate to describe their effects on oilseed rape (Schnug and Haneklaus, 1994).

Factors influencing plant response

Sulphur deficiency occurs most commonly on coarse-textured soils and those with low organic matter (Thomas, 1984). As with nitrogen, high soil moisture and low soil temperatures will prevent sulphate mineralization by microbial activity.

Effect of sulphur on quality

As mentioned above, sulphur is essential for glucosinolate biosynthesis, and the influence of sulphur fertilization on glucosinolate content has been widely debated. The main difference from nitrogen is that only excess sulphur increases glucosinolate content (Mailer and Wratten, 1987; Marquard *et al.*, 1991). A strong relationship between leaf sulphur content at the beginning of flowering and final seed glucosinolate content has been observed (Evans *et al.*, 1991).

Previous results from Schnug (1987) indicated that seed glucosinolate content related more to sulphur content of younger

leaves at the stem elongation stage in single-low cultivars than in double-low cultivars. A 0.1% increase in foliar sulphur content is followed by a rise of 7.5 μmol of glucosinolates g^{-1} seed in single-low cultivars and by a rise of 1.5 μm in double-low cultivars. It is recognized in France that an application of more than 75 SO_3 units ha^{-1} can increase glucosinolate content (Merrien *et al.*, 1987), although this is of less significance with double-low cultivars and concern about the relationship between sulphur fertilization and quality should decline now that these are more widely grown (Grant and Bailey, 1993).

Magnesium

Deficiency symptoms

Magnesium is required for chlorophyll production and for numerous enzymatic functions. Deficiency is denoted by chlorotic marbling of the leaves and stem, which are also tinged orange to purple.

Crop requirements

Magnesium affects oilseed rape nutrition and the amount in the crop can rise to 28 kg

Table 3.4. Optimal soil content for minor elements.

Element	Optimal content (ppm)	Comment
Iron–EDTA	40–200	Interaction with pH
Iron–DTPA	20–150	
Copper–EDTA	1.7–2.5	Interaction with organic matter
Copper–DTPA	0.5–10.0	
Copper–oxalic	1–5	
Zinc–EDTA	2–5	Interaction with pH
Manganese–EDTA	14–16	
Manganese–DTPA	10–80	
Boron (hot water)	0.3–0.8	Interaction with pH
Molybdenum (ammonium acetate)	0.1–1.0	

Source: Merrien and Pouzet (1989).
DTPA, diethylenetriaminopentaacetic acid; EDTA, ethylenediaminotetraacetic acid.

magnesium or 46 kg MgO ha^{-1}. About 50% of this amount is removed with the seeds. Magnesium deficiency is more frequent on light acid soils, where minimum content is about 0.1%, compared to 0.15% on heavy soils. Calcium affects magnesium absorption and the magnesium : calcium ratio in the soil should be taken into consideration when deciding application rates.

Magnesium sulphate provides a convenient form for application of both sulphur and magnesium to crops. An application of 140 kg ha^{-1} provides 40 kg MgO ha^{-1} and 75 kg SO$_3$ ha^{-1}.

Minor elements

Comprehensive references for soil and plant contents have been published (Merrien and Pouzet, 1989) and are summarized in Tables 3.4 and 3.5. Deficiency symptoms are described by Grant and Bailey (1993).

Haneklaus and Schnug (1991) have described methods for diagnosing nutrient deficiencies based on foliar analysis. Results showing the relationship between yield and leaf content of minor elements have determined threshold levels for each element and are presented as a computer program named PIPPA (Schnug, 1990).

Table 3.5. Optimal content of minor elements in leaves at stem elongation.

Element	Optimal content (ppm)
Copper	4.5
Zinc	37.5
Iron	100–130
Manganese	40
Boron	20–25
Molybdenum	0.5–0.7

Source: Merrien and Pouzet (1989).

Results obtained by Krauze et al. (1991) in Poland showed that oilseed rape (winter cultivar with high erucic acid content) is very sensitive to boron deficiency and one application of 0.2–0.4 kg boron ha^{-1} increased seed yields by 0.4–0.6 tonne ha^{-1}. Boron also increased seed oil content and slightly reduced erucic acid content. Higher doses could be phytotoxic and placement with the seed should be avoided (Grant and Bailey, 1993).

Barszczak et al. (1991b) also studied the toxic effect of aluminium and manganese on B. napus and found that cultivars differ in their tolerance to these elements.

Table 3.6. Effect of triapenthenol (350 g ha^{-1}) on yield.

Year	Cultivar	Lodging score[1]	Yield increase[2] (100 kg ha^{-1})
1987	Bienvenu	2	−1.5
1987	Bienvenu	2	−0.5
1987	Darmor	2	+1.0
1987	Darmor	2	−0.5
1987	Darmor	3	+1.2
1987	Bienvenu	4	+5.9
1987	Darmor	5	+6.0
1988	Doublol	4	+1.6
1988	Doublol	5	+8.8
1988	Doublol	6	+9.4
1988	Bienvenu	5	+7.5

Source: Merrien and Pouzet (1989).
[1] Lodging score recorded on untreated plot: 1 = none, 5 = severe.
[2] Yield increase: difference between treated and untreated plots.

GROWTH REGULATORS

Growth regulators were developed for cereals to reduce crop height and improve standing ability. Such effects have been sought in rapeseed crops with the intention of reducing lodging, improving efficiency of nitrogen utilization, controlling excessive vegetative growth resulting from heterosis in hybrids, controlling stem elongation during autumn, and reducing canopy height so as to facilitate fungicide or insecticide applications after flowering, as well as direct combining (Daniels and Scarisbrick, 1986).

Two kinds of chemicals have been tested: anti-gibberellins and anti-auxins and are considered here. In the rapeseed plant, gibberellins are mainly produced at the beginning of stem elongation and auxins are produced during stem extension and up to the time when flower buds are visible (de Bouillé et al., 1987). Experiments over several years and in various countries indicate that application of

growth regulator may be particularly effective in reducing lodging. Thus results from Merrien and Pouzet (1989) show that growth regulators did not improve yield in the absence of lodging (Table 3.6). Compensation between yield components is probably the explanation: growth regulators increase pod number per plant with a subsequent decrease in seed number per pod. Nevertheless, growth regulators are still of interest as a means of expediting harvest, especially when lodging occurs.

Observations by the author have shown that reducing crop canopy height with growth regulators can facilitate application of plant protection chemicals after flowering. The shorter height permits ground rather than aerial spraying to control late diseases or insects. Other observations (Musnicki et al., 1987; Lerhmann et al., 1991) show that plant growth regulators reduce height and lodging without affecting seed quality or delaying maturity.

Growth regulators can be used on winter rapeseed in the autumn to limit stem elongation and thus susceptibility to frost (Musnicki et al., 1987). Susceptibility to frost is related to the average distance between two stem nodes and plant growth regulators can therefore be useful on early-sown crops or when a mild autumn encourages growth (Lucas, 1994). Application at the four-leaf stage can reduce vegetative growth, but the cost of the chemical and autumn application is generally uneconomic. It is difficult to decide whether an application is justified in the spring as it must be done before stem elongation and future growth, including the risk of lodging, can be predicted. The decision has therefore to be based on cultivar susceptibility, soil conditions and the amount of nitrogen used.

Application of a growth regulator can be harmful to a crop, especially when the temperature is low at time of application (Merrien and Maisonneuve, 1987). The conditions for the absorption of the active

Table 3.7. Influence of rainfall on the efficiency of triapenthenol.

Year	Cultivar	Rain in 8 days prior to application (mm)	Rain in 20 days after application (mm)	Height reduction (%)[1]
1986	Bienvenu	14	104	43
1986	Bienvenu	7.0	38	17
1987	Darmor	27	18	19
1987	Darmor	19.5	10.8	17
1987	Bienvenu	13.0	7.3	15
1987	Bienvenu	27.5	11.6	13
1987	Darmor	24.2	12.9	10
1987	Darmor	4.0	22.0	6
1987	Bienvenu	2.0	30.0	4
1988	Doublol	4.6	90.4	47
1988	Ceres	4.6	90.4	43
1988	Doublol	17	76.5	37
1988	Bienvenu	27	63.5	32

Source: Merrien and Pouzet (1989).
[1] Height reduction: Height of untreated plots − Height of treated plots expressed as percentage of height of untreated plots.

ingredient are also important. Thus, for triapenthenol, the more the rain from 8 days before to 20 days after application, the greater the effect in reducing canopy height (Table 3.7).

Hybrids have now been registered in Europe but, although observations by the author have shown experimental winter hybrids 15–20% taller than conventional cultivars, this is unlikely to be a feature of hybrids currently under development as breeders are seeking dwarf hybrids (Renard, 1991). Further improvement of cultivars with resistance to lodging is possible and studies have shown that both shearing and bending stresses are related to the stem cross-section area (Skubisz and Tys, 1987; Skubisz, 1991).

Finally, growth regulator application is very sensitive to the economic climate. When favourable, European farmers can use growth regulators to reduce lodging and to facilitate late applications of agrochemicals and harvest but, when unfavourable, farmers are likely to restrict nitrogen applications and avoid chemical control

of late diseases or insects, making plant growth regulators uneconomic. For example, with the new Common Agricultural Policy in France, less than 10% of the rapeseed acreage is sprayed now compared with more than 25% previously when rapeseed production was more profitable.

IRRIGATION

Water requirements

Rapeseed plants require large amounts of water for photosynthesis and translocation of nutrients. Around flowering time, the daily evapotranspiration of *B. rapa* is as high as 8 mm. There is a linear relationship between yield and evapotranspiration in spring rapeseed and Thomas (1984) obtained yield increases of 6 kg ha^{-1} for each additional millimetre of water. Crops are especially sensitive to drought at the time of germination and at the pod elongation stage. Drought at germination inhibits seed imbibition, which delays emergence with deleterious consequences for subsequent growth and delayed harvest. This is particularly serious when there is sufficient water to initiate germination but not enough to support initial seedling growth. It often leads to poor establishment and frequently to the need to resow.

Water management

Irrigation is not widely used on rapeseed due either to absence of water supplies or the cost of equipment, which may be better justified on more profitable crops. However, rapeseed is quite sensitive to water shortage and irrigation could be valuable where water is available. Substantial areas of spring rapeseed are irrigated in western Canada and winter crops are irrigated in the Mediterranean area, where irrigation facilities are available. However, water supplies are becoming restricted in the latter area owing to more extensive use with more profitable crops and to increased competition for water between agriculture and other activities.

Irrigation for spring rapeseed crops

Clarke and Simpson (1978) registered a positive effect of irrigation for all yield components of spring *B. napus*. Water was highly restricted in this dryland experiment: the rainfed control yielded less than 1 tonne ha^{-1} while irrigated plots yielded more than 2.5 tonnes ha^{-1}. It is not abnormal that all yield components should be affected by such severe stress. Early-flowering *B. napus* cultivars have potential to avoid drought effects to some degree as they flower before the dry summer peak (MacPherson *et al.*, 1987). However, D.I. McGregor (1994, personal communication) suggests that this is not true under dryland conditions in Canada as even early-flowering *B. rapa* cultivars may not always avoid drought.

Experiments with spring rapeseed were undertaken in France under controlled conditions (Mingeau, 1974), when the growth cycle was divided into eight periods during which water stress treatments were applied. The control plants were irrigated to achieve maximum evapotranspiration and compared with those given only half that quantity of water (Table 3.8). The results indicate that the period of seed development is the most sensitive (Table 3.8a). At that time, number of seeds per pod is determined by pod number and supplies of water, carbon and mineral nutrients. If there is a water shortage, the number of seeds per pod will be reduced and compensation, normally achieved by increased seed weight, will be incomplete (Table 3.8b). Due to the long duration of flowering in the field, these two components are largely determined during the same period (Mingeau, 1974).

Mendham and Salisbury (p. 53, Chapter 2) confirm that water supply is critical at the early pod development stage but observe that early water stress during pod development affected pod number, while late water stress affected seed number per pod. Water stress has not apparently shown any major effect on seed quality (Mingeau, 1974), but stress during flowering can reduce oil content (Table 3.8c).

Irrigation for winter oilseed rape

Irrigation can ensure good germination and establishment in a drought, after which oilseed rape has good drought tolerance due to specialized root adaptation (Balestrini *et al.*, 1983): under dry conditions, the secondary root system is transformed into short tuberized roots which elongate normally when water becomes available. This phenomenon has also been observed in the field by Potfer (1987).

Irrigation just prior to flowering can increase pod number but the negative compensation between pod number and number of seeds per pod makes it unnecessary unless drought is severe (MacPherson

Table 3.8. Irrigation experiment with spring oilseed rape.

Treatment[1]	Plant age (days)	Stage of development	Drought intensity (%)
1	1–113	Emergence–harvest	0
2	16–29	3–7 leaves	44
3	30–43	Stem elongation–first buds	52
4	44–57	First buds–first flowers	51
5	58–71	Flowering	53
6	72–85	End flowering–oil content 30%	51
7	86–99	Oil content from 30 to 49%	52
8	100–113	Moisture content from 50 to 25%	52

(a) Dry matter distribution at harvest.

Treatment[1]	Dry matter (g per plant)					
	Roots	Stems	Leaves	Pods	Seeds	Total
1	6.28ab	17.66ab	5.62	9.31	11.93a	50.80a
2	6.78a	17.53ab	5.39	9.48	11.64a	50.82a
3	5.63bc	16.19c	5.27	9.78	11.48ab	48.35bc
4	5.04c	12.98e	4.96	8.71	10.59cd	42.28e
5	6.01b	13.96d	5.22	8.40	9.48e	43.07e
6	6.68ab	17.71a	5.38	7.68	8.95e	46.40d
7	6.30ab	17.01b	5.20	8.99	10.32d	47.82cd
8	6.56ab	17.40ab	5.65	9.44	11.01bc	50.06ab

(b) Yield components as percentage of control.

Treatment[1]	No. of pods per plant	No. of seeds per pod	No. of seeds per plant	Mean seed weight
1	100	100	100	100
2	98	99	97	100
3	103	95	98	98
4	96	96	92	96
5	87	90	78	102
6	75	85	64	116
7	98	95	93	94
8	101	99	100	92

(c) Effect of water stress on oil and protein content.

Treatment[1]	Oil content (% of seed)	Protein content (% of seed)
1	50.00	17.71
2	50.25	17.67
3	49.25	18.64
4	47.50	19.84
5	47.40	20.14
6	47.25	20.55

Table 3.8. (contd).

Treatment[1]	Oil content (% of seed)	Protein content (% of seed)
7	47.40	19.45
8	48.20	18.69

Source: Mingeau (1974).
Note: figures in a column followed by the same letter are not significantly different at
$P = 0.05$.
[1] Treatment 1, plants irrigated to achieve maximum evapotranspiration; other treatments, half
the amount of water applied to Treatment 1 for the period of water restriction.

Table 3.9. Effect of irrigation on the yield of winter oilseed rape at three trial sites in France.

	Herault		Cher		Rhône	
	Dry	Irrigated	Dry	Irrigated	Dry	Irrigated
Yield (t ha^{-1})	2.38	2.78	2.51	3.11	1.85	2.99
Mean seed weight (mg)	3.87	3.81	4.08	3.56	3.31	3.46
No. of seeds m^{-2} ('000s)	61.5	73.0	61.5	87.4	56.0	86.5
Oil content as % of dry matter	50.3	50.5	46.0	47.2	38.6	42.5
Glucosinolate content (μmol g^{-1} seed at 9% moisture content)	15.7	15.1	14.1	12.3	11.9	16.6
Amount of water as irrigation (mm) during spring		70		142		80

Source: Merrien (1991).
Dry, rainfed plots.

et al., 1987). As with spring rape, the most sensitive stage after flowering is at pod elongation (Table 3.9) (Merrien, 1991). Similar results were obtained in the United States where, under water or heat stress, efficacy of water application was higher during flowering and at the time of flower fertilization (Hang and Gilliland, 1991).

There appears to be no evidence of interactions between water nutrition and agronomic practices, except between water and nitrogen as mentioned above. Of course, the microclimate in irrigated crops may facilitate the development of diseases and sprinkling irrigation after flowering can contribute to lodging but, as far as we know, these hypotheses are not supported by experimental evidence.

HARVESTING

Harvesting is a critical operation: losses can be heavy due to the small seeds and because the growth habit prevents all seeds in a crop maturing at the same time. Furthermore, early harvesting can reduce seed quality and late harvesting can enhance pod shattering. From a practical point of view, the crop is mature when all the seeds are black and when the seed moisture content is less than 15%. Harvesting at high moisture content depends on the availability of a drier as the content has to be reduced to 9% for storage. In the absence of a drier it will be necessary to allow the crop to attain this level in the field. If harvest takes place when moisture content is above 15%, chlorophyll content will be higher and drying costs greater. In some countries, such as Canada, early frosts prevent seeds maturing and seed quality is poor due to higher residual chlorophyll content (Ward *et al.*, 1991). So it is important that agronomic decisions on choice of cultivar, sowing date, sowing rate, fertilizers, irrigation, etc. are made with date of maturity in mind.

Direct combining

Direct combining is possible when the crop matures evenly and early enough, or after desiccation. Vertical knives on the combine should be used when the crop is lodged or leaning heavily. Forward speed should be low (about two thirds of that for cereal crops) and the cutting height should be as high as possible to avoid an unduly high throughput of straw.

Desiccants

Chemical desiccation, used to limit seed loss and seed quality deterioration due to weathering, has been studied in Canada, Denmark (Flengmark, 1983), Britain (Bowerman, 1984) and Poland. It appears that glyphosate, diquat and dimethipin have no harmful effect on yield or quality, and had no effect on chlorophyll content in Polish experiments (Ciesielski and Czeslaw, 1987) and, in Canadian studies, gluphosinate ammonium was effective in decreasing the proportion of green seeds (Bahri, 1991).

Swathing

Swathing is also effective in reducing crop losses and almost all the Canadian crop is harvested in this way. Swathing allows the crop to mature earlier and more evenly. It also reduces shattering due to winds in the standing crop which can be a hazard in some countries but, as with desiccation,

costs cannot always be justified but may be considered an insurance to prevent loss of the whole crop in windy seasons.

The effect of the chemical pod sealant di-1-*p*-menthene (spodnam) was tested in Poland (Szot and Tys, 1991) where it appeared to contribute to pod resistance to splitting but its efficacy varied with cultivar.

Seed losses and damage at harvest

Seed losses

Polish studies (Szot *et al.*, 1991) showed that seed losses at harvest can be as high as 0.5 tonne ha^{-1}. The main cause is bad combine harvester setting and there is little difference between cultivars except for susceptibility to shattering. Szot *et al.* (1991) studied the physics of shattering and their conclusions are pertinent for breeders as lines could be characterized as susceptible to pod shattering. Breeding for shattering resistance could be an asset but resistant varieties which do not lean over as crops ripen may be more vulnerable to strong winds. The best compromise would be cultivars with stems which lean to form a relatively weatherproof canopy together with high resistance to pod shattering (Szot and Tys, 1987; Tys and Bengtsson, 1991). Bowerman (1984) investigated seed losses in relation to method of harvest and found little difference in yields whether crops were swathed or desiccated. The highest proportion of losses occurred usually at the cutterbar.

Seed damage

The same Polish team also described seed quality deterioration at harvest or during storage (Stepniewski *et al.*, 1991; Szot, 1987; Szot *et al.*, 1991) and found that harvesting and drying are major contributors to seed damage (78–96% depending on year). The mechanical characteristics of seed have been studied in the laboratory (Dobrzanski, 1991; Dobrzanski and

Stepniewski, 1991). Seeds are more susceptible to damage at higher temperatures or moisture contents (Stepniewski and Dobrzanski, 1991). High moisture content also predisposes seed to permanent deformation, but differences between cultivars were observed and large seeds are more resistant to mechanical forces than smaller seeds (Szot, 1987).

STORAGE

Damp or green seeds, or impurities in the seeds, can be a problem especially when weather conditions at harvest are unpredictable. Ventilation of such seed in store is essential in order to prevent heating. It is recommended that rapeseed is dried to less than 9% moisture content and cooled to less than 10 °C as it will otherwise deteriorate: both fungi and insect activity will increase and germination ability will be impaired.

DEVELOPMENT OF MANAGEMENT SYSTEMS

One of the aims of agronomic research is to optimize crop management systems.

Brassica oilseeds have to fit into farming systems which may have constraints with respect to optimum management. For example, livestock producers must allocate time for care of their animals and arable farmers may find available labour limiting at peak periods. Thus it may not be possible to adhere to optimum timing for fertilizer application, insect control or other production requirements.

Threshold levels are available in most countries for the economic control of pests and diseases, for fertilizer application, desiccation and swathing, seed drying and storage provision. Their practical application will be influenced by supplementary data on climate, soil and other relevant local factors. Such data are available in Britain, Canada and most rapeseedgrowing countries.

Electronic data systems have been devised to assist decision-making on the farm, notably those of Leterme *et al.* (1991) in France and Sykes (1991) in Australia. Both are primarily intended for advisers and are important for training purposes, but have potential for the development of highly sophisticated assistance for determining the timing and relevance of the various crop treatments.

REFERENCES

Almond, J.A., Dawkins, T.C.K. and Askew, M.F. (1986) Aspects of crop husbandry. In: Scarisbrick, D.H. and Daniels, R.W. (eds) *Oilseed Rape.* Collins, London, pp. 127–175.

Anon. (1994) *La Culture du colza d'hiver.* CETIOM, Paris, 36 pp.

Auld, D.L. and Mahler, K.A. (1991) Production of canola and rapeseed in the U.S. In: McGregor, D.I. (ed.) *Proceedings of the Eighth International Rapeseed Congress*, Saskatoon, Canada. Organizing Committee, Saskatoon, pp. 978–983.

Baddesha, H.S. and Pasricha, N.S. (1991) Interaction of green manuring and fertilizer-N in rapeseed mustard crops. In: McGregor, D.I. (ed.) *Proceedings of the Eighth International Rapeseed Congress*, Saskatoon, Canada. Organizing Committee, Saskatoon, pp. 558–566.

Bahri, R.W. (1991) Glufosinate ammonium (HOE 039866) as a harvest aid in Canada. In: McGregor, D.I. (ed.) *Proceedings of the Eighth International Rapeseed Congress*, Saskatoon, Canada. Organizing Committee, Saskatoon, pp. 1860–1865.

Balestrini, S., Vartanian, N. and Rollier, M. (1983) Variabilité génétique dans les réactions adaptives du colza à la sécheresse. In: *Proceedings of the Sixth International Rapeseed Congress*, Paris, pp. 64–71.

Barber, S.J., Rakow, G. and Downey, R.K. (1991a) Field emergence and yield of certified seed lots of cvs Westar and Tobin canola on the Canadian prairie. In: McGregor, D.I. (ed.) *Proceedings of the Eighth International Rapeseed Congress*, Saskatoon, Canada. Organizing Committee, Saskatoon, pp. 659–664.

Barber, S.J., Rakow, G. and Downey, R.K. (1991b) Laboratory and growth room seed vigor testing of certified canola seed. In: McGregor, D.I. (ed.) *Proceedings of the Eighth International Rapeseed Congress*, Saskatoon, Canada. Organizing Committee, Saskatoon, pp. 727–732.

Barszczak, Z., Barszcsak, T., Gorczinski, J. and Kot, A. (1991a) Effect of moisture, nitrogen doses and soil acidity on seed yield, chemical composition and thousand seed weight of some winter oilseed rape cultivars. In: McGregor, D.I. (ed.) *Proceedings of the Eighth International Rapeseed Congress*, Saskatoon, Canada. Organizing Committee, Saskatoon, pp. 1181–1185.

Barszczak, Z., Barszcsak, T., Gorczinski, J. and Kot, A. (1991b) Sensibility of genotypes of winter rape seedlings to higher concentrations of Al and Mn ions in soil. In: McGregor, D.I. (ed.) *Proceedings of the Eighth International Rapeseed Congress*, Saskatoon, Canada. Organizing Committee, Saskatoon, pp. 1193–1196.

Belayneh, H. and Alemayehu, N. (1987) Comparative performance of Ethiopian mustard (*B. carinata*) and Argentina rapeseed (*B. napus*) under improved and traditional farming practices. In: *Proceedings of the Seventh International Rapeseed Congress*, Poznan, Poland. The Plant Breeding and Acclimatization Institute, Poznan, pp. 1044–1048.

Bilsborrow, P.E., Evans, E.J. and Zhao, F.J. (1993) The influence of spring nitrogen on yield, yield components and glucosinolate content of autumn-sown oilseed rape (*Brassica napus*). *Journal of Agricultural Science* 120, 219–224.

Booth, E.J., Walker, K.C. and Schnug, E. (1991) The effect of site, foliar sulphur and nitrogen application on glucosinolate content and yield of oilseed rape (*Brassica napus* L.). In: McGregor, D.I. (ed.) *Proceedings of the Eighth International Rapeseed Congress*, Saskatoon, Canada. Organizing Committee, Saskatoon, pp. 567–572.

Bowerman, P. (1984) Comparison of harvesting methods of oilseed rape. *Aspects of Applied Biology* 6, 157–165.

Bowerman, P. (1991) Establishment of winter oilseed rape in the presence of straw of the previous wheat crop. In: McGregor, D.I. (ed.) *Proceedings of the Eighth International Rapeseed Congress*, Saskatoon, Canada. Organizing Committee, Saskatoon, pp. 1224–1228.

Cedell, T. (1983) Direct drilling of oilseed rape – experiences of experimental work in Sweden. In: *Proceedings of the Sixth International Rapeseed Congress*, Paris, France, pp. 719–725.

Christensen, J.V. and Drabble, J.C. (1984) Effect of row spacing and seeding rate on rapeseed yield in Northwest Alberta. *Canadian Journal of Plant Science* 64, 1011–1013.

Ciesielski, F. and Czeslaw, M. (1987) A new plant maturity regulator for winter rape with a wide spectrum of action. In: *Proceedings of the Seventh International Rapeseed Congress*, Poznan, Poland. The Plant Breeding and Acclimatization Institute, Poznan, pp. 962–968.

Clarke, J.M. and Simpson, G.M. (1978) Influence of irrigation and seeding rates on yield and yield components of *Brassica napus* cv. Tower. *Canadian Journal of Plant Science* 58, 731–737.

Clossais-Besnard, N. and Larher, F. (1991) Glucosinolates accumulation pattern during the development of single low and double low oilseed rape (*Brassica napus* L.). In: McGregor, D.I. (ed.) *Proceedings of the Eighth International Rapeseed Congress*, Saskatoon, Canada. Organizing Committee, Saskatoon, pp. 1714–1719.

Daniels, R.W. and Scarisbrick, D.H. (1986) Plant growth regulators for oilseed rape (1986). In: Scarisbrick, D.H. and Daniels, R.W. (eds) *Oilseed Rape*. Collins, London, pp. 176–194.

Daun, J.K., Clear, K.M. and Candlish, V.E. (1983) Agronomic factors associated with high chlorophyll levels in rapeseed grown in western Canada. In: *Proceedings of the Sixth International Rapeseed Congress*, Paris, France, pp. 132–137.

de Bouillé, P., Sotta, B., Miginiac, E. and Merrien, A. (1987) Hormones endogènes et développement de l'inflorescence principale du colza *Brassica napus*, cultivar Bienvenu. In: *Proceedings of the Seventh International Rapeseed Congress*, Poznan, Poland. The Plant Breeding and Acclimatization Institute, Poznan, pp. 645–650.

Debouzie, D., Thioulouse, J. and Pouzet, A. (1987) Recherches sur la conduite de la culture du colza d'hiver en France: nouvelles méthodes et premiers résultats. In: *Proceedings of the Seventh International Rapeseed Congress*, Poznan, Poland. The Plant Breeding and Acclimatization Institute, Poznan, pp. 665–670.

Delorme, S. (1993) *Etude de la qualité des torteaux de colza et de tournesol.* Mémoire de maîtrise. CETIOM, Paris, France, 26 pp.

Dobrzanski, B. (1991) Theoretical model of large deformation of compressed rapeseed. In: McGregor, D.I. (ed.) *Proceedings of the Eighth International Rapeseed Congress*, Saskatoon, Canada. Organizing Committee, Saskatoon, pp. 1261–1266.

Dobrzanski, B. and Stepniewski, A. (1991) Modulus of elasticity of rapeseed shell. In: McGregor, D.I. (ed.) *Proceedings of the Eighth International Rapeseed Congress*, Saskatoon, Canada. Organizing Committee, Saskatoon, pp. 1256–1260.

Donald, D., Sharp, G.S., Atkinson, D. and Duff, F.I. (1993) Effect of nitrogen and sulphur fertilization on the yield and composition of winter oilseed rape. *Communications in Soil Science and Plant Analysis* 24, 813–826.

Evans, E.J., Bilsborrow, P.E., Zhao, F.J. and Syers, J.K. (1991) The sulphur nutrition of winter oilseed rape in northern Britain. In: McGregor, D.I. (ed.) *Proceedings of the Eighth International Rapeseed Congress*, Saskatoon, Canada. Organizing Committee, Saskatoon, pp. 542–546.

Farman, C.D., Henman, A.P. and Warry, P.J. (1989) *Oilseed Rape Manual.* National Agricultural Centre, Kenilworth, 136 pp.

Flengmark, P. (1983) Methods of harvesting spring oilseed rape. In: *Proceedings of the Sixth International Rapeseed Congress*, Paris, France, pp. 669–674.

Geisler, G. and Kullmann, A. (1991) Changes of dry matter, nitrogen content and nitrogen efficiency in oilseed rape in relation to nitrogen nutrition. In: McGregor, D.I. (ed.) *Proceedings of the Eighth International Rapeseed Congress*, Saskatoon, Canada. Organizing Committee, Saskatoon, pp. 1175–1180.

Gerath, N. and Schweiger, W. (1991) Improvement of the use of nutrients in winter rape – a strategy of economically and ecologically responsible fertilizing. In: McGregor, D.I. (ed.) *Proceedings of the Eighth International Rapeseed Congress*, Saskatoon, Canada. Organizing Committee, Saskatoon, pp. 1197–1201.

Grant, C.A. and Bailey, L.D. (1993) Fertility management in canola production. *Canadian Journal of Plant Science* 73, 651–670.

Greenwood, D.J., Lemaire, G., Gosse, G., Cruz, P., Draycott, A. and Neeteson, J.J. (1990) Decline in percentage of N in C3 and C4 crops with increasing plant mass. *Annals of Botany* 66, 425–436.

Haneklaus, S. and Schnug, E. (1991) Evaluation of the nutritional status of oilseed rape plants by leaf analysis. In: McGregor, D.I. (ed.) *Proceedings of the Eighth International Rapeseed Congress*, Saskatoon, Canada. Organizing Committee, Saskatoon, pp. 536–541.

Hang, A.N. and Gilliland, G.C. (1991) Water requirement for winter rapeseed in central Washington. In: McGregor, D.I. (ed.) *Proceedings of the Eighth International Rapeseed Congress*, Saskatoon, Canada. Organizing Committee, Saskatoon, pp. 1235–1240.

Heikkinen, M.K. and Auld, D.L. (1991) Harvest index and seed yield of winter rapeseed grown at different plant populations. In: McGregor, D.I. (ed.) *Proceedings of the Eighth International Rapeseed Congress*, Saskatoon, Canada. Organizing Committee, Saskatoon, pp. 1229–1234.

Henry, J.L. and MacDonald, K.B. (1978) The effect of soil and fertilizer nitrogen and moisture stress on yield, oil and protein content of rape. *Canadian Journal of Soil Science* 58, 303–310.

Holmes, M.R.J. (1980) *Nutrition of the Oilseed Rape Crop.* Applied Science Publishers, Barking, Essex, UK.

Jasinska, Z., Kotechi, A., Malarz, W., Musnicki, Cz., Jodkowski, M., Budzynski, W., Wrobel, E. and Sikora, B. (1987) The influence of sowing date and sowing rates on the development and yield of winter rape varieties. In: *Proceedings of the Seventh International Rapeseed Congress*, Poznan, Poland. The Plant Breeding and Acclimatization Institute, Poznan, pp. 886–892.

Kjellstrom, C. (1987) Stand density experiments with *Eruca sativa.* In: *Proceedings of the Seventh International Rapeseed Congress*, Poznan, Poland. The Plant Breeding and Acclimatization Institute, Poznan, pp. 930–934.

Kjellstrom, C. (1991) Growth and distribution of the root system in *Brassica napus.* In: McGregor, D.I. (ed.) *Proceedings of the Eighth International Rapeseed Congress*, Saskatoon, Canada. Organizing Committee, Saskatoon, pp. 722–726.

Krauze, A., Bowszys, T., Bobrzecka, D. and Rotkiewicz, D. (1991) Effect of foliar boron fertilization on yield and quality of winter rape. In: McGregor, D.I. (ed.) *Proceedings of the Eighth International Rapeseed Congress*, Saskatoon, Canada. Organizing Committee, Saskatoon, pp. 547–553.

Lehrmann, P., Despeghel, J.P. and Guguin, N. (1991) Plant growth regulators effect on double low winter rapeseed varieties. In: McGregor, D.I. (ed.) *Proceedings of the Eighth International Rapeseed Congress*, Saskatoon, Canada. Organizing Committee, Saskatoon, pp. 584–589.

Leterme, P. (1991) Modélisation du fonctionnement du colza d'hiver: utilisation pour la conduite de la culture. In: McGregor, D.I. (ed.) *Proceedings of the Eighth International Rapeseed Congress*, Saskatoon, Canada. Organizing Committee, Saskatoon, pp. 671–676.

Leterme, P., Debouzie, D. and Wagner, D. (1991) Présentation d'une base de donnés relative aux itinéraires techniques du colza d'hiver. In: McGregor, D.I. (ed.) *Proceedings of the Eighth International Rapeseed Congress*, Saskatoon, Canada. Organizing Committee, Saskatoon, pp. 1218–1223.

Lewis, C.E. and Knight, C.W. (1987) Yield response of rapeseed to row spacing and rates of seeding and N-fertilization in interior Alaska. *Canadian Journal of Plant Science* 67, 53–57.

Lucas, J.L. (1994) *Intérêt chez le colza d'hiver d'un semis précoce ou d'un apport d'azote automnal suivi d'une application d'un régulateur de croissance au stade B4*. Mémoire ingénieur. CETIOM, Paris, France, 62 pp.

MacPherson, H., Scarth, R., Rimmer, S.R. and MacVetty, P.B.E. (1987) The effect of drought stress on yield determination in oilseed rape. In: *Proceedings of the Seventh International Rapeseed Congress*, Poznan, Poland. The Plant Breeding and Acclimatization Institute, Poznan, pp. 822–827.

Mailer, R.J. and Wratten, N. (1987) Glucosinolate variability in rapeseed in Australia. In: *Proceedings of the Seventh International Rapeseed Congress*, Poznan, Poland. The Plant Breeding and Acclimatization Institute, Poznan, pp. 671–675.

Malhi, S.S., Nyborg, M., Jahn, H.G. and Penny, D.C. (1988) Yield and nitrogen uptake of rapeseed (*Brassica campestris*) with ammonium and nitrate. *Plant Soil* 105, 231–239.

Marquard, R., Demes, H. and Gaudchau, M. (1991) Influence of sulphur supply on glucosinolate content of rapeseed. In: McGregor, D.I. (ed.) *Proceedings of the Eighth International Rapeseed Congress*, Saskatoon, Canada. Organizing Committee, Saskatoon, pp. 1708–1713.

Mendham, N.J. and Scott, R.K. (1975) The limiting effect of plant size at inflorescence initiation on subsequent growth and yield of oilseed rape (*Brassica napus*). *Journal of Agricultural Science* 84, 487–502.

Mendham, N.J., Shipway, P.A. and Scott, R.K. (1981) The effects of delayed sowing and weather on growth, development and yield of winter oilseed rape (*Brassica napus*). *Journal of Agricultural Science* 96, 389–415.

Merrien, A. (1991) *Irrigation du colza d'hiver*. Dossier technique. CETIOM, Paris, France, 17 pp.

Merrien, A. and Maisonneuve, C. (1987) Usage de régulateurs de croissance sur colza: Résultats expérimentaux. In: *Proceedings of the Seventh International Rapeseed Congress*, Poznan, Poland. The Plant Breeding and Acclimatization Institute, Poznan, pp. 948–955.

Merrien, A. and Pouzet, A. (1989) *Agronomie colza*. Cahier technique. CETIOM, Paris, France, 56 pp.

Merrien, A., Ribaillier, D., Agbo, P. and Devineau, J. (1987) Impact de la fertilisation soufrée sur le teneur en glucosinolates des graines de colza: conséquences agronomiques. In: *Proceedings of the Seventh International Rapeseed Congress*, Poznan, Poland. The Plant Breeding and Acclimatization Institute, Poznan, pp. 907–916.

Merrien, A., Champolivier, L., Devineau, J., Estragnat, A. and Jung, L. (1991) Valorisation de l'azote par une culture de colza d'hiver. In: McGregor, D.I. (ed.) *Proceedings of the Eighth International Rapeseed Congress*, Saskatoon, Canada. Organizing Committee, Saskatoon, pp. 1186–1192.

Mingeau, M. (1974) Comportement du colza de printemps à la sécheresse. *Informations Techniques*, Paris, France, 36, 1–11.

Morall, R.A.A., Turkington, T.K., Kaminski, D.A., Thompson, J.R., Gugel, R.K. and Rude, S.V. (1991) Forecasting *Sclerotinia* stem rot of spring rapeseed by petal testing. In: McGregor, D.I. (ed.) *Proceedings of the Eighth International Rapeseed Congress*, Saskatoon, Canada. Organizing Committee, Saskatoon, pp. 483–488.

Munir, M. and Rahman Khan, A. (1987) Status of rapeseed and mustard crops in Pakistan. In: *Proceedings of the Seventh International Rapeseed Congress*, Poznan, Poland. The Plant Breeding and Acclimatization Institute, Poznan, pp. 1038–1043.

Murphy, G.M. and Pascale, N.C. (1991) Cultivation areas of winter and spring rapeseed in Argentina. In: McGregor, D.I. (ed.) *Proceedings of the Eighth International Rapeseed Congress*, Saskatoon, Canada. Organizing Committee, Saskatoon, pp. 1288–1293.

Musnicki, C., Mrowczynski, M., Tobola, P. and Cichy, H. (1987) Growth regulators in winter oilseed rape cultivation. In: *Proceedings of the Seventh International Rapeseed Congress*, Poznan, Poland. The Plant Breeding and Acclimatization Institute, Poznan, pp. 940–947.

Nyborg, M. (1961) The effect of fertilizers on emergence of cereal grains, flax and rape. *Canadian Journal of Soil Science* 41, 89–98.

Nyborg, M. and Hennig, A.M.F. (1969) Field experiments with different placements of fertilizers for barley, flax and rapeseed. *Canadian Journal of Soil Science* 49, 79–88.

ONIDOL-CETIOM (1993) *Enquête qualité colza d'hiver*. Brochure. ONIDOL-CETIOM, Paris, France, 15 pp.

Palleau, J.P. (1989) *Contribution à l'élaboration d'un diagnostic agronomique en culture de colza: intérêt de la mesure de la vitesse de croissance au printemps*. Mémoire ingénieur. CETIOM, Paris, France, 66 pp.

Palleau, J.P. and Tittonel, E.D. (1991) Effet d'un apport d'azote à un stade repère de la fleur; répercussions sur les composantes du rendement. In: McGregor, D.I. (ed.) *Proceedings of the Eighth International Rapeseed Congress*, Saskatoon, Canada. Organizing Committee, Saskatoon, pp. 628–633.

Porter, P.M., Bathke, G.R. and Robinson, D.M. (1991) Agronomic practices for canola grown in South Carolina, USA. In: McGregor, D.I. (ed.) *Proceedings of the Eighth International Rapeseed Congress*, Saskatoon, Canada. Organizing Committee, Saskatoon, pp. 617–621.

Potfer, J.P. (1987) Etude du système racinaire du colza de printemps en conditions de sécheresse. Unpublished compte rendu d'expérimentation. CETIOM, Paris, France, 28 pp.

Pouzet, A. (1985) Raisonnement de la fertilisation azotée du colza d'hiver en France. *Bulletin GCIRC* 2, 62–68.

Pouzet, A. and Rollier, M. (1983) Possibilités d'implantation du colza d'hiver par semis direct. In: *Proceedings of the Sixth International Rapeseed Congress*, Paris, France, pp. 845–853.

Pouzet, A., Raimbault, J., Estragnat, A. and Gosse, G. (1983a) Analyse de la croissance automnale du colza d'hiver (cv. Jet Neuf): influence de la date de semis, de la densité de semis et de l'apport d'azote à l'automne. In: *Proceedings of the Sixth International Rapeseed Congress*, Paris, France, pp. 699–713.

Pouzet, A., Sauzet, G. and Raimbault, J. (1983b) Influence du type de sol, de la profondeur de semis, de la pluviométrie et des techniques culturales sur le taux de levée au champ du colza d'hiver. In: *Proceedings of the Sixth International Rapeseed Congress*, Paris, France, pp. 836–844.

Pouzet, A., Estragnat, A., Gilly, J.M., Raimbault, J. and Sauzet, G. (1984) Eléments pour le raisonnement de la fertilisation azotée d'automne pour le colza d'hiver. *Informations Techniques* 89, 3–19.

PROLEA (1993) *Ecobilan du Diester: Evaluation comparée des filières Gazole et Diester*. PROLEA, 23 pp.

Reau, R., Paillard, C. and Wagner, D. (1994) *Azote et colza d'hiver: Propositions de règles pour raisonner la fertilisation*. Dossier technique. CETIOM, Paris, France, 36 pp.

Renard, M. (1991) Breeding: closing summary. In: McGregor, D.I. (ed.) *Proceedings of the Eighth International Rapeseed Congress*, Saskatoon, Canada. Organizing Committee, Saskatoon, pp. 1937–1942.

Rollier, M. (1970) Le colza et l'azote. *Oléagineux* 25, 91–96.

Rollier, M. and Pouzet, A. (1983) Eléments pour le raisonnement de la fertilisation azotée du colza de printemps en France. In: *Proceedings of the Sixth International Rapeseed Congress*, Paris, France, pp. 737–749.

Romagosa, I. and Fox, P.N. (1993) Genotype × environment interaction and adaptation. In: Hayward, M.D., Bosemark, N. and Romagosa, I. (eds) *Plant Breeding: Principles and Prospects*. Chapman and Hall, London, pp. 373–390.

Schnug, E. (1987) Relations between sulphur supply and glucosinolate content of O- and OO-oilseed rape. In: *Proceedings of the Seventh International Rapeseed Congress*, Poznan, Poland. The Plant Breeding and Acclimatization Institute, Poznan, pp. 682–686.

Schnug, E. (1990) Professional interprofessional program for plant analysis of oilseed rape and cereals. In: *Proceedings of the Inaugural Conference of the European Society of Agronomy*, Paris, France, p. 35.

Schnug, E. and Haneklaus, S. (1994) Regulation of glucosinolate biosynthesis in oilseed rape by nutritional factors. *Bulletin GCIRC* 10, 72–78.

Scott, N.M. (1981) Evaluation of sulphate status of soils by plant and soil tests. *Journal of Science, Food and Agriculture* 32, 193–199.

Skubisz, G. (1991) The variability of mechanical properties of winter rape stems during plant vegetation period. In: McGregor, D.I. (ed.) *Proceedings of the Eighth International Rapeseed Congress*, Saskatoon, Canada. Organizing Committee, Saskatoon, pp. 1795–1800.

Skubisz, G. and Tys, J. (1987) An evaluation of the mechanical properties of winter rape stems. In: *Proceedings of the Seventh International Rapeseed Congress*, Poznan, Poland. The Plant Breeding and Acclimatization Institute, Poznan, pp. 862–867.

Soper, R.J. (1971) Soil tests as a means of predicting response of rape to added N, P and K. *Agronomy Journal* 63, 564–566.

Stepniewski, A. and Dobrzanski, B. (1991) Higher temperature of rapeseeds as a factor affecting their mechanical resistance. In: McGregor, D.I. (ed.) *Proceedings of the Eighth International Rapeseed Congress*, Saskatoon, Canada. Organizing Committee, Saskatoon, pp. 1246–1250.

Stepniewski, A., Szot, B. and Kushwaha, R.L. (1991) Decreasing the quality of rapeseed during postharvest handling. In: McGregor, D.I. (ed.) *Proceedings of the Eighth International Rapeseed Congress*, Saskatoon, Canada. Organizing Committee, Saskatoon, pp. 1267–1271.

Sun, W.C., Pan, Q.Y., An, X. and Yang, Y.P. (1991) *Brassica* and *Brassica*-related oilseed crops in Gansu, China. In: McGregor, D.I. (ed.) *Proceedings of the Eighth International Rapeseed Congress*, Saskatoon, Canada. Organizing Committee, Saskatoon, pp. 1130–1135.

Sykes, J.D. (1991) Canola Check – a management support system for growers. In: McGregor, D.I. (ed.) *Proceedings of the Eighth International Rapeseed Congress*, Saskatoon, Canada. Organizing Committee, Saskatoon, pp. 1214–1217.

Sykes, J.D. and Mailer, R.J. (1991) The effect of nitrogen on yield and quality of canola. In: McGregor, D.I. (ed.) *Proceedings of the Eighth International Rapeseed Congress*, Saskatoon, Canada. Organizing Committee, Saskatoon, pp. 554–557.

Szczygielski, T. and Owczarek, E. (1987) Response of new winter rape varieties to the sowing density. In: *Proceedings of the Seventh International Rapeseed Congress*, Poznan, Poland. The Plant Breeding and Acclimatization Institute, Poznan, pp. 868–878.

Szot, B. (1987) An evaluation of the mechanical properties and the susceptibility to damage of winter rape seeds. In: *Proceedings of the Seventh International Rapeseed Congress*, Poznan, Poland. The Plant Breeding and Acclimatization Institute, Poznan, pp. 850–855.

Szot, B. and Tys, J. (1987) Characterization of the strength properties of winter rape siliques in the aspect of their cracking susceptibility. In: *Proceedings of the Seventh International Rapeseed Congress*, Poznan, Poland. The Plant Breeding and Acclimatization Institute, Poznan, pp. 856–861.

Szot, B. and Tys, J. (1991) The influence of Spodnam DC preparation on agrophysical properties of rape silique and seed losses at maturation and harvest. In: McGregor, D.I. (ed.) *Proceedings of the Eighth International Rapeseed Congress*, Saskatoon, Canada. Organizing Committee, Saskatoon, pp. 1272–1276.

Szot, B., Tys, J. and Bilanski, W.K. (1991) Studies of rapeseed resistance to mechanical damage. In: McGregor, D.I. (ed.) *Proceedings of the Eighth International Rapeseed Congress*, Saskatoon, Canada. Organizing Committee, Saskatoon, pp. 1251–1255.

Thomas, P. (1984) *Canola Growers Manual*. Canola Council of Canada Publication, Winnipeg, Canada.

Topinka, A.R.C., Downey, R.K. and Rakow, G.F.W. (1991) Effect of agronomic practices on the overwintering of winter canola in southern Alberta. In: McGregor, D.I. (ed.) *Proceedings of the Eighth International Rapeseed Congress*, Saskatoon, Canada. Organizing Committee, Saskatoon, pp. 665–670.

Tys, J. and Bengtsson, L. (1991) Estimation of rape silique resistance to cracking and rape seed shattering resistance for some selected varieties and lines of spring rape. In: McGregor, D.I. (ed.) *Proceedings of the Eighth International Rapeseed Congress*, Saskatoon, Canada. Organizing Committee, Saskatoon, pp. 1848–1853.

Van de Putte, B. and Messéan, A. (1994) Méthodologie de l'évaluation variétale: une nouvelle approche. *Bulletin GCIRC* 10, 86–93.

Vullioud, P. (1993) Densité de semis en culture de colza d'automne. *Bulletin GCIRC* 9, 125–133.

Ward, K.A., Scarth, R., McVetty, P.B.E. and Daun, J.K. (1991) Genotypic and environmental effects on seed chlorophyll levels in canola (*Brassica napus*). In: McGregor, D.I. (ed.) *Proceedings of the Eighth International Rapeseed Congress*, Saskatoon, Canada. Organizing Committee, Saskatoon, pp. 1241–1245.

Williams, I.H., Darby, R.J., Leach, J.E. and Rawlinson, C.J. (1991) An analysis of factors affecting yield in winter rape. In: McGregor, D.I. (ed.) *Proceedings of the Eighth International Rapeseed Congress*, Saskatoon, Canada. Organizing Committee, Saskatoon, pp. 612–616.

4 Weeds and Their Control

J. Orson
ADAS Cambridge, UK

WEEDS

Weed type

Five groups of weeds can occur in field crops:

- perennial dicotyledonous (broadleaved)
- biennial dicotyledonous
- annual dicotyledonous
- perennial monocotyledonous (usually grasses)
- annual monocotyledonous

Perennial weeds have storage organs such as swollen roots or stems, the latter being above or below ground, from which new growth emerges. They can also develop from seeds, as do all annual weeds although the latter can perennate with regular cutting, which prevents seed set. Biennials also develop from seed and can occur in autumn crops where they germinate in the autumn and are vernalized over the winter to flower and set seed the following summer. They may also perennate with regular cutting.

Weed species occurring in *Brassica* oilseeds

The weed flora in a crop will be determined by the time of drilling; the species present in the soil, either as storage organs of perennial weeds or seeds of annual, biennial and perennial weeds; and the state of dormancy of those seeds, which can be influenced by their age and position in the soil profile (Roberts, 1982). Weeds may also get established in crops sown in late summer or early autumn from seeds distributed by wind or other means.

There is a reasonable consensus (Lutman, 1991) about the commonest species in autumn-sown rape in northern Europe: apart from volunteer cereals, these are: common chickweed (*Stellaria media*), annual meadow-grass (winter grass in Australia, *Poa annua*), mayweeds (*Matricaria* spp.) and speedwells (*Veronica* spp.). Species commonly occurring in spring-sown crops include: charlock (wild mustard in Canada, *Sinapis arvensis*), common chickweed, mayweeds, annual meadow-grass and black bindweed (wild buckwheat in Canada, *Fallopia convolvulus*). The feature of these species is that they have no specific emergence pattern or, if they have, emerge primarily in the autumn in winter-sown crops or in the spring in spring-sown crops. Biennials regularly occur in early-sown winter oilseed rape. They often originate from wind-blown seed from plants in neighbouring vegetation. Those commonly occurring include spear thistle (*Cirsium vulgare*) and hemlock (*Conium maculatum*) (Froud-Williams and Chancellor, 1987). Rapeseed in Canada is spring sown but weeds which germinate in the autumn may survive the winter and seedbed cultivations if specific control measures are not employed. Examples include the annual stinkweed (field pennycress in Britain, *Thlaspi arvense*) and the perennial creeping thistle (Canada or Californian thistle, *Circium arvense*). Other commonly occurring weeds are wild oats (*Avena fatua*), black bindweed/wild buckwheat, fat hen (lamb's quarters in Canada, *Chenopodium album*), redroot pigweed in Canada (amaranthus in the UK, redroot amaranth in Australia, *Amaranthus retroflexus*), charlock/wild mustard and green foxtail in Canada (*Setaria viridis*) (Thomas, 1992).

The weed species present in the soil seedbank are determined by many factors, including:

- previous cropping
- seeds shed in previous crops
- soil texture
- soil pH
- cultivation practices

Previous cropping

The time of crop establishment has a profound effect on the weed species likely to grow and set seed. Annual weeds must germinate at the same time as the crop, or at least well before a full crop canopy has developed, in order to survive and shed seed before harvest. The period of germination varies: some species germinate predominantly at certain times of the year, while others can germinate at almost any time. The latter, particularly if they have a short life cycle, will predominate in a rotation involving crops with drilling dates at different times of the year, while, if the rotation is limited to crops sown at the same time of the year, the weed flora will be dominated by species having the same germination period and able to shed seed in the growing crop. Where the rotation includes a 1-year fallow, the soil seedbank is more likely to comprise species which have a high proportion of seeds with sufficient dormancy to survive.

Seeds shed in previous crops

One objective of weed control is to prevent an unreasonable return of weed seeds to the soil and control measures are rarely fully effective, particularly for broadleaved weeds in broadleaved crops, such as brassicas. In addition, financial and environmental constraints mitigate against control of weed populations which do not present a threat to current and future crops. Hence prevention of seed return to the soil seedbank will depend on financial and environmental considerations as well as the ease of weed control in these crops.

Weed seeds vary in their ability to survive in the soil. Some species contain a proportion of dormant seed capable of surviving many years. Estimates vary but weeds which can be difficult to control in *Brassica* oilseed crops, such as volunteer oilseed rape, charlock/wild mustard and poppies (*Papaver* spp.), are reported to survive 15 to 20 years or more. So the soil may contain significant numbers of these weed seeds which, together with related weeds, could cause increasing problems where brassicas are grown regularly in a rotation. On the other hand, seeds of species with short persistence will only be influenced by cropping practices in the recent past (Cussans *et al.*, 1987).

Soil texture

Some species are favoured by light soils, others by heavy soils. Soil texture has a direct effect on the seedbank by influencing the establishment of species and survival of weed plants, survival of seeds and, indirectly, by influencing seed behaviour. Weeds present on specific soil types also reflect the crops commonly grown (Roberts, 1982). Seeds of some species survive longer in heavy soils where soil temperatures are more stable and waterlogging enhances carbon dioxide concentration which induces or enforces dormancy.

Soil pH

Some weed species flourish in soils with a wide pH range, while others have more specific requirements. In countries, notably in parts of northern Europe, where most agricultural soils now have a pH of 6.5 or more due to the addition of lime or chalk, acid-loving species like corn spurrey (*Spergula arvensis*) have largely disappeared (Roberts, 1982).

Cultivation practices

Every operation disturbing the soil has the potential to control some weeds and damage perennial plants, either directly or by stimulating germination of seeds which fail to establish or are subsequently controlled by cultivations or non-selective herbicides. However, predation by insects, birds and rodents and mortality of freshly shed seed are maximized by leaving seed exposed on the soil surface. Hence, depending on weed species and the position of the seeds, shallow cultivations may or may not enhance the loss of viable seeds in the surface layers of the soil.

Inversion tillage (mouldboard ploughing) is the traditional method of weed control but some weed species are encouraged. These tend to be those having seeds with marked dormancy and longevity. Examples are knotgrass (*Polygonum aviculare*), charlock/wild mustard, common poppy (*Papaver rhoeas*) and the *Brassica* spp. Some weeds are encouraged by non-inversion tillage, particularly perennials and some annual weeds. The latter tend to be those possessing a short generation time, short dormancy and/or short longevity in the soil and copious seed production. These include some annual grasses and cereal volunteers in addition to broadleaved weeds which occur commonly in *Brassica* oilseeds, such as the sowthistles (*Sonchus* spp.), groundsel (*Senecio vulgaris*), shepherd's purse (*Capsella bursa-pastoris*), cleavers (*Galium aparine*) and the mayweeds. Some species, such as common chickweed and the speedwells, commonly occur whatever the type of cultivation (Cussans *et al.*, 1987).

Problems caused by weeds

Weeds are acknowledged to be the most important limiting factor in rapeseed production in Canada and their control worldwide is commonly one of the highest growing costs in *Brassica* oilseeds.

Weeds can:

- reduce quantity and quality of seed harvested
- delay harvest and reduce harvesting efficiency
- contaminate harvest crop seed samples – resulting in increased heating, spoilage, dockage, transportation and cleaning costs.
- affect incidence of pests, viruses and diseases in *Brassica* oilseeds or following crops
- shed seed to cause problems in subsequent crops
- necessitate increased cultivation practices which may reduce seedbed moisture and soil structure

Reduce yield of seed harvested

The fundamental principles of the competitive effects of weeds have been described by Harper (1977). Plants in close proximity compete for limited resources of light, nutrients and moisture, favouring some but at the expense of others. Total production of plant material per unit area increases up to a maximum with plant density but production per plant decreases. Hence the growth of *Brassica* plants will be influenced by the vigour and density of neighbouring plants and, conversely, the vigour and density of the crop will influence weed growth.

Competition begins when environmental resources, principally water, nutrients and light, cease meeting the needs of two or more plants in an area. The critical time for weed competition resulting in yield loss is often during the period of very rapid dry matter accumulation creating a greater demand on resources (Zimdahl, 1980). In *Brassica* oilseeds competition leading to yield reduction often starts with stem elongation in autumn-sown crops but is at an earlier growth stage in spring-sown crops.

Weed competition reduces crop growth and leaf area resulting in increased flower, pod and seed abortion. Weeds causing problems in *Brassica* oilseeds are those which grow tall and are shade tolerant. In addition, autumn-sown crops have a long establishment phase prior to rapid growth in the spring, and so any weed emerging at the same time as the crop and growing rapidly during this period can significantly reduce yield and maybe reduce the ability of the crop to survive the winter.

In autumn-sown rapeseed crops in northern Europe the most competitive weeds include volunteer cereals, charlock/wild mustard and common chickweed. Cleavers is more tolerant of shade than most weeds but can grow rapidly through the canopy as pods and seeds develop and also reduce crop yields appreciably by shading (Lutman, 1991). In spring crops in Canada, fat hen/lamb's quarters and charlock/wild mustard can be very com-

petitive throughout the life cycle of the crop (Thomas, 1992).

Competition leading to lower yields commences at different growth stages according to:

- time of sowing
- density and spatial arrangement of the crop
- species and density of the weeds
- relative vigour of crop and weeds
- availability of moisture
- availability of nutrients
- relative time of emergence of crops and weeds

Weed competition is generally more intense, necessitating earlier control measures where:

- the crop is thin and lacks vigour
- weed density is high and the weeds are vigorous
- moisture and nutrient supply favours weed growth more than crop growth
- weeds significantly reduce moisture supply to crops

It follows that there can be no general guidelines for timing control measures or defining densities of specific weeds needed to justify their removal. Spring rapeseed offers a striking example. Crops are often dense and very vigorous given sufficient rainfall in early and mid-season in parts of northern Europe and, consequently, weeds are smothered. Crops are able therefore to withstand high weed numbers without prejudicing yields (Lutman, 1991). In Canada, moisture is usually the limiting factor and many crops are grown at lower plant densities, thus increasing the potential for weed competition (Blackshaw *et al.*, 1987). Weeds in this situation need to be controlled early. This is strongly supported by research data from Canada which suggests that it is weeds emerging with the crop which cause most damage to yield and that weed control in the early stages of crop development is one of the most important requirements for growing high-yielding *Brassica* oilseed crops.

Attempts have been made to predict weed competition based on weed numbers, particularly in areas where the weather conditions are more predictable such as in areas of Canada. Variable autumn weather conditions in the maritime climate zones of northern Europe have led to great variations in the number of weeds required to reduce yields significantly. In one study in the UK the number of common chickweed plants required to reduce yields of autumn-sown oilseed rape by 5% varied between 1.4 and 464 plants m^{-2} (Bowerman *et al.*, 1994).

Delay harvest and reduce harvesting efficiency

The presence of green weeds in the crop just before and at harvest may delay and reduce efficiency of harvest. Patches of weeds in fields can cause variable ripening, thus making decisions on method and timing of harvest more difficult. These problems can be particularly acute in northern Europe with cleavers and mayweed species.

The use in some countries of diquat, glufosinate or glyphosate to desiccate reduces weed effects on harvesting, which would otherwise be slowed down as well as increasing the risk of seed loss. However, although the presence of weeds at harvest is likely to impair harvesting efficiency and lead to increased drying, storage, dockage and seed cleaning costs, it is not clear when this effect is critical and at what density the effects of weeds become significant.

Contamination of harvest seed samples

The presence of weeds in the crop at harvest can physically contaminate the sample and/or lower quality of the harvested sample leading to serious downgrading or rejection. Surveys in Denmark have highlighted the risk of *Cruciferae* weeds

increasing oil impurities in harvested samples (Jensen, 1990). *Cruciferae* weeds, such as charlock/wild mustard, contain high levels of erucic acid and contamination can affect levels in the harvested crop. Where rape is grown for industrial purposes, shed seed may pose problems with volunteers in future crops. As the genotype of the embryo determines the level of erucic acid, any pollination of a low erucic acid crop by pollen from a high erucic acid source could raise the level of erucic acid in the seed harvested. Thus erucic acid levels in a low erucic acid cultivar can be increased in the harvested crop by the presence of high erucic acid volunteers or charlock/wild mustard seed.

Physical contamination with seeds of *Cruciferae* and cleavers and with the flowering heads of weeds such as mayweeds can incur extra cleaning charges and reduced grading, particularly when the contaminants have a similar seed size to that of the harvested crop.

Effect of weeds on pests and diseases of *Brassica* oilseeds

Weeds can be hosts of fungi pathogenic to brassicas, particularly those closely related, such as runch (*Raphanus raphanistrum*), charlock/wild mustard and shepherd's purse. However, there is some doubt about the economic importance of weeds as a source of crop infection as the pathotypes involved may be less virulent (Lutman, 1991).

The presence of weeds in *Brassica* crops will increase humidity at the base of the crop, which favours development of pathogens (see Table 4.1). For instance, Kust *et al.* (1988) found in Germany that grey mould (*Brassica cinerea*) was more severe in rapeseed crops when chickweed was present. Higher humidity will favour slugs but may also increase the number of beneficial ground beetles (Schwerin, 1989).

Weeds in *Brassica* oilseed crops may pose a threat to neighbouring crops as, for

Table 4.1. Host ranges of *Brassica* oilseed crop diseases.

Diseases	Hosts
Sclerotinia sclerotiorum (stem rot)	Very wide host range
Leptosphaeria maculans (leaf spot/canker)	Cruciferae
Verticillium dahliae (wilt disease)	Very wide host range
Alternaria brassicae (dark/leaf pod spot)	Cruciferae
Pyrenopeziza brassicae (light leaf spot)	Brassicae
Botrytis cinerea (grey mould)	Wide host range
Peronospora parasitica (downy mildew)	Cruciferae
Erysiphe cruciferarum (powdery mildew)	Cruciferae
Pseudocercosporella capsellae (white leaf spot)	Cruciferae

Source: Lutman (1991).

example, volunteer cereals may harbour foliage diseases and virus vectors. Volunteer cereals may also help root and stem-base diseases, such as eyespot (caused by *Pseudocercosporella herpotrichoides*) and take-all (caused by *Gaeumannomyces graminis*), to survive and infect subsequent cereal crops (Kayser and Heitifuss, 1990).

Shed seed to cause weed problems in subsequent crops

Weed control in any crop will affect incidence of weeds in subsequent crops and demands for the whole rotation should be taken into account when making decisions for individual crops. This means considering not only the potential for weeds to increase in following crops but also their ease of control in other crops in the rotation.

Rapeseed usually provides an opportunity for the control of annual and perennial grass weeds. Effective herbicides for the selective control of annual grass weeds and selective suppression of perennial grass weeds are registered for usage in the major *Brassica* oilseed crops in many countries. Crops with weed populations well below a level likely to compete with the crop may be treated with such a herbicide to reduce grass weed populations in subsequent crops.

It may be desirable to control weeds which set copious amounts of seed and survive in the soil for several years even though they are expensive to control in *Brassica* oilseed crops and may not pose a threat to the current crop. Weeds such as poppies shed extremely high numbers of seeds in northern Europe with the potential for even higher numbers in the long term, including the next rapeseed crop, despite effective control in other crops in the rotation. A similar situation exists with stinkweed in Canada.

Conversely, it may be worth controlling apparently innocuous levels of weeds in other crops in the rotation in order to reduce the cost of weed control or improve the quality of the *Brassica* oilseed crop. For example, broadleaved weeds, including volunteer rapeseed, are normally controlled effectively in cereals. Hence, very low populations of rapeseed volunteers likely to set seed in cereals are often treated, particularly when of a cultivar which might have an adverse effect on quality of subsequent *Brassica* oilseed crops. However, the advent of herbicide-tolerant rapeseed cultivars should allow the control of such volunteers but these cultivars may prove to be of advantage in this respect only for a short term if they become volunteer problems for the future.

WEED CONTROL

Weeds are still controlled by cultural means in some parts of the world and this was the method of control in all countries prior to the advent of effective herbicides. The cultural methods include rotation, time of drilling, inversion tillage, between-crop management, stale seedbeds and hand and mechanical weeding. Herbicides have allowed farmers to intensify rotations with crops more suited to their land and market, rather than for their contribution to weed control within a rotation. Moreover, the availability of effective herbicides led to the adoption, in many situations, of non-inversion tillage and of optimal drilling dates for yield.

With the adoption of minimal non-inversion tillage associated with reduced diversity of autumn-sown crops in parts of northern Europe, weed numbers, particularly annual grass weeds, increased but the number of weed species declined as some broadleaved species were effectively controlled with herbicides. Consistent drilling dates through rotations also helped to reduce the number of weed species. Those surviving were those which produced copious amounts of seed and/or were incompletely controlled by herbicides. The resulting selection pressure and annual exposure of young plants from freshly shed seed to herbicides are associated with herbicide resistance in black-grass (*Alopecurus myosuroides*) in England (Moss and Clarke, 1994). Similarly, herbicide resistance in some broadleaved weeds to sulphonylurea herbicides in North America has been associated with non-inversion tillage. In the UK, problems with weeds were among the main reasons for the retreat from farming methods totally dependent on herbicides in favour of a more integrated approach to weed control using both cultural, notably a return to inversion tillage, and chemical methods.

Cultural weed control

Drilling dates

Using different drilling dates throughout the rotation will discourage weed species which readily germinate at a particular time of the year. Weed species which germinate throughout the year, such as common chickweed and annual meadow-grass in northern Europe, are less affected. Unfortunately, some *Cruciferae* weed species and volunteer *Brassica* oilseeds, which are difficult to control in *Brassica* crops, can germinate in the autumn or spring and, consequently, are not often controlled by growing crops in the rotation with different drilling dates.

Inversion tillage

Non-inversion tillage is adopted to reduce costs of establishment, to reduce moisture loss during primary cultivations or to minimize erosion. It may result in inferior yields to inversion tillage in some climates and soil types. In addition, surface straw may cause problems in the establishment of *Brassica* oilseeds shortly after harvesting cereals. Non-inversion tillage encourages high numbers of weeds which can present significant problems where crops are grown intensively and the time interval between harvest of one crop and sowing the next is short. However, in many countries there is a significant time interval between harvest and sowing the succeeding crop in the following spring. In addition, fallows or setaside may also provide an opportunity for cultural and chemical control measures to reduce weeds with limited dormancy and longevity, which are those encouraged by non-inversion tillage.

Inversion tillage buries seed and thus reduces weed numbers, especially of those species which cannot then emerge and also have limited dormancy and longevity. Poppies and some *Cruciferae*, including volunteer rapeseed, tend to have prolonged dormancy and are among those species that tend to dominate where inversion tillage is the sole method of primary tillage. Additionally, some weeds are more dormant when buried by inversion tillage. There is evidence to suggest that inversion tillage encourages the dormancy of shed *Brassica* oilseeds (Lutman, 1991).

Mechanical weed control

Mechanical weed control using tractor-drawn machines can be effective provided the weeds are small and the weather conducive to shallow cultivation. Unfortunately disturbance of the soil can stimulate germination of weed seeds which are more likely to establish with moist conditions after mechanical weeding, particularly when the crop does not fully cover the soil. However, these will be significantly later emerging than the crop and may either be uncompetitive or smothered by the rapidly expanding canopy of the crop.

The most effective mechanical methods are those which result in the shallow burial of seedling weeds. However, this may be impracticable in terms of crop safety with tractor-based equipment when both the crop and weed plants are small, unless the crop is grown in sufficiently wide rows to allow inter-row cultivation. In many situations, it may be impossible to carry out mechanical weeding at a stage when weeds are most susceptible due to unsuitable weather, crop or soil conditions.

As weeds develop, so do differences in their susceptibility to mechanical disturbance. The most susceptible broadleaved weeds tend to have a fibrous root system, while the least susceptible are those with tap or branched roots. Plants of the *Brassicae* have a branched root system, and *Brassica* oilseed plants are relatively tolerant of mechanical weeding but, unfortunately, so are the weed members of the genus. Weed scientists are currently investigating whether the use of very low doses of herbicides will make subsequent mechanical

weeding more effective and reliable (Blair and Green, 1993).

In the future, machine recognition technology may produce sophisticated mechanisms which will allow the precise guidance of mechanical weeders. This will enable more complete control of weeds between the rows and avoid damage to the crop, provided the row widths necessary do not compromise yield, agronomic requirements or environmental demands unacceptably.

Between-crop management

Often there may be little opportunity to practise between-crop management in rotations because of the short time interval between crops. But, in areas where crops occupy the land for a relatively short period of the year, there will be more opportunity to control weeds between crops. This often reduces weed numbers to a level where non-inversion tillage can be adopted.

Shallow cultivation after harvest can, if the soil is moist, encourage germination of some weed species but can also enforce dormancy in others (Cussans *et al.*, 1987). It can also reduce loss of freshly shed seed from the soil surface through insect, small mammal and bird predators and mortality. Clearly, it is important to understand the reaction of the target species to shallow cultivations. Weed species with little dormancy often dominate where non-inversion tillage is adopted and freshly shed seed will often readily germinate given appropriate soil management and weather conditions after harvest. The resulting plants can then be effectively controlled by shallow cultivations or with herbicides in the autumn or spring (Thomas, 1992).

The main objective of management between crops is to achieve a level of control of weeds which are difficult or expensive to control in the following crop. In Canada, winter annuals like stinkweed do not germinate until late in the autumn. The weed forms a rosette and is not killed completely by winter frosts or non-inversion tillage in the autumn or seedbed cultivations in the spring. Hence it is often sprayed with 2,4-D or MCPA in the late autumn. Unfortunately some populations are now resistant to these phenoxy herbicides. Care must be taken over the depth of tillage. Too deep tillage will bury the seed where it may live longer and too deep seedbed cultivations will bring seed to the surface layers where it will germinate.

Stale seedbeds

Preparing a seedbed in advance of the drilling date allows weeds to germinate which can then be killed with further cultivation or a non-selective herbicide before sowing. This technique is less commonly practised now, particularly where weather conditions are less predictable, as any delay may result in sowing after the optimum date. Moreover, the seedbed can be dried out by repeated cultivations (Roberts, 1982).

Fallows

Keeping land weed-free for a season is not usually an economically viable alternative compared with other forms of weed control but is practised in some parts of the world for the purpose of water conservation or to meet political requirements, e.g. setaside in the European Union. Preventing the shedding of viable seeds for a year or more will particularly reduce the infestation of weed species with limited dormancy in the subsequent crop. This is especially useful in parts of the world where fallows are commonly used in rotations and where *Brassica* oilseeds are established by non-inversion tillage to reduce erosion. Non-inversion tillage tends to encourage those weeds with short dormancy and longevity, or those with a short generation time, which are those most affected by preventing seed shed in a 1-year fallow.

Preventive control measures

Preventive control measures include:

- use of weed-free seed
- ensuring farm machinery is clean
- covering harvested produce and screenings during transport
- restricting animal movement from weedy fields
- prevention of seed set close to cropped land
- control of spread of weeds through manure

Chemical weed control

Two basic types of herbicide are used in *Brassica* oilseed production.

Non-selective herbicides

Non-selective herbicides, such as glufosinate, glyphosate and paraquat, kill emerged vegetation and are used prior to or shortly after drilling *Brassica* oilseeds to control those susceptible weeds which have emerged after primary cultivation, between crop management or stale seedbeds. They are usually most effective applied prior to cultivations when it seems unlikely that these will control the weeds present. Herbicides applied shortly after cultivations may be less effective where soil protects partially buried weeds. In some countries glufosinate and glyphosate are registered for use prior to harvest for the control of weeds, including perennial weeds, or to desiccate the crop. Glufosinate- and glyfosate-resistant cultivars of oilseed rape have been developed which allow the use of non-selective herbicide post-emergence of the crop.

Selective herbicides

Selective herbicides are intended to suppress or kill weeds with no or only minimal effect on the crop. These can be further subdivided as follows:

Foliage-acting herbicides Foliage-acting herbicides mainly enter through the leaves and stems.

1.   Contact herbicides affect the point of contact or part of plant in the immediate vicinity.
2.   Translocated herbicides are transported within the plant.

Soil-acting herbicides Soil-acting herbicides mainly enter through the roots or emerging shoots and may control weeds ranging in growth stage from germination to quite large plants. Some herbicides, particularly those which enter through the emerging shoots, may only control weeds at or shortly after germination, while others, particularly those entering through the roots, often control weeds pre- and post-emergence. Most soil-acting herbicides exert an effect on weeds germinating for a period after application. Such treatments are termed residual.

Each herbicide is described according to its time of application:

1.   pre-sowing herbicides – applied after or during seedbed cultivations but before the crop is drilled. An outstanding example is trifluralin which is widely used in *Brassica* oilseeds throughout the world.
2.   pre-emergence herbicides – applied after the crop is sown but before it emerges.
3.   post-emergence herbicides – applied after the crop has emerged, weeds may or may not have emerged. These include the specific foliage-acting broad-spectrum grass-weed herbicides used widely in world *Brassica* oilseed production.

Herbicides are usually applied overall but band-spraying along the rows of the crop or directed sprays between the crop rows may be possible in some situations (Roberts, 1982).

Integrated weed control through the rotation

As for many broadleaved crops, there is not a comprehensive range of herbicides for selective control of broadleaved weeds in *Brassica* crops. This reinforces the need to consider broadleaved weed control over the whole rotation rather than just for the current crop. However, *Brassica* oilseeds often provide an opportunity for very effective annual grass-weed control. Hence the following should be considered in all crops in the rotation:

• which weed species cannot be controlled easily and at low cost by cultural or chemical means in one or more of the crops grown or likely to be grown in the rotation?
• how competitive are the weeds which cannot be controlled easily and what is their likely effect on harvest or crop quality?
• how long can the seeds of the weeds survive in the soil if they are not easily controlled in one or more crops in the rotation?
• what additional specific cultural methods can be employed on the weeds not easily controlled otherwise?

Weeds difficult to control with selective herbicides in brassicas include many of the *Cruciferae*, poppies (*Papaver* spp.) and cleavers. Herbicide-tolerant cultivars are being developed by plant breeders and biotechnologists (see pp. 159–160, 184) which may overcome the problems of controlling these weeds, although volunteers derived from these crops will be tolerant of the same herbicide.

Seeds of poppies and some *Cruciferae* species, including volunteer rapeseed, can survive in the soil over many years and burial by deep non-inversion tillage or inversion tillage encourages their survival. They can germinate in substantial numbers in autumn- and spring-sown crops, and are often difficult or impossible to control with selective herbicides in *Brassica* oilseeds. Every opportunity should be taken to control them in other crops in the rotation and between crops or in fallows/setaside. Shallow cultivation may encourage germination but may not control emerged plants and non-selective herbicides applied to the foliage, like paraquat and glyphosate, are not very effective on some *Brassicae*. Hence herbicides, such as MCPA and 2,4-D, have been used to control emerged weeds between crops or in fallows, where safety of the following crop is not compromised. The more frequently *Brassica* crops are grown on the same field, the more rigorous the requirements for control of some of the *Cruciferae* and poppies in all crops in the rotation. Selective control with foliage- or soil-acting herbicides is comparatively cheap and effective in cereals.

Weeds which have a short dormancy and do not survive long in the soil are usually significantly reduced by inversion tillage and are encouraged by shallow non-inversion tillage. Where inversion tillage is carried out every year a short-term view may be taken with weeds which survive only a few years in the soil and may be difficult to control in *Brassica* oilseeds but control measures need to increase in rigour in the three or four crops prior to the *Brassica* oilseed crop. A longer-term view has to be taken with non-inversion tillage where high populations of these species can thrive. However, in situations where there are relatively long intervals between the harvest of one crop and the sowing of the next, and particularly where there are also additional significant gaps in cropping due to fallows or setaside, weed populations may be contained even where non-inversion tillage is adopted throughout the rotation. This may involve an initial shallow cultivation of non-cropped land to promote germination and subsequent weed control with further shallow cultivations to en-courage further germination. If these subsequent cultivations do not control the emerged weeds the use of foliage-acting herbicide may be necessary. Broadleaved weeds in this category may be cheaply and easily controlled in cereal crops within the rotation. Hence the opportunity is often taken to reduce potential weed populations in *Brassica* oilseeds, where they may be difficult and expensive to control, by controlling even low populations in cereal crops.

On the other hand, most annual grasses, which are encouraged by non-inversion tillage, are usually easily controlled with herbicides in most *Brassica* oilseed crops. In this situation uncompetitive populations may be treated with herbicides in *Brassica* oilseeds to reduce populations in later cereal crops in the rotation.

Control of perennial weeds through the rotation is often based on glyphosate applied either preharvest for some crops or postharvest, or on the use of effective herbicides for specific species. Thus MCPA may be applied in cereals or on uncropped land, or clopyralid applied to a range of crops for the control of creeping/Canadian/Californian thistle. It is prudent to apply these herbicides when the weed growth stage and weather conditions are conducive to herbicide activity. Specific foliage-acting broad-spectrum grass-weed selective herbicides will provide some control of perennial grass weeds in *Brassica* oilseed crops, which can also be controlled by chopping up the perennating organs with cultivators and destroying regrowth with repeat cultivations or non-selective herbicides. Perennial weeds are encouraged by non-inversion tillage.

Clearly there may be competing priorities in integrating cultural and chemical weed control with the agronomic requirements of the crop. A strategy of management needs to be defined for each field within an economic agronomic framework.

The need to discourage herbicide

resistance in weeds should also be taken into account in an integrated approach to weed control. Some populations of some broadleaved weeds occurring in Canada are resistant to the phenoxy herbicides or the sulphonylureas, and some green foxtail populations are resistant to trifluralin and ethalfluralin. In addition, there is a worldwide problem with the resistance of some annual grass weeds, notably green foxtail, the ryegrasses (*Lolium* spp.), wild oats and black-grass to the specific broad-spectrum foliage-acting selective grass-weed herbicides which belong to the aryloxyphenoxypropionate ('fops') and cyclohexanedione ('dims') groups.

The occurrence of herbicide resistance in weeds is often associated with non-inversion tillage. There is usually a requirement to rotate herbicides with different modes of action, to use herbicide mixtures and to maximize cultural control within a strategy to discourage resistance. The introduction of herbicide-resistant crops may not only reduce the need to have an integrated weed control strategy across the rotation but also reduce, at least in the short term, concerns about the presence of herbicide resistance to selective herbicides in *Brassica* oilseeds.

Chemical weed control in rapeseed crops

Rapidly growing and vigorous crops not only delay and reduce the need for chemical weed control but will also improve the weed control achieved with herbicides in all crops. This is particularly true with *Brassica* oilseed crops.

In some countries, herbicides are likely to have label recommendations for the major *Brassica* oilseed crops and may only provide limited chemical weed control options in the minor ones. Some herbicides may not be available in some countries if the crop area does not justify marketing. Chemical control measures can usually be applied to rapeseed crops over a wide range of dates and often more than one herbicide may be required to control the spectrum of weed species present or expected.

Combinations of herbicides, as 'tank mixes' or 'product mixes', can reduce cost of application but may require compromises in timing, which may result in some herbicides not performing to their optimum. It is essential that herbicides are used according to the product label. Care should be taken to ensure that herbicide mixtures are physically and biologically compatible. When using residual herbicides, it is important to check recropping possibilities in case of crop failure. This is particularly important when conditions are not conducive to crop establishment.

Pre-sowing and pre-emergence herbicides

The use of non-selective herbicides, such as paraquat, diquat, glyfosate and glufosinate will control susceptible weeds which may not be controlled by seedbed cultivations. This is particularly useful for weeds that cannot be controlled by selective herbicides in the subsequent crop and for weeds likely to be too big for their control by selective herbicides.

Pre-sowing herbicides, such as trifluralin and ethalfluralin, are commonly used where cheap and early weed control is required. (Resistance in green foxtail to these herbicides has been noted in Canada.) They are volatile and subject to degradation in bright sunlight and hence are incorporated into the soil surface, usually during seedbed preparation, to maximize activity. They enter the weeds in the initial stages of shoot and root growth, wholly or partly in the gaseous phase. This explains why they may be more reliable in drier climates than other soil-acting herbicides which enter the plant from soil solution. Napropamide enters the plant from soil solution and is incorporated pre-sowing to reduce degradation by light. Due to its persistence and requirement for soil moisture, it is used in winter oilseed rape.

Pre-weed emergence soil-acting herbicides, such as the pre-emergence herbicides propachlor and metazachor, are best applied when the soil is moist. This is for two reasons: first, to ensure crop emergence as, should it fail, the choice of alternative crops may be severely limited; and, second, to enhance availability of the herbicide for weed control. They may also be used after crop emergence, but before the weeds have emerged with propachlor; metazachor will control some seedling weeds. The moist soil requirement limits their worldwide adoption and they are usually used in autumn-sown oilseed rape. There may be recommendations for incorporation to enhance activity in drier climates.

Tebutam is a soil-acting, preemergence herbicide which should be applied as soon as possible after sowing winter oilseed rape as it inhibits germination of weeds. It is also available as a product mixture with clomazone, another soil-acting herbicide, to broaden the range of annual grass and broadleaved weeds controlled.

Labels of soil-acting herbicides may specify doses according to soil texture and indicate seedbed requirement. Soil-acting herbicides applied to the soil surface tend to be most effective with a fine, consolidated and moist soil, so that they are either more likely to make contact with emerging weeds for shoot-uptake herbicides, or the weed roots are more likely to be nearer the surface for root-uptake herbicides. Moisture is essential for uptake of most soil-acting herbicides, which will be more readily available when applied directly to a moist soil surface. Rainfall after application is also important to achieve optimum control with soil-acting root-uptake herbicides as they need to be moved into the root zone. This must occur before the herbicide has degraded too far. The product label may indicate a restriction

according to the organic matter status of the soil, as this adsorbs herbicides and makes them less available. Commonly, soil-acting herbicides are not recommended for use in soil with more than 10% organic matter.

Post-emergence herbicides

Post-emergence selective herbicides are commonly used for the control of annual grass weeds and have largely replaced TCA, applied either pre-drilling or pre-emergence of the crop. Foliage-acting broadleaved herbicides may be marginal in their selectivity and it is very important that the crop is at a recommended growth stage and stress-free. In some cases, the amount of leaf wax is critical to crop safety and rapidly growing plants, having less leaf wax, may be more susceptible to a particular herbicide.

Annual grass and broadleaved weeds are generally more susceptible to foliage-acting herbicides when growing rapidly and well supplied with moisture. Hence applications made under such conditions will help to ensure maximum weed control, provided crop safety is assured. Guidance is usually given on the product label giving the length of time when rain after application will reduce weed control.

The foliage-acting herbicides include benazolin, clopyralid and ethametsulfuron. Benazolin is used in some countries for its activity on specific weeds, notably common chickweed, while clopyralid is used on a world scale for the control of *Compositae* weeds, many of which are associated with non-inversion tillage.

Ethametsulfuron is used in Canada for the control of problem weeds such as stinkweed, charlock/wild mustard and flixweed (*Descurainia sophia*). Residues can cause problems in following crops and hence crop rotation guidelines should be followed. In addition, this herbicide belongs to the sulphonylurea chemical group to which herbicide resistance in some weed species has developed. Rotating this group with other chemical groups within the rotation is recommended.

Cyanazine is foliage- and soil-acting on annual broadleaved weeds and is used in some countries for the control of speed-wells, common chickweed and charlock/wild mustard. It is marginal in crop safety in some conditions and has been recommended only on triazine-tolerant varieties in Canada where it was used to control stinkweed, wild mustard and redroot pigweed/amaranth. There are populations of charlock/wild mustard which are resistant to cyanazine.

Pyridate is a contact foliage-acting herbicide which offers control of a range of annual broadleaved weeds, including cleavers and speedwells, in spring and winter oilseed rape. A relatively large minimum-growth stage of the crop is required to ensure crop safety and good spray coverage of the weed is necessary. It works more effectively when the weather is mild and the weeds are actively growing. It is also product-mixed with the foliage-acting herbicide picloram to broaden the range of annual broadleaved weeds controlled. Picloram is very persistent in the soil and hence usage of the product mixture is recommended for winter oilseed rape.

Quinmerac is primarily soil-acting with some foliage action. It controls a narrow range of annual broadleaved weeds, including cleavers, and pre-emergence and post-emergence applications may be recommended. A product mixture with metazachlor broadens the range of weeds controlled in oilseed rape.

Rapeseed varieties resistant to the non-selective herbicide, glufosinate, have been introduced in Canada. These offer the possibility of the control of weeds closely related to oilseed rape as well as grass weeds. There have also been introductions of cultivars resistant to the non-selective herbicide, glyphosate, and imazethapyr, which is selective in other crops. There is some concern amongst weed scientists about the implications of such releases, particularly where the herbicide may have to be used for the control of volunteer rapes in other parts of the rotation.

Specific foliage-acting broad-spectrum grass-weed herbicides (the 'fops' and 'dims' mentioned above), such as sethoxydim, fluazifop-butyl and quizalofop-ethyl, are used widely to control annual grass weeds and suppress perennial grass weeds. Their activity is influenced by moisture supply to the weed and lower than the maximum recommended dose can often be used in moist conditions. They do not control, but sometimes suppress, annual meadow-grass/winter grass, and they vary in their ability to control certain species or large weeds of the same species. Timing can be a problem when the different target weed species emerge at different times. In this situation and where early weed competition within the crop is occurring a sequence of two lower than maximum recommended doses may be used.

Some soil-acting herbicides are used post-emergence of crop and weeds in order to maximize crop safety and weed control achieved in autumn-sown crops. Propyzamide and carbetamide control a range of annual broadleaved and grass weeds including common chickweed, volunteer cereals and wild oats. Carbetamide is also product-mixed with dimefuron, another soil-acting herbicide, to broaden the range of annual broadleaved weeds controlled in winter oilseed rape. This objective can also be achieved with propyzamide as a product mixture with the soil-acting herbicide, diuron. Isoxaben is a soil-acting herbicide for use pre-emergence of weeds. It is recommended for application to relatively large and advanced winter oilseed rape crops for the control of susceptible annual broadleaved weeds, including *Brassica* weeds, which may germinate after application.

Unfortunately, herbicide resistance in

green foxtail, ryegrasses, black-grass and wild oats to the 'fops' and 'dims' group of chemicals has been noted (S.R. Moss, Rothamsted, 1994, personal communication) and their use may have to be rationed within a weed control strategy to discourage resistance. There are recommendations for the use of herbicides, from other chemical groups, for the control of wild oats. The control of black-grass by trifluralin, propyzamide and carbetamide in winter oilseed rape is not reduced by the presence of herbicide resistance in this species in the UK (Moss and Clarke, 1994). There are populations of this weed which are cross-resistant to all selective herbicides commonly used in cereals.

A minimum growth stage for crop safety may be specified for post-emergence herbicides, as well as calendar dates to ensure residual action of the soil-acting herbicides only recommended for post-emergence use. Labels of selective foliage-acting grass-weed herbicides may specify minimum and maximum weed growth stages for effective use and all post-emergence herbicides specify maximum weed size.

Preharvest herbicides

Recommendations for preharvest applications of glufosinate and glyphosate on rapeseed operate in some countries for use as a desiccant and/or the control of weeds, particularly perennial weeds. Time of application often coincides with flowering of perennial broadleaved weeds, which is when they are most sensitive to glyphosate. At crop maturity perennial grass weeds may have a large green shoot to rhizome ratio, which assists effective weed control. Glyphosate is best applied when the target weeds are actively growing, soil is moist and air humidity high.

Selecting the appropriate dose of herbicides

The manufacturer's recommended dose generally provides reliable control of a range of weeds in a range of conditions from one application. Lower than recommended doses of some herbicides may provide reliable weed control and may be appropriate in certain situations where:

- conditions are conducive to herbicide activity
- target weed species are particularly susceptible to the herbicide.
- later herbicide applications are possible before weeds are likely to compete with the crop, or where a previous herbicide application has already damaged older weed plants present
- weeds are smaller than the maximum size quoted in the manufacturer's recommendation for a foliage-acting herbicide
- full weed control is not required
- herbicide mixtures are being used which offer the possibility of reducing recommended doses of all components that may control the target species

Herbicide dose–response curves are helpful in determining lower than recommended rates as efficacy falls off gradually initially as dose is reduced. However, there is not a standard dose response, even for specific weeds to specific herbicides, because of variation due to weather and soil. Pesticide scientists are now trying to identify the appropriate dose for specific situations (Orson *et al.*, 1994).

APPLICATION OF HERBICIDES

Alternatives to the conventional sprayer, such as controlled droplet application (CDA) and electrostatics, have been developed, but do not provide an opportunity to reduce doses because:

- the conventional sprayer is as efficient

as other methods in applying foliage-acting herbicides to the parts of the plant where activity is greatest: leaf bases for grass-weed herbicides and the younger leaves for many broadleaved herbicides (Merrit, 1980)

• any improvement over conventional sprayers in distribution of soil-applied herbicides on the soil surface may not improve weed control

However, conventional nozzles produce drops which can drift. Although drift usually involves only a small amount of the spray applied, it should be kept to a minimum. Low-drift conventional nozzles offer some improvement, as do air-assisted sprayers (Rutherford and Miller, 1993).

Farmers like to use the minimum spray volume possible to achieve maximum work rates. There is limited evidence that some foliage-acting grass-weed herbicides and glyphosate may give improved weed con-trol through conventional nozzles at 50 to $100 \, l \, ha^{-1}$, when compared to $200 \, l \, ha^{-1}$. This is probably because these herbicides are easily translocated, do not damage the plant around the point of contact of the spray droplet, and the increased concentration in the lower volume assists penetration into the plant. Most herbicides work more robustly through conventional nozzles at higher volumes, usually around $200 \, l \, ha^{-1}$. However, applications at around $100 \, l \, ha^{-1}$ can provide similar results for both soil-acting and foliage-acting herbicides where the weeds are small and the crop canopy is open. These lower volumes are commonly used throughout the world when weed control is carried out either pre-sowing, pre-emergence or early post-emergence of the crop. In similar situations, controlled droplet application at a total volume of $40 \, l \, ha^{-1}$ can provide similar levels of control to conventional nozzles at $80–100 \, l \, ha^{-1}$.

REFERENCES

Blackshaw, R.E., Anderson, G.W. and Dekker, J. (1987) Interference of *Sinapis arvensis* L. and *Chenopodium album* L. in spring rapeseed (*Brassica napus* L.). *Weed Research* 27, 207–213.

Blair, A.M. and Green, M.R. (1993) Integrating chemical and mechanical weed control to reduce herbicide use. *Brighton Crop Protection Conference* 3, 985–990.

Bowerman, P., Lutman, P.J.W., Whytock, G and Palmer, M. (1994) Weed control requirements for oilseed rape under the reformed CAP. In: *Aspects of Biology* No. 40: *Arable Farming under CAP Reform*. AAB, Wellesbourne, Warwickshire, pp. 361–368.

Cussans, G.W., Moss, S.R. and Wilson, B.J. (1987) Straw disposal techniques and their influence on weeds and weed control. *Brighton Crop Protection Conference – Weeds* 1, 97–106.

Froud-Williams, R.J. and Chancellor, R.J. (1987) A survey of weeds in oilseed rape in central southern England. *Weed Research* 27, 187–194.

Harper, J.L. (1977) *Population Biology of Plants*. Academic Press, London, 892 pp.

Jensen, H.A. (1990) Thresholds for weeds in spring oilseed rape. *Danske Plantevaernskonferenz – Ukrudt* 4, 129–139.

Kayser, A. and Heitifuss, R. (1990) Influence of weeds on the infection of winter oilseed rape (*Brassica napus* L. var. *oleifera* Metzger) with fungal pathogens. In: *Proceedings of the IOBC Conference: Disease, Pests and Integrated Control in Winter Oilseed Rape*, Paderborn.

Kust, G., Wahmhoff, W. and Heitifuss, R. (1988) Untersuchungen sur Erfassung der artspecifischen Koncurrenz ausgewählter Unkrauter in Winterraps als Grünlage für die Erarbeitung von Schadensschwellen. *Mitteilungen aus der Biologischen Bundesanstalt für Land- und Forstwirtschaft* (245). 46 Deutsche Pflanzenschutz-Tagung.

Lutman, P.J.W. (1991) *Weeds in Oilseed Crops*. HGCA Oilseeds Research Review No. OS2. HGCA, London.

Merrit, C.M. (1980) The influence of application variables on the biological performance of foliage-acting herbicides. *BCPC Monograph* 24, 35–44.

Moss, S.R. and Clarke, J.H. (1994) Guidelines for the prevention and control of herbicide-resistant black grass (*Alopecurus myosuroides* Huds.). *Crop Protection* 13, 230–234.

Orson, J.H., Blair, A.M. and Clarke, J.H. (1994) Weed control requirements under the reformed CAP. In: *Aspects of Applied Biology 40. Arable Farming under CAP Reform*. AAP, Wellesbourne, Warwick, pp. 297–306.

Roberts, H.A. (ed.) (1982) *Weed Control Handbook: Principles*. Blackwell Scientific Publications, Oxford.

Rutherford, I and Miller, P. (1993) *Evaluation of Sprayer Systems for Applying Agrochemicals to Cereal Crops*. Project Report No. 81. HGCA, London.

Schwerin, C.W. (1989) Auswirkung der Herbicidan wendung und Werunkrautung auf Laufkäfer (Col. *Carabidae*) im Winterraps. *EWRS Newsletter* 45, 17.

Thomas, P. (1992) *Canola Growers' Manual*. Canola Council of Canada, Winnipeg, Canada.

Zimdahl, R.L. (1980) *Weed–Crop Competition*. International Plant Protection Centre, Oregon State University, Corvallis, Oregon, USA, pp. 5–140.

5 Diseases

S.R. Rimmer and L. Buchwaldt
Department of Plant Science, University of Manitoba, Canada

INTRODUCTION

There are numerous diseases of *Brassica* oilseed crops that may cause production losses to a greater or lesser extent in different areas of the world. This review considers the major diseases of these crops caused by fungal, bacterial or viral pathogens from a global perspective and, where possible, a comparative approach is taken. Our discussion relates to diseases of *Brassica napus* L., both winter and spring cultivars, *Brassica rapa* L. and *Brassica juncea* Czern. These three species account for more than 99% of world *Brassica* oilseed production. Diseases will be presented in terms of area of importance, causal organism, life cycle and epidemiology, and control. Economic considerations have greatly influenced control practices for individual diseases in different areas of the world and will be referred to as they arise. Due to space limitations, the literature citations will be restricted to recent key papers and reviews where more complete references may be obtained. Descriptions and photographs of symptoms of diseases of *Brassica* oilseed crops may be found in Davies (1986), Smith *et al.* (1988), Martens *et al.* (1988) and Pauls and Rawlinson (1992).

SCLEROTINIA STEM ROT

Stem rot is the most important disease of oilseed rape in central China (Liu *et al.*, 1990), and is also a major cause of crop loss in France, Germany (Krüger and Stoltenberg, 1983) and other parts of Europe. It occurs sporadically in Canada and may cause large yield losses in some years (see Morrall *et al.*, 1991). The causal agent is the ascomycete *Sclerotinia sclerotiorum* (Lib.) de Bary. This pathogen causes disease on many other dicotyledonous plants including sunflower, field beans, peas, carrots and lettuce.

Symptoms

Generally, the first symptoms of sclerotinia stem rot are visible after flowering because infection is related to abscised petals. Lesions on leaves are greyish, irregularly shaped and often associated with adhering petals. The lesions on stems are white or almost bleached, sometimes with darker

Fig. 5.1. Stems of oilseed rape with sclerotinia stem rot caused by *Sclerotinia sclerotiorum.*

rings showing the stepwise growth of the fungus, often with a sharp line between infected and healthy tissue (Fig. 5.1). Girdled stems are weaker and have a tendency to lodge at the point of infection. Stem-infected plants ripen prematurely and can be seen as brown patches in the field. At the end of the season, black sclerotia can be found inside the infected part of the stems, in infected pods, and under very humid conditions even on the outside of infected tissue.

Occasionally, grey mould (*Botrytis cinerea*) is associated with stem rot, covering the infected areas with brown to grey fluffy mycelium.

Epidemiology

Sclerotinia sclerotiorum survives for several years in soil in the form of sclerotia. These are dark, melanized resting bodies usually about 0.5–0.8 cm in length and 0.3–0.4 cm in width. Sclerotia in the top 3–5 cm of the soil may germinate to form numerous apothecia on the soil surface. Cool and prolonged periods of moisture are required for apothecial formation and these conditions most commonly occur when the soil is covered by a crop canopy. Sclerotia may also germinate myceliogenically but this is not considered significant in the epidemiology of stem rot of *Brassica* oilseed crops. Ascospores are discharged into the air from apothecia with sudden changes in relative humidity and are dispersed primarily by wind. Since ascospores are the only source of inoculum in the field, stem rot is primarily a monocyclic disease.

The role of petals in the infection of *Brassica* oilseed crops is well established (Krüger; 1975, Lamarque, 1983). After abscission, many petals, on which ascospores may have been deposited, lodge on leaves or in the leaf axils. Utilizing the petals as a source of nutrients, the fungus invades the leaf or leaf axil tissue and, subsequently, the stem is colonized.

The pathogen produces cell-wall-degrading enzymes, primarily polygalacturonases, and oxalic acid, which probably aids infection by sequestering Ca^{2+} from the pectate complexes of the cell walls and reducing the pH of the tissues. Polygalacturonases are known to function optimally at low Ca^{2+} concentration and low pH (Godoy *et al.*, 1990).

Sclerotinia sclerotiorum is homothallic and occurs as reproductively isolated clones which may be identified by DNA fingerprinting. Some clones are present over large geographical regions although clonal heterogeneity occurs in single fields infected with stem rot (Kohli *et al.*, 1992). Why some clones are prevalent on *Brassica* oilseeds and whether some clones are more adapted to different host species remain to be determined.

Sclerotia are formed within the infected portions of the stem and are released at harvest when the crop is mechanically threshed with a combine and either remain in the field or contaminate the harvested seed. Severe crop losses are due to high levels of stem infections. Percentage seed loss can be predicted and has been estimated as half of the disease incidence (DI) of infected stems (Morrall *et al.*, 1984).

Control

Because of the wide host range of *S. sclerotiorum*, the longevity of sclerotia and its dependence on prolonged moist conditions, this disease has been difficult to control economically. Chemical control is effective when applied in the period of early flowering to full bloom, and a number of chemicals are, or have been, used including benomyl, iprodione, vinclozolin and sumisclex. Stem rot is only a problem in certain years when environmental conditions are favourable for epidemic development and, consequently, much effort has been devoted to the development of prediction systems to avoid unnecessary chemical treatments. These include the check-list type of risk assessment to ascertain the likelihood of a

stem rot problem in a given field in a given year. This type of prediction system has been used in Scandinavia and in Canada (Thomas, 1984). Morrall and co-workers (1991) have developed and patented a prediction system based on the percentage of petal infestation following a specified sampling procedure in the field. Petals are plated on an antibiotic-containing medium and the percentage of petals contaminated with *S. sclerotiorum* is determined. Risk of economic crop loss is then evaluated as low, moderate or high based on this assessment and on current weather conditions. Petal testing kits, which include an instruction manual, agar media, viable sclerotia as a control, etc., have been used successfully in western Canada by farmers for determining the need for fungicide application.

Evidence of variation for resistance to stem rot in *Brassica* oilseed crops has been reported. Brun *et al.* (1987b) compared different inoculation techniques and demonstrated significant differences in stem rot severity among a number of *B. napus* accessions. Norin 9, a Japanese accession, showed good resistance to stem rot. Inserting toothpicks infested with mycelium of the pathogen into the main stem was shown to be a reliable method for evaluation of resistant materials. Sedun *et al.* (1989) compared the rate of lesion expansion on stems of various *Brassica* species as a measure of disease resistance. Lesion expansion was slower on *Brassica carinata* and *B. napus* compared to *B. nigra*, *B. juncea* and *B. rapa*, which showed the most rapid lesion expansion. It is interesting that *B. carinata*, when inoculated in the leaf axil, frequently exhibited premature leaf abscission before stem infection could become established and consequently escaped stem rot. Liu *et al.* (1990) reported high heritability of sclerotinia resistance in *B. napus*, which was controlled by nuclear genes and unlinked to the low erucic acid trait.

Because of the significant role played by petals in infection of stem rot it may be possible to avoid stem rot by growing apetalous cultivars. An apetalous mutant of *B. napus* selected by G. Buzza (Pacific Seeds Pty, Australia) was substantially free of stem rot compared to the normal petalous cultivar Westar (Downey and Rimmer, 1993). Fu *et al.* (1990) also observed that apetalous lines were unaffected by stem rot and found that four recessive genes control this trait in *B. napus*. Transformation of *B. napus* with a gene for oxalate oxidase isolated from barley has been reported (Thompson *et al.*, 1993). The gene was expressed and degradation of oxalic acid occurred. It remains to be seen whether this gene will provide resistance to stem rot under field conditions.

In Europe where intensive oilseed rape production is widespread the availability of satisfactory fungicide control has probably inhibited a concerted effort to select for resistance to this disease. It is perhaps not too surprising, then, that most progress in selection for resistance to stem rot has been made in China, where chemical control is unavailable for most farmers due to the expense of suitable fungicides. Under drier conditions stem rot occurs sporadically and in these situations, for example in western Canada, field selection for resistance has not been effective.

STEM CANKER (BLACK LEG)

Stem canker caused by *Leptosphaeria maculans* (Desm.) Ces. & de Not. is a serious disease of *Brassica* crops and has caused yield loss in vegetable brassicas for many years. The importance of the disease on oilseed rape is more recent and often associated with the increase in area of production. Epidemics in France, Great Britain and Germany occurred subsequent to the rapid expansion of cultivation of *B. napus* in the period 1965–1980 (see Davies, 1986; Gugel and Petrie, 1992 for references). In Australia, losses were so extensive that production was virtually

eliminated until resistant cultivars were developed. In Canada, since 1975, the disease has spread throughout the provinces of Saskatchewan and Manitoba. In 1987, a stem canker survey conducted in Manitoba indicated an overall yield reduction of 9.6% (see Gugel and Petrie, 1992). Although relationships between yield loss and disease severity have been described (see Hall, 1992), differences among methods of disease assessment and among individual cultivars render such relationships unreliable. Stem canker is not important in China and India, possibly due to the widespread practice of harvesting rape stems for fuel, thus limiting an important source of inoculum, and the fact that, in India, the major oilseed *Brassica* crop is *B. juncea* which is resistant to the disease.

A recent symposium on stem canker reviewed the biology of the pathogen, economic importance and crop loss, epidemiology, genetics of resistance and host–pathogen relations, and physiology of host–pathogen interactions (Gugel and Petrie, 1992; Hall, 1992; Pedras and Seguin-Swartz, 1992; Rimmer and van den Berg, 1992; Williams, 1992). These reviews are useful sources of recent literature on this disease.

The causal fungus, an ascomycete, *L. maculans* (anamorph *Phoma lingam* Tode ex Fries), occurs as two distinct sympatric populations, which differ in pathogenicity, cultural characteristics and physiological characters. One population of isolates, referred to as aggressive or virulent, is strongly pathogenic and causes severe stem canker on oilseed rape. Another group of isolates is usually described as non-aggressive or weakly virulent. Non-aggressive isolates produce a red-brown or yellow-brown pigment in Czapek-Dox medium while aggressive isolates generally produce no pigment. Other important differences are that only aggressive isolates produce sirodesmins, which are phytotoxic secondary metabolites (Koch *et al.*, 1989)

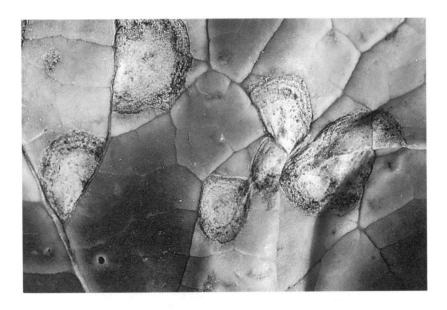

Fig. 5.2. Leaf lesions on *Brassica napus* caused by *Leptosphaeria maculans*.

and, although a number of workers have successfully crossed isolates among aggressive types, *in vitro* mating between either aggressive and non-aggressive or non-aggressive and other non-aggressive isolates has not been achieved (Petrie and Lewis, 1985; Mengistu *et al.*, 1993). There is some evidence for pathogenic specialization within aggressive isolates of *L. maculans*. Koch *et al.* (1991) distinguished three pathotypes of aggressive isolates based on their virulence to the *B. napus* cultivars, Westar, Glacier and Quinta. These were designated pathogenicity group (PG) 2, PG3 and PG4. Non-aggressive isolates were designated PG1.

Symptoms

The fungus causing stem canker also causes symptoms on leaves in the vegetative growth period. Leaf lesions are greyish or yellow-brown with small, black pycnidia (Fig. 5.2).

Typical stem canker symptoms are found after flowering. Diseased plants develop a corky dry rot at the base of the stem and, on the stems, pale necrotic lesions with a darker margin are formed (Fig. 5.3). When the stem is girdled the plants tend to lodge at ground level. Pycnidia are produced in the infected area and inside the root collar a large part of the tissue develops a black discoloration.

Epidemiology

Stem canker is primarily a stubble-borne disease on *Brassica* oilseed crops, though infected seed may play an important role as a primary source of infection in some instances. Generally the primary inoculum is ascospores released into the air from pseudothecia formed on *Brassica* or other cruciferous residues. To serve as a source of infection, residues must be on the soil surface to allow dispersal of inoculum. Ascospore inoculum may be dispersed and cause stem canker 1.5 to 8 km from the source (see Hall, 1992 for references). The timing of ascospore release varies with different regions. In Canada, pseudothecia

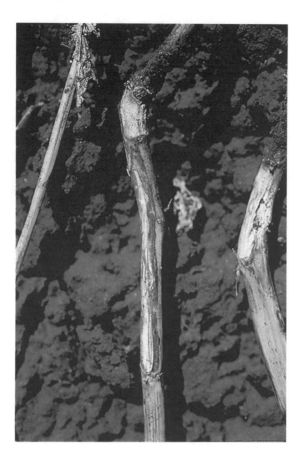

Fig. 5.3. Basal stem canker symptoms on *Brassica napus* caused by *Leptosphaeria maculans*.

of aggressive isolates on the previous year's crop residues do not mature and release ascospores until late June, July and August at the time of flowering or during later stages of crop development. In subsequent years following a diseased crop, matured pseudothecia may release ascospores earlier in the cropping season in May and June with the result that severe seedling infection may occur (G.A. Petrie, Saskatoon, 1994, personal communication). In Europe, ascospore discharge is usually most pronounced in the autumn to coincide with seedling development of winter rape. In Australia, following summer, ascospore discharge coincides with the first

rains, the time when planting or crop growth commences. In fact McGee and Emmett (1977) reported delaying planting until August can reduce losses due to canker.

On spring rape, in Canada for instance, the disease may be considered monocyclic, as primary infection of seedlings from the ascospore inoculum results in latent (asymptomatic) infection of the leaf petiole and stem. It is these latent infections which eventually progress and cause the characteristic cankering and girdling of the basal stem, the major cause of crop loss. The significance of latent infection for chemical control of canker in winter oilseed rape was first realized by Nathaniels and Taylor (1983). Subsequently, it was shown that the infection pathway involves a bio-trophic growth phase of the pathogen where leaf infection is followed by a systemic infection of the lamina and petiole which eventually leads to stem lesions (Hammond et al., 1985; Hammond and Lewis, 1986).

Xi et al. (1991b) observed that on a susceptible cultivar (Westar) the period of latent infection is much shorter than on a resistant cultivar (Cresor). Moreover, species of Brassica with the B genome (B. nigra, B. juncea and B. carinata) have been described as highly resistant, but are frequently found to have latent infections. Whether or not this results in ascocarp development on the dead stem and root tissues is not known, but presumably there is potential for this to occur. Wild or weedy cruciferous species vary in susceptibility to L. maculans and a similar process may occur here. Weeds such as Raphanus raphanistrum have been considered as possible sources of infection (e.g. in Australia) even though this species is not very susceptible to canker. Thus latent infections on resistant hosts may result in subsequent formation of primary inoculum on the plant residues. The extent to which this occurs is an area where more research is urgently needed.

Pycnidiospores (asexual spores) may serve as primary inoculum as pycnidia also often occur on residues. Dispersal of pycnidiospores is due to rain splash, and thus dispersal only occurs over short distances. Secondary spread of the disease is mainly due to pycnidiospores formed in pycnidia on leaves, stems and pods of in-fected plants, splashed during rain periods. In spring rape, secondary spread by pycnidiospores is not likely to result in extensive damage unless it occurs early in crop development. With winter rape the opportunity for secondary infection to reach damage thresholds is obviously much greater. Wounding of plants by insects,

farm equipment or other means increases the risk of severe infection.

Seed infection may initiate canker in Brassica oilseed crops. Low but detectable disease levels have occurred at 0.1% or less seed infection (Wood and Barbetti, 1977). Seed infection is most significant for importing the disease into regions where it is not yet present. Seed testing procedures for L. maculans in brassicas have been established by the International Seed Testing Association (Neergaard, 1977).

Control

Stem canker is most effectively controlled by the use of resistant cultivars, though other management practices are also useful in supplementing resistance. Reduction of primary inoculum levels may be achieved by crop rotation and techniques which lead to rapid decomposition of infected residues, control of volunteers and cruciferous weeds, seed treatments and fungicide applications. In western Canada decom-position of residues does not occur in frozen soils in the winter months. Similarly, lit-tle decomposition occurs in Australia in the hot and dry summer period. Thus, in these climates, crop residues may take a con-siderable time to decompose; three- to four-year crop rotations are necessary and strategic planting of crops as far from infected stubble as possible is important. In Europe, residues decompose more quickly in the temperate climate and deep ploughing of the stubble hastens this pro-cess. In China, where land is usually flooded after oilseed rape cropping to pro-duce rice, the warm wet soils lead to rapid decomposition of the stubble. This may be part of the reason why this disease is of no consequence to oilseed rape production in China.

Although seed-dressing with fungicides has not been demonstrated to control stem canker effectively when ascospore inocu-lum is present, it may be essential in regions where the disease is not endemic. However, it should be mentioned that, in Australia, Ballinger *et al.* (1988a, b) showed that flutriafol, applied as a fertilizer dressing, significantly reduced levels of stem canker in areas where the disease was prevalent. This chemical was not effective in trials in western Canada (Xi *et al.*, 1991a).

Small, but economic, increases in seed yield have been obtained in Europe with fungicide applications in the late autumn and early spring (Rawlinson *et al.*, 1984). Results have sometimes been inconsistent but prochloraz is recommended for stem canker control in many parts of Europe.

Considerable variation for resistance has been reported among oilseed rape (*B. napus*) cultivars and breeding lines. Extensive discussion of the genetics and methodology of breeding for resistance to stem canker has recently been published (Rimmer and van den Berg, 1992; Downey and Rimmer, 1993). Resistance in many European winter rape cultivars is derived from the French cultivar Jet Neuf, whereas resistance in spring cultivars of *B. napus* occurs in the French cultivar Cresor, in Australian cultivars, e.g. Maluka, Taparoo, and other more recent cultivars. Many Australian cultivars derive their resistance from Japanese sources, e.g. Chisaya, Chikuzen and Mutu natane. For methodology and comparison of techniques for selection for resistance see Gretenkort and Ingram (1993) and McNabb *et al.* (1993).

Mapping of resistance genes to restriction fragment length polymorphisms (RFLP) linkage maps has been described (Ferreira *et al.*, 1993; Landry, 1993). Ferreira and co-workers used doubled haploid-derived F_2 plants from the cross of Major × Stellar and identified a single major gene for cotyledon resistance to a PG2 isolate of *L. maculans*. Four other genomic regions were associated with quantitative measurements of resistance of cotyledon and stem inoculated plants. Two other genomic

regions, different from regions associated with resistance of greenhouse inoculated plants, were associated with field measurements of resistance.

Resistance to stem canker has been observed in only a few accessions of *B. rapa* and the rather weak resistance that has been found seems to be largely restricted to winter oilseed accessions (Rimmer and van den Berg, 1992). Species with the B genome (*B. nigra*, *B. juncea* and *B. carinata*) generally appear to be highly resistant to stem canker. However, the apparent lack of symptoms in cotyledon and leaves is often associated with subsequent extensive colonization of root or basal stem tissues (Gugel *et al.*, 1990; Keri, 1991). Keri observed that over 80% of 250 accessions of *B. juncea* inoculated with Canadian isolates of *L. maculans* on the cotyledons were subsequently found to have root infection. Furthermore, Ballinger *et al.* (1991) reported some isolates of the pathogen in Australia were virulent to *B. juncea*.

If the mechanism of resistance is to extend the period of latent infection, as has been observed with Cresor, beyond the time when significant crop damage can occur, yet allow subsequent survival of the pathogen in crop residues, then it would seem that selection pressure for adaptation of the pathogen for greater virulence will not occur, at least not very quickly. Thus resistance may be durable as seems to be the case with the resistance of Jet Neuf. Further exploration of this phenomenon seems worthwhile, not only for control of stem canker, but for control of other diseases of other crops.

ALTERNARIA LEAF AND POD SPOT (BLACK SPOT, DARK LEAF SPOT, ALTERNARIA BLIGHT)

Alternaria diseases of *Brassica* oilseed crops are most often caused by *Alternaria brassicae* Berk. and occasionally by *Alternaria bras-*

Fig. 5.4. Alternaria black spot on leaves of oilseed rape.

sicicola Schwein and *Alternaria raphani* Groves and Skolko, although these two species are mostly associated with vegetable brassicas (*Brassica oleracea, B. rapa*) and radish (*Raphanus sativus*), respectively. *Alternaria brassicae* occurs at high incidence but low severity in most temperate areas such as western Canada (Petrie *et al.*, 1985) and England (Evans *et al.*, 1984; Humpherson-Jones, 1983) mainly because *B. napus* cultivars are relatively resistant, although more severe epidemics have occurred in years with warm, wet weather conditions late in the season as reported in former West Germany (Daebeler and Amelung, 1988), Poland (Stankova, 1972) and Australia (Stovold *et al.*, 1987). In India,

however, alternaria blight is a serious disease, which often causes severe yield losses, mainly because cultivars of *B. juncea* (the predominant *Brassica* oilseed in this region) and *B. rapa* are highly susceptible (Singh and Bhowmik, 1985).

Symptoms

Symptoms of black spot can be found on all parts of the plant, leaves, stems and pods. On leaves, the spots are more or less concentric, brown and necrotic, often surrounded by a chlorotic halo if the leaf is still green (Fig. 5.4). Spots on stems are elongate and almost black. On pods, spots are dark and circular, sometimes developing into larger necrotic, brown areas with a sunken middle. Severe infection may cause the pods to open and shatter the seed.

Epidemiology

Alternaria brassicae is mainly stubble-borne, but seed infection also occurs and is particularly important in causing seedling rot in India and other areas where seed is sown into warm soil. In one study the ratio of seed-borne infection to seedling mortality was 1 : 0.88 (Chahal and Kang, 1979). In contrast, seed-borne infection does not usually reduce plant stands sufficiently to affect yields in temperate regions.

Alternaria brassicae survives as a saprophyte on dead plant material and is able to cause infection the following year (Humpherson-Jones, 1989). In one study, a gradient of number of leaf spots was recorded 100–1000 m from a previously infected oilseed rape field, probably caused by wind-borne conidia from infected plant debris (Daebeler and Amelung, 1988). In India, infected plant debris is also a source of inoculum. In one experiment the highest number of leaf infections was recorded in *B. juncea* planted in soil where plant debris was buried in the top 7.5 cm (Tripathi and Kaushik, 1984), plant debris at 15 and 22.5 cm caused little and no infection, respectively. *Alternaria brassicae* does not survive on the surface under Indian conditions, probably because of the high summer air temperatures. *Alternaria brassicae* has been shown to produce microsclerotia and chlamydospores (Tsuneda and Skoropad, 1977), but their involvement in long-term survival has not yet been determined.

Infection incidence and conidial concentration remain low until flowering, probably restricted by suboptimal temperatures and presence of mainly young tissue, which is less susceptible. Infection starts at the lower leaves and moves upwards in the canopy and increases rapidly as the pods mature (Marchegay *et al.*, 1990). Secondary conidia are formed in lesions on leaves and stems and their release is correlated with rising temperatures and lower relative humidity, with consequent drying of the host surface. Conidia are dispersed by wind (Mridha and Wheeler, 1987). Conidia are also spread within the canopy by runoff rainwater and in rain-splashed droplets (Louvet and Billotte, 1964).

Alternaria brassicicola infects directly through the epidermis or through stoma, whereas *A. brassicae* infects only through stoma (Changsri and Weber, 1963). Conidia require free water or above 95% relative humidity for germination and infection (Louvet and Billotte, 1964). From the literature it appears that *Alternaria* species have adapted to different climatic conditions. Generally, *A. brassicola* requires higher temperature for a longer period for germination (25–31°C for 9 h) than *A. brassicae* (21–28°C for 3 h) and higher temperature for disease development (23–25°C for 3 days for *A. brassicicola* compared with 19–23°C in *A. brassicae*) and for production of secondary conidia (23–31°C for 3 days for *A. brassicicola* and 15–23°C for *A. brassicae*) (Degenhardt *et al.*, 1982). The high temperature requirement of *A. brassicicola* restricts its development in northern parts of North America, Europe and other temperate regions.

The fungus attacks the inflorescent stems and pods resulting in reduced seed yield, seed shattering and poor seed quality. In years favourable for *A. brassicae*, 25% of rapeseed lots with up to 9% seed infection have been found in England (Humpherson-Jones, 1985) and incidences of 14–22% have been observed in some areas of New South Wales in Australia (Stovold *et al.*, 1987). In India, 41–59%, 23–51% and 3–19% of rapeseed or mustard seed carried *A. brassicae* depending on the climatic region (Chahal and Kang, 1979). *Alternaria brassicae* can infect seed internally, as a result of early infection of the seed. This is mainly a problem in Indian rapeseed and mustard production. The fungus can also occur as a contaminant on the seed surface which probably occurs during harvesting. However, seed infection is substantially reduced within a few months of storage at room temperature 20–22°C in Australia (Stovold et al., 1987), 40 ± 5°C in India (Tripathi and Kaushik, 1984). A good correlation between disease severity on pods and yield loss has been found using single plants: with 5 to 20% infected pod area the relative yield was 81% of uninfected plants, with 20 to 50% it was 78% and above 50% infection the relative yield was 68% (Daebeler and Amelung, 1988). Degenhardt et al. (1974) also found a good correlation between disease severity, yield per plant and 1000 kernel weight. Effect on oil quality varies from not measurable (Stovold et al., 1987) to slightly reduced (Degenhardt et al., 1974) and up to 35% reduction (Ansari et al., 1988). A key for assessment of percentage leaf and pod area with black spot has been published (Conn et al., 1990).

Control

Dispersal of spores during harvest of infected crops and from infected debris left in the field constitutes an important source of inoculum for nearby host crops. This has been a problem in England, where vegetable and Brassica oilseed crops are grown in the same area (Humpherson-Jones, 1989) and in other parts of Europe where winter and spring rape are both grown (Daebeler and Amelung, 1988; Marchegay et al., 1990). In order to escape disease, distance to infected crops both in time and space is an important preventive control measure.

Seed treatment with the dicarboximide fungicide, iprodione or the systemic fungicide, fenpropimorph, controls effectively both A. brassicae and A. brassicicola (Maude et al., 1984). Iprodione is the most consist-

ently effective fungicide used for control of alternaria in the field. A single application at petal fall controls the disease persistently until harvest and increases seed yield significantly (Cox et al., 1981; Evans and Gladders, 1981). However, good coverage at the base of the crop is important for optimum control. In England, crops for commercial seed production are routinely sprayed, and this practice reduced the disease from 8.9% in 1981 to 0.1% in 1984 (Humpherson-Jones, 1985).

Most Brassica oilseed cultivars are susceptible to alternaria, and disease resistance has not featured in control strategies so far. However, several reports have shown that some cultivars or breeding lines are less susceptible, primarily in B. napus but also in B. juncea and B. rapa (Bhowmik and Munde, 1987; Brun et al., 1987a). In Poland, four out of 65 winter cultivars of B. napus were less susceptible than the average, when compared under field conditions (Stankova, 1972).

Several mechanisms seem to be involved in reduced susceptibility. Thickness and physical features of the wax layer are important factors conferring physical resistance to A. brassicae (Tewari and Skoropad, 1976). Young tissue is more resistant to A. brassicae than older tissue, and senescing tissue is very susceptible. This is likely to be related to the thickness of the epicuticular wax layer, which is reduced as the plant ages. Reduction in glucosinolates with leaf age may also be implicated (Porter et al., 1991), together with increase in leaf exudates with age (Kohle and Hoffmann, 1988). Production of phytoalexin may influence susceptibility at the species level (Conn et al., 1988).

Reduced number of stomata per cm^2, resulting in fewer lesions per leaf, reduced lesion size and spore production, together with longer latent period, have been shown in two partly resistant rapeseed and mustard cultivars (Saharan and Kadian, 1983), and could explain the lower

Fig. 5.5. Light leaf spot on leaves of oilseed rape (courtesy of NIAB, Cambridge, UK).

infection rate measured on plants grown in the greenhouse.

LIGHT LEAF SPOT

Light leaf spot of brassicas is caused by the discomycete *Pyrenopeziza brassicae* Sutton and Rawlinson whose anamorphic state is *Cylindrosporium concentricum* Grev. This disease is perhaps the most important disease of winter oilseed rape in the United Kingdom and has become increasingly problematic in France (McCartney *et al.*, 1987). Rawlinson *et al.* (1978) and Courtice *et al.* (1988) have described and reviewed the pathogen, including biology, infection processes, genetics and inoculation techniques. Many *Brassica* crops are affected including oilseed rape and vegetable brassicas, especially cauliflower, broccoli, Brussels sprouts, turnip, etc. Variation among isolates of the pathogen for aggressiveness and for resistance among and within cultivars of *B. napus* (Rawlin-

son *et al.*, 1978; Maddock *et al.*, 1981) has been demonstrated.

Symptoms

Early symptoms appear on the upper and lower leaf surfaces as 1 mm small, white colonies (acervuli). Soon after, a white film develops and individual colonies are no longer recognizable. When the infected leaf tissue turns brown and necrotic, the white spore masses are situated in the green tissue at the margin of the expanding lesion (Fig. 5.5). Infected stems have superficial fawn lesions with black speckling at the edges, later splitting the epidermis. Severe infection can kill or stunt autumn-planted crops and cause damage to buds, flowers and pods the following year.

Epidemiology

Primary inoculum may be ascospores produced from minute apothecia on decaying oilseed rape tissues (primarily leaf tissue), conidia dispersed by rain splash and, perhaps, infected seed. It is likely that ascospore inoculum is the most important source of primary infection where the disease is prevalent. Ascospore dispersal is by wind and thus there is potential for long-distance spore movement (McCartney and Lacey, 1991). In the United Kingdom on winter oilseed rape, infection is frequently initiated in the autumn and under average conditions the latent period of infection is in the range of 20–30 days (Rawlinson et al., 1978). Rain-splashed conidia are responsible for secondary infections.

Control

The use of resistant cultivars is the most effective control. Many European cultivars of *B. napus* have resistance or moderate resistance to this disease. Reduced leaf surface wax due to herbicide, surfactant or mechanical damage increases susceptibility to infection due to increased leaf surface wettability. Similarly, cultivar differences in susceptibility and resistance may be partly due to differences in leaf surface waxes. Crop rotation to allow decomposition of oilseed rape residues and planting as far away as possible from the previous year's crops are also helpful. Control may also be achieved by application of fungicides (carbendazim in combination with other fungicides) in the autumn, although results have been inconsistent.

WHITE LEAF SPOT

White leaf spot is caused by the ascomycete *Mycosphaerella capsellae* Inman & Sivanesan (anamorph *Pseudocercosporella capsellae* (Ell. & Ev.) Deighton). The disease has a worldwide distribution and has been severe in some fields in the United Kingdom and France (see Inman et al., 1991b).

Symptoms

Leaf lesions are white to greyish with a faint purple-brown border. Stem and seed pod infections are associated with elongate, grey to purple speckled spots containing the fruiting bodies.

Epidemiology

Little is known about the epidemiology of white leaf spot and what has been described relates principally to the United Kingdom. Ascomata are produced on the current year's crop residues and apparently do not survive overwinter. From October to January, ascomata release ascospores which are aerially dispersed following a period of wetting by rain or dew and serve as the source of primary inoculum. Ascospores permit dispersal potentially over much greater distances than are possible by conidia which are splash-dispersed (Fitt et al., 1989). Since ascomata do not overwinter, secondary spread is dependent on conidia. The latent period of infection varies with temperature and ranges from 18–36 days at 5°C to 6.5–11 days at 20°C (Inman et al., 1991a).

Control

The disease is rarely severe enough to cause yield loss and is managed currently by crop rotation. Avoiding planting in the autumn adjacent to the previous crop will reduce risk of primary infection from ascospores.

WHITE RUST (WHITE BLISTER, STAGHEAD)

White rust is an important disease of *B. rapa* and *B. juncea* in India and Canada. The disease is caused by the oomycete *Albugo candida* (Pers. ex Hook.) Kuntze. It caused considerable yield losses of *B. rapa* in western Canada during the 1970s (see Downey and Rimmer, 1993).

Fig. 5.6. White rust pustules on leaves of oilseed turnip rape (*Brassica rapa*) caused by *Albugo candida*.

Albugo candida occurs as a number of pathotypes characterized by their pathogenicity to species (and genotypes within species) of crucifers. At least ten pathotypes of *A. candida* have been identified and classified, including race 2 on *B. juncea*, (Pound and Williams, 1963), race 7 on *B. rapa* (Petrie, 1988) and race ACcar on *B. carinata* (Liu and Rimmer, 1992).

Albugo candida pathotypes are virulent on many genotypes of their homologous host species. However, some may also cause disease on some genotypes of closely related species (heterologous hosts) (Petrie, 1988). No pathotypes are known whose homologous host is *B. napus*, but some genotypes of this species are susceptible to isolates from *B. rapa* (Fan *et al.*, 1983), *B. juncea* (Verma and Bhowmik, 1989) and *B. oleracea* and *B. carinata* (Liu and Rimmer, 1992). Isolates from *B. rapa* and *B. juncea* in Canada are homothallic, whereas isolates from *B. oleracea* and from *B. carinata* are heterothallic (Liu, 1992).

Symptoms

White-rust-infected plants may be covered with white, chalk-like blisters. On the upper surface of the leaves, the infected areas are bleached and thickened, due to increased host cell division and cell size; on the lower surface, a mycelial mat is formed (Fig. 5.6). The disease also affects stems and flowering parts. Brown and split flecks are frequently found on stems and seed pods in which whitish conidia are formed. Systemic infection of meristems and inflorescence give rise to so-called 'stagheads', which are enlarged and deformed peduncles and seed pods (Fig. 5.7). On mature plants, stagheads are hard, brown and frequently infected with other fungi, e.g. *Alternaria* spp. White rust often occurs in association with downy mildew.

Epidemiology

In western Canada, primary inoculum consists of oospores, the sexual state of

Fig. 5.7. Hypertrophy of inflorescence of *Brassica rapa* caused by *Albugo candida*.

the organism, which may be soil-borne or carried with the seed as a contaminant. Oospores are known to survive for long periods under dry conditions, e.g. as an admixture with seed. Though little information is available concerning their persistence in soil, it is assumed that they may also survive for long periods as dormant, resting spores in this environment. Under moist conditions, oospores germinate to form many motile zoospores. More rarely, *in vitro*, they may germinate to form a germ tube, though the epidemiological significance of this in the field is not known. Infection of young seedlings occurs and zoosporangia which serve as secondary inoculum are formed in white, raised pustules on the abaxial and to a lesser extent on the adaxial surface of cotyledons or leaves, and also on stems or pods. In other climates, where cruciferous hosts grow all the year round, zoosporangia may serve as primary inoculum on oilseed *Brassica* crops.

Zoosporangia may be dispersed by wind or rain splash to cause secondary infections on nearby plants. Secondary infections occur due to the germination of zoosporangia to form six to ten zoospores. These swim to stomata where they encyst. Cysts germinate to form a penetration hypha which enters the stomata and infects adjacent cells. Zoospores germinate well in the presence of free water at temperatures of 10–15°C.

Infection of the flowering meristem results in hypertrophy of the stem and floral parts, often called stagheads, in which little or no seed are produced. Oospores may be formed in any infected plant tissue but are present in large numbers in stagheads. Stagheads break when the crop is harvested and oospores are released to contaminate the harvested seed or are blown out of he combine to contaminate the soil.

Control

European and Canadian cultivars of *B. napus* are resistant to indigenous populations of the pathogen. However, care should be employed when oriental accessions of *B. napus* are used in breeding programmes as some Japanese and Chinese accessions are susceptible to isolates from other *Brassica* oilseed crops. Resistance appears to be controlled by one or more independent dominant genes (Fan *et al.*, 1983; Verma and Bhowmik, 1989). Although isolates from *B. oleracea* are virulent on some genotypes of *B. napus* in greenhouse tests (I.R. Crute, personal communication), there have been no reports of problems with white rust on oilseed rape in the UK where these isolates occur naturally.

Resistance in *B. juncea* to race 2 has been described and appears to be controlled by a single dominant gene (Tiwari *et al.*, 1988). This resistance has been utilized in condiment mustard cultivars, e.g. Domo and Cutlass. A variant of race 2 virulent to the resistance gene in Cutlass has been identified and screening of *B. juncea* accessions for resistance to this new pathotype has not been successful (G. Rakow, Saskatoon, 1994, personal communication). Thus, in a period of about ten years, resistance in this species has been overcome and the outlook for development of cultivars

with resistance to the new pathotype is not optimistic. In *B. rapa* resistance to race 7 was identified in wild accessions from Mexico and incorporated into the Canadian cultivar, Tobin. Because of the outcrossing nature of this species it is very difficult to obtain homozygosity for resistance in a cultivar. With Tobin only about 50–70% of the population was resistant. Using recurrent selection procedures, the cultivar, Reward, was developed at the University of Manitoba with 90% resistant plants. However, since the release of Tobin, a new pathotype of race 7 has emerged virulent on Tobin and may have been responsible for heavy local infections in Alberta (Conn and Tewari, 1991). Resistance to this new pathotype is present at low frequency in the population of Tobin and is now being utilized. It is doubtful, however, that this resistance will be effective for more than a short period of time. Other sources of resistance are urgently needed in this species also.

Oospore contamination of seed may be controlled with metalaxyl though this chemical is not registered for use as a seed treatment of *Brassica* oilseed crops in Canada. If this chemical were to be used, the risk of development of fungicide resistance would have to be considered. Indeed, it is fairly easy to obtain metalaxyl-insensitive isolates of races 2 and 7, and such isolates are currently being used in our laboratory to study the genetics of virulence and other traits in *A. candida*.

Sanitation practices such as crop rotation, control of volunteers and cruciferous weeds and pathogen-free seed should also be employed where the disease is prevalent.

DOWNY MILDEW

Downy mildew is a common disease of *Brassica* oilseed crops throughout the world, and is caused by the oomycete *Peronospora*

Fig. 5.8. Downy mildew on the abaxial side of a leaf of *Brassica napus* caused by *Peronospora parasitica*.

parasitica Pers. ex Pers. Fr. Although seedlings may be severely attacked by the disease, there is little evidence that significant seed yield loss occurs, at least in *B. napus* (see Davies, 1986). There is evidence that pathotypes of *P. parasitica* are adapted to *Brassica* crops as with *A. candida* (Sherriff and Lucas, 1990).

Symptoms

Systemic invasion of seedlings leads to shrivelling and, subsequently, the plant may be killed. On older leaves, the first symptoms are chlorotic areas on the upper surface. When conditions are moist, a profuse, downy growth of white-grey sporangiophores is found on the abaxial surface within the infected area (Fig. 5.8), (in contrast to light leaf spot, where white colonies appear on both leaf surfaces at the margin of the infected area). Eventually, necrotic lesions develop, which may dry out a large part of the leaf.

Epidemiology

Infection is favoured by cool (15 to 20°C) wet weather. Oospores are produced by this organism and are probably responsible for primary infection but conclusive evidence for this has not been demonstrated. Sporangia are produced on the underside of infected leaves and are responsible for secondary spread of the disease. The latent period of infection is very short (3–4 days under favourable environmental conditions). On many crucifers, including *B. rapa* and *B. juncea*, downy mildew is frequently found in association with white rust. Evidence suggests that prior infection with *A. candida* may predispose the infected tissues to infection by *P. parasitica*, perhaps to pathotypes normally avirulent to the host (Bains and Jhooty, 1985).

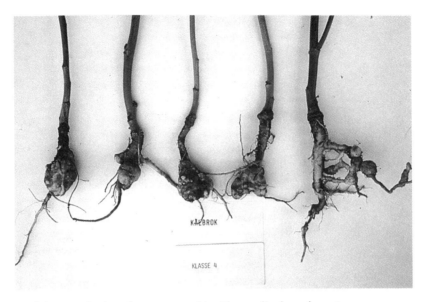

Fig. 5.9. Club root of oilseed rape caused by *Plasmodiophora brassicae*.

Control

Generally, this disease is not considered of sufficient economic importance to warrant control measures unless seedlings are attacked so severely as to reduce plant stands significantly below optimum densities. In regions where this occurs regularly, seed treatment with fungicides containing metalaxyl can be employed. Resistance to pathotypes adapted to *B. napus* in the UK has been identified in the French spring rape cultivar, Cresor (Lucas *et al.*, 1988).

CLUB ROOT

Club root caused by *Plasmodiophora brassicae* Woronin has a broad host range including almost all cruciferous species. The disease is widespread and has been known for over one hundred years as a pathogen on vegetable and fodder crops, especially in transplanted crops of *B. oleracea* which are highly susceptible. Club root problems of *B. napus* and *B. rapa* occur in humid, temperate regions of Europe, especially in Scandinavia, eastern Canada, in parts of New Zealand and Australia (see Buczacki, 1979).

Plasmodiophora brassicae induces gall formation on roots; only rarely are the plants killed, but malfunction of the root system results in stunting and wilt symptoms under even slight water stress.

Symptoms

The growth of club-root-infected plants is stunted compared to healthy plants in the field. When these plants are pulled up, the root system is more or less deformed, due to formation of galls on the taproot and lateral roots (Fig. 5.9). Early in the growing season the swellings are firm and the tissue is white. At maturity, however, the galls turn brown and soft as the tissue decomposes and a large portion of the roots may remain in the ground when the plants are pulled up. Deformed roots may also be caused by the cabbage gall weevil, but, when these gall are cut open, tunnels made by the larvae are visible.

Epidemiology

Plasmodiophora brassicae is a soil-borne pathogen, which survives as resting spores for more than 10 years. The resting spores germinate with zoospores spontaneously or in the presence of host plant exudates (Macfarlane, 1970). The zoospores have flagella and are motile as long as the soil is wet. The life cycle of *P. brassicae* is not completely understood, but the description by Tommerup and Ingram (1971) is accepted. Zoospores locate roots of the host, encyst and penetrate the root hair cells. Inside the root hair, the uninucleate amoeba changes into a multinucleate plasmodium, which then cleaves to form zoosporangia. From these secondary zoospores are released which migrate further into the root tissue and invade cortex cells to form secondary binucleate plasmodia. During this process, the host cells divide and grow rapidly (hyperplasia) resulting in formation of root galls, characteristic of the disease. It is believed that enzymes produced by the pathogen degrade glucobrassicins normally present in the roots, leading to excessive production of auxins, which result in gall formation (Butcher *et al.*, 1974). Secondary zoospores may also be released into the soil, where they reinfect other parts of the root system. At a later stage, secondary plasmodia differentiate into thick-walled resting spores, which remain in the soil after the host is dead and disintegrated.

Several methods for assessment of resting spores in soil have been developed, using sedimentation and microscopy counts (Buczacki and Ockendon, 1978) and fluorescence microscopy (Takahashi and Yamaguchi, 1989).

Control

The selection and breeding of club root resistance has been hampered by a lack of resistance sources and by limited knowledge about physiological specialization in the pathogen. Studies of host resistance and pathogenic variation had been attempted for a long time, when Buczacki *et al.* (1975) suggested the use of a European club root differential (ECD) set comprised of five accessions from each of *B. rapa*, *B. napus* and *B. oleracea*. These differentials are now generally accepted as standards. Studies using the European club root differential set have been summarized (Crute *et al.*, 1983; Toxopeus *et al.*, 1986) and showed that the strongest level of resistance exists in *B. rapa* accessions which only rarely were infected, while *B. napus* accessions exhibited a complete range of disease reactions, suggesting a simple gene-for-gene relationship in these two species; however, the majority of *B. oleracea* accessions were susceptible, suggesting a non-differential reaction in this species. Genetic analysis of the number of genes involved has shown that there are most likely four genes determining resistance to club root in *B. napus* (Gustafsson and Falt, 1985). Recently, an assessment of more than 500 *B. napus* spring and winter lines found three resistant and seven fairly resistant lines (Rod and Havel, 1992).

Crop rotation, application of lime and drainage are standard recommendations for club root control. Much research has been carried out in order to determine the complex relationships between club root severity, inoculum level, soil type, moisture content and the effects of pH and calcium imposed by liming. Hamilton and Crete (1978) showed that increasing soil moisture resulted in higher disease severity and that the disease was much more severe on organic soil compared with mineral soil, thus land drainage partly works by reducing spore germination and zoospore mobility. Several reports have shown a synergistic effect of pH and calcium on club root control (Myers and Campbell, 1985); increasing both factors suppressed pathogen inoculum and enhanced host resistance to infection (Webster and Dixon, 1991). To obtain consistent control by lime application it is important to incorporate finely ground lime thoroughly into the soil, thereby raising the pH above 7.2 and limiting pockets in the soil which have lower pH (Dobson *et al.*, 1983). Application of calcium cyanamide has a similar inhibitory effect on *P. brassicae*, by reducing the inoculum density, and it has several side-effects on other microorganisms and plant growth (Williamson and Dyce, 1989). Growing of club-root-resistant *B. rapa* cultivars may have a sanitary effect on the soil, since part of the resting spore population will germinate without being able to infect and complete the life cycle on these cultivars (Yoshikawa, 1993).

VERTICILLIUM WILT

Verticillium wilt caused by *Verticillium dahliae* Kleb. has been reported to cause severe damage to oilseed rape in southern Sweden (Jönsson, 1978) and in Germany (Krüger, 1989). In France, this pathogen has been implicated, along with *L. maculans* and other agents, as a possible cause of 'pieds sec', a syndrome which results in premature senescence of plants of *B. napus*. The disease has not been reported in Canada though isolates of *V. dahliae* pathogenic on sunflower and potato are common. The disease is not important in China or India.

Verticillium dahliae causes disease on a broad host range of cultivated and wild dicotyledonous plants and a number of cruciferous weed species are hosts. There is evidence that isolates attacking oilseed rape are diploid whereas most common isolates of *V. dahliae* are haploid (Jackson and Heale, 1985). RFLP analysis has also provided evidence that isolates pathogenic to brassicas are distinct from strains attacking other hosts (Okoli *et al.*, 1994).

Symptoms

Symptoms caused by verticillium wilt are related to infection of the vascular system and are most noticeable after stem elongation. Only parts of the vascular system become infected, often resulting in wilting of one side of the stem (Fig. 5.10) or, occasionally, one half of a leaf. The affected tissue turns yellow to brown in contrast to the green healthy tissue on the opposite side. Later the stems have a grey, frayed appearance because masses of microsclerotia (Fig. 5.11) have formed underneath the epidermis, causing it to peel off.

Epidemiology

Verticillium wilt is a monocyclic disease and the source of primary inoculum is microsclerotia which are produced on infected plants and may survive in soil for many years (Berg, 1984). Microsclerotia germinate myceliogenically and infection of the roots occurs directly or via wounds. Invasion of the vascular tissues follows where the pathogen grows and sporulates. Upward movement of the pathogen is due to conidia carried in the transpiration stream. The pathogen grows out of the vascular tissues when the plant senesces to form microsclerotia both outside and within the stem. Seed infection may also occur when pods are infected.

Control

Control is mainly by crop rotation and control of volunteer and weed hosts. Most damage has been reported when rape has been grown in an area for many years or when rape is planted frequently in the rota-

Fig. 5.10. Stem of oilseed rape with partial browning caused by verticillium wilt.

tion. No commercially grown cultivars are resistant but Jönsson (1978) reported that variation for resistance to verticillium wilt occurred in both summer and winter *B. napus* lines, but that *B. rapa* lines were susceptible. The disease cannot be controlled by fungicides.

SEEDLING BLIGHTS, DAMPING-OFF AND ROOT ROTS

Seedling blight and root rot in *Brassica* oilseed crops are frequently complex in aetiology. A number of fungal species can cause seedling blight and some of these pathogens are frequently associated with the seed. Pathogen species involved include *Rhizoctonia solani* (anastomosis groups (AG) 2-1 and 4), *Fusarium* spp., *Pythium* spp., *Phytophthora megasperma*, *Alternaria* spp., *L. maculans* and others. Interaction with root-feeding insects may occur.

Severe root rot, known as brown girdling root rot, is the major cause of crop loss of *B. rapa* in northern Alberta in Canada (Martens *et al.*, 1988). *Rhizoctonia solani* AG 2-1 appears to be the most commonly isolated pathogen from infected plants (Gugel *et al.*, 1987).

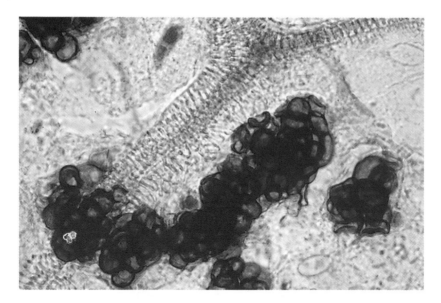

Fig. 5.11. Microsclerotia of *Verticillium dahliae* (×400) from oilseed rape.

Symptoms

Damping-off and seedling blights occur shortly after emergence when the seedlings collapse following rapid infection at the ground level. In some cases the root collar is darkly coloured and shrunken (Fig. 5.12). Brown girdling root rot usually appears at the onset of flowering, when brown lesions develop on the taproot 5–7 cm below the soil surface and on the main lateral roots. The infected areas turn dark brown, and girdling of the taproot occurs and may continue upwards but never into the stem. The plants remain turgid as long as there is a partially functional root system, but often the plants ripen prematurely.

Control

Seed treatments can alleviate seedling blight and damping-off problems but have rarely been shown to affect final yields. *Brassica* oilseed crops can compensate for large variations in stand and, unless stands are severely reduced, yields are often unaffected. Pathogens such as *R. solani* are soil-inhabiting so inoculum reduction by crop rotation may be helpful.

Bacterial Diseases

The major bacterial diseases of crucifers are black rot, soft rot, bacterial leaf spot and aster yellows. None of these diseases are considered very important on *Brassica* oilseed crops. Black rot, caused by *Xanthomonas campestris* pv. *campestris*, and soft rot, caused by *Erwinia carotovora*, are very significant diseases of vegetable brassicas in warm humid climates. Bacterial leaf spot, caused by *Pseudomonas syringae* pv. *maculicola*, has been reported to be a problem on oilseed *B. rapa* in the Tibetan plateau of China (Stinson *et al.*, 1982). It is favoured by very cold (<10°C) and wet environmental conditions. It may also occur on *B. juncea* and Indian oilseed types of this species have been reported to be susceptible (K. Helms, Canberra, 1994, personal

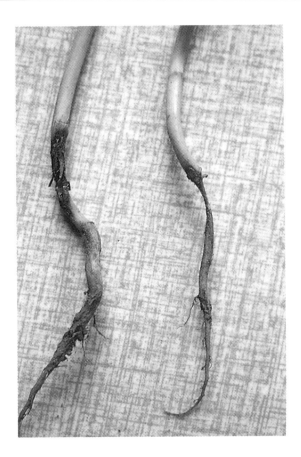

Fig. 5.12. Seedling blight of oilseed rape caused by one or more soil-borne fungal pathogens.

communication). Aster yellows is a common disease in western Canada but disease incidence is usually low (<1%). Phyllody of flower parts occurs and infected plants produce no seed. The disease is caused by a mycoplasma-like organism. It is transmitted by leafhoppers and the incidence of disease is related to the time of migration and abundance of these vectors into western Canada from the USA. No control measures are feasible for this disease.

VIRUS DISEASES

At least 11 distinct viruses are known to infect *Brassica* crops. These are: beet western yellows, broccoli necrotic yellows, cauliflower mosaic, cucumber mosaic, radish mosaic, radish yellow edge, ribgrass mosaic, turnip crinkle, turnip mosaic, turnip rosette and turnip yellow mosaic. *Brassica* oilseed crops are affected by a number of these virus diseases and, in some areas of China, considerable damage may occur, especially where oilseed rape is grown adjacent to areas of intensive market-garden production of vegetable brassicas. The most important virus disease in China is mosaic caused, primarily, by turnip mosaic virus (TuMV). Cucumber mosaic virus (CuMV) and ribgrass mosaic virus may also be involved (R. Stace-Smith, Vancouver, 1994, personal communication). Affected seedlings may be killed; surviving infected seedlings are stunted, produce little or no seed, while necrotic rings and spots occur on stems and leaves.

Turnip mosaic virus has a wide host range including a number of weed species (Shattuck, 1992). The virus is transmitted by aphids in a non-persistent manner. In China, transmission of the virus occurs on oilseed rape in the autumn when the seedlings are still in the seedbeds or shortly after transplanting. Severe damage to seedlings can occur if seed is planted at the time of the aphid vector migration. Numerous strains of TuMV occur and it has been very difficult to find resistance or tolerance to the most common strains present in China. Shattuck (1992) has reviewed the biology, epidemiology and control of TuMV. CuMV is also non-persistently transmitted by aphids. Beet western yellows virus has been reported to cause some damage in the United Kingdom (Gilligan et al., 1980).

Symptoms of turnip mosaic virus and cucumber mosaic virus

Diagnosis of virus-infected brassicas in the field is impossible as symptoms of many virus infections are very similar and mixed infections may occur. Diagnosis is dependent on serological tests of infected plant material. Turnip mosaic virus alone causes a characteristic irregular mosaic and crinkling effect of the leaves, and infected plants are also smaller. Cucumber mosaic virus is recognized by a lightening of the vein tissue, a heavy mosaic pattern and deformed leaves.

Control

In China, where oilseed rape is mainly transplanted from seedbeds to fields, severe disease can sometimes be avoided by delaying planting or by protection of the seedbeds with insecticides. Resistance is considered to be the only real hope for con-

sistently effective control and programmes for development of resistance in oilseed rape have been initiated. *Brassica rapa* is highly susceptible and *B. napus* is generally less susceptible.

SUMMARY AND CONCLUSIONS

Diseases continue to threaten stable production of *Brassica* oilseed crops in many areas of the world. In Canada the main diseases are sclerotinia stem rot and stem canker. In China, stem rot and viral diseases cause substantial yield losses. Apart from the widespread occurrence of stem rot and stem canker, a variety of other diseases of local importance are found in Europe, including verticillium wilt in Sweden and Germany, club root in Scandinavian countries and light leaf spot in the UK. In India, alternaria blight is especially devastating in northern regions where *B. rapa* is grown, while simultaneous infection of white rust and downy mildew is very common on both *B. rapa* and *B. juncea*. Stem canker was an enormous problem for the production of oilseed rape in Australia, until resistant cultivars were developed.

Utilization of disease resistance has had appreciable economic and environmental value. Within *B. napus*, there is a considerable variation for resistance to stem canker and light leaf spot but only a limited number of genotypes have some level of resistance to alternaria and stem rot. Breeding for club root and verticillium resistance is carried out in parts of the world where these diseases are a problem.

Brassica juncea is naturally highly resistant to stem canker, but few sources of resistance to white rust (race 2) are available and, if this species is more widely grown as an oilseed in drier areas of the world, white rust may well become more important. Because of the outcrossing nature of *B. rapa* it is more difficult to obtain resistant cultivars. Good resistance to white rust has been obtained in Canadian cultivars but resistance to other diseases such as stem canker and alternaria is lacking.

It seems that genetic variation for resistance to the most important diseases can be found in *B. napus* but utilization is dependent on the level of resistance available, the ease with which it can be transferred to and selected in adapted cultivars, and the economic importance of the disease. For some diseases, intraspecific crossing or genetic transformation may be the only recourse to incorporate resistance into susceptible species.

A few diseases are primarily controlled by chemical application. In India, where alternaria blight is a serious problem, both seed treatment and foliar application are used routinely; in other parts of the world, Europe for example, sprays for alternaria are occasionally used in years with warm and wet weather conditions. Sclerotinia stem rot occurs sporadically in many countries and can cause severe yield losses. Thus considerable efforts have been applied to the development of prediction systems to assist farmers in making appropriate management decisions. Farmers require sophisticated management skills to produce *Brassica* oilseed crops in the modern world. Management of diseases is only one but an important component of this management process. Continued improvement of the technology for disease management and the producer's skills in utilizing new developments will be essential in the future for both increased production and environmental protection.

REFERENCES

Ansari, N.A., Khan, M.W. and Muheet, A. (1988) Effect of *Alternaria* blight on oil content of rape seed and mustard. *Current Science, India* 57, 1023–1024.

Bains, S.S. and Jhooty, J.S. (1985) Association of *Peronospora parasitica* with *Albugo candida* on *Brassica juncea* leaves. *Phytopathologiche Zeitschrift* 112, 28–31.

Ballinger, D.J., Salisbury, P.A., Dennis, J.L., Kollmorgen, J.F. and Potter, T.D. (1988a) Evaluation of fungicides, applied at sowing, for control of blackleg in rapeseed. *Australian Journal of Experimental Agriculture* 28, 511–515.

Ballinger, D.J., Salisbury, P.A., Kollmorgen, J.F., Potter, T.D. and Coventry, D.R. (1988b) Evaluation of rates of flutriafol for control of blackleg in rapeseed. *Australian Journal of Experimental Agriculture* 28, 517–519.

Ballinger, D.J., Salisbury, P.A. and Kadkol, G.P. (1991) Race variation in *Leptosphaeria maculans* and the implications for resistance breeding in Australia. In: McGregor, D.I. (ed.) *Proceedings of the Eighth International Rapeseed Congress*, Saskatoon, Canada. Organizing Committee, Saskatoon, pp. 226–231.

Berg, G. (1984) *Verticillium dahliae*: an investigation of root injuries in winter-oilseed-crops in Sweden. *Groupe Consultatif International de Recherche sur la Colza Bulletin* 1, 19.

Bhowmik, T.P. and Munde, P.N. (1987) Identification of resistance in rapeseed and mustard against *Alternaria brassicae* (Berk.) Sacc. and some resistant sources. *Beitrage zur Tropishen Landwirtschaft und Veterinärmedizin* 25, 49–53.

Brun, H., Plessis, J. and Renard, M. (1987a) Resistance of some crucifers to *Alternaria brassicae* (Berk.) Sacc. In: *Proceedings of the Seventh International Rapeseed Conference*, Poznan, Poland. The Plant Breeding and Acclimatization Institute, Poznan, pp. 1222–1227.

Brun, H., Tribodet, M., Renard, M., Plessis, J. and Tanguy, X. (1987b) A field study of rapeseed (*Brassica napus*) resistance to *Sclerotinia sclerotiorum*. In: *Proceedings of the Seventh International Rapeseed Congress*, Poznan, Poland. The Plant Breeding and Acclimatization Institute, Poznan, pp. 1216–1221.

Buczacki, S.T. (1979) *Plasmodiophora brassicae*. Commonwealth Mycological Institute, Descriptions of Pathogenic Fungi and Bacteria No. 621.

Buczacki, S.T. and Ockendon, J.G. (1978) A method for the extraction and enumeration of resting spores of *Plasmodiophora brassicae* from infested soil. *Annals of Applied Biology* 88, 363–367.

Buczacki, S.T., Toxopeus, H., Johnston, P., Dixon, T.D. and Hobolt, L.A. (1975) Study of physiologic specialisation in *Plasmodiophora brassicae*: proposals for attempted rationalisation through an international approach. *Transactions of the British Mycological Society* 65, 295–303.

Butcher, D.N., El-Tigani, S. and Ingram, D.S. (1974) The role of indol glucosinolates in the club root disease of the Cruciferae. *Physiological Plant Pathology* 4, 127–140.

Chahal, A.S. and Kang, M.S. (1979) Some aspects of seed borne infection of *Alternaria brassicae* in rape and mustard cultivars in the Punjab. *Indian Journal of Mycology and Plant Pathology* 9, 51–55.

Changsri, W. and Weber, G.F., (1963) Three *Alternaria* species pathogenic on certain cultivated crucifers. *Phytopathology* 53, 643–648.

Conn, K.L. and Tewari, J.P. (1991) Survey of *Alternaria* blackspot and sclerotinia stem rot of canola in Central Alberta in 1991. *Canadian Plant Disease Survey* 71, 96–97.

Conn, K.L., Tewari, J.P. and Dahiya, J.S. (1988) Resistance to *Alternaria brassicae* and phytoalexin-elicitation in rapeseed and other crucifers. *Plant Science* 56, 21–25.

Conn, K.L., Tewari, J.P. and Awasthi, R.P. (1990) A disease assessment key for *Alternaria* blackspot in rapeseed and mustard. *Canadian Plant Disease Survey* 70, 19–22.

Courtice, G.R.M., Ilott, T.W., Ingram, D.S., Johnstone, K., Sawczye, M.C. and Skidmore, D.I. (1988) *Pyrenopeziza brassicae*, cause of light leaf spot of *Brassica* spp. *Advances in Plant Pathology* 6, 225–231.

Cox, T.W., Swash, D. and Paviot, J. (1981) The control of *Alternaria brassicae* and *Sclerotinia sclerotiorum* on oilseed rape with iprodione. In: *Proceedings of the 1981 British Crop Protection Conference – Pests and Diseases*, Brighton, UK, pp. 513–520.

Crute, I.R., Phelps, K., Barnes, A., Buczacki, S.T. and Crisp, P. (1983) The relationship between genotypes of three *Brassica* species and collections of *Plasmodiophora brassicae*. *Plant Pathology* 32, 405–420.

Daebeler, F. and Amelung, D. (1988) Auftreten und Bedeutung der *Alternaria*-rapsschwarze im Winter-raps. *Nachrichtenblatt fur den Pflanzenschutz in der DDR* 42, 196–199.

Davies, J.M.L. (1986) Diseases of oilseed rape. In: Scarisbrick, D.H. and Daniels, R.W. (eds) *Oilseed Rape*. Collins, London, pp. 195–236.

Degenhardt, K.J., Skoropad, W.P. and Kondra, Z.P. (1974) Effects of *Alternaria* blackspot on yield, oil content and protein content of rapeseed. *Canadian Journal of Plant Science* 54, 795–799.

Degenhardt, K.J., Petrie, G.A. and Morrall, R.A.A. (1982) Effects of temperature on spore germination and infection of rapeseed by *Alternaria brassicae, A. brassicicola* and *A. raphani*. *Canadian Journal of Plant Pathology* 4, 115–118.

Dobson, R.L., Gabrielson, R.L., Baker, A.S. and Bennett, L. (1983) Effects of lime particle size and distribution and fertilizer formulation on clubroot disease caused by *Plasmodiophora brassicae*. *Plant Disease* 67, 50–52.

Downey, R.K. and Rimmer, S.R. (1993) Agronomic improvement in oilseed brassicas. *Advances in Agronomy* 50, 1–66.

Evans, E.J. and Gladders, P.G. (1981) Diseases of winter oilseed rape and their control. East and South-East England. In: *Proceedings of the 1981 British Crop Protection Conference – Pests and Diseases*, Brighton, UK, pp. 505–512.

Evans, E.J., Gladders, P.G., Davies, J.M.L., Ellerton, D.R., Hardwick, N.V., Hawkins, J.H., Jones, D.R. and Simkin, M.B. (1984) Current status of diseases and disease control of winter oilseed rape in England. *Aspects of Applied Biology* 6, 323–334.

Fan, Z., Rimmer, S.R. and Stefansson, B.R. (1983) Inheritance of resistance to *Albugo candida* in rape (*Brassica napus* L). *Canadian Journal of Genetics and Cytolology* 25, 420–424.

Ferreira, M.E., Teutonica, R. and Osborn, T.C. (1993) Mapping trait loci in oilseed *Brassica*. In: McGregor, D.I. (ed.) *Proceedings of the Eighth Crucifer Genetics Workshop*, Saskatoon, Canada. Organizing Committee, Saskatoon, p. 10 (Abstr.).

Fitt, B.D.L., Dhua, U., Lacey, M.E. and McCartney, H.A. (1989) Effects of leaf age and position on splash dispersal of *Pseudocercosporella capsellae*, cause of white leaf spot on oilseed rape. *Aspects of Applied Biology* 23, 457–464.

Fu, S., Lu, Z., Chen, Y., Qi, C., Pu, H. and Zhang, J. (1990) Inheritance of apetalous character in *Brassica napus* L and its potential in breeding. In: Chu, Q.R. and Fang, G.H. (eds)

Proceedings of the Symposium China International Rapeseed Sciences, Shanghai, China. Shanghai Scientific and Technical Publishers, Shanghai, pp. 7–8.

Gilligan, C.A., Pechan, P.M., Day, R. and Hill, S.A. (1980) Beet western yellows virus on oilseed rape (*Brassica napus* L.). *Plant Pathology* 29, 53.

Godoy, G., Steadman, J.R., Dickman, M.B. and Dam, R. (1990) Use of mutants to demonstrate the role of oxalic acid in pathogenicity of *Sclerotinia sclerotiorum* on *Phaseolus vulgaris*. *Physiological and Molecular Plant Pathology* 37, 179–191.

Gretenkort, M.A. and Ingram, D.S. (1993) A comparison of the disease reaction of stems and detached leaves of soil and *in vitro* grown plants and regenerants of oilseed rape to *Leptosphaeria maculans* and protocols for selection for novel disease resistance. *Journal of Phytopathology* 137, 89–104.

Gugel, R.K. and Petrie, G.A. (1992) History, occurrence, impact, and control of blackleg of rapeseed. *Canadian Journal of Plant Pathology* 14, 36–45.

Gugel, R.K., Yitbarek, S.M., Verma, P.R., Morrall, R.A.A. and Sadasivaiah, R.S. (1987) Etiology of the rhizoctonia root rot complex of canola in the Peace River region of Alberta. *Canadian Journal of Plant Pathology* 9, 119–128.

Gugel, R.K., Seguin-Swartz, G. and Petrie, G.A. (1990) Pathogenicity of three isolates of *Leptosphaeria maculans* on *Brassica* species and other crucifers. *Canadian Journal of Plant Pathology* 12, 75–82.

Gustafsson, M. and Falt, A.-S. (1985) Genetic studies on resistance to clubroot in *Brassica napus*. *Annals of Applied Biology* 108, 409–415.

Hall, R. (1992) Epidemiology of blackleg of oilseed rape. *Canadian Journal of Plant Pathology* 14, 46–55.

Hamilton, H.A. and Crete, R. (1978) Influence of soil moisture, soil pH, and liming sources on the incidence of clubroot, the germination and growth of cabbage produced in mineral and organic soils under controlled conditions. *Canadian Journal of Plant Science* 58, 45–53.

Hammond, K.E. and Lewis, B.G. (1986) The timing and sequence of events leading to stem canker disease in populations of *Brassica napus* var. *oleifera* in the field. *Plant Pathology* 35, 551–564.

Hammond, K.E., Lewis, B.G. and Musa, T.M. (1985) A systemic pathway in the infection of oilseed rape plants by *Leptosphaeria maculans*. *Plant Pathology* 34, 557–565.

Humpherson-Jones, F.M. (1983) The occurrence of *Alternaria brassicicola, Alternaria brassicae* and *Leptosphaeria maculans* in brassica seed crops in south-east England between 1976 and 1980. *Plant Pathology* 32, 33–39.

Humpherson-Jones, F.M. (1985) The incidence of *Alternaria* spp. and *Leptosphaeria maculans* in commercial brassica seed in the United Kingdom. *Plant Pathology* 34, 385–390.

Humpherson-Jones, F.M. (1989) Survival of *Alternaria brassicae* and *Alternaria brassicicola* on crop debris of oilseed rape and cabbage. *Annals of Applied Biology* 115, 45–50.

Inman, A.J., Fitt, B.D.L. and Evans, R.L. (1991a) Aspects of the biology and epidemiology of *Pseudocercosporella capsellae*, the cause of white leaf spot on oilseed rape. In: McGregor, D.I. (ed.) *Proceedings of the Eighth International Rapeseed Congress*, Saskatoon, Canada. Organizing Committee, Saskatoon, pp. 1652–1657.

Inman, A.J., Sivanesan, A., Fitt, B.D.L. and Evans, R.L. (1991b) The biology of *Mycosphaerella capsellae* sp. nov., the teleomorph of *Pseudocercosporella capsellae*, cause of white leaf spot of oilseed rape. *Mycological Research* 95, 1334–1342.

Jackson, C.W. and Heale, J.B. (1985) Relationship between DNA content and spore volume in sixteen isolates of *Verticillium lecanii* and two new diploids of *V. dahliae* (= *V. dahliae* var. *longisporum* Stark). *Journal of General Microbiology* 131, 3229–3236.

Jönsson, R., (1978) Resistensförädling mot kransmögel, *Verticillium dahliae*, i raps och rybs. *Sveriges Utsädesförenings Tidskrit* 88, 165–177.

Keri, M. (1991) Resistance of *Brassica juncea* Czern. & Coss. to blackleg disease caused by *Leptosphaeria maculans* Ces. & de Not. MSc Thesis, University of Manitoba, Winnipeg, Canada, 123 pp.

Koch, E., Badawy, H.M.A. and Hoppe, H.H. (1989) Differences between aggressive and non-aggressive single spore lines of *Leptosphaeria maculans* in cultural characteristics and phytotoxin production. *Journal of Phytopathology* 124, 52–62.

Koch, E., Song, K., Osborn, T.C. and Williams, P.H. (1991) Relationship between pathogenicity and phylogeny based on restriction fragment polymorphism in *Leptosphaeria maculans*. *Molecular Plant–Microbe Interactions* 4, 341–349.

Kohle, H. and Hoffmann, G.M. (1988) Untersuchungen zur Physiologie des *Alternaria*-Befalls von Raps. *Zeitschrift für Pflanzenkrankheiten und Pflanzenschutz* 96, 225–238.

Kohli, Y., Morrall, R.A.A., Anderson, J.B. and Kohn, L.M. (1992) Local and trans-Canadian clonal distribution of *Sclerotinia sclerotiorum* on canola. *Phytopathology* 82, 875–880.

Krüger, W. (1975) Die Beeinflussung der Apothezien – und Ascosporen Entwicklung des Rapskrebserregers *Sclerotinia sclerotiorum* (Lib.) de Bary durch Umweltfaktoren. *Zeitschrift für Pflanzenkrankheiten und Pflanzenschutz* 82, 101–108.

Krüger, W. (1989) Untersuchungen zur Verbreitung von *Verticillium dahliae* Kleb. und anderen Krankheits- und Schaderregern bei raps in der Bundesrepublik Deutschland. *Nachrichtenblatt Deutsche Pflanzenschutzdienst* 41, 49–56.

Krüger, W. and Stoltenberg, F. (1983) Bekämpfung von Rapskranheiten. II. Massnahmen zur Befallsverringerung von *Sclerotinia sclerotiorum* unter Berüksichtigung 8 konomischer Faktoren. *Phytopathologiche Zeitschrift* 108, 114–126.

Lamarque, C. (1983) Conditions climatiques qui favorisent le processus naturel de la contamination du colza par le *Sclerotinia sclerotiorum*. In: *Proceedings of the Sixth International Rapeseed Congress*, Paris, France, pp. 903–907.

Landry, B.S. (1993) Towards map-based cloning in *Brassica*. *Proceedings of the Eighth Crucifer Genetics Workshop*, Saskatoon. National Research Council, Canada, p. 11 (Abstr.).

Liu, C.Q., Du, D.Z., Zou, C.S. and Huang, Y.J. (1990) Initial studies on tolerance to *Sclerotinia sclerotiorun* (Lib.) de Bary in *Brassica napus* L. In: Chu, Q.R. and Fang, G.H. (eds) *Proceedings of the Symposium China International Rapeseed Sciences*, Shanghai, China. Shanghai Scientific and Technical Publishers, Shanghai, pp. 70–71.

Liu, Q. (1992) A methodology for genetic studies with *Albugo candida*. PhD Thesis, University of Manitoba, Winnipeg, Canada, 96 pp.

Liu, Q. and Rimmer, S.R. (1992) Inheritance of resistance in *Brassica napus* to an Ethiopian isolate of *Albugo candida* from *Brassica carinata*. *Canadian Journal of Plant Pathology* 13, 197–201.

Louvet, J. and Billotte, J.M. (1964) Influence des facteurs climatiques sur les infections du Colza par l'*Alternaria brassicae* et conséquences pour la lutte. *Annales Epiphytologie (Paris)* 15, 229–243.

Lucas, J.A., Crute, I.R., Sherriff, C. and Gordon, P.L. (1988) The identification of a gene for race-specific resistance to *Peronospora parasitica* (downy mildew) in *Brassica napus* var. *oleifera* (oilseed rape). *Plant Pathology* 37, 538–545.

McCartney, H.A. and Lacey, M.E. (1991) Spread of light leaf spot (*Pyrenopeziza brassicae*) in oilseed rape crops in the United Kingdom. In: McGregor, D.I. (ed.) *Proceedings of the Eighth International Rapeseed Congress*, Saskatoon, Canada. Organizing Committee, Saskatoon, pp. 454–459.

McCartney, H.A., Lacey, M.E. and Rawlinson, C.J. (1987) The perfect stage of *Pyrenopeziza brassicae* on oilseed rape and its agricultural implications. In: *Proceedings of the Seventh International Rapeseed Congress*, Poznan, Poland. The Plant Breeding and Acclimatization Institute, Poznan, pp. 1262–1267.

Macfarlane, I. (1970) Germination of resting spores of *Plasmodiophora brassicae*. *Transactions of the British Mycological Society* 55, 97–112.

McGee, D.C. and Emmett, R.W. (1977) Black leg (*Leptosphaeria maculans* (Desm.) Ces. et de Not.) of rapeseed in Victoria: crop losses and factors which affect disease severity. *Australian Journal of Agricultural Research* 28, 47–51.

McNabb, W., van den Berg, C.G.J. and Rimmer, S.R. (1993) Comparison of inoculation methods for selection of plants resistant to *Leptosphaeria maculans* in *Brassica napus*. *Canadian Journal of Plant Science* 73, 1199–1207.

Maddock, S.E., Ingram, D.S. and Gilligan, C.A. (1981) Resistance of cultivated brassicas to *Pyrenopeziza brassicae*. *Transactions of the British Mycological Society* 76, 371–382.

Marchegay, P., Thorin, N. and Schiavon, M. (1990) Effets des facteurs climatiques sur l'émission aérienne des spores d'*Alternaria brassicae* (Berk.) Sacc. et sur l'épidémiologie de l'alternariose dans un culture de colza. *Agronomie* 10, 831–839.

Martens, J.W., Seaman, W.L. and Atkinson, T.G. (eds) (1988) *Diseases of Field Crops in Canada*, revised edn. Canadian Phytopathological Society, Harrow, Ontario, Canada, 160 pp.

Maude, R.B., Humpherson-Jones, F.M. and Shuring, C.G. (1984) Treatments to control *Phoma* and *Alternaria* infections of brassica seeds. *Plant Pathology* 33, 525–535.

Mengistu, A., Rimmer, S.R. and Williams, P.H. (1993) Protocols for *in vitro* sporulation, ascospore release, sexual mating, and fertility in crosses of *Leptosphaeria maculans*. *Plant Disease* 77, 538–540.

Morrall, R.A.A., Dueck, J. and Verma, P.R. (1984) Yield losses due to sclerotinia stem rot in western Canada. *Canadian Journal of Plant Pathology* 6, 265 (Abstr.).

Morrall, R.A.A., Turkington, T.K., Kaminski, D.A., Thomson, J.R., Gugel, R.K. and Rude, S.V. (1991) Forecasting sclerotinia stem rot of spring rapeseed by petal testing. In: McGregor, D.I. (ed.) *Proceedings of the Eighth International Rapeseed Congress*, Saskatoon, Canada. Organizing Committee, Saskatoon, pp. 483–488.

Mridha, M.A.U. and Wheeler, B.J. (1987) Spore liberation and disease development in winter oilseed rape by *Alternaria brassicae* under field conditions. In: *Proceedings of the Seventh International Rapeseed Congress*, Poznan, Poland. The Plant Breeding and Acclimatization Institute, Poznan, p. 253 (Abstr.).

Myers, D. F. and Campbell, R.N. (1985) Lime and the control of clubroot of crucifers: effect of pH, calcium, magnesium and their interaction. *Phytopathology* 75, 670–673.

Nathaniels, N.Q.R. and Taylor, G.S. (1983) Latent infection of winter oilseed rape by *Leptosphaeria maculans*. *Plant Pathology* 32, 23–31.

Neergaard, P. (1977) *Seed Pathology* (Vols 1 & 2). Wiley, New York.

Okoli, C.A.N., Carder, J.H. and Barbara, D.J. (1994) Restriction fragment length polymorphisms (RFLPs) and the relationships of some host-adapted isolates of *Verticillium dahliae*. *Plant Pathology* 43, 33–40.

Pauls, V.H. and Rawlinson, C.J. (1992) *Diseases and Pests of Rape*. Th. Mann Gelsenkirchen-Buer, Germany.

Pedras, M.S.C. and Seguin-Swartz, G. (1992) The blackleg fungus: phytotoxins and phytoalexins. *Canadian Journal of Plant Pathology* 14, 67–75.

Petrie, G.A. (1988) Races of *Albugo candida* (white rust and staghead) on cultivated Cruciferae in Saskatchewan. *Canadian Journal of Plant Pathology* 10, 142–150.

Petrie, G.A. and Lewis, P.A. (1985) Sexual compatibility of isolates of the rapeseed blackleg fungus *Leptosphaeria maculans* from Canada, Australia and England. *Canadian Journal of Plant Pathology* 7, 253–255.

Petrie, G.A., Mortensen, K. and Dueck, J. (1985) Blackleg and other diseases of rapeseed in Saskatchewan, 1978 to 1981. *Canadian Plant Disease Survey* 65, 35–41.

Porter, A.J.R., Morton, A.M., Kiddle, G., Doughty, K.J. and Wallsgrove, R.M. (1991) Variation in the glucosinolate content of oilseed rape (*Brassica napus*) leaves. I. Effect of leaf age and position. *Annals of Applied Biology* 118, 461–467.

Pound, G.S. and Williams, P.H. (1963) Biological races of *Albugo candida*. *Phytopathology* 53, 1146–1149.

Rawlinson, C.J., Sutton, B.C. and Muthyalu, G. (1978) Taxonomy and biology of *Pyrenopeziza brassicae* sp. nov. (*Cylindrosporium concentricum*), a pathogen of winter oilseed rape (*Brassica napus* ssp. *oleifera*). *Transactions of the British Mycological Society* 71, 425–439.

Rawlinson, C.J., Muthyalu, G. and Cayley, G.R. (1984) Fungicide effects on light leaf spot, crop growth and yield of winter oil-seed rape. *Journal of Agricultural Science, Cambridge* 103, 613–628.

Rimmer, S.R. and van den Berg, C.G.J. (1992) Resistance of oilseed *Brassica* species to blackleg caused by *Leptosphaeria maculans*. *Canadian Journal of Plant Pathology* 14, 56–66.

Rod, J. and Havel, J. (1992) Assessment of oilseed rape resistance to clubroot (*Plasmodiophora brassicae*). *Test of Agrochemicals and Cultivars 13, Annals of Applied Biology* (Supplement) 120, 106–107.

Saharan, G.S. and Kadian, A.K. (1983) Analysis of components of horizontal resistance in rape seed and mustard cultivars against *Alternaria brassicae*. *Indian Phytopathology* 36, 503–507.

Sedun, F.S., Seguin-Swartz, G. and Rakow, G.F.W. (1989) Genetic variation in reaction to *Sclerotinia sclerotiorum* in *Brassica* species. *Canadian Journal of Plant Science* 69, 229–232.

Shattuck, V.I. (1992) The biology, epidemiology, and control of turnip mosaic virus. *Horticultural Review* 14, 199–238.

Sherriff, C. and Lucas, J.A. (1990) The host range of isolates of downy mildew, *Peronospora parasitica*, from *Brassica* crop species. *Plant Pathology* 39, 77–91.

Singh, A. and Bhowmik, T.P. (1985) Persistence and efficacy of some common fungicide against *Alternaria brassicae*, the causal agent of leaf blight of rapeseed and mustard. *Indian Phytopathology* 38, 35–38.

Smith, I.M., Dunez, J., Lelliot, R.A., Phillips, D.H. and Archer, S.A. (eds) (1988) *European Handbook of Plant Diseases*. Blackwell Scientific Publications, Oxford, 583 pp.

Stankova, J. (1972) Varietal variability of winter rape with regard to its inclination to dark leaf spot and the factors influencing the development of this disease. *Rostlinna vyroba (Praha)* 18, 625–630.

Stinson, P., Downey, R.K. and Rimmer, S.R. (1982) *Rapeseed Improvement in China*. Report to International Development Research Centre, Ottawa, Canada, 77 pp.

Stovold, G.E., Mailer, R.J. and Francis, A. (1987) Seed-borne levels, chemical seed treatment and effect on seed quality following a severe outbreak of *Alternaria brassicae* on rapeseed in New South Wales. *Plant Protection Quarterly* 2, 128–131.

Takahashi, K. and Yamaguchi, T. (1989) Assessment of pathogenicity of resting spores of *Plasmodiophora brassicae* in soil by fluorescence microscopy. *Annals of the Phytopathological Society of Japan* 55, 621–628.

Tewari, J.P. and Skoropad, W.P. (1976) Relationship between epicuticular wax and blackspot caused by *Alternaria brassicae* in three lines of rapeseed. *Canadian Journal of Plant Science* 56, 781–785.

Thomas, P. (1984) *Canola Growers' Manual*. Canola Council of Canada, Winnipeg, Canada.

Thompson, C., Dunwell, J.M., Johnstone, C.E., Lay, V., Ray, J., Schmitt, M., Watson, H. and Nisbet, G. (1993) Degradation of oxalic acid by transgenic canola plants expressing oxalate oxidase. In: *Proceedings of the Eighth Crucifer Genetics Workshop*, Saskatoon. National Research Council, Canada, p. 31 (Abstr.).

Tiwari, A.S., Petrie G.A. and Downey, R.K. (1988) Inheritance of resistance to *Albugo candida* race 2 in mustard [*Brassica juncea* (L.) Czern.]. *Canadian Journal of Plant Science* 68, 297–300.

Tommerup, I.C. and Ingram, D.S. (1971) The life-cycle of *Plasmodiophora brassicae* Woron. in *Brassica* tissue and in intact roots. *New Phytologist* 70, 327–332.

Toxopeus, H., Dixon, G.R. and Mattusch, P. (1986) Physiological specialization in *Plasmodiophora brassicae*: an analysis by international experimentation. *Transactions of the British Mycological Society* 87, 279–287.

Tripathi, N.N. and Kaushik, C. (1984) Studies on the survival of *Alternaria brassicae*, the causal organism of leaf spot of rapeseed and mustard. *Madras Agriculture Journal* 71, 237–241.

Tsuneda, A. and Skoropad, W.P. (1977) Formation of microsclerotia and chlamydospores from conidia of *Alternaria brassicae*. *Canadian Journal of Botany* 55, 1276–1281.

Verma, U. and Bhowmik, T.P. (1989) Inheritance of resistance to a *Brassica juncea* pathotype of *Albugo candida* in *B. napus*. *Canadian Journal of Plant Pathology* 11, 443–444.

Webster, M.A. and Dixon, G.R. (1991) Calcium, pH and inoculum concentration influencing colonization by *Plasmodiophora brassicae*. *Mycological Research* 95, 64–73.

Williams, P.H. (1992) Biology of *Leptosphaeria maculans*. *Canadian Journal of Plant Pathology* 14, 30–35.

Williamson, C.J. and Dyce, P.E. (1989) The effect of calcium cyanamide on the reaction of swede cultivars to populations of *Plasmodiophora brassicae*. *Plant Pathology* 38, 230–238.

Wood, P.McR. and Barbetti, M.J. (1977) The role of seed infection in the spread of blackleg of rape in Western Australia. *Australian Journal of Experimental Agriculture and Animal Husbandry* 17, 1040–1044.

Yoshikawa, H. (1993) Studies on breeding of clubroot resistance in cole crops. *Bulletin of the National Research Institute of Vegetables, Ornamental Plants and Tea*, Series A, 7, 1–165.

Xi, K., Kutcher, H.R., Westcott, N.D., Morrall, R.A.A. and Rimmer, S.R. (1991a) Effect of seed treatment and fertilizer coated with flutriafol on blackleg of canola (oilseed rape) in western Canada. *Canadian Journal of Plant Pathology* 13, 336–346.

Xi, K., Morrall, R.A.A., Gugel, R.K. and Verma, P.R. (1991b) Latent infection in relation to the epidemiology of blackleg of spring rapeseed. *Canadian Journal of Plant Pathology* 13, 321–331.

6 Insect Pests

B. Ekbom
Swedish University of Agricultural Sciences, Uppsala

INTRODUCTION

One of the most important limiting factors for production of *Brassica* oilseeds is the complex of insect pests associated with these plants. The necessary insecticide input to secure acceptable production levels may not only be high in any given year but is often essential every season. The insect pests of *Brassica* oilseeds are primarily crucifer specialists. One particular chemical characteristic of the genus *Brassica* and the family *Brassicaceae* is the presence of a group of secondary compounds, the glucosinolates. Most economically important herbivores of *Brassica* oilseeds use the glucosinolates or their fission products as attractants, feeding stimuli or oviposition stimuli. These same secondary compounds act as feeding deterrents or toxins for herbivores not adapted to crucifers.

The recent and dramatic increase in arable areas devoted to the production of *Brassica* oilseeds has provided crucifer specialists with an enormous resource for feeding and reproduction (Lamb, 1989). Presence of spring (annual) and winter (biennial) varieties of the crop in the same area will enhance the temporal availability of host plants for insect pests. While recent plant breeding has drastically reduced the glucosinolate levels in the seed, these secondary compounds still occur in sufficient quantities to act as attractants and stimuli for specialist insects.

An excellent and recent review of entomology in *Brassica* oilseeds focused on interactions between the plant and insect pests, mainly the effects the pests have on plant growth and production (Lamb, 1989). The aim of this review is to describe principal insect pests, their relative importance and measures for their control.

THE PESTS

Table 6.1 lists most insect pests mentioned in the literature. Some of these pests are referred to only briefly in the literature and are unlikely to occur as major pests except in isolated cases. In the following section information about pests that are of major economic importance in one or more geographical areas is summarized. Some groups described are rarely economically important; however, they occur in many areas and could potentially cause problems.

141

Table 6.1. Pests of *Brassica* oilseed crops.

Pest	Common name	Plant part attacked	Area	References
Hemiptera				
Lygus elisus		Bud, flower, pod, seed	Canada	Butts and Lamb (1991)
Lygus lineolaris		Bud, flower, pod, seed	Canada	Butts and Lamb (1991)
Lygus rugulipennis	Tarnished plant bug	Seedlings	Europe	Wallenhammar (1992)
Homoptera				
Brevicoryne brassicae	Cabbage aphid	Leaf, stem, bud, flower, pod	Europe New Zealand Australia	Lamb (1989)
Lipaphis erysimi	Mustard aphid	Leaf, stem, bud, flower, pod	India	Sekhon and Åhman (1992)
Myzus persicae	Green peach aphid	Leaf, stem, bud, flower, pod	Europe	Hill *et al.* (1991)
Coleoptera				
Meligethes aeneus	Pollen beetle	Bud, flower	Europe	Lamb (1989)
Phyllotreta cruciferae	Flea beetle	Seedlings	N. America India	Lamb (1989)
Phyllotreta striolata	Flea beetle	Seedlings	N. America Europe	Lamb (1989)
Phyllotreta undulata	Small striped flea beetle	Seedlings	Europe	Lamb (1989)
Psylliodes chrysocephala	Cabbage stem flea beetle	Stem, leaf	Europe	Lamb (1989)
Entomoscelis americana	Red turnip beetle	Leaf	Canada	Lamb (1989)
Ceutorhynchus assimilis	Cabbage seed weevil	Pod	Europe	Lamb (1989)
Ceutorhynchus quadridens	Cabbage stem weevil	Stem	Europe	Lamb (1989)
Ceutorhynchus napi	Cabbage weevil	Stem	Europe	Lerin (1993)
Ceutorhynchus sulcicollis		Stem	Europe	Björkman (1975)
Baris coerulescens	Crucifer weevil	Root	Europe	Lerin and Koubaiti (1991)
Lepidoptera				
Pieris brassicae	Large cabbage white butterfly	Leaf	India	Sekhon and Åhman (1992)
Mamestra configurata	Cabbage moth	Leaf, pod	Canada	Lamb (1989)
Plutella xyostella	Diamondback moth	Leaf	India, Canada New Zealand Australia	Sekhon and Åhman (1992)
Diacrisia obliqua		Leaf	India	Sekhon and Åhman (1992)
Agrotis flammatra	Cutworm (dart moth)	Leaf	India	Sekhon and Åhman (1992)
Diptera				
Dasineura brassicae	Brassica pod midge	Pod	Europe	Lamb (1989)

Table 6.1. (contd).

Pest	Common name	Plant part attacked	Area	References
Delia radicum	Cabbage root fly	Root	Canada, Europe	Lamb (1989)
Phytomyza horticola	Chrysanthemum leaf miner	Leaf	India	Sekhon and Åhman (1992)
Hymenoptera				
Athalia proxima		Leaf	India	Sekhon and Åhman (1992)
Athalia rosae	Turnip sawfly	Leaf	Europe	Mörner and Ekbom (1987)
Thysanoptera				
Several species of thrips	Thrips	Leaf, stem, bud, flower, pod	Canada	Burgess and Weegar (1988)

Source: Lamb (1989).

The order of presentation for the pests follows, where possible, the growth stages of the plants. The first group injures seedlings, while the last damages pods and seeds. The damage caused by each group is presented along with a short summary of the insects' biology. Finally control measures are outlined, including methods available at present as well as strategies with future potential.

Flea beetles, *Phyllotreta* spp.

Several species of *Phyllotreta* spp. occur as pests in different areas of the world (Lamb, 1984; Ekbom, 1991; Medhin and Mulatu, 1992). All the species cause the same type of damage. The adult beetles invade newly germinated spring *Brassica* oilseed fields and feed upon the small plants. Damage is typically described as feeding holes, often distributed fairly evenly on both cotyledons and leaves. Entire seedlings may be eaten when beetle populations are high. Damage is particularly severe when the weather is warm and dry. Resulting damage is expressed not only as lower seed yields but also lower oil content in the seed and delayed maturation, causing higher chlorophyll levels. *Phyllotreta* flea beetles are

of economic importance only on spring varieties of *Brassica* oilseeds.

Flea beetles overwinter as adults away from the fields, in wooded areas, and fly to the fields as the spring weather warms up. After a period of feeding they lay eggs close to the base of the plants. Larvae feed on the roots of the plants. Larval feeding has not been shown to cause any significant additional damage to the crop. Pupation occurs in the soil and the new generation emerges in the late summer. The new adults feed on the leaves of maturing summer *Brassica* oilseeds and will also, later in the season, move to winter crops if they are available. Feeding damage in the late summer and autumn rarely causes yield losses.

It is difficult to predict exactly when flea beetle attacks will take place. Enormous damage can be caused during a very short time period when weather conditions are favourable for the beetles. For this reason insecticide seed dressings or in-furrow granules are the most important control measures (Weiss *et al.*, 1991). Spraying can be an alternative if the crop is monitored intensively. Attempts at biological control using an introduced parasitoid in Canada (Wylie, 1988) and studies on natural parasitism in Sweden (Ekbom, 1991) have shown that control by natural enemies is not a viable option in managed systems. Recently progress has been made in identifying chemical traits that stimulate and deter feeding behaviour by the flea beetles (Bodnaryk and Lamb, 1991; Palaniswamy and Lamb, 1992). Host plant resistance may become an important alternative to chemical control in the future (Peng *et al.*, 1992).

Cabbage stem flea beetle, *Psylliodes chrysocephala*

This pest (Fig. 6.1) is dependent on biennial, or winter, forms of *Brassica* oilseeds. The insect uses the crop as its overwintering habitat. Adults will chew holes in leaves, as do flea beetles, but it is the larvae which cause economically important damage. Newly hatched larvae eat their way into leaf stalks and later into the base and stem of the plant. The injuries can provide points of entry for pathogens. When infestations are heavy the growth point of the plant may be damaged or the poor condition of the plant will cause the plant to die during the winter. The cabbage stem flea beetle is one of the most important pests of winter varieties of oilseed in Europe.

During the late summer adults of the cabbage stem flea beetles move into newly sown winter *Brassica* oilseeds. Eggs are laid in the soil near the base of the plants. Larvae overwinter in leaf stalks and stems. If the winter is mild, oviposition and larval development may continue until early spring, and the larvae will also resume their development when the plants begin growing again. Adults emerge in late spring or early summer. The summer is generally spent in aestivation.

Psylliodes chrysocephala does not occur in economically important numbers every year. The pest is monitored in Austria (Berger, 1991), the United Kingdom (Lane and Walters, 1993) and Sweden (Nilsson, 1990). Larval densities are estimated by sampling plants in the autumn and sometimes in the very early spring. If the average number of larvae exceeds two per plant then seed dressing for crops in the following season is recommended. Spraying with a pyrethroid in the autumn may be necessary if beetles are very numerous. In Sweden a spray recommendation is given when the threshold of two beetles per metre row is passed. In England three to five larvae per plant or about 60% scarred leaf stalks is the treatment threshold. There are at present no alternatives to chemical control. Recently studies have been initiated on the host plant spectrum of the stem beetle (Bartlet and Williams, 1991). Insect pathogenic fungi are also being studied as a possible biological control agent against *P. chrysocephala* (Butt *et al.*, 1992).

Stem weevils

The larvae of several weevils belonging to the genus *Ceutorhynchus* mine the stems of plants. Direct damage occurs when the larvae are very numerous. The stems will split (Lerin, 1993) or weaken, causing plants to bend and break (Björkman, 1975). Indirect damage can occur at lower population levels, as injuries caused by oviposition or the small larvae boring into the stems can serve as entry points for several plant-pathogenic fungi (Broschewitz *et al.*, 1993). These weevils are almost exclusively pests in winter forms of *Brassica* oilseeds.

Both *Ceutorhynchus napi* and *Ceutorhynchus quadridens* occur as important pests on the European continent (Berger, 1991; Büchs, 1993; Lerin, 1993). Their biology is rather similar. The weevils overwinter as adults and move into the winter oilseed crop in the early spring. Oviposition takes place on leaf petioles or at the base of the plant. On hatching, the larvae eat their way into the stem where they feed until shortly before pupation. Last-instar larvae leave the stems and pupate in the soil. Another stem weevil, *Ceutorhynchus*

Fig. 6.1. Cabbage stem flea beetle, *Psylliodes chrysocephala*.

sulcicollis, with a slightly different biology is sometimes a problem in Sweden (Björkman, 1975). This weevil overwinters in the crop and migration takes place in early autumn; oviposition can begin earlier as no spring migration is necessary.

Stem weevils are mainly monitored using yellow water traps (Büchs, 1993); however, other trapping methods used directly in the crop canopy are available (Björkman, 1975). Using monitoring methods it is possible to determine the actual need for an insecticide application as well as the optimal timing for an application. Action thresholds based on yellow water traps (Berger, 1991) or other methods (Björkman, 1975) are available. Effective control can be achieved by spraying in the early spring, two to three weeks after detection of activity peaks (Büchs, 1993).

Aphids

Three species of aphids, namely *Lipaphis erysimi* (Fig. 6.2), *Brevicoryne brassicae* and *Myzus persicae*, can be of economic importance on *Brassica* oilseeds. *Lipaphis erysimi* and *B. brassicae* are crucifer specialists while *M. persicae* is extremely polyphagous. *Lipaphis erysimi* is the most devastating insect pest in India (Sekhon and Åhman, 1992) where it can cause losses of up to 50% in seed yield. *Brevicoryne brassicae*, which is a common pest on vegetable brassicas in temperate zones, appears only occasionally as a pest in oilseed crops. All three aphid species cause direct damage when population levels are high, and suction feeding on the plant can decrease plant vigour and cause deformation. *Myzus persicae* is also important as a virus vector of beet western yellows virus (BWYV) (Hill *et al.*, 1991; Lane and Walters, 1993). Aphids occur on all varieties and species of *Brassica* oilseeds.

In many areas of the world aphids reproduce parthenogenetically and almost continuously during the entire year. Sexual reproduction is recorded from more northern areas. Winter mortality of eggs and parthenogenetic forms is important in limiting the pest potential of aphids in temperate zones. When aphid populations become large and colonies are crowded, winged forms of the aphids appear and migrate to new host plants. Aphid reproduction rates are strongly regulated by

Fig. 6.2. Aphid, *Lipaphis erysimi*.

temperature and many generations are possible in warm climates.

Chemical control of aphids is possible. However, with repeated insecticide use against a multivoltine species the risk of development of pesticide resistance by the pest is high. Recent research on aphid control has been focused on breeding for aphid resistance (Sekhon and Åhman, 1992). Resistance or tolerance has been found in several *Brassica* spp. and cultivars as well as in related crucifers. Mechanisms of resistance include both plant morphological and anatomical characteristics as well as biochemical factors in plants.

Lepidoptera

There are a variety of lepidopteran pests which occur sporadically on *Brassica* oilseeds. The larvae feed mainly on the leaves

of the plant. Defoliation at the beginning of flowering can result in economic loss; however, leaf loss is less significant as the plant matures (Pechan and Morgan, 1985). Crucifer specialists such as *Plutella xylostella* and *Pieris brassicae* are more important pests on vegetable crops where leaf damage directly affects the market value of the crop. *Mamestra configurata*, which is a polyphagous species, is a major defoliator in Canada where the larvae will move from the leaves to pods and stems (Bracken, 1987).

Lepidopteran pests will have one or more generations per year, depending on the climate area. Crucifer specialists are attracted to the plants by volatiles (Palaniswamy *et al.*, 1986) and lay their eggs on the leaves. Newly hatched larvae begin feeding on the leaves and move within the plant as their feeding resource

Fig. 6.3. Pollen beetle, *Meligethes aeneus*.

is depleted. Pupation takes place in the soil.

Because leaf damage is generally marginal for *Brassica* oilseed production, control of lepidopteran pests is seldom necessary. Chemical control can be used if a large influx of insects is detected, but extremely high population levels are necessary before economic damage occurs.

Pollen beetles, *Meligethes* spp.

Pollen beetles (Fig. 6.3) have long been the most important insect pest of *Brassica* oilseeds in Scandinavia (Nilsson, 1987; Hokkanen, 1989). They have been of lesser importance in the UK and on the European continent but have become more significant with the higher proportion of spring crops grown in recent years. The adult beetles feed on pollen, often from open flowers. However, they also feed directly on buds, eating through the bud to the developing flower parts. Eggs are laid inside buds and larvae will eat stamens during the first part of their development. At high population levels larvae will also attack the stem of the plant. When pollen beetle numbers are low, damage may be confined to bud and flower abortion, but as plants may abort up to 50 or 60% of their buds, without insect attack, small or moderate loss of buds and flowers due to insect damage will not necessarily severely affect yield (Williams and Free, 1979). Development of more side shoots often compensates for serious damage to the main shoot. Damaged plants will have an extended flowering period and maturation will be uneven and delayed. Fewer pods per stalk and blind stalks will also occur. Both annual and biennial varieties of the crop are attacked by the beetles.

Fig. 6.4. Seed weevil, *Ceuthorhynchus assimilis.*

Pollen beetles overwinter as adults in wooded and protected areas, and feed on pollen from early-flowering plants in the spring. They migrate to the crop at bud formation in *Brassica* oilseeds for further feeding, mating and oviposition. Eggs are laid first in winter varieties and, later, as the optimal stage for oviposition is past, the beetles will seek out spring varieties. Oviposition can occur during a two-month period. Development from eggs to last-instar larvae takes about three to four weeks depending on temperature. Final-instar larvae fall to the ground and pupate. The new generation of pollen beetles emerges from the middle of July into August. The new adults feed on pollen in autumn-flowering plants before migrating to over-wintering sites.

Chemical control of pollen beetles is often necessary to ensure yields. Economical thresholds for both winter and spring varieties are in use in Scandinavia (Nilsson, 1987) and the UK (Lane and Walters, 1993). More beetles are tolerated by winter varieties and the threshold also increases as the plants mature. Pyrethroids are the most commonly used chemicals. Potential for alternative control measures does exist. Several parasitoids are common and cultivation methods, such as avoiding ploughing, can increase parasitoid numbers (Nilsson, 1985). An insect-pathogenic protozoon (Lipa and Hokkanen, 1992) and the fungus, *Beuveria bassiana*, are being studied as potential control methods (Vänninen *et al.*, 1989). The impact of natural enemies is, however, probably very marginal. This is most likely due to the intensive use of insecticides against pollen beetles which can also destroy many potential biological control agents.

Seed weevil, *Ceuthorhynchus assimilis*, and pod midge, *Dasineura brassicae*

The seed weevil (Fig. 6.4) occurs in both Europe and North America. The pod midge is, on the other hand, restricted to

Europe. Both pests lay their eggs in pods and their larvae feed on developing seeds. Presence of pod midge larvae in pods will also cause pods to open before seed development is complete, resulting in loss of pod contents. Adult seed weevils will not only lay eggs in pods but feed on the pods as well. The feeding damage, in the form of small holes in the pod, creates openings which are used by the midge for oviposition (Åhman, 1987). While the pod weevil is a pest of economic importance even when only biennial varieties are available, pod midge importance is enhanced by close proximity of spring varieties (Mörner and Ekbom, 1987).

The pod weevil has one generation a year. Adults leave overwintering sites in the late spring at flowering and feed on the crop. Eggs are laid in newly formed pods and larval development in the pod takes 3 to 5 weeks. Larvae eat their way out of the pods and drop to the soil for pupation. The pod midge overwinters as larvae in cocoons in the soil. In late spring the larvae move to the soil surface and pupate. Adults emerge and fly to host plants for oviposition. Midges do not feed as adults and live only a few days. Eggs are laid in pods and development from egg to last-instar larvae takes about 2 weeks. Pods split and larvae fall to the ground for pupation. The pod midge usually completes three generations per year. The second generation usually coincides with pod formation in spring varieties. Because of the short oviposition period for each generation successful reproduction is extremely dependent on warm, calm weather. The midge is unable to build up large populations unless both winter and spring varieties are available.

Economic thresholds and monitoring methods exist for recommendations for insecticide treatment of these pod pests (Williams, 1990; Hansen, 1993). In the UK at least two weevils per plant at flowering in both winter and spring oilseed crops are required to justify economic treatment

(Lane and Walters, 1993). The thresholds are lowered to one weevil per plant when the pod midge is known to be a problem. Earlier spraying against pollen beetles will most likely have a detrimental effect on pod pests as well. Studies are currently being carried out on oviposition-deterring pheromones (ODP) of *C. assimilis* (Ferguson and Williams, 1989). This pheromone may have potential in an integrated pest management programme against oilseed pests. Pheromones of the pod midge have also been studied (Williams and Martin, 1986); these might aid in monitoring programmes by using baited traps. There are also several parasitoids of pod midge and the potential of these enemies has been evaluated in simulation models (Axelsen, 1992).

Miridae, *Lygus* spp.

Mirid bugs (Miridae, *Lygus* spp.) have recently been recorded as pests of *Brassica* oilseeds. This is interesting as well as disturbing because insects of *Lygus* spp. are generalist herbivores. They are found on a wide variety of plant families. It has been suggested that plant breeding which changes the chemical composition of the crop could affect insect responses (Lamb, 1989). However, no radical changes in the pest complex have been observed following the introduction of double-low varieties. *Lygus elisus* and *Lygus lineolaris* have been reported as pests of economic significance in Canada (Butts and Lamb, 1991). These species feed on pods and decrease yields by sucking nutrients. In Sweden *Lygus rugulipennis* has been noted to cause damage to seedlings of summer *Brassica* oilseeds. The bugs suck at the growth point and kill the main shoot (Varis, 1972). Plants have a typical U-shaped form (Wallenhammar, 1992). The economic importance is not clear, but up to 80 or 90% of the plants in a field with symptoms have been recorded.

Mirid bugs overwinter as adults, probably in field verges and nearby vegetation. Warm weather is advantageous for reproduction. It is not known to what extent mirid bugs actually use the oilseed crop as reproductive sites, but large numbers of nymphs can sometimes be found in the crop, indicating that oviposition has occurred in the crop. Both the overwintering adults and the new generation can cause damage.

Little information is available, at present, as to economic thresholds and recommendations for chemical control. Large populations are presumably necessary to cause economically significant damage and therefore some type of monitoring should be used. These mirid bugs are very polyphagous and it may be difficult to find alternative methods for control.

PERSPECTIVES FOR THE FUTURE OF PEST CONTROL IN *BRASSICA* OILSEEDS

A high level of insecticide input for insect pest control in *Brassica* oilseeds is at present an unavoidable prerequisite for ensuring high yields and good quality. The potential of natural enemies is probably severely hampered by intensive use of chemicals. In a relatively isolated area of Finland where insecticides had never been used in oilseed crops it was found that the pollen beetle, *Meligethes* spp., was kept at low population levels by parasitism rates of about 80% (Hokkanen, 1989). Infection in pollen beetles by the protozoan, *Nosema meligethi*, is fairly common in Eastern European countries while a survey of thousands of beetles in Western European countries has not revealed a single infection (Lipa and Hokkanen, 1992). Protozoan infections weaken the beetles and are probably removed from the beetle population by insecticides. Cultivation methods may also be detrimental to parasitoids of the insect pests (Nilsson, 1985). Many parasitoids overwinter in the soil and intensive cultivation will destroy large

numbers of them. The action of polyphagous predators may potentially be improved by changing agricultural practices such as intercropping. Intercropping may also serve to deflect or confuse insect pests and reduce feeding and reproductive activities. Another alternative could be to use trap crops to draw invading insects away from the crop (Hokkanen *et al.*, 1986; Büchi, 1990; Ekbom and Borg, 1993).

Where only one form of *Brassica* oilseed is grown (annual or biennial) pest problems are generally somewhat smaller. Some pests, such as pod midge, are dependent on the presence of both forms to build up high population levels. The occurrence of annual and biennial crops in close proximity to each other seems to enhance pest problems (Mörner and Ekbom, 1987). The temporal availability of host plants is extended to cover virtually the entire year. Host plants at a suitable development stage will be accessible for a protracted time period for reproduction of pollen beetles and the pod weevil. Regional planning may be important in the future to avoid inflating pest problems.

One of the most promising alternatives to chemical control for the future is the possibility of developing varieties with resistance or tolerance to insect pests. Chemical aspects of insect–plant interactions of crucifers and their herbivores have been studied since Verschaeffelt (1911) first discovered that larvae of *Pieris brassicae* could be stimulated to feed on non-hosts by applying extracts from cabbage. A large body of knowledge concerning both stimulants and deterrents for a wide range of crucifer specialist insects is accumulating. Specific host plant resistance mechanisms for the cabbage stem beetle (Bartlet and Williams, 1991), *Phyllotreta* flea beetles (Bodnaryk and Lamb, 1991; Palaniswamy and Lamb, 1992; Peng *et al.*, 1992), pollen beetles (Charpentier, 1985; Åhman, 1993), pod midges (Åhman, 1993) and aphids (Sekhon and Åhman, 1992) are being investigated. Important chemical components which determine cabbage root fly oviposition choices (Roessingh *et al.*, 1992) are currently being identified. As the chemical puzzle is solved for individual insects, synthesis of this knowledge may lead to some general conclusions about chemical-mediated insect–plant relationships. It may be possible to change the plants so that insect attack and damage can be minimized. Recent advances in biotechnological methods of plant breeding, especially for *Brassica* oilseeds, may allow entomologists to specify oilseed plants designed for insect resistance.

These prospects are, however, still distant possibilities. For the time being insecticides will remain the basis of insect pest control in *Brassica* oilseeds. Important research for determining economic thresholds and developing good monitoring methods for the insect pest complex is continuing. Rational use of chemicals within integrated pest management systems can result in effective control of pests and minimize harmful insecticide side-effects to the environment.

REFERENCES

Åhman, I. (1987) Oviposition site characteristics of *Dasineura brassicae*. *Zeitschrift für Angewandte Entomologie* 104, 85–91.

Åhman, I. (1993) A search for resistance to insects in spring oilseed rape. *IOBC/WPRS Bulletin* 16(5), 36–46.

Axelsen, J. (1992) The population dynamics and mortalities of the pod gall midge (*Dasineura brassicae*) in winter rape and spring rape (*Brassica napus*) in Denmark. *Journal of Applied Entomology* 114, 463–471.

Bartlet, E. and Williams, I.H. (1991) Factors restricting the feeding of the cabbage stem flea beetle. *Entomologia Experimentalis et Applicata* 60, 233–238.

Berger, H.K. (1991) Oilseed rape in Austria – pest problems. *IOBC/WPRS Bulletin* 14(6), 14–20.

Björkman, I. (1975) Trials with trapping methods and control of *Ceutorhynchus sulcicollis*. *Växtskyddsnotiser* 39(6), 129–134.

Bodnaryk, R.P. and Lamb, R.J. (1991) Mechanisms of resistance to the flea beetle *Phyllotreta cruciferae* in mustard seedlings, *Sinapis alba*. *Canadian Journal of Plant Science* 71, 13–20.

Bracken, G.K. (1987) Relation between pod damage caused by larvae of bertha armyworm, *Mamestra configurata* and yield loss shelling and seed quality in Canola. *Canadian Entomologist* 119, 365–369.

Broschewitz, B., Steinbach, P. and Goltermann, S. (1993) The effect of insect larval damage upon the attack of winter oilseed rape by *Phoma lingam* and *Botrytis cinerea*. *Gesunde Pflanzen* 45(3), 106–110.

Büchi, R. (1990) Investigations on the use of turnip rape as trap plant to control oilseed rape pests. *IOBC/WPRS Bulletin* 13(4), 32–39.

Büchs, W. (1993) Investigations on the occurrence of pest insects in oilseed rape as a basis for the development of action thresholds, concepts for prognosis and strategies for the reduction of the input of insecticides. *IOBC/WPRS Bulletin* 16(9), 216–234.

Burgess, L. and Weegar, H.H. (1988) Thrips in canola crops in Saskatchewan. *Canadian Entomologist* 120, 815–819.

Butt, T.M., Barrisever, M., Drummond, J., Schuler, T.H., Tillemans, F.T. and Wilding, N. (1992) Pathogenicity of the enomogenous, hyphomycete fungus, *Metarhizium anisopliae* against the chrysomelid beetles *Psylliodes chrysocephala* and *Phaedon cochleariae*. *Biocontrol Science and Technology* 2(4), 327–334.

Butts, R.A. and Lamb, R.J. (1991) Seasonal abundance of three *Lygus* species in oilseed rape and alfalfa in Alberta. *Journal of Economic Entomology* 84(2), 450–456.

Charpentier, R. (1985) Host plant selection by the pollen beetle *Meligethes aeneus*. *Entomologia Experimentalis et Applicata* 38, 277–285.

Ekbom, B.S. (1991) The effect of abiotic and biotic factors on flea beetle populations in spring rape in Sweden. *IOBC/WPRS Bulletin* 14(6), 21–27.

Ekbom, B. and Borg, A. (1993) Predators, *Meligethes* and *Phyllotreta* in unsprayed oilseed rape. *IOBC/WPRS Bulletin* 16(9), 175–184.

Ferguson, A.W. and Williams, I.H. (1989) Oviposition-deterring pheromone of the cabbage seed weevil (*Ceutorhynchus assimilis*). *Aspects of Applied Biology* 23, 339–342.

Hansen, L.M. (1993) Monitoring of *Brassica* pod midge (*Dasineura brassicae*). *IOBC/WPRS Bulletin* 16(9), 139–143.

Hill, S.A., Lane, A. and Hardwick, N.V. (1991) The incidence and importance of beet western yellows virus in oilseed rape. *IOBC/WPRS Bulletin* 14(6), 36–45.

Hokkanen, H.M.T. (1989) Biological and agrotechnical control of the rape blossom beetle *Meligethes aeneus*. *Acta Entomologica Fennica* 53, 25–29.

Hokkanen, H., Granlund, H., Husberg, G.-B. and Markkula, M. (1986) Trap crops used successfully to control *Meligethes aeneus*, the rape blossom beetle. *Annales Entomologici Fennici* 52, 115–120.

Lamb, R.J. (1984) Effect of flea beetles, *Phyllotreta* spp., on the survival, growth, seed yield and quality of canola, rape and yellow mustard. *Canadian Entomologist* 116, 269–280.

Lamb, R.J. (1989) Entomology of oilseed *Brassica* crops. *Annual Review of Entomology* 34, 211–229.

Lane, A. and Walters, K.F.A. (1993) Recent incidence and cost effective control of pests of oilseed rape in England and Wales. *IOBC/WPRS Bulletin* 16(9), 185–192.

Lerin, J. (1993) Influence of the growth rate of oilseed rape on the splitting of the stem after an attack of *Ceutorhynchus napi*. *IOBC/WPRS Bulletin* 16(9), 160–163.

Lerin, J. and Koubaiti, K. (1991) Biology of *Baris coerulescens*, a pest of winter rape. *IOBC/WPRS Bulletin* 14(6), 28–35.

Lipa, J.J. and Hokkanen, H.M.T. (1992) *Nosema meligethi* in populations of *Meligethes* spp. in Europe. *Biocontrol Science and Technology* 2, 119–125.

Medhin, T.G. and Mulatu B. (1992) Insect pests of noug, linseed and *Brassica*. In: *First National Oilseeds Workshop*, Addis Ababa, Ethiopia, 3–5 December 1991. Institute of Agricultural Research, Addis Ababa, pp. 174–177.

Mörner, J. and Ekbom, B. (1987) Cultivation structure and occurrence of pest in oilseed crops. *Växtskyddsrapporter, Jordburk* 46.

Nilsson, C. (1985) Impact of ploughing in emergence of pollen beetle parasitoids after hibernation. *Zeitschrift für Angewandte Entomologie* 100, 302–308.

Nilsson C. (1987) Yield losses in summer rape caused by pollen beetles (*Meligethes* spp.). *Swedish Journal of Agricultural Research* 17, 105–111.

Nilsson, C. (1990) Yield losses in winter rape caused by cabbage stem flea beetle larvae (*Psylliodes chrysocephala*). *IOBC/WPRS Bulletin* 13(4), 53–56.

Palaniswamy, P. and Lamb, R.J. (1992) Host preferences of the flea beetles *Phyllotreta cruciferae* and *P. striolata* for crucifer seedlings. *Journal of Economic Entomology* 85(3), 743–752.

Palaniswamy, P., Gillott, C. and Slater, G. (1986) Attraction of diamondback moths, *Plutella xylostella* by volatile compounds of Canola, white mustard and faba bean. *Canadian Entomologist* 118, 1279–1285.

Pechan, P.A. and Morgan, D.G. (1985) Defoliation and its effects on pod and seed development in oilseed rape (*Brassica napus*). *Journal of Experimental Botany* 36, 458–468.

Peng, C., Weiss, M.J. and Anderson, M.D. (1992) Flea beetle response, feeding and longevity on oilseed rape and Crambe. *Environmental Entomology* 21(3), 604–609.

Roessingh, P., Städler, E., Fenwick, G.R., Lewis, J.A., Kvist Nielsen, J., Hurter, J. and Ramp, T. (1992) Oviposition and tarsal chemoreceptors of the cabbage root fly are stimulated by glucosinolates and host plant extracts. *Entomologia Experimentalis et Applicata* 65, 267–282.

Sekhon, B.S. and Åhman, I. (1992) Insect resistance with special reference to mustard aphid. In: Labana, K.S., Banga, S.S. and Banga S.K. (eds) *Breeding Oilseed Brassicas*. Narosa Publishing House, New Delhi, India, pp. 206–221.

Vänninen, I., Husberg, G.-B. and Hokkanen, H.M.T. (1989) Occurrence of entomopathogenic fungi and entomoparasitic nematodes in cultivated soils in Finland. *Acta Entomologica Fennica* 53, 65–71.

Varis, A.-L. (1972) The biology of *Lygus rugulipennis* and the damage caused by this species to sugar beet. *Annales Agriculturae Fenniae* 11, 1–56.

Verschaeffelt, E. (1911) The causes determining the selection of food in some herbivorous insects. *Proceedings of the Academy of Sciences, Amsterdam* 13, 536–542.

Wallenhammar, A.-C. (1992) Damage by mirids in summer oilseed rape. *Svensk Frötidning* 11, 20–21.

Weiss, M.J., McLeod, P., Schatz, B.G. and Hanson, B.K. (1991) Potential for insecticidal management of flea beetle on Canola. *Journal of Economic Entomology* 84(5), 1597–1603.

Williams, I.H. (1990) Monitoring *Dasineura brassicae* by means of pheromone traps. *IOBC/WPRS Bulletin* 13(4), 40–45.

Williams, I. and Free, J.B. (1979) Compensation of oilseed rape (*Brassica napus*) plants after damage to their buds and pods. *Journal of Agricultural Science, Cambridge* 92, 53–59.

Williams, I.H. and Martin, A.P. (1986) Evidence for a female sex pheromone in the brassica pod midge *Dasineura brassicae*. *Physiological Entomology* 11, 353–356.

Wylie, H.G. (1988) Release in Manitoba, Canada of *Townesilitus bicolor*, a European parasite of *Phyllotreta* spp. *Entomophaga* 33, 25–32.

7 Plant Breeding

G.C. Buzza

Pacific Seeds Pty Ltd, Toowoomba, Queensland, Australia

INTRODUCTION

Plant breeding is not a science or an art, but a technology. The aim of plant breeding is to produce a product and not, as in the case of science, to advance knowledge. This product is a cultivar or hybrid, which is a more or less uniform population of the species used to grow a crop. Of course plant breeders use science to help develop new variation but it is inaccurate to call a breeder a scientist. Much of the world's plant breeding was done before the development of modern science and in fact goes back to the origin of agriculture. The first plant breeders were the people who decided, after gathering the harvest, to retain seed to sow the next season's crop. By selecting seed they were changing the population of the original species to make it more suitable for their purposes. Because of this it has been claimed that plant breeding may be the world's second oldest profession!

To develop a product, the plant breeder draws on many scientific disciplines, with all of which he or she should be familiar. Apart from the obvious fields of genetics, agronomy and biometrics, the breeder should have a knowledge of plant pathology, plant physiology, biochemistry and other fields that relate to the relevant crop species. But the real skill of the breeder comes from organizing the resources, which are always limited, to produce products efficiently.

This chapter aims to show how this process has led to the improvement of *Brassica* oilseed cultivars and where we are heading.

ORIGIN OF *BRASSICA* OILSEED

The different *Brassica* species and relatives in closely related genera are described in Chapter 1. Nowadays three species dominate, *Brassica napus, Brassica rapa* and *Brassica juncea*, but other species, *Brassica carinata* and *Brassica tournefortii*, are still grown. A related species, *Eruca sativa*, is widely cultivated on the Indian subcontinent and *Sinapis alba* may be eventually converted into an oilseed crop. Reference is made to the much quoted 'triangle of U' (U, 1935) in Chapter 1, which neatly describes the relationship between the genomes of six important

Brassica species. In recent years the use of molecular markers has given new insight into the relationships of species in the subtribe *Brassicae* (Song and Osborn, 1991; Warwick and Black, 1993), which indicates that the taxonomy of *Brassica* spp. and related genera needs revision.

Pre-Mendelian Breeding

Plant breeding started with the ancient beginnings of agriculture, first by selecting which species to cultivate and then by domesticating them. Presumably the diploid species, *Brassica oleracea*, *B. nigra* and *B. rapa*, were cultivated first and the amphidiploid species, *B. carinata*, *B. juncea* and *B. napus*, arose spontaneously where the two parent species were cultivated together. This is similar to the development of hexaploid wheat. As with the domestication of other species there has been selection for seed size and seed retention. There may also have been selection for self-compatibility, as all three amphidiploids are self-compatible while all three parental diploids are usually self-incompatible. The exceptions are *B. alboglabra* (a vegetable in the $n = 9$ *B. oleracea* group) and yellow sarson, an Indian subspecies of *B. rapa*. It is not surprising that the groups showing the most domesticated traits come from areas where *Brassica* oilseeds have been cultivated longest. Yellow sarson has large seeds and shatter-resistant pods. *Brassica juncea* and *B. carinata* are also relatively shatter-resistant, compared with the more recent species *B. napus* which has little resistance to shattering. There is much more genetic diversity in *B. rapa* and *B. juncea* than in *B. napus*. This diversity reflects the more ancient cultivation which has allowed their wider dispersal and has led to selection (either conscious or unconscious) for adaptation to a wider range of environments and for different characteristics.

Modern Breeding

Modern plant breeding began with the identification and multiplication of the best landraces. The simplest procedure was mass selection, that is, selection and bulking of individual plants from heterogeneous landraces to produce a more uniform cultivar. Many of the European winter *B. napus* cultivars of the first half of the 20th century were bred in this way.

With advances in the understanding of genetics and the development of statistics, plant breeders devised faster and more efficient means to produce improved cultivars and hybrids. This is a continuing process and, with recent advances in tissue culture, genetic fingerprinting, gene mapping and transformation, plant breeding methods continue to change.

What the plant breeder does to produce a new cultivar is seldom reported. Most publications describe the science behind plant breeding, the inheritance of traits, their measurement, statistical procedures and so on. In practice, the plant breeder's job is to breed a cultivar that is an improvement on existing ones. This means improving one or more traits which can be generally classified into four groups: yield *per se*, seed quality, resistance to pests and diseases and improved agronomic characteristics. In order to improve any of these traits the first job is to find genetic variation. Next the breeder needs to know something about the heritability of the trait, i.e. whether the phenotype is a good reflection of its genotype. Then, often one of the most difficult tasks is to devise a quick, cheap and accurate method of measuring the trait. Finally, the breeder must combine the desirable traits into a single cultivar or hybrid. Modern plant breeding still takes a long time and is expensive. New technology allows the plant breeder to make more changes to *Brassica* oilseeds and to do so more efficiently, but still requires the production, growing and evaluation of new lines over several seasons.

BREEDING FOR QUALITY

The seed of *Brassica* oilseeds is used for two main products – oil and meal. The oil can be used for human use (edible oil, margarine) or as an industrial oil. The meal, which is high in protein, can be used for human or animal consumption. It has also been used as a fertilizer. The chemistry of the seed is described in Chapter 10.

Oil quality is determined by its fatty acid composition, while meal quality is determined by the levels of antinutritional factors (particularly glucosinolates) and the proportions of protein and fibre. Quality can also refer to the proportion of oil and protein fractions in the seed, particularly oil as this is the most valuable component.

One of the most spectacular plant breeding achievements has been the improvement in quality of the old rapeseed cultivars. The changes from high to low erucic acid content of the oil and from high to low content of glucosinolates in the meal has resulted in a change in status of the crop: from low quality to high quality for both oil and meal. To emphasize this improvement the term 'canola' has been introduced to refer to seed and seed products of low erucic and low glucosinolate (double-low) cultivars.

Oil quality

Oil from *Brassica* seeds normally contains high levels of erucic acid. Cultivars with little or no erucic acid were first identified in *B. napus*, then in *B. rapa* and more recently in *B. juncea*. Because of nutritional concerns, particularly in the Western world, breeders produced cultivars with low erucic acid levels (Sauer and Kramer, 1983). As erucic acid content is determined by the embryo genotype and is controlled

by one gene locus in *B. rapa* and two gene loci in both *B. napus* and *B. juncea* (Kirk and Hurlstone, 1983), it was possible to change the fatty acid profile dramatically by introducing these genes into adapted cultivars. This led to the first low erucic acid rapeseed (LEAR) cultivars: Oro (*B. napus*), Span (*B. rapa*) (Downey *et al.*, 1975) and Zem1 (*B. juncea*) (Kirk and Oram, 1981). In China, the change with *B. napus* and *B. rapa* is still in progress. Although low erucic genes are available in *B. juncea* nearly all commercial cultivars are still high erucic (Love *et al.*, 1990a, b). Breeders need efficient selection methods and the development of the 'half seed' technique enabled breeders to select for low erucic acid content in early generations (Downey and Harvey, 1963). This technique involves germination of the seed, removal of one cotyledon for analysis and, if low in erucic acid content, the remainder of the seed is grown on. The cotyledon can be analysed by gas chromatography (Daun *et al.*, 1983) or by paper chromatography (Thies, 1971). The latter method is cheap, sufficiently accurate to detect the presence or absence of erucic acid, and has been particularly useful for breeders.

Further changes to fatty acid composition have not been so easy. It has been difficult to find genotypes which confer low linolenic or high oleic status and, even when found, inheritance has been more complex (Brunklaus-Jung and Robbelen, 1987; Diepenbrock and Wilson, 1987; Pleins and Friedt, 1989). The environment, particularly temperature, also affects linolenic level and measurement needs to be done by gas chromatography or by a fast screening method (McGregor, 1974). Various breeders have made claims for increasing oleic acid, which seems to be controlled by several genes, from about 60% to more than 80% (Auld *et al.*, 1992). When low erucic genes became available most breeders stopped breeding for high erucic acid content. By about 1980 low erucic cultivars were superior in perfor-mance to the old high erucic acid ones and breeders selecting for low linolenic or high oleic genotypes have had to compete with them. The low linolenic cultivars released so far have been lower yielding than their conventional counterparts. It is not yet clear whether this is due to a pleiotropic effect of the low linolenic genes or is simply due to the modest breeding effort so far.

Demand for higher levels of erucic acid for the industrial oil market has encouraged breeders to try to produce cultivars with levels greater than 50%. It had been thought that *Brassica* spp. were not capable of producing trierucin, the triacylglycerol with erucic acid esterified to all three hydroxyl positions of the glycerol, because erucic acid could not be put in the middle position of the triacylglycerol. However, recent work has shown that triacylglycerol with erucic acid in the middle position is produced in one line of *Brassica oleracea* (Taylor *et al.*, 1994). This gene should at least be readily transferable to *B. napus*.

Genotypes with total saturated fatty acid content lower than the usual 6–7% have been sought but little variation has been found. Specialist vegetable oils have been developed by transgenic methods and lines with high stearic (Knutson *et al.*, 1992) or lauric acid (Voelker *et al.*, 1992) content are being evaluated. As canola (usually *B. napus*) can be transformed by transgenic genes with relative ease (Maloney *et al.*, 1989) it is likely to be used as the starting-point for the development of specialized oils.

Meal quality

Meal quality has been improved drama-tically by plant breeding. The major problem was the presence of a series of compounds collectively called glucosinol-ates which hydrolyse when the seed is crushed to produce isothiocyanates. These reduced the amounts of meal that can be incorporated in livestock feed.

In 1968 the *B. napus* cultivar Bronow-ski was found to have much lower levels of glucosinolates than other cultivars. It was used to produce new cultivars with the low glucosinolate trait, first in *B. napus* spring cultivars (Kondra and Stefansson, 1970; Stefansson and Kondra, 1975) and later in *B. rapa* spring cultivars and *B. napus* winter cultivars. The incorporation of this trait into new cultivars was a dramatic demonstration of the potential of conven-tional breeding. As with other traits, suc-cess in breeding came not only from having genetic variation to work with but also from having good methods to measure the trait.

Many different techniques have been developed for glucosinolate analysis; some measure individual glucosinolates while others measure total glucosinolate levels (see Chapter 11). Plant breeding has made extensive use of the cheap, quick methods such as the glucose test tape which is used for urine sugar analysis (Lein, 1970; McGregor and Downey, 1975). This is sufficiently accurate to identify high gluco-sinolate selections for discard and, if used with care, low glucosinolate cultivars can be bred using only this method of analysis. However, for a quantitative measure there are a number of methods which measure total glucosinolate content (Brzezinski and Mendelewski, 1984; Pinkerton *et al.*, 1993) or individual glucosinolates (Sang and Truscott, 1984).

Despite the success in breeding low glucosinolate cultivars, the inheritance of this trait in *B. napus* and *B. rapa* has only recently been studied extensively (Gland *et al.*, 1981; Magrath *et al.*, 1993). After finding the trait in Bronowski it soon became clear that the mother-plant geno-type determined the glucosinolate level in the seeds (De March *et al.*, 1989). Several genes control the level in *B. napus* so that in an F_2 population, derived from a cross between high glucosinolate and low glucosinolate parents, usually less than 2% of plants are classified as low glucosinolate and the frequency is often much lower. This is why it has been important for breeders to have cheap and quick screening methods.

After the success with *B. napus* and *B. rapa* there has been considerable effort to find or produce a *B. juncea* line with low glucosinolate levels (Palmer *et al.*, 1988; Love *et al.*, 1991). Although one earlier report (Cohen *et al.*, 1983) was later shown to be false, it now seems that lines have been produced with this trait using interspecific crossing. However the frequency of low glucosinolate plants in F_2 populations seems even lower than for *B. napus* (G.F.W. Rakow, Saskatoon, 1994, personal communication) indicating that more genes are involved and, consequently, that it will be more difficult to transfer this trait into adapted cultivars.

Seed fractions

Oil is the most valuable fraction of the seed. Although oil content is strongly influenced by the environment, particularly temperature, moisture stress and soil nitrogen, there is also genetic variation in *B. napus* (Grami *et al.*, 1977), *B. rapa* (De Pauw and Baker, 1978) and *B. juncea*. Selection for oil content has led to slow but steady improvement. Non-commercial germplasm often has low oil content and the range is small in commercial cultivars. Efficient methods of analysis such as nuclear magnetic resonance (Madsen, 1976), near-infrared reflectance (Tkachuk, 1981) and various methods based on solvent extraction (Bengtsson, 1985; Raney *et al.*, 1987) have allowed large numbers of samples to be processed. As there is a negative correlation between yield and oil content, breeders must often choose which should have priority in applying selection pressure. Unless there is a price premium for

oil content, breeders generally select to meet a minimum standard.

Protein content is expressed on a whole-seed or oil-extracted meal basis, the latter being more important as protein affects meal quality. Since, in addition to protein content, breeders must also select for oil content, many select for the sum of oil and protein. In effect this means selecting for low proportions of the other seed constituents, i.e. fibre and carbohydrates.

Yellow seed coat has been shown to be associated with low fibre content and therefore higher oil and protein (Stringam *et al.*, 1974). It is not clear whether this applies to all sources of the trait. Yellow seed also produces a more visually appealing meal. Spring cultivars of *B. rapa* with either 'semi-yellow' or pure yellow seed have been bred in Canada and Europe and there is an expectation that future cultivars will be yellow. Both yellow- and brown-seeded types of *B. juncea* are available and preference is determined largely from local demand.

The search for true yellow-seeded genotypes of *B. napus* has been frustrating. Although partial yellow or 'yellow-brown' seeds have been produced a true yellow-seeded type has been elusive. Recent work, using genes from other species in the triangle of U, seems more promising (Chen *et al.*, 1988; Rashid *et al.*, 1994; and G.F.W. Rakow, Saskatoon, 1993, personal communication). The inheritance of a yellow seed coat depends on the species and source of genes but there is no general agreement (Stringam, 1980; Abraham and Bhatia, 1986; Shirzadegan, 1986; Chen and Heneen, 1992).

BREEDING FOR DISEASE RESISTANCE

Some diseases cause problems in most *Brassica*-growing areas, while others are of only local importance. Each species has different disease problems. Plant breeding offers one method for controlling diseases (see Chapter 5) and has obvious advantages if successful. As with other traits, the breeder's task is firstly to find sources of disease resistance and effective ways to screen for resistant genotypes. The trait has then to be transferred into a useful cultivar or hybrid.

Sclerotinia

This disease, caused by *Sclerotinia sclerotiorum*, is a problem in most *Brassica* oilseed-growing areas. The wide host range of the fungus and failure to find sources of resistance in many species pose a difficult challenge for *Brassica* breeders. Although there have been reports of differences in susceptibility between conventional cultivars (Sedun *et al.*, 1989), there seems little prospect of success and two other approaches are being investigated.

Infection mainly occurs as a result of the fungus entering the plant by using senescing petals which have dropped on to leaves or are lodged in leaf axils as a medium for penetration by the ascospores. The development of apetalous cultivars offers a means of blocking this route. Although no apetalous cultivars have been released, several sources of the apetalous trait are being studied (D. Lydiate, Norwich, 1994, personal communication).

Another approach is to introduce genes from distant species by transformation. When the fungus invades the plant, oxalic acid is released which causes necrosis. Attempts are being made to control the disease by introducing a gene for oxalate oxidase (Thompson *et al.*, 1993).

Stem canker (black leg)

This disease, sometimes called 'phoma' because of the asexual stage of the fungus *Phoma lingam*, is caused by *Leptosphaeria maculans*. It is an important disease of *B.*

napus in most countries, China being the main exception. Breeders have been able to identify sources of resistance and to incorporate them into commercial cultivars (Mithen and Lewis, 1988; Rimmer and van den Berg, 1992).

New resistant cultivars of winter *B. napus* have effectively kept this disease under control in Europe, while the eruption of more virulent races in Canada has necessitated more breeding effort. Sources of resistance from Europe and Australia are being used and should give good control. Apart from finding resistance within the *B. napus* species, other resistance sources are being transferred from *B. juncea* and *B. sylvestris* (a wild form of *B. rapa*) (Roy, 1984; Crouch *et al.*, 1994). These sources may confer greater tolerance or even immunity against leaf necrosis, although control of stem canker is the main objective.

Methods of screening genotypes for resistance have been developed using natural and artificial inoculation (Bansal *et al.*, 1994). Most cultivars have been bred using field screening, usually by growing plants on infected stubble or in areas where a high incidence of the disease occurs. Until recently it has been difficult to get a good correlation between resistance to artificial inoculation of young plants in a greenhouse or growth room and adult plant resistance in the field. Methods of artificial screening now seem able to do so and offer an opportunity to screen seedlings at any time of the year (McNabb *et al.*, 1993).

White rust

This disease, caused by *Albugo candida*, is a serious disease of *B. juncea* and *B. rapa* but most cultivars of *B. napus* are resistant to the prevalent races of the fungus (Fan *et al.*, 1983). In Canada, newer cultivars of *B. rapa* are resistant or tolerant of the predominant races but it is premature to claim that the disease is under control. In India, white rust is a major disease of

B. juncea and all commercial cultivars are susceptible, although the degree of susceptibility varies. There is enough variation in *B. juncea* and *B. rapa* to allow selection for tolerance (Tiwari *et al.*, 1988). Enduring sources of resistance will probably have to be found in other species such as *B. napus* and *B. carinata*.

Methods of screening using artificial seedling inoculation have proved very useful for breeders in Canada (T. Huskowska, Winnipeg, 1993, personal communication). Field screening is effective in India, particularly when the screening nursery is sown later than the main crops, as this gives high levels of infection.

Other diseases

There are many other diseases of oilseed *Brassica* spp. and their importance varies with site and season (see Chapter 5). In China, virus diseases (especially turnip mosaic virus) are difficult to breed against (Stobbs *et al.*, 1989) but progress has been made. In India, *Alternaria* is a major disease of *B. juncea* but no good sources of resistance are available within the species. In Canada, a root rot complex, known as brown girdling root rot, causes problems for *B. rapa* and resistance has been hard to incorporate. In Europe there are a number of foliar diseases which vary in importance each year (Rawlinson and Muthyalu, 1979). When there are several important diseases it is very difficult for the breeder to incorporate resistance to them all in one cultivar, particularly if the inheritance is complex.

BREEDING FOR AGRONOMIC TRAITS

The breeder develops a cultivar or hybrid for a particular environment. Part of that environment is the cropping system and cultural practices adopted by the farmer. This means that, when selecting for agro-nomic traits, the breeder selects not only to maximize yield but also to reduce growing costs by selecting for traits which make the crop easier and cheaper to manage.

Winter hardiness

Winter cultivars of *B. napus* grown in parts of Europe and China are subjected to very low temperatures which damage or kill plants. Apart from phenological differences between cultivars, which affect the time a cultivar is exposed, there are differences in winter survival. In general, attempts to develop screening methods based on morphological or physiological traits have been unsuccessful and so breeders continue to screen lines in the field under harsh conditions. As the climate varies from year to year this is a difficult trait to work on.

Herbicide resistance

Herbicides are used to control grasses and some broadleaved species in *Brassica* crops but it is difficult to use herbicides on broadleaved weeds that are closely related to oilseed brassicas. The discovery of a *B. rapa* weed with resistance to the triazine group of herbicides led to the release of a number of resistant commercial cultivars (Beversdorf *et al.*, 1980). However, the triazine resistance is due to a cytoplasmic mutant which also reduces the yield and oil content of the cultivar (Grant and Beversdorf, 1985a).

Transgenic cultivars have now been developed with resistance to glufosinate ammonium (or Basta) (Oelck *et al.*, 1991) and glyphosate (or Roundup) (Kishore and Shah, 1991; Parker *et al.*, 1991), herbicides which are used to control a wide spectrum of weeds and are considered total herbicides. This development results from the insertion of a single gene, which therefore behaves as a single dominant gene. It is expected that cultivars with these traits will be released if regulatory hurdles can be overcome.

Apetalous flowers

The bright yellow petals of *Brassica* crops are a striking feature. But when compared with wheat, which remains green at anthesis, the efficiency of the *Brassica* crop for intercepting light must be questioned. Apetalous genotypes have been identified (Buzza, 1983; Fu, S., *et al.*, 1990) and light interception has been shown to be more efficient (Mendham *et al.*, 1991; Rao *et al.*, 1991). The inheritance of a number of sources of the apetalous trait is being studied but the most useful sources seem to have complex inheritance (D. Lydiate, Norwich, 1993, personal communication). This will make it more difficult to transfer to commercial cultivars.

Height

Many *Brassica* oilseed crops are more than 180 cm high and are therefore taller than most annual cereal crops. Cereal breeders have reduced height and increased yield, resulting in improved harvest index and thus an easier-to-harvest crop. This has also permitted the application of higher rates of nitrogen fertilizer without causing lodging. *Brassica* breeders have attempted to select for shorter genotypes but commercial cultivars have shown little reduction in height.

Recently a gene which reduces the height of crops to 100 cm or less was reported (M. Renard, Le Rheu, 1993, personal communication). Inheritance is codominant which would allow heterozygous cultivars (F_1 hybrids) to be intermediate in height. Height reduction in some crops (e.g. sunflower) is associated with reduced yield so dwarfness may not always be an economic character.

Lodging resistance

Breeders select for lodging resistance by evaluating plots in the field. There are clear differences between cultivars but the inheritance of the trait is unknown. Although there is a correlation between height and lodging it is by no means the only factor.

Breeders have sometimes selected for 'tabling', i.e. leaning over of the crop to form a canopy well clear of the ground. This sort of trait depends on the eye of the breeder and cannot easily be defined.

Maturity

Growth cycles show a very wide range from the short-season *B. rapa* cultivars to the very long-season winter forms of *B. napus*. The time from sowing to flowering is controlled by a number of genes influenced by daylength and vernalization (Thurling and Das, 1977). The range available to the breeder in each species varies (Myers *et al.*, 1982).

Brassica rapa is the earliest species to flower followed by *B. juncea* and then *B. napus*. The species overlap in maturity and selection is possible within each species for a wide range of maturities to suit different environments.

Perhaps the main breeding challenge for this trait is to select early cultivars of *B. napus* for the shorter growing season in Canada and India.

Shattering resistance

Seed loss through shattering at harvest time is a problem in *B. napus* but less so in *B. rapa* and *B. juncea*. Although there are differences between *B. napus* cultivars there is no good source of resistance to shattering within the species (Agnihotri *et al.*, 1990). Attempts have been made to introduce resistance genes from other species (Kadkol *et al.*, 1986) but inheritance is complex. Moreover it is not easy to screen for the trait (Kadkol *et al.*, 1985) and it has proved difficult to transfer interspecifically.

BREEDING FOR YIELD

Plant breeders can breed for yield *per se* or breed for yield by selecting against those factors which decrease yield. The latter has sometimes been called deficiency breeding as it seeks to correct weaknesses in a cultivar: it covers selection for disease resistance, lodging resistance and so on.

Selection for yield *per se* is usually done by measuring the trait in small plots. In most breeding programmes it is customary initially to measure a large number of lines at a few sites and then, after selection, to test fewer lines at more sites before deciding which to release.

An alternative approach is to select for an ideotype which the breeder has decided *a priori* should give higher yields. This ideotype is a list of desirable traits for a plant type which is expected to give higher yields because of greater efficiency in plant structure and in allocation of resources within the plant (Thurling, 1991). One such ideotype is described in Chapter 2.

In practice most breeders adopt their own combination. Breeding for yield *per se* has led to small but steady increases in yield over many years, while it is argued that the ideotype approach may lead to bigger increases in yield if and when successful.

BREEDING CULTIVARS

Until the recent development of synthetics (seed produced from mixtures of two or more cultivars or relatively uniform inbred lines) and hybrids, commercial cultivars have been open-pollinated and several different methods have been used in their development.

Mass selection

The simplest method of breeding has been to select from landraces which have developed over centuries of cultivation. As both *B. napus* and *B. juncea* are self-fertile and are mostly self-pollinated (depending on insect activity), it is possible to derive relatively uniform lines by selecting individual plants, bulking them and multiplying the seed in isolation. For the self-incompatible *B. rapa* enough single-plant selections need to be bulked to prevent inbreeding depression. This method can be further improved in *B. napus* and *B. juncea* by pure-line breeding. By selecting individual plants and enforcing self-pollination (by bagging plants) for several generations a uniform line can be produced. Many early cultivars were developed by some form of mass selection or pure-line breeding.

Pedigree breeding

This is the most widely used method for developing cultivars in self-pollinated crops and has been used to produce many cultivars of *B. napus* and *B. juncea*. Although there are many variations of pedigree breeding, the essential idea is the same: parental lines are chosen with a view to combining the desirable traits from each. After crossing the parents by hand the F_1 generation is grown to increase seed. The F_2 generation displays a wide range of variation which may include the desirable genetic combinations from each parent. Single plants are selected and bagged (to ensure self-pollination) and the seed used to grow single rows in the F_3 generation. Plants are 'selfed' in each generation and this enforced inbreeding leads to the production of uniform inbred lines. Single plants or rows of plants in each generation can be measured or evaluated for desirable traits.

Eventually in the F_6 or later generations the breeder will have produced an inbred line with the desired combination of traits from the parents. An example of this method is that used by the author to develop the cultivar Marnoo. After making the cross and growing the F_1 generation, a large F_2 generation was grown and plants selected and measured for glucosinolate content, retaining only 1–2% of the plants with an acceptable low glucosinolate content. Seed of these plants were selected for zero erucic acid content using the half-seed technique. After selfing the F_3 generation, rows of F_4 lines were evaluated for disease resistance and oil content. The best lines were then evaluated in subsequent generations for yield and agronomic traits.

This simple example illustrates dilemmas posed for the breeder. Which trait should be selected and in which generation? Which method should be used to measure each trait, a cheap less accurate one or another with more accuracy but measuring fewer lines? How much selection pressure should be applied to each trait? Clearly some compromise must be made because of limited resources. Because there is no right answer and each breeder's decision will be influenced by the importance of the traits and resources available, breeders seldom publish their breeding methods in detail.

Backcross breeding

This method is usually used to transfer a simply inherited trait from a poorly adapted donor line into a well-adapted cultivar. It was used to transfer the low erucic acid and low glucosinolate traits into adapted cultivars in many parts of the world. For example, the canola-quality genes were first found in spring cultivars of *B. napus*, Liho for low erucic acid and Bronowski for low glucosinolate content. These traits were moved into adapted *B. napus* germplasm by backcrossing and little of the germplasm of the donor parents

remains in current cultivars (Stefansson, 1983).

This method is being used to transfer a range of new traits, such as different fatty acid compositions, seed colour and herbicide resistances, which are the subject of current developments.

Recurrent selection

Although some form of inbreeding is usually employed for developing cultivars of *B. napus* and *B. juncea*, recurrent selection is often used for self-incompatible species like *B. rapa*. There are different methods of recurrent selection but the same basic ideas apply. First, a population is developed which contains a level of diversity. This can be achieved by handcrossing or by mixing a number of lines and allowing them to intercross for several generations. Second, single plants are selected and some seed of each line is retained, while some is planted in progeny rows. These are evaluated and, after selecting the best progeny rows, the breeder goes back to the seed retained prior to drilling those rows, which is mixed for a further round of intercrossing before repeating the process. In this way the population is improved for the selected traits and eventually becomes sufficiently uniform for release as a cultivar.

Dihaploid breeding

There is a low frequency of naturally occurring haploid plants which can be found by the careful observer (Thompson, 1969; Stringam and Downey, 1973). Techniques have been developed using colchicine to double the chromosome number and produce dihaploid lines which are in fact completely homozygous diploid lines. This method was used to produce the cultivar Maris Haplona (Thompson, 1979a). However, the ability to produce plants, first by anther culture in *B. napus* (Wenzel *et al.*, 1977), *B. rapa* (Keller and Armstrong,

1979) and *B. juncea* (George and Rao, 1982), and later by microspore culture (Thomas and Wenzel, 1975; Keller and Armstrong, 1978; Sharma and Bhojwani, 1985; Chuong *et al.*, 1988) has given the breeder a new tool to develop cultivars.

The technology of using microspore culture to produce haploid and dihaploid lines in *B. napus* is now well developed and can be used with any genotype. It is a routine technique in many breeding programmes and also works with *B. juncea*. It has not been as successful with *B. rapa* (S. Kelly, Winnipeg, 1993, personal communication). Because most forms of *B. rapa* are self-incompatible and suffer from inbreeding depression, microspore culture is inappropriate for developing new cultivars. Recent work to introduce a self-compatible gene may allow this technique to be exploited.

At first glance it would seem that dihaploid breeding could replace pedigree breeding as a means of combining traits from a cross. Microspores from the F_1 plants could produce many dihaploid lines from which to screen for the desired combination of traits. Selected lines are completely homozygous and can be increased for further evaluation. Such lines have fewer problems passing the uniformity and stability criteria for plant breeders' rights. Moreover, this stage should be reached earlier than with the pedigree method, which is often taken to F_6 or later generations before segregation is considered to have no further influence on uniformity and stability.

Compared with pedigree breeding, dihaploid breeding has the disadvantage that selections cannot be made during the development of dihaploids and the number of lines that can be produced is often limited by resources. Of course it is possible to combine with the pedigree method, for example, by selecting initially for simple traits in early generations before putting selected lines through micro-

spore culture to produce pure breeding lines.

Development of synthetics

Synthetic cultivars of *B. napus* and *B. rapa* have been developed. These partly exploit the heterosis available and can be regarded as intermediate between cultivars and F_1 hybrids. The term 'synthetic' has several interpretations and is slightly different in *B. napus* and *B. rapa*.

Synthetics of *B. napus* are developed from two or more relatively uniform cultivars or inbred lines (Schuster and Böhm, 1983). The lines are mixed in equal proportions (although not always) and grown in isolation. As *B. napus* is self-compatible and the degree of outcrossing is dependent on insect pollination, there will be varying degrees of crossing between the component lines. The seed of the next generation (Syn 1) will be a mixture of the component parental lines and all possible F_1 hybrids between them. Thus a three-parent synthetic would include the original three parent lines and the three possible F_1 hybrids. The process can be continued by sowing 'Syn 1' seed to produce 'Syn 2' seed and so on. As the degree of outcrossing is variable it is impossible to be sure in advance what proportions of the components will be in the commercial seed. This has led to debates about which synthetic generation should be used and what constitutes a synthetic cultivar (Leon, 1991). Although synthetic *B. napus* cultivars were marketed in Europe they were often not very uniform and this method of breeding is no longer used in *B. napus*.

Synthetics of *B. rapa* are usually composed of two, or at most three, parental lines. *B. rapa* has an advantage in making synthetics as it is self-incompatible, which makes the outcome of mixing more predictable. The 'Syn 1' seed from a two-component synthetic will be composed of 25% of each of the component parts and 50%

of the F_1 hybrid from the cross between the two parents. This assumes an equal propensity to cross between lines as within a line. The exploitation of heterosis using this system has been demonstrated by the first *B. rapa* synthetics (Hysyn 100 and Hysyn 110) which were registered in Canada in 1994.

HYBRID BREEDING

A number of studies have shown that there is considerable heterosis for yield in all three oilseed *Brassica* species, *B. napus* (Schuster and Michael, 1976; Grant and Beversdorf, 1985b; Lefort-Buson and Dattee, 1985; Brandle and McVetty, 1989b), *B. rapa* (Sernyk and Stefansson, 1983; Schuler *et al.*, 1992) and *B. juncea* (Singh, 1973; Larik and Hussain, 1990; Pradham *et al.*, 1993). The problem has been to find a pollination control system which will allow the production of F_1 hybrids. Great interest has been shown in the development of different systems but so far success has been limited.

Cytoplasmic male sterility (CMS)

To be accurate cytoplasmic male sterility (CMS) should really be called cytoplasmic genetic male sterility but is usually called CMS. Hybrids based on CMS systems have been very successful in sunflower and sorghum, which have some features in common with oilseed brassicas. As with sunflower and sorghum, to use a CMS system effectively it is an advantage if the crop has a low seeding rate and can readily self-pollinate but can also readily cross-pollinate. It is also helpful if inbreeding depression is not too severe (Brandle and McVetty, 1989a). On these criteria both *B. napus* and *B. juncea* conform well but *B. rapa* does not do so in all respects.

A CMS system is composed of three lines, usually referred to as the A line (male-sterile or female parent line), B line (maintainer line) and R line (restorer or male parent line). The A and B lines have the same nuclear genotype but different cytoplasms. The A line has a cytoplasm conferring male sterility in the absence of nuclear genes which can override this cytoplasmic trait. The B line has a cytoplasm which allows normal pollen production. The R line has a nuclear gene or genes which overcome the effect of the 'male-sterile' cytoplasm and restores normal fertility. The cytoplasm of the R line may be similar to that of the A line or B line but it is better in practice to have the 'male-sterile' cytoplasm as this allows checking for the presence of the restorer gene. Figure 7.1 shows how this system is used to produce commercial F_1 hybrid seed.

The problem in oilseed *Brassica* breeding has been to find an effective CMS system. Several systems are available but each has problems which researchers are trying to correct (Fan *et al.*, 1986). In most cases the 'male-sterile' cytoplasm comes from another species (Rawat and Anand, 1979; Batra *et al.*, 1990). The interaction between the foreign cytoplasm and the host species can be associated with a reduction in yield. Apart from causing male sterility (due to incompatibility between the 'foreign' mitochondria and nuclei of the host species), the plant may show chlorosis (due to an interaction between the nucleus and the foreign chloroplast), distorted petals (making bee pollination less effective) and poor nectary function.

With some systems male sterility is not complete for all genotypes and in all environments and this partial male fertility means that the line is not effective as either an A line or an R line. Some systems also have problems in not having effective restorer lines. A recent description of the various CMS systems for brassicas shows the limitations of each (Banga, 1993).

At the time of writing, only one system has been used to produce commercial

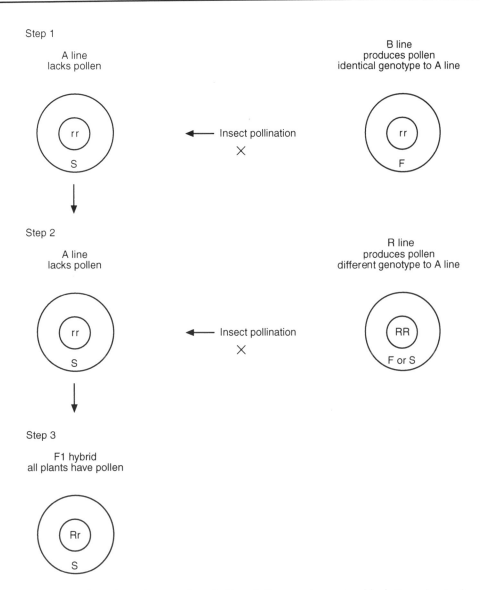

Fig. 7.1. Hybrid seed production. *Step 1*: Foundation seed crossing block to maintain A line supplies. *Step 2*: Hybrid seed crossing block to produce hybrid planting seed. *Step 3*: Commercial canola production. *Key*: S, 'sterile' cytoplasm; F, 'fertile' cytoplasm; RR, homozygous for dominant restorer gene; rr, homozygous for maintainer gene (or recessive for restorer gene); Rr, heterozygous for restorer gene.

hybrids. This is the *Polima* system developed in China (Fu, T.D., *et al.*, 1990). With this system it is difficult to get complete male sterility under all environments (Fan and Stefansson, 1986; Burns *et al.*, 1991) and the cytoplasm is thought to cause lower yields (McVetty *et al.*, 1990). The floral morphology, particularly the reduction in petal size, of the male-sterile lines means that seed set is often reduced due to poor pollination (McVetty *et al.*, 1989). Despite this, there has been sufficient heterosis to make it economical to produce hybrids in China (Niu and Zhao, 1990), Australia and Canada.

Another CMS system is close to commercialization in an incomplete form. The '*ogura*' system modified in France by protoplast fusion to correct problems of chlorosis and poor nectary function (Mesquida *et al.*, 1987, 1991) gives good floral morphology in the male-sterile lines and complete male sterility under all environments. Protoplast fusion produces a cybrid, a hybrid cytoplasm with components from both donors. This modified *ogura*, now called the INRA-*ogura* system, uses a restorer gene transferred from *Raphanus*.

This restorer gene was translocated into a chromosome of the C genome which means it can be used for *B. napus* but will need to be translocated further for use in *B. rapa* and *B. juncea*. The R gene was linked to genes for high glucosinolate content but this linkage is now thought to be broken. However, this system is being tested without using restorer lines by making so-called mixed or 'composite' hybrids (Renard *et al.*, 1987). These consist of a high proportion (usually 80%) of a sterile F_1 hybrid mixed with a lower proportion (usually 20%) of fertile plants to act as pollinators. The pollinator is made up of male-fertile plants, usually the parent genotypes of the male-sterile hybrid. In some environments it seems feasible for farmers to sow such a mixed hybrid and achieve sufficient pollination to get good seed set. It is suggested that even when restorer lines

are available it may be advantageous to have a proportion of sterile hybrid in the crop as the reduction in pollen may allow more energy to be diverted into seed yield.

Self-incompatibility

The use of self-incompatibility (SI) genes in breeding hybrid vegetable *Brassica* crops (both *B. oleracea* and *B. rapa*) led to efforts to try this method with oilseed brassicas (Thompson, 1979b; Gowers, 1980; Schuler *et al.*, 1992; Banks and Beversdorf, 1994). This is not easy in vegetable crops where self-pollination caused by incomplete self-incompatibility needs to be detected by 'grow-outs', i.e. growing on samples, or electrophoresis. One of the major problems of self-incompatibility systems is the effort and cost involved in increasing the self-incompatible lines by bud pollination, increased CO_2 or other methods. The economics of seed production makes this feasible for vegetable crops but is more difficult to justify for field crops. Despite these problems self-incompatibility-based hybrids have been released in Canada and are under test in Europe.

Transgenic systems

Brassica napus is relatively easy to transform (Maloney *et al.*, 1989; Boulter *et al.*, 1990) and is often used as a test species to try new ideas in genetic engineering (see Chapter 8). One method of pollination control using transgenic plants is well advanced (Mariani *et al.*, 1992; De Block and Debrouwer, 1993). In this system the gene conferring male sterility is linked to a gene conferring herbicide tolerance. The 'female' line is composed of 50% male-sterile, herbicide-resistant plants and 50% normal plants which can be removed by spraying the herbicide. The male line has an antisense gene which restores fertility in the hybrid.

Other hybrid systems

Attempts to produce hybrids using chemical hybridizing agents (CHAs) have not been successful and are unlikely to be so. Brassicas have indeterminate flowering and these agents usually have a narrow window of time of application to cause male sterility, so it is difficult to ensure male-sterile flowers for the whole of the flowering time (Banga and Labana, 1983).

Although sources of genetic male sterility (GMS) have been described (Toakari, 1970) it is not usually an economic method for hybrid seed production because of the need to remove male-fertile plants from the female line. One source of genetic male sterility is being developed to overcome this necessity (Li *et al.*, 1988) but the genetics is complex.

LOGISTICS OF BREEDING

The technology of breeding oilseed brassicas continues to change. Some improvements have come from developing new sources of germplasm, some from the development of new screening techniques and some from new breeding techniques, such as hybrid breeding and tissue culture. One aspect often overlooked is the organization of the breeding programme itself. As we have seen, plant breeders are concerned with finding or producing variation for a number of traits, combining these traits and then screening large numbers of plants or lines to find desirable combinations. This is a continual process with progress being made in each generation.

Generation interval

Breeders' lines or hybrids for each crop can be evaluated in one normal growing season each year. For winter *B. napus* in Europe this takes up almost the whole year, from August to July. However, for the spring type and other species or in areas where the growing season is shorter, it is possible to grow an extra generation outside the 'normal' growing season. This can be in growth rooms, greenhouses, or in 'off-season' or 'contra-season' (other hemisphere) nurseries. Apart from advancing another generation of inbreeding, this contra-season can speed up the breeding programme if a similar site (daylength, precipitation, temperature) can be found where it is possible to select for desirable traits.

Breeders of spring *B. napus* and spring *B. rapa* in Canada and northern Europe have used off-season sites in southern USA and Mexico, and contra-season sites in New Zealand, Australia and Chile. In China the winter *B. napus* crop is grown from September to May which leaves only a short time in summer to grow another generation. However, some breeders have used cooler regions in southwest China or nurseries at higher altitudes to get another generation even though the off-season is atypical and the following normal planting is delayed. Breeders in India have a difficulty in finding an off-season site within the country. The normal growing season is 'rabi' from October to March. In the other two seasons, summer (hot and dry) and kharif (hot and wet), there are only isolated areas in the mountains where *B. juncea* can be grown relatively normally. Regulations prevent an easy flow of seed to sites outside India which would make breeding easier.

Information technology

Plant breeding involves handling data on a large number of lines. Data from the chemical analyses of *Brassica* lines have involved increased use of computers both in collecting and in processing data directly from analytical instruments.

The organization of large breeding programmes is now more computerized and this will certainly increase. In future, data generated by restriction fragment length polymorphism (RFLP) analysis or similar technology will increase the need to correlate this with laboratory, greenhouse and field data.

Breeding and biotechnology

Brassicas have been among the first agricultural species to receive attention from biotechnology (see Chapter 8). Although *B. napus* seems likely to be used as a vehicle or tool to make new products this section is concerned with how biotechnology can help the breeder produce better oilseed cultivars. From this perspective biotechnology can help in two ways: first, by generating new sources of variation by the transfer of genes from other species; second, by improving efficiency in allowing the breeder to analyse the genotype rather than the phenotype.

New variation

New traits continue to be added to those introduced into brassicas (mainly *B. napus*) by transformation. The traits most likely to be introduced into commercial cultivars in the near future include herbicide resistances, changes in fatty acid composition, pollination control genes and disease resistances (see Chapter 8). Apart from providing variation not present in the species, from a breeding perspective these introduced genes should be easy to manipulate. Transformed genes are introduced in a 'construct' which contains several genes, usually a promoter to limit expression to a particular growth stage or plant part, the gene for the trait and a selectable marker. If only one insertion is needed then this construct behaves as a dominant gene at a single locus and can be moved easily from one line to another. In the case of brassicas, if the construct is inserted into the A genome, the gene can be used by all three

oilseed species, *B. rapa, B. napus* and *B. juncea*.

Genetic fingerprinting

Restriction fragment length polymorphism (RFLP) maps of the three main oilseed *Brassica* species have been produced by several groups and give a good cover of the genomes (Landry *et al.*, 1991; Chyi *et al.*, 1992). Many ideas have been put forward by *Brassica* researchers and other groups on how this technology can help the breeder. Although much new information has been generated, providing a new understanding of the relationship between genomes, cytoplasms and cultivars (Demeke *et al.*, 1992; McGrath and Quiros, 1992), breeders are now investigating how to use it to produce better cultivars and hybrids.

The relationship between cultivars or lines within a species can be shown by principal component analysis or by dendrograms. This gives breeders a measure of the genetic distance between lines and cultivars and an insight into the ancestry and 'phylogenetic' relationships between cultivars. This information is being used to test the hypothesis that heterosis is likely to be greater between more distantly related groups and that the gene pool can be divided into heterotic groups (Lefort-Buson *et al.*, 1986, 1987a, b).

Trait tagging using isozymes, RFLP markers or random amplified polymorphic DNA markers (RAPDs) has been proposed as a way of following the introgression of a gene which is difficult to measure in the phenotype (see Chapter 8). For *Brassica* breeders this offers the possibility of pyramiding genes for fatty acid composition or disease resistance. This should prove useful where one gene masks the effect of another.

Backcrossing by selecting for the genetic background of the recurrent parent is much quicker than the conventional method. It also enables the breeder to move the background of the donor parent in a controlled way so that isogenic lines can be produced for evaluation. With the introgression of transgenic traits and the demand for this to be done more quickly, this technique is becoming more widely used.

Regulation and breeding

Regulations for the registration of cultivars and hybrids are often necessary but sometimes have a negative effect. Plant breeders are confronted with a range of regulations which vary greatly in different parts of the world. These include quarantine regulations which inhibit the movement of seed, regulations for the registration of cultivars, regulations for the development of transgenic plants, plant variety rights and patent law. Although these affect all plant breeders, some have a particular impact on *Brassica* breeding.

Quarantine and import/export rules are, in general, not a serious problem. Brassicas are widely grown around the world and most diseases are widespread and therefore quarantine control is not widely used. Until recently, India imposed severe restrictions on seed movement and even the present regulations make it difficult for breeders to cooperate internationally.

Regulations for the registration of cultivars and hybrids vary from no compulsory registration requirements (Australia and USA) to mandatory testing for up to 4 years to ensure that the new cultivar meets set standards (UK). Although standards are intended to protect the quality of the crop, to reduce the risk from diseases or to protect the farmer from buying seed of an inferior cultivar, the effect of mandatory testing and compulsory registration is to increase the time required to develop a new cultivar and release it to the farmer. This also often entails a considerable cost.

In Canada, regulations for quality specifications are changing but the emphasis is on meeting minimum standards not only for yield, but also for oil and protein content and fatty acid composition. This has forced breeders to put more emphasis on quality, presumably to help crushers and exporters, and less emphasis on yield for the farmer. In the European Union, concern about the environment (see Chapter 9) and the excess use of agrochemicals (see Chapter 3) has resulted in a pricing structure which encourages growers to grow low-input cultivars, i.e. emphasis is on the deficiency breeding mentioned in 'Breeding for Yield' above. In India and China emphasis is on yield rather than on quality.

Regulations for breeding with transgenic plants or genetically modified organisms (GMOs) are usually on the ultra-cautious side so as to allay public concerns; although plant breeders have transferred genes from other species and genera for some time, transformation is likely to produce further changes. *Brassica* breeding is at the forefront of this technology and oilseed *Brassica* cultivars will be among the first transgenic plants to be grown commercially. This places a burden on *Brassica* research workers but, as politicians feel their way and try to grope with the issues, we must hope that logic will prevail and sensible, workable regulations be put in place.

Regulations and laws covering plant variety rights (PVR) and patent law have had some effect on *Brassica* breeding. Patents have been issued to cover *Brassica* transformation by *Agrobacterium*, products of protoplast fusion and fatty acid profiles, to mention but three (Kishore and Shah, 1991). While plant variety rights are considered by *Brassica* breeders to give good protection on the principle of 'reward for effort' and still allow the 'breeders' privilege' of using other breeders' germplasm, the use of patent legislation to lock up technology could have a negative effect on plant breeding unless it moves from its present confused state.

Conclusion

Plant breeding has played an important role in the dramatic increase in the production of oilseed brassicas. It has adapted the three major species to different regions of the world, increased yields, improved the quality of the oil and meal and increased resistances to diseases.

Brassicas are at the forefront of the biotechnological revolution which will introduce even more dramatic changes. However, 'designer genes' need to be carried in an adapted cultivar and it will be the combination of biotechnology with plant breeding that will deliver the new products.

References

Abraham, V. and Bhatia, C.R. (1986) Development of strains with yellow seedcoat in Indian mustard (*Brassica juncea* Czern. & Coss.). *Plant Breeding* 97, 86–88.

Agnihotri, A., Shivanna, K.R., Raina, S.N., Lakshmikumaran, M., Prakash, S. and Jagannathan, V. (1990) Production of *Brassica napus* × *Raphanobrassica* hybrids by embryo rescue: an attempt to introduce shattering resistance into *B. napus*. *Plant Breeding* 105, 292–299.

Auld, D.L., Heikkinen, M.K., Erickson, D.A., Sernyk, J.L. and Romero, J.E. (1992) Rapeseed mutants with reduced levels of polyunsaturated fatty acids and increased levels of oleic acid. *Crop Science* 32, 657–662.

Banga, S.S. (1993) Heterosis and its utilisation. In: Labanda, K.S, Banga, S.S. and Banga, S.K. (eds) *Breeding Oilseed Brassicas*. Springer Verlag, Berlin, pp. 21–43.

Banga, S.S. and Labana, K.S. (1983) Production of F$_1$ hybrids using ethrel-induced male sterility in Indian mustard (*Brassica juncea* (L.) Coss.). *Journal of Agricultural Science, Cambridge* 101, 453–455.

Banks, P.R. and Beversdorf, W.D. (1994) Self-incompatibility as a pollination control mechanism for spring oilseed rape, *Brassica napus*. *Euphytica* 75, 27–30.

Bansal, V.K., Kharbanda, P.D., Stringam, G.R., Thiagarajah, M.R. and Tewari, J.P. (1994) A comparison of greenhouse and field screening methods for blackleg resistance in doubled haploid lines of *Brassica napus*. *Plant Disease* 78, 276–281.

Batra, V., Prakash, S. and Shivanna, K.R. (1990) Intergeneric hybridisation between *Diplotaxis siifolia*, a wild species, and crop brassicas. *Theoretical and Applied Genetics* 80, 537–541.

Bengtsson, L. (1985) Some experiences of using different analytical methods in screening for oil and protein content in rapeseed. *Fette Seifen Anstrichmittel* 87, 262–265.

Beversdorf, W.D., Weiss-Lerman, J., Erickson, L.R. and Souza Machado, V. (1980) Transfer of cytoplasmically-inherited triazine resistance from bird's rape to cultivated oilseed rape (*Brassica campestris* and *B. napus*). *Canadian Journal of Genetics and Cytology* 22, 167–172.

Boulter, M.E., Croy, E., Simpson, P., Shields, R., Croy, R.R.D. and Shirsat, A.H. (1990) Transformation of *Brassica napus* L. (oilseed rape) using *Agrobacterium tumefaciens* and *Agrobacterium rhizogenes* – a comparison. *Plant Science* 70, 91–99.

Brandle, J.E. and McVetty, P.B.E. (1989a) Effects of inbreeding and estimates of additive genetic variance within seven summer oilseed rape cultivars. *Genome* 32, 115–119.

Brandle, J.E. and McVetty, P.B.E. (1989b) Heterosis and combining ability in hybrids derived from oilseed rape cultivars and inbred lines. *Crop Science* 29, 1191–1195.

Brunklaus-Jung, E. and Robbelen, G. (1987) Genetical and physiological investigations on mutants for polyenoic fatty acids in rapeseed (*Brassica napus* L.). *Plant Breeding* 98, 9–16.

Brzezinski, W. and Mendelewski, P. (1984) Determination of total glucosinolate content in rapeseed meal with thymol reagent. *Zeitschrift für Pflanzenzüchtung* 93, 177–183.

Burns, D.R., Scarth, R. and McVetty, P.B.E. (1991) Temperature and genotypic effects on the expression of *pol* cytoplasmic male sterility in summer rape. *Canadian Journal of Plant Science* 71, 655–661.

Buzza, G.C. (1983) The inheritance of an apetalous character in Canola (*Brassica napus*). *Cruciferae Newsletter* 8, 11–12.

Chen, B.Y. and Heneen, W.K. (1992) Inheritance of seed colour in *Brassica campestris* L. and breeding for yellow-seeded *B. napus* L. *Euphytica* 59, 157–163.

Chen, B.Y., Heneen, W.K. and Jonsson, R. (1988) Resynthesis of *Brassica napus* L. through interspecific hybridization between *B. alboglabra* Bailey and *B. campestris* L. with special emphasis on seed colour. *Plant Breeding* 101, 52–59.

Chuong, P.V., Pauls, K.P. and Beversdorf, W.D. (1988) High-frequency embryogenesis in male sterile plants of *Brassica napus* through microspore culture. *Canadian Journal of Botany* 66, 1676–1680.

Chyi, Y.S., Hoenecke, M.E. and Sernyk, J.L. (1992) A genetic linkage map of restriction fragment length polymorphism loci for *Brassica rapa* (syn. *campestris*). *Genome* 35, 746–757.

Cohen, D.B., Knowles, P.F., Theis, W. and Robbelen, G. (1983) Selection of glucosinolate free lines of *Brassica juncea*. *Zeitschrift für Pflanzenzühtung* 91, 169–172.

Crouch, J., Lewis, B. and Mithen, R. (1994) The effect of A genome substitution on the resistance of *Brassica napus* to *Leptosphaeria maculans*. *Plant Breeding* (in press).

Daun, J.K., Mazur, P.B. and Marrek, C.J. (1983) Use of gas liquid chromatography for monitoring the fatty acid composition of Canadian rapeseed. *Journal of American Oil Chemists' Society* 60, 1751–1754.

De Block, M. and Debrouwer, D. (1993) Engineered fertility control in transgenic *Brassica napus* L.: histochemical analysis of anther development. *Planta* 189, 218–225.

De March, G., McGregor, D.I. and Seguin-Swartz, G. (1989) Glucosinolate content of maturing pods and seeds of high and low glucosinolate summer rape. *Canadian Journal of Plant Science* 69, 929–932.

Demeke, T., Adams, R.P. and Chibbar, R. (1992) Potential taxonomic use of random amplified polymorphic DNA (RAPD): a case study in *Brassica*. *Theoretical and Applied Genetics* 84, 990–994.

De Pauw, R.M. and Baker, R.J. (1978) Correlations, heritabilities, and components of variation of four traits in *Brassica campestris*. *Canadian Journal of Plant Science* 58, 685–690.

Diepenbrock, W. and Wilson, R.F. (1987) Genetic regulation of linolenic acid concentration in rapeseed. *Crop Science* 27, 75–77.

Downey, R.K. and Harvey B.L. (1963) Methods for breeding for oil quality in rape. *Canadian Journal of Plant Science* 43, 271–275.

Downey, R.K., Stefansson, B.R., Stringam, G.R. and McGregor, D.I. (1975) Breeding rapeseed and mustard crops. In: Harapiak, J.P. (ed.) *Oilseed and Pulse Crops in Western Canada: A Symposium*. Western Cooperative Fertilizers, Calgary, Canada, pp. 157–183.

Fan, Z. and Stefansson, B.R. (1986) Influence of temperature on sterility of two cytoplasmic male sterility systems in rape (*Brassica napus* L.). *Canadian Journal of Plant Science* 66, 229–227.

Fan, Z., Rimmer, S.R. and Stefansson, B.R. (1983) Inheritance of resistance to *Albugo candida* in rape (*Brassica napus* L.). *Canadian Journal of Genetics and Cytology* 25, 420–424.

Fan, Z., Stefansson, B.R. and Sernyk, J.L. (1986) Maintainers and restorers of three male sterility inducing cytoplasms in rape (*Brassica napus* L.). *Canadian Journal of Plant Science* 66, 229–234.

Fu, S., Lu, Z., Chen, Y., Qi, C., Pu, H. and Zhang, J. (1990) Inheritance of apetalous character in *Brassica napus* L. and its potential in breeding. In: Chu, Q.R. and Fang, G.H. (eds) *Proceedings of the Symposium of China International Rapeseed Sciences*, Shanghai, China. Shanghai Scientific and Technical Publishers, Shanghai, 7 pp.

Fu, T.D., Yang, G. and Yang, X. (1990) Studies on 'three-line' Polima cytoplasmic male sterility developed in *Brassica napus* L. *Plant Breeding* 104, 115–120.

George, L. and Rao. P.S. (1982) *In vitro* induction of pollen embryos and plantlets in *Brassica juncea* through anther culture. *Plant Science Letters* 26, 111–116.

Gland, A., Robbelen, G. and Thies, W. (1981) Variation of alkenyl glucosinolates in seeds of *Brassica* species. *Zeitschrift für Pflanzenzüchtung* 87, 96–110.

Gowers, S. (1980) The production of hybrid oilseed rape using self-incompatibility. *Cruciferae Newsletter* 5, 15–16.

Grami, B., Baker, R.J. and Stefansson, B.R. (1977) Genetics of protein and oil content in summer rape: heritability, number of effective factors, and correlations. *Canadian Journal of Plant Science* 57, 937–943.

Grant, I. and Beversdorf, W.D. (1985a) Agronomic performance of triazine-resistant single-cross hybrid oilseed rape (*Brassica napus* L.) *Canadian Journal of Plant Science* 65, 889–892.

Grant, I. and Beversdorf, W.D. (1985b) Heterosis and combining ability estimates in spring-planted oilseed rape (*Brassica napus* L.). *Canadian Journal of Genetic Cytology* 27, 472–478.

Kadkol, G.P., Halloran, G.M. and Macmillan, R.H. (1985) Evaluation of *Brassica* genotypes for resistance to shatter. II. Variation in siliqua strength within and between accessions. *Euphytica* 34, 915–924.

Kadkol, G.P., Halloran, G.M. and Macmillan, R.H. (1986) Inheritance of siliqua strength in *Brassica campestris* L. I. Studies of F$_2$ and backcross populations. *Canadian Journal of Plant Pathology* 28, 365–373.

Keller, W.A. and Armstrong, K.C. (1978) High frequency production of microspore-derived plants from *Brassica napus* anther cultures. *Zeitschrift für Pflanzenzüchtung* 80, 100–108.

Keller, W.A. and Armstrong, K.C. (1979) Stimulation of embryogenesis and haploid production in *Brassica campestris* anther cultures by elevated temperature treatments. *Theoretical and Applied Genetics* 55, 65–67.

Kirk, J.T.O. and Hurlstone, C.J. (1983) Variation and inheritance of *Brassica juncea*. *Zeitschrift für Pflanzenzüchtung* 90, 331–338.

Kirk, J.T.O. and Oram, R.N. (1981) Isolation of erucic acid free lines of *Brassica juncea*. *Journal of Australian Institute of Agricultural Science* 47, 51–52.

Kishore, G.M. and Shah, D. (1991) Glyphosate-tolerant 5-enolpyruvyl 3-phosphoshikimate synthase. *United States Patent* 4, 971, 908; *Biotechnology Advances* 9, 89; *Plant Breeding Abstracts* 61(11), No. 9768.

Knutson, D.S., Thompson, G.A., Radke, S.E., Johnson, W.B., Knauf, V.C., and Kridl, J.C. (1992) Modification of *Brassica* seed oil by antisense expression of a stearoyl-acyl carrier protein desaturase gene. *Proceedings of the National Academy of Sciences of the United States of America* 89, 2624–2628.

Kondra, Z.P. and Stefansson, B.R. (1970) Inheritance of the major glucosinolates of rapeseed (*Brassica napus*) meal. *Canadian Journal of Plant Science* 50, 643–647.

Landry, B.S., Hubert, N., Etoh, T., Harada, J.J. and Lincoln, S.E. (1991) A genetic map for *Brassica napus* based on restriction fragment length polymorphisms detected with expressed DNA sequences. *Genome* 34, 543–552.

Larik, A.S. and Hussain, M. (1990) Heterosis in Indian mustard *Brassica juncea* (L.) Coss. *Pakistan Journal of Botany* 22(2), 168–171.

Lefort-Buson, M. and Dattee, Y. (1985) Étude de l'hétérosis chez le colza oléagineux d'hiver (*Brassica napus* L.) I. Comparaison de deux populations, l'une homozygote et l'autre hétérozygote. *Agronomie* 5(2), 101–110.

Lefort-Buson, M., Guillot-Lemoine, B. and Dattee, Y. (1986) Heterosis and genetic distance in rapeseed (*Brassica napus* L.). Use of different indicators of genetic divergence in a 7 × 7 diallel. *Agronomie* 6(9), 839–844.

Lefort-Buson, M., Dattee, Y. and Guillot-Lemoine, B. (1987a) Heterosis and genetic distance in rapeseed (*Brassica napus* L.): use of kinship coefficient. *Genome* 29, 11–18.

Lefort-Buson, M., Guillot-Lemoine, B. and Dattee Y. (1987b) Heterosis and genetic distance in rapeseed (*Brassica napus* L.): crosses between European and Asiatic selfed lines. *Genome* 29, 413–418.

Lein, K.A. (1970) Quantitativ Bestimmungsmethoden für Samenglucosinolate in *Brassica*-Arten und ihre Anwendung in der Züchtung von Glucosinolatarmen Raps. *Zeitschrift für Pflanzenzüchtung* 63, 137–154.

Leon, J. (1991) Heterosis and mixing effects in winter oilseed rape. *Crop Science* 31, 281–284.

Li, Qian, Y., Wu, Z. and Stefansson, B.R. (1988) Genetic male sterility in rape (*Brassica napus* L.) conditioned by interaction of genes at two loci. *Canadian Journal of Plant Science* 68, 1115–1118.

Love, H.K., Rakow, G., Raney, J.P. and Downey, R.K. (1990a) Development of low glucosinolate mustard. *Canadian Journal of Plant Science* 70, 419–424.

Love, H.K., Rakow, G., Raney, J.P. and Downey, R.K. (1990b) Genetic control of 2-propenyl and 3-butenyl glucosinolate synthesis in mustard. *Canadian Journal of Plant Science* 70, 425–429.

Love, H.K., Rakow, G., Raney, J.P. and Downey, R.K. (1991) Breeding improvements towards canola quality *Brassica juncea*. In: McGregor, D.I. (ed.) *Proceedings of the Eighth International Rapeseed Congress*, Saskatoon, Canada. Organizing Committee, Saskatoon, pp. 164–169.

McGrath, J.M. and Quiros, C.F. (1992) Genetic diversity at isozyme and RFLP loci in *Brassica campestris* as related to crop type and geographical origin. *Theoretical and Applied Genetics* 83, 783–790.

McGregor, D.I. (1974) A rapid and sensitive spot test for linolenic acid levels in rapeseed. *Canadian Journal of Plant Science* 54, 211–213.

McGregor, D.I. and Downey, R.K. (1975) A rapid and simple assay for identifying low glucosinolate rapeseed. *Canadian Journal of Plant Science* 55, 191–196.

McNabb, W.M., van den Bert, C.G.J. and Rimmer, S.R. (1993) Comparison of inoculation methods for selection of plants resistant to *Leptosphaeria maculans* in *Brassica napus*. *Canadian Journal of Plant Science* 73, 1199–1207.

McVetty, P.B.E., Pinnisch, R. and Scarth, R. (1989) The significance of floral characteristics in seed production of four summer rape cultivar A-lines with *pol* cytoplasm. *Canadian Journal of Plant Science* 69, 915–918.

McVetty, P.B.E., Edie, S.A. and Scarth, R. (1990) Comparison of the effect of *nap* and *pol* cytoplasms on the performance of intercultivar summer oilseed rape hybrids. *Canadian Journal of Plant Science* 70, 117–126.

Madsen, E. (1976) Nuclear magnetic resonance spectrometry as a quick method of determination of oil content in rapeseed. *Journal of American Oil Chemists' Society* 53, 467.

Magrath, R., Herron, C., Giamoustaris, A. and Mithen, R. (1993) The inheritance of aliphatic glucosinolates in *Brassica napus. Plant Breeding* 111, 55–72.

Maloney, M.M., Walker, J. and Sharma, K. (1989) High efficiency transformation of *Brassica napus* using *Agrobacterium* vectors. *Plant Cell Reports* 8, 238–242.

Mariani, C., Gossele, V., De Beuckeleer, M., De Block, M., Goldberg, R.B., De Greef, W. and Leemans, J. (1992) A chimaeric ribonuclease-inhibitor gene restores fertility to male sterile plants. *Nature (London)* 357, 384–387.

Mendham, N.J., Rao, M.S.S. and Buzza, G.C. (1991) The apetalous flower character as a component of a high yielding ideotype. In: McGregor, D.I. (ed.) *Proceedings of the Eighth International Rapeseed Congress*, Saskatoon, Canada. Organizing Committee, Saskatoon, pp. 596–600.

Mesquida, J., Renard, M., Pellen-Delourme, R., Pelletier, G. and Morice, J. (1987) Influence des sécrétions nectarifers des lignées mâles stériles pour la production de semences hybrides F1 de colza. In: *Les Colloques INRA,* 22–23 April 1987, Ste Sabine, France, pp. 269–280.

Mesquida, J., Pham-Delegue, M.H., Marilleau, R., Le Metayer, M. and Renard, M. (1991) La sécrétion nectarifère des fleurs de cybrides mâles-stériles de colza d'hiver (*Brassica napus* L). *Agronomie* 11, 217–227.

Mithen, R.F. and Lewis, B.G. (1988) Resistance to *Leptosphaeria maculans* in hybrids of *Brassica oleracea* and *Brassica insularis. Journal of Phytopathology* 123, 253–258.

Myers, L.F., Christian, K.R. and Kirchner, R.J. (1982) Flowering responses of 48 lines of oilseed rape (*Brassica* spp.) to vernalization and daylength. *Australian Journal of Agricultural Research* 33, 927–936.

Niu, Y. and Zhao, Y. (1990) Today and tomorrow for introduced hybrid rapeseed in Anhui province. In: Chu, Q.R. and Fang, G.H. (eds) *Proceedings of the Symposium of China International Rapeseed Sciences*, Shanghai, China. Shanghai Scientific and Technical Publishers, Shanghai, 58 pp.

Oelck, M.M., Phan, C.V., Eckes, P., Donn, G., Rakow, G. and Keller, W.A. (1991) Field resistance of canola transformants (*Brassica napus* L.) to ignite (Phosphinotricin). In: McGregor, D.I. (ed.) *Proceedings of the Eighth International Rapeseed Congress*, Saskatoon, Canada. Organizing Committee, Saskatoon, pp. 292–297.

Palmer, M.V., Sang, J.P., Oram, R.N., Tran, D.A. and Salisbury, P.A. (1988) Variation in seed glucosinolate concentrations of Indian mustard (*Brassica juncea* (L.) Czern. + Coss). *Australian Journal of Experimental Agriculture* 28, 779–782.

Parker, G.B., Mitchell, S.H., Hart, J.L., Padgette, S.R., Fedele, M.J., Barry, G.F., Didier, D.K., Re, D.B., Eichhilzt, D.A., Kishore, G.M. and Delannay, X. (1991) Development of canola genetically modified to tolerate Roundup herbicide. *Agronomy Abstracts of the Annual Meeting of the American Society of Agronomy, Denver, Colorado, USA,* 199.

Pinkerton, A., Randall, P.J. and Wallace, P.A. (1993) Determination of total glucosinolates in oilseed rape by X-ray spectrometric analysis for oxidised sulphur (S^{6+}). *Journal of Science and Food of Agriculture* 61, 79–86.

Pleins, S. and Friedt, W. (1989) Genetic control of linolenic acid concentration in seed oil of rapeseed (*Brassica napus* L.). *Theoretical and Applied Genetics* 78, 793–797.

Pradham, A.K., Sodhi, Y.S., Mukhopadhyay, A. and Pental, D. (1993) Heterosis breeding in Indian mustard (*Brassica juncea* L. Czern & Coss): analysis of component characters contributing to heterosis for yield. *Euphytica* 69, 219–229.

Raney, J.P., Love, H.K., Rakow, G.F.W. and Downey, R.K. (1987) An apparatus for rapid preparation of oil and oil-free meal from *Brassica* seed. *Fat Science Technology* 6, 235–237.

Rao, M.S.S., Mendham, N.J. and Buzza, G.C. (1991) Effect of the apetalous flower character on radiation distribution in the crop canopy, yield and its components in oilseed rape (*Brassica napus*). *Journal of Agricultural Science, Cambridge* 117, 189–196.

Rashid, A., Rakow, G. and Downey, R.K. (1994) Development of yellow seeded *Brassica napus* through interspecific crosses. *Plant Breeding* 112, 127–134.

Rawat, D.S. and Anand, I.J. (1979) Male sterility in Indian mustard. *Indian Journal of Genetic Plant Breeding* 39, 412–415.

Rawlinson, C.J. and Muthyalu, G. (1979) Diseases of winter oilseed rape: occurrence, effects and control. *Journal of Agricultural Science, Cambridge* 93, 593–606.

Renard, M., Mesquida, J., Pellan-Delourme, R., Pelletier, G. and Morice, J. (1987) Pollination des cybrids mâle stériles dans un système de culture mixte de colza. In: *Variabilité génétique cytoplasmique et stérilité mâle cytoplasmique*. INRA, Paris, France, pp. 281–292.

Rimmer, S.R. and van den Berg, C.G.J. (1992) Resistance of oilseed *Brassica* spp. to blackleg caused by *Leptosphaeria maculans*. *Canadian Journal of Plant Pathology* 14, 56–66.

Roy, N.N. (1984) Interspecific transfer of *Brassica juncea*-type high blackleg resistance to *Brassica napus*. *Euphytica* 33, 295–303.

Sang, J.P. and Truscott, R.J.W. (1984) Liquid chromatographic determination of glucosinolates in rapeseed as desulfoglucosinolates. *Journal of the Association of Official Analytical Chemists* 67, 829–833.

Sauer, F.D. and Kramer, J.K.G. (1983) The problem associated with the feeding of high erucic acid rapeseed oils and some fish oils to experimental animals. In: Kramer, J.K.G., Sauer, F.D. and Pigden, W.J. (eds). *High and Low Erucic Acid Rapeseed Oils: Production, Usage, Chemistry and Toxicological Evaluation*. Academic Press, New York, pp. 254–292.

Schuler, T.J., Hutcheson, D.S. and Downey, R.K. (1992) Heterosis in intervarietal hybrids of summer turnip rape in western Canada. *Canadian Journal Plant Science* 72, 127–136.

Schuster, W and Böhm, J. (1983) Untersuchungen über den Einfluss von Standort, Jahr, Stickstoffdüngung und Bienenflug auf die Saatgutwert unterschiedlicher Winterrapssorten. In: *Proceedings of the Sixth GCIRC International Congress on Rapeseed*, Paris, France, pp. 390–393.

Schuster, W. and Michael, J. (1976) Untersuchungen über Inzuchtdepressionen und Heterosis effekte bei Raps (*Brassica napus oleifera*). *Zeitschrift für Pflanzenzüchtung* 77, 56–66.

Sedun, F.S., Seguin-Swartz, G. and Rakow, G.F.W. (1989) Genetic variation in reaction to sclerotinia stem rot in *Brassica* species. *Canadian Journal of Plant Science* 69, 229–232.

Sernyk, J.L. and Stefansson, B.R. (1983) Heterosis in summer rape (*Brassica napus* L.). *Canadian Journal of Plant Science* 63, 407–413.

Sharma, K.K. and Bhojwani, S.S. (1985) Microspore embryogenesis in anther cultures of two Indian cultivars of *Brassica juncea* (L) Czern. *Plant Cell Tissue Organ Culture* 4, 235–239.

Shirzadegan, M. (1986) Inheritance of seed color in *Brassica napus* L. *Zeitschrift für Pflanzenzüchtung* 96, 140–146.

Singh, S.P. (1973) Heterosis and combining ability estimates in Indian mustard, *Brassica juncea* (L) Czern and Coss. *Crop Science* 13, 497–499.

Song, K. and Osborn, T.C. (1991) Origins of *Brassica napus*: new evidence based on nuclear and cytoplasmic DNAs. In: McGregor, D.I. (ed.) *Proceedings of the Eighth International Rapeseed Congress*, Saskatoon, Canada. Organizing Committee, Saskatoon, pp. 324–327.

Stefansson, B.R. (1983) The development of improved rapeseed cultivars. In: Kramer, J.K.G., Sauer, F.D. and Pigden, W.J. (eds) *High and Low Erucic Acid Rapeseed Oils: Production, Usage, Chemistry, and Toxicological Evaluation*. Academic Press, New York, pp. 144–159.

Stefansson, B.R. and Kondra, Z.P. (1975) Tower summer rape. *Canadian Journal of Plant Science* 55, 343–344.

Stobbs, L.W., Hume, D. and Forrest, B. (1989) Survey of canola germplasm for resistance to turnip mosaic virus. *Phytoprotection* 70, 1–6.

Stringam, G.R. (1980) Inheritance of seed color in turnip rape. *Canadian Journal of Plant Science* 60, 331–335.

Stringam, G.R. and Downey, R.K. (1973) Haploid frequencies in *Brassica napus*. *Canadian Journal of Plant Science* 53, 229–231.

Stringam, G.R., McGregor, D.I. and Pawlowski, S.H. (1974) Chemical and morphological characteristics associated with seed coat colour in rapeseed. In: *Proceedings of the Fourth International Rapeseed Congress*, Giessen, Germany. GCIRC, Giessen.

Taylor, D.C., MacKenzie, S.L., McCurdy, A.R., McVetty, P.B.E., Giblin, E.M., Pass, E.W., Stone, S.J., Scarth, R., Rimmer, S.R. and Pickard, M.D. (1994) Stereospecific analyses of seed triacylglycerols from high-erucic acid *Brassicaceae*: detection of erucic acid at the *sn*-2 position in *Brassica oleracea* L. genotypes. *Journal of American Oil Chemists' Society* 71, 163–167.

Thies, W. (1971) Schnelle und einfache Analysen der Fettsäurezusammensetzung in einzelnen Raps-Kotyledonen. *Zeitschrift für Pflanzenzüchtung* 65, 181–202.

Thomas, E. and Wenzel, G. (1975) Embryogenesis from microspores of *Brassica napus*. *Zeitschrift für Pflanzenzüchtung* 74, 77–81.

Thompson, C., Dunwell, J.M., Johnstone, C.E., Lay, V., Ray, J., Schmitt, M., Watson, H. and Nisbet, G. (1993) Degradation of oxalic acid by transgenic canola plants expressing oxalate oxidase. In: *Proceedings of the Eighth Crucifer Genetics Workshop*, Saskatoon, Canada, 31 pp.

Thompson, K.F. (1969) Frequencies of haploids in spring oilseed rape (*Brassica napus*). *Heredity* 24, 318–319.

Thompson, K.F. (1979a) Superior performance of two homozygous diploid lines from naturally occurring polyhaploids in oilseed rape (*Brassica napus*). *Euphytica* 28, 127–135.

Thompson, K.F. (1979b) Application of recessive self-incompatibility to production of hybrid rapeseed. In: *Proceedings of the Fifth International Rapeseed Conference*, Malmo, Sweden. Organizing Committee, Malmo, pp. 56–59.

Thurling, N. (1991) Application of the ideotype concept in breeding for higher yield in the oilseed brassicas. *Field Crops Research* 26, 201–219.

Thurling, N. and Das, L.D.V. (1977) Variation in pre-anthesis development of spring rape (*Brassica napus* L.). *Australian Journal of Agricultural Research* 28, 597–607.

Tiwari, A.S., Petrie, G.A. and Downey, R.K. (1988) Inheritance of resistance to *Albugo candida* Race 2 in mustard (*Brassica juncea* (L.) Czern.). *Canadian Journal of Plant Science* 68, 297–300.

Tkachuk, R. (1981) Oil and protein analysis of whole rapeseed kernels by near infrared reflectance spectroscopy. *Journal of American Oil Chemists' Society* 58, 819.

Toakari, Y. (1970) Monogenic recessive male sterility in oil rape induced by irradiation. *Zeitschrift für Pflanzenzüchtung* 64, 242–247.

U, N. (1935) Genome analysis in *Brassica* with special reference to the experimental formation of *B. napus* and its peculiar mode of fertilization. *Japanese Journal of Botany* 7, 389–452.

Voelker, T.A., Worrell, A.C., Anderson, L., Bkeibaum, J., Fan, C. and Hawkins, D.J. (1992) Fatty acid biosynthesis redirected to medium chains in transgenic oilseed plants. *Science* 257, 72–73.

Warwick, S.I. and Black, L.D. (1993) Molecular relationships in subtribe *Brassicinae* (*Cruciferae*, tribe *Brassiceae*). *Canadian Journal of Botany* 71, 906–918.

Wenzel, G., Hoffman, F. and Thomas, E. (1977) Anther culture as a breeding tool in rape. *Zeitschrift für Pflanzenzüchtung* 78, 149–155.

8 Biotechnology

D.J. Murphy and R. Mithen
John Innes Centre, Norwich, UK

INTRODUCTION

For the purposes of this chapter, the term 'biotechnology' will be defined according to its newer, rather than more classical, meaning as follows:

> The utilisation of a combination of conventional and novel techniques, many of the latter of which are based on molecular biology, to alter the characteristics of biological organisms in order that they may be better utilised for a variety of applications.

With respect to the *Brassica* oilseeds, and indeed any other crop, modern biotechnology will have a major impact in two areas. Firstly, it provides a new range of techniques enabling the efficient selection of favourable variants in plant breeding programmes. Secondly, it provides the opportunity to improve germplasm by increasing its diversity beyond conventional genetic limitations. Due to the relative ease of genetic transformation, *Brassica* oilseed crops have been amongst the first to be subject to the full range of modern biotechnological methods.

TECHNIQUES

Synthetic amphidiploids, wide crosses and genome manipulation

Following the elucidation of the genetic relationships of the cultivated *Brassica* species (see Chapter 1), there have been successful programmes which have developed synthetic amphidiploid lines by interspecific hybridization of the relevant diploid parents. This has been particularly useful for *Brassica napus* in which there is little variation for several important agronomic traits. The synthetic forms of *B. napus* are sexually compatible with natural forms of *B. napus* and have been used as a means to introduce novel characters into oilseed rape breeding programmes. Early attempts at resynthesis relied upon obtaining hybrids by undertaking a large number of pollinations so that the few embryos which did not abort following fertilization could be recovered as viable seed. Crosses were often undertaken between tetraploids developed by doubling the chromosomes of the diploid species with colchicine. The success of these methods has been considerably enhanced through the use of tissue culture methods which enable embryos to be rescued from developing ovaries prior to abortion and grown *in vitro*. Moreover, the cultivation of ovaries *in vitro* prior to embryo rescue increases the number of viable embryos which can be recovered (Inomata, 1977; Mithen and Herron, 1991). These techniques are now routinely used to develop synthetic *B. napus* lines from crosses between any two geno-types of *Brassica rapa* and *Brassica oleracea*, and have been extended to include wild forms of both species (Mithen and Herron, 1991; Crouch *et al.*, 1994). Although the emphasis within resynthesis programmes has been on *B. napus*, due to its economic importance, synthetic lines of *Brassica juncea* and *Brassica carinata* can also be produced. In addition to embryo rescue techniques, protoplast fusion has been used to develop synthetic amphidiploid lines (Schenck and Röbbelen, 1982; Sunberg and Glimelius, 1986). Through these tech-niques, variations in several characters have been introduced into breeding pro-grammes including alteration of flower-ing time (Akbar, 1987), disease resistance (Crouch *et al.*, 1994), glucosinolate con-tent (Magrath *et al.*, 1994; Parkin *et al.*, 1994), oil content (Olsson, 1986) and self-incompatibility (Ozminkowski and Jourdan, 1993).

Despite the extensive variation in the wild and cultivated forms of *B. rapa* and *B. oleracea* (which remains to be fully evaluated and exploited), there are several characters which scientists have sought to introduce from other species within the *Brassicaceae*. Intergeneric hybrids have been obtain with an extensive range of species either through sexual crossing and embryo rescue techniques or by somatic fusion of protoplasts, including *Eruca sativa* (Agnihorti *et al.*, 1990), *Diplotaxis* spp. (Delourme *et al.*, 1989; Batra *et al.*, 1990), *Sinapis alba* (Ripley and Arnison, 1990), *Raphanus sativus* (Lelivet *et al.*, 1993)

and *Moricandia arvensis* (Takahata and Takeda, 1990), and it seems likely that hybrids can be developed between *Brassica* species and most other species within the *Brassicaceae*. Symmetrical protoplast fusion results in hybrids possessing the complete genomes of both parents, typi-cally resulting in abnormal and infertile plant types. A more promising approach is asymmetrical protoplast fusion. Prior to fusion, the protoplasts from the donor species are irradiated with UV or X-rays to disrupt their nuclei, while the proto-plasts of the recipient species are treated with iodoacetic acid which disrupts cyto-plasmic organelles. Following fusion, only hybrid cells containing the entire nuclear genome of the recipient species and orga-nelles and variable amounts of nuclear DNA from the donor are able to regene-rate. It is important that the DNA from the donor integrates into the recipient genome to produce a genetically stable line which can then be used within a breeding programme. Several techniques such as flow cytometry and the use of molecular markers can assist the selection of suitable hybrids with the minimum amount of donor DNA. These techniques have been used in attempts to introduce resistance to *Leptosphaeria maculans* from *B. nigra* (Sjodin and Glimelius, 1989), aspects of intermediate C3/C4 metabo-lism from *Moricandia* spp. (Murata, O'Neill and Mathias, personal communication) and nervonic acid from *Thlaspi perfoliatum* (Fahleson *et al.*, 1994) into *B. napus*.

Induced variation

Genetic variation can be induced in breed-ing lines by using chemical mutagens such as ethanemethylene sulphonate (EMS), or by radiation, to produce a large number of mutations in a population of as many as 100,000 seeds. With such large numbers of seeds, a rapid and simple screening method is necessary for the selection of the desired phenotype in the M_1 generation. Since mutagenesis is non-selective, each plant in the M_1 generation will contain many other mutations in addition to the desired one. Hence, an extensive back-crossing programme is necessary in order to obtain plants containing single-gene mutations.

Chemical mutagenesis has been used in *B. napus* in order to modify the fatty acid composition of the seed oil, although the range of variation observed did not lie dramatically outside the kind of profiles that can be obtained from classical breeding programmes. Nevertheless, the potential of this method for altering seed fatty acid composition has recently been demonstrated in linseed. Linseed oil can contain up to 70% α-linolenic acid and is therefore oxidation-prone and unsuitable for edible applications. An EMS-based mutagenesis programme resulted in the selection of a double-recessive mutant line containing only 2% α-linolenic acid and up to 70% of the high-grade edible fatty acid, linoleic acid (Green, 1986).

Another source of induced variation that has been characterized in *Brassica* oil crops is that caused by tissue culture, i.e. somaclonal variation (Sacristan, 1982). Although this variation has been shown to be heritable, its causes are not yet fully understood. It is possibly due to the activation of latent transposable elements, or to chromosomal rearrangements (Lee and Phillips, 1988). Somaclonal variation has been used successfully in the production of novel vegetable crops, but its use for the production of new *Brassica* oilseed breeding lines has yet to be reported.

Molecular markers

Molecular marker technology is likely to have a profound effect on future plant breeding programmes, and is well suited to being exploited within breeding programmes for oilseed *Brassica* crops. Several

different types of markers are available. Isozyme markers rely on protein separation techniques such as electrophoresis and isoelectric focusing. Preparation is relatively easy and costs are low, enabling large numbers of plants to be scored. However the number of isozyme markers is generally low and they are not being widely applied within breeding programmes. In contrast, restriction fragment length polymorphism (RFLP) markers are now being applied in several *Brassica* breeding programmes. Several independent *Brassica* RFLP maps have been produced, each of which may contain several hundred loci (e.g. Slocum *et al.*, 1990; Landry *et al.*, 1991; Lydiate *et al.*, 1993). Many of the RFLP maps of *B. napus* are based upon a recombinant homozygous doubled haploid population derived from the microspore culture of F_1 hybrids.

RFLPs have several advantages over other marker systems.

1.   Many polymorphic loci, which are distributed throughout the genome, can be detected within crosses between oilseed *Brassica* cultivars.

2.   Markers are co-dominant which enables heterozygotes to be detected in backcrossing programmes so that recessive traits can be selected without the need for test crosses.

3.   There are no environmental effects on the occurrence of RFLPs and tissue can be selected from any part of the plant. This enables, for example, alleles regulating oil quality to be selected by examining the genotype of young seedlings without having to wait for flowering and seed setting.

The most important application of RFLPs may be the rapid introgression of traits from exotic sources. For example, once disease resistance genes have been introduced into *B. napus* via wide crosses it is necessary to introgress only the resistant genes into an agronomically acceptable background. By conventional methods this may take many years, particularly if the resistance genes are linked to undesirable characters. However, with the use of RFLP markers, genotypes in backcrossing programmes can be selected which have the desired alleles without the need for excessive phenotypic screening, which may be particularly problematic if the trait is recessive so that test crosses would be required. Several important genes have now been located on *Brassica* RFLP maps such as the genes which determine aliphatic glucosinolate composition (Magrath *et al.*, 1994; Parkin *et al.*, 1994) and those which regulate various aspects of fatty acid quality (D.J. Lydiate, personal communication). In addition to selecting for a particular trait, RFLPs are efficient at selecting against the alien background.

Frequently, important agronomic traits are inherited in a quantitative manner and may be determined by several loci, termed quantitative trait loci (QTLs), which cannot be resolved through Mendelian analysis. However, through RFLP analysis it is possible to identify linkages between QTLs and RFLPs so that the number and relative positions of loci which determine complex traits can be elucidated. The RFLP markers can then be used to select for specific alleles at these loci within breeding programmes. Several approaches to identifying QTLs can be adopted. Interval mapping based on maximum likelihood methods (Lander and Botstein, 1989) can locate QTLs between marker loci but requires a detailed map, whereas other approaches such as regression analysis and ANOVA can detect linkage between markers and QTLs even if they cannot be positioned on a map, which may provide sufficient information for their exploitation within breeding programmes. Selective genotyping, in which only a small proportion of the segregating population is 'genotyped' may also provide a rapid means to derive markers for complex traits (Darvasi

and Soller, 1992). As with the initial development of linkage maps, the availability of recombinant homozygous lines developed through microspore culture greatly assists the genetic analysis of complex traits. QTLs have been identified for a range of morphological traits in *B. oleracea* (Kennard *et al.*, 1994) and several research programmes are currently identifying QTLs for agronomic traits in *B. napus.*

A variety of polymerase chain reaction (PCR)-based marker systems are also being developed. The process of PCR involves the amplification of specific DNA sequences that lie between pairs of oligonucleotide primers. Random amplified polymorphic DNA markers (RAPDs) use arbitrary short oligonucleotides (usually 10-mers) as primers. The amplified DNA is then separated by electrophoresis and visualized with ethidium bromide. While the technological investment required for RAPDs is less than that for RFLPs, their usefulness may be limited by other factors such as reproducibility and by problems with their use in complex genomes such as *Brassica*. Sequence analysis of amplified bands can lead to the development of longer primers with greater specificity which are more reliable. Other specific primers may consist of short repeats which may amplify microsatellite loci, which can be integrated into RFLP maps (Bell and Ecker, 1994).

Gene identification and isolation

Genetic engineering involves the transfer of copies of a gene or parts of a gene from a donor organism to a recipient organism. At present, it is only possible to transfer relatively small numbers of genes at a time using transformation techniques. This limits the use of transformation to the manipulation of genes involved in simple genetic traits, e.g. genes encoding enzymes responsible for fatty acid modification in seeds. In order to isolate the gene controlling a particular trait, it is first necessary to identify the relevant gene. This often requires a sophisticated knowledge of plant biochemical pathways and physiological processes. For example, the enzymes responsible for the formation of very-long-chain fatty acids such as erucic acid, are acyl-CoA-dependent elongases. However, the presence of elongases alone is not sufficient to ensure high levels of erucic acid in the seed oil. An additional class of enzyme, termed acyltransferases, is responsible for the assembly of newly synthesized fatty acids on to a glycerol backbone to form the triacylglycerols that constitute the seed oil (Murphy, 1993). Therefore, research groups involved in using a biotechnological approach to produce very high erucic rapeseed cultivars have identified an acyltransferase gene as an additional target for isolation and subsequent transformation.

The classical approach to gene isolation involved the purification of the protein (normally an enzyme) encoded by the gene. The purified protein could then be partially sequenced and/or antibodies raised against it. In these respective cases, specific oligonucleotide or antibody probes would then be available in order to screen gene libraries, leading to the isolation of the gene of interest. Such a method has been used to isolate a soluble n-9-stearate desaturase gene from several *Brassica* species (Slocombe *et al.*, 1992; Knutzon *et al.*, 1992). This gene is responsible for the formation of the most common fatty acids found in edible rapeseed oil, i.e. oleic, linoleic and linolenic. It is therefore an important gene in programmes involved in the modification of *Brassica* seed oil content.

In many cases, however, it is very difficult to purify the target protein and more indirect methods for gene isolation must be used. If the target protein belongs to a wider family of proteins, some of whose members have already been isolated and

their genes cloned from other organisms, it may be possible to isolate it using sequence homology via PCR. This technique has considerably facilitated the isolation and manipulation of genes over the past few years. For example, PCR has recently been used in our laboratories and elsewhere for the cloning of part of the acetyl-CoA carboxylase gene from rapeseed (R. Deka and S. Rawsthorne, personal communication). This gene is responsible for the first committed step of fatty acid biosynthesis and is therefore potentially a key regulatory point in determining the rate of fatty acid synthesis, and hence the amount of oil, that is produced in a seed.

In cases where protein purification or the use of homologies with existing related proteins is not possible, it is necessary to resort to more indirect methods of gene isolation, often using the model plant *Arabidopsis thaliana* as an intermediate system. This species is a member of the *Brassicaceae* and is therefore a close relative of the *Brassica* oilseeds. *Arabidopsis* is exceptionally amenable to a molecular genetic manipulation due to its small genome size, its ease of transformation and the large number of cloned genes and molecular markers that are now available. For example, the location of several fatty acid desaturase mutants in *Arabidopsis* were mapped with respect to RFLP markers, with sufficient precision to allow for isolation of the relevant genes by the technique of chromosome walking (Arondel *et al.*, 1992). The isolation of the n-15-linoleate desaturase gene from *Arabidopsis* was quickly followed by the isolation of its homologue from rapeseed (Arondel *et al.*, 1992), hence opening up the prospect of using molecular genetics to produce a premium-grade seed oil entirely free of the oxidation-prone fatty acid, α-linolenic acid.

Another powerful approach to the isolation of genes whose protein products are unknown or unavailable involves the use of gene tagging. One method, termed DNA tagging, involves the random insertion of a foreign piece of DNA, such as T-DNA from *Agrobacterium*, into the host, i.e. *Arabidopsis*, genome. Often only a single gene will be disrupted and by screening for the required phenotype (such as an altered seed fatty acid profile), it is possible to isolate the gene encoding the desired character by using probes based on the T-DNA sequence. This method has been used recently to isolate an n-12-oleate desaturase gene from *Arabidopsis* (Okuley *et al.*, 1994). Once again, the *Arabidopsis* gene was then used to isolate the homologous oleate desaturase from rapeseed, thus opening up the prospect of producing a very high oleic seed oil, for either edible or industrial use.

Brassica transformation

Genetic modification by plant transformation is relatively far advanced in *Brassica* species compared with most other major food crops. This is largely due to the availability of numerous tissue culture techniques for the regeneration of whole plants via organogenesis and embryogenesis.

The three principal transformation methods used successfully in *Brassica* oilseeds involve: (i) *Agrobacterium* infection; (ii) microprojectile bombardment and (iii) direct DNA uptake into protoplasts. Of the three methods, *Agrobacterium* infection of tissue explants has proved to be the most effective. In recent years, hypocotyl and cotyledonary explants have been used as sources of tissue for *Brassica* transformation via *Agrobacterium* infection. At the time of writing, the most efficient *Brassica* transformation method, with reliable transformation efficiencies of 10–40% in *B. napus*, is that of Moloney *et al.* (1989) which uses cotyledonary petioles. This system routinely produces shoots from over 80% of explants with *B. napus, B. juncea* and *B.*

oleracea. Other methods of *Brassica* transformation have recently been reviewed (Moloney and Holbrook, 1993). Although most efforts in *Brassica* oilseed transformation have focused upon *B. napus*, increasing attention is now being paid to other important *Brassica* oilseeds such as *B. juncea* and *B. rapa* and it is likely that routine high-efficiency transformation methods will soon be available for all of the economically important *Brassica* oilseeds.

Transgene expression

Gene expression in eukaryotes is regulated primarily by the 5′ non-coding region of the gene, i.e. the promoter. The tissue specificity (i.e. varying between different tissue types) and the developmental specificity (i.e. varying between different stages of development in the same tissue) of gene expression are regulated by several promoter elements acting together in a manner which has yet to be fully elucidated. In some *Brassica* gene promoters, the elements responsible for tissue specificity are relatively close, i.e. within 200–400 base pairs, to the transcriptional start site of the gene (Stayton *et al.*, 1991). However, in other *Brassica* genes, controlling elements have been identified as much as 2000 base pairs (bp) upstream of the transcriptional start site (Plant *et al.*, 1994).

Although the factors regulating transgene expression are relatively poorly understood, it has been possible to use *Brassica* seed-specific promoters, of 800–2000 bp, to obtain modest to high levels of expression of transgenes in a seed-specific and temporally regulated manner. Much of this work has involved the use of seed storage protein gene promoters such as napin or cruciferin (Stayton *et al.*, 1991). More recently, other seed-specific promoters such as oleosin and acyl carrier protein have been used (Knutzon *et al.*, 1992). An interesting alternative technique has involved the development of chimeric gene promoters, which combine seed-specific elements from endogenous *Brassica* promoters with elements from other non-plant genes to ensure high levels of tissue-specific expression (Comai *et al.*, 1990). In the future it will be important to build upon progress in the elucidation of the basic mechanisms of endogenous gene expression in order to facilitate the design of appropriate [promoter + gene] cassettes in order to achieve optimal modes of transgene expression.

Gene amplification/inhibition

While transformation enables novel genes to be inserted into oilseed crops, phenotypic changes may also be effected by the amplification and suppression of exisiting genes. This involves the insertion of complete or partial copies of genes from the host plant back into the same species. In the case of gene amplification, the objective is to obtain enhanced levels of gene expression by inserting additional complete copies of the relevant gene into the host plant. The expectation is that the additional copies will be expressed, leading to increased formation of the gene product. If the gene product is an enzyme that is not already present in excess, elevated levels of the enzyme can lead to increased throughput in the relevant metabolic pathway, leading to the desired phenotypic change. Examples of pathways that could be manipulated in this way include fatty acid formation, glucosinolate catabolism, triacylglycerol accumulation and storage protein deposition. However, it has been found that the insertion of additional, complete or partial, copies of a gene into a plant can sometimes lead to the inhibition of the endogenous gene activity, a phenomenon termed 'co-suppression'. The occurrence of co-suppression, the mechanism of which remains obscure, makes it difficult to consider gene amplification via the insertion of extra gene copies as a general strategy for crop improvement at the present time.

The deliberate suppression of gene activity is often a desirable goal in crop improvement. For example, in the *Brassica* oilseeds, the presence of certain classes of glucosinolates and certain fatty acids may adversely affect seed quality, particularly for edible use. It is possible to inhibit the accumulation of such undesirable products by suppressing the expression of genes involved in their biosynthesis. By far the most important method in current use is that involving antisense genes. This involves the insertion of a partial or complete copy of the target gene in a reverse, or antisense, orientation. Expression of the antisense gene leads to the suppression of the expression of the equivalent sense gene in the host plant. Perhaps the most dramatic example of the use of antisense technology in *Brassica* oilseeds involves the n-9-stearoyl ACP desaturase gene, responsible for the conversion of stearic acid to oleic acid in developing seeds. A partial copy of the n-9 desaturase gene from *B. rapa* was inserted in the antisense orientation into *B. napus* plants under the control of a seed-specific promoter (Knutzon *et al.*, 1992). The result was a significant inhibition of the endogenous n-9 desaturase gene expression, leading to the accumulation of a seed oil containing up to 40% stearic acid, whereas normal rapeseed oil contains less than 5% of this fatty acid. Antisense methods have also been used to reduce the accumulation of napin (one of the two major seed storage proteins) by as much as 90% in *B. napus* (Hayakawa and Murase, personal communication).

GOALS FOR CROP IMPROVEMENT

Modifying agronomic traits

Disease resistance
Biotechnological approaches to enhancing

disease resistance involve either exploitation of natural forms of resistance or genetic
engineering approaches, such as the introduction of chitinases, glucanases and other
antifungal proteins. The former approach
may involve the introduction of novel resistance genes from wild species (e.g. Crouch
et al., 1994) and the subsequent introgression of genes through the use of
molecular markers, or attempts to clone
resistance genes. The approach to gene
cloning that is most likely to be successful
is to exploit *Arabidopsis*. For example,
attempts are being made to clone resistance
genes to *Peronospora parasitica* which have
been mapped in *Arabidopsis* (Parker et al.,
1993), and which could be used as probes
to isolate homologous genes from *Brassica*.

Chitinases, which hydrolyse the β-
(1,4)-glycoside in chitin (a major component of fungal cell walls), are often induced
in plants following fungal attack and it is
thought that they are involved in plant
defence. Constitutive expression of a bean
chitinase in *B. napus* led to a reduction in
susceptibility to *Rhizoctonia solani* (Brogile
et al., 1991). Lines which contain chitinases
that are specifically induced in *Brassica* by
pathogens such as *Leptosphaeria maculans*
(Rasmussen et al., 1992) may give a higher
degree of protection. Another strategy has
been the introduction of a gene for oxalate
oxidase in order to reduce susceptibility to
infection by *Scelerotinia sclerotorium* which
relies upon the production of oxalic acid
in the infection process (Thompson et al.,
1993). Likewise, genes which may reduce
the susceptibility of *Brassica* to insect attack
such as trypsin inhibitors and *Bacillus
thuringensis* (Bt) toxins have been introduced
into breeding lines. How effective transgenic lines will be in a field situation
remains to be tested.

Hybrids

Hybrid rapeseed has been demonstrated
to be higher yielding than inbred lines
(Brandle and McVetty, 1989). Self-incompatibility, reliant on the introduction of
S alleles from synthetic lines, and cytoplasmic male sterility (CMS) are currently
being exploited in several breeding programmes as a means of developing hybrid
cultivars. Several systems of CMS are
available, but none has proven completely
satisfactory (Banga, 1993). Recently a
novel method of producing hybrid cultivars
has been developed which involves transgenic parental lines with *barnase* (bacterial
RNAase) genes driven by a tapetum
specific promoter (*TA29*). Expression of
the *barnase* gene in the tapetal cells of rapeseed flowers results in male sterility.
Another gene called *barstar* encodes an
inhibitor of *barnase*. Therefore, crosses between male-sterile lines containing a
TA29-barnase gene and male-fertile lines
containing a *TA29-barstar* gene result in
hybrids which contain both genes but are
male-fertile due to the inhibitory action of
barstar on *barnase*, both of which are
expressed only in the tapetal cells. The
TA29-barstar gene thus acts as a dominant
restorer of male fertility. Additionally, the
TA29-barstar and *TA29-barnase* genes are
linked to the *bar* gene which gives resistance
to the herbicide, bialophos, and enables the
selection of individuals which contain the
transgenes from progeny derived from
hybrid lines. This system provides a viable
method for developing hybrid cultivars and
maintaining lines which contain the male
restorer gene (Mariani et al., 1992). The
first field trials of transgenic rapeseed
hybrids produced using the *barnase/barstar*
system were scheduled to be held in the UK
in 1994. In addition to *barnase/barstar*,
numerous additional hybrid *Brassica* systems are being developed, some of which
involve the use of biotechnological techniques. It is likely that, within the next
decade, the majority of new *Brassica* oilseed
cultivars will be hybrids, due to their
superior yields and the necessity to protect
newly developed breeding lines.

Other agronomic characters

It may be desirable to manipulate a range of quantitative traits, such as oil and protein content, flowering time, plant architecture, winter hardiness and drought tolerance, to develop cultivars for particular products and for specific environments. Molecular marker technology and genetic analysis will be particularly valuable in elucidating and manipulating QTLs which regulate these characters. Transgenic approaches may also be possible, particularly through exploitation of cloned genes from *Arabidopsis*.

Oil modification

Metabolic pathways

Brassica seed oils are made up of triacylglycerols, comprising a glycerol moiety esterified to three fatty acyl residues, the chemical nature of which determines the properties and end use of the oil. The main ways in which fatty acids can vary are with respect to their chain length and their functionality. Interesting functionalities in fatty acids include double bonds, hydroxyl groups, epoxy groups and hydroperoxide groups. Seed fatty acids are synthesized according to the biochemical pathway shown in outline form in Fig. 8.1. This figure shows the large range of possible fatty acids that can be synthesized and incorporated into a seed oil. In most plant species, however, a relatively restricted range of fatty acids is present in the seed oil. For example in the 'double-low' varieties of rapeseed, over 98% of the seed oil is made up of only five C_{16} and C_{18} fatty acids, i.e. palmitic, stearic, oleic, linoleic and α-linolenic acids. High erucic cultivars of rapeseed can accumulate very-long-chain fatty acids, such as 20 : 1 and 22 : 1, in addition to the C_{16} and C_{18} fatty acids also present in double-low varieties. However, *Brassica* oilseeds produced by conventional breeding are not able to accumulate other potentially useful fatty acids such

as short- or medium-chain fatty acids, hydroxy fatty acids or epoxy fatty acids. A major contribution of modern biotechnology to *Brassica* oilseed improvement will be to enable the transfer of genes responsible for the formation of these exotic fatty acids to create novel transgenic varieties. Gene technology can also be used to manipulate the ratios of existing fatty acids in order to enhance the usefulness of *Brassica* seed oils.

Edible oils

Zero erucic acid cultivars of *Brassica* oilseeds normally contain 60–80% oleic acid, 10–30% polyunsaturated fatty acids and only 5–10% saturated fatty acids. This fatty acid profile gives an excellent salad oil that is comparable to that of olive oil, but the presence of relatively high levels of polyunsaturates reduces the oxidative stability and limits the utility of rapeseed oil for cooking or margarine production, unless it is first hydrogenated. The recent availability of cloned genes encoding the *Brassica* oleate desaturase opens up the prospect of using antisense technology to reduce or even eliminate polyunsaturates in rapeseed oil (Okuley *et al.*, 1992). Similar methods have already been used to down-regulate the stearate desaturase, resulting in the production of a new rapeseed variety containing as much as 40% stearic acid in its seed oil (Knutzon *et al.*, 1992). This new high stearic oil could be blended with conventional high oleic rapeseed oil for margarine production and may also be used for the manufacture of cocoa butter substitutes.

Industrial oils

Although edible oils currently represent the major market outlet for *Brassica* oilseeds, the prevalence of agricultural surpluses in many developed countries has focused attention on the possible industrial use of *Brassica* seed oils. Several examples of fatty acids which can be manipulated in or

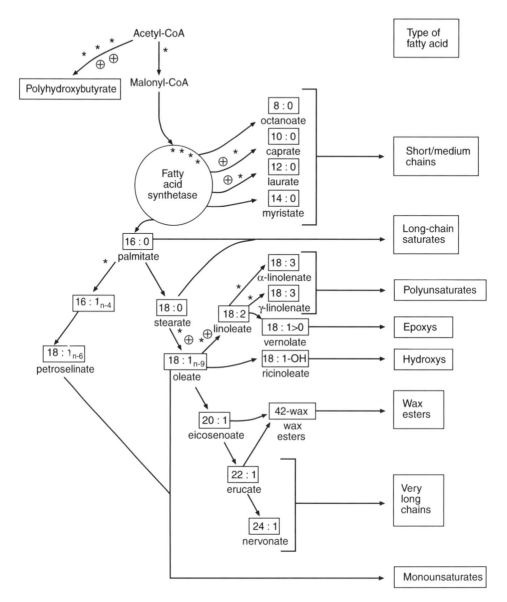

Fig. 8.1. Pathway of fatty acid biosynthesis in oilseeds. In plants, fatty acids are formed from acetyl-CoA via the stepwise addition of malonyl-CoA units on a multi-enzyme complex, termed fatty acid synthase (FAS). This normally involves eight cycles through FAS to produce 16 : 0 (palmitate). In some plants, however, premature chain termination by acyl-ACP thioesterases leads to the accumulation of short- and medium-chain acyl groups. In the *Umbelliferae*, palmitate can be converted into $18 : 1_{n-6}$ (petroselinate) via a $16 : 1_{n-4}$ intermediate. More commonly, palmitate is elongated to 18 : 0 (stearate) and desaturated to $18 : 1_{n-9}$ (oleate) which has been termed the 'central substrate' of long-chain fatty acid metabolism. Oleate can undergo a variety of modifications to produce polyunsaturates, epoxy acids, wax esters, hydroxy acids and very-long-chain

added to *Brassica* oilseeds are given below.

Erucic acid (C_{22 : 1}). Although this is already a major industrial feedstock, the best available varieties of high erucic rapeseed contain only 45–50% erucic acid in their seed oil. The reason for this is that erucic acid is effectively excluded from the C_2 position of the triacylglycerol molecule, as the relevant acyltransferase fails to recognize erucic acid as an efficient substrate. Several laboratories in Europe and North America are attempting to rectify this problem by cloning genes for acyltransferases which do recognize erucic acid efficiently, e.g. those from *Limnanthes* spp., and inserting these into rapeseed. If successful, these efforts may result in the production of transgenic rapeseed varieties containing over 90% erucic acid which, as outlined in Chapter 12, would greatly stimulate the market for erucic-derived oleochemical feedstocks. An alternative route may be to use the recently described *B. oleracea* genotypes that can accumulate up to 35% erucic acid on the C_2 position (Taylor *et al.*, 1994) in order to create resynthesized *B. napus* lines with enhanced erucic levels which can then be backcrossed into elite *B. napus* cultivars.

Lauric acid (C_{12 : 0}). Lauric acid is an important industrial feedstock in the detergent and surfactant industry, as well as having some edible applications. New transgenic rapeseed varieties containing over 30% lauric acid have recently been produced in the USA (Voelker *et al.*, 1992). This lauric-containing rapeseed has an additional thioesterase gene obtained from the California Bay plant, which normally accumulates short-chain fatty acids in its seed oil. The widespread commercial use of this novel rapeseed variety will probably depend upon a further increase in the proportion of lauric acid from 30% up to the values of 50% and beyond that are present in competitive sources of lauric acid such as coconut and palm kernel oil. Current economic conditions notwithstanding, the real benefit of this development is to diversify the sources of lauric acid, which may be beneficial for lauric processors if their traditional sources of supply become less accessible.

Ricinoleic acid (C_{18 : 1-OH}). This is a 12-hydroxylated derivative of oleic acid which constitutes up to 90% of castor seed oil and has many medium- to high-value industrial uses in products ranging from cosmetics and pharmaceuticals to polymers and high-grade lubricants. Several groups in the USA and Europe are attempting to clone the oleate hydroxylase gene from castorbean for transfer into rapeseed in order to produce a high ricinoleic oilseed which can be grown in the temperate climatic zones of Europe and North America, where castorbean cultivation is not possible (Murphy, 1994).

Petroselinic acid (C_{18 : 1,n-6}). This is an isomer of oleic acid with the double bond in the n-6 rather than the n-9 position. A desaturase gene believed to be responsible for petroselinic acid formation has been cloned by groups in the USA and UK from coriander (Cahoon *et al.*, 1992). This gene has been inserted into rapeseed with the aim of producing transgenic plants containing a high petroselinic seed oil. Petroselinic acid has the potential to be

acids. The *Brassica* oilseeds only produce a small fraction of this wide range of fatty acids.

During the past few years, many of the genes encoding enzymes of fatty acid biosynthesis and modification (*) have been cloned from various species and in several cases transgenic *Brassica* oilseed plants containing novel fatty acids (⊕) have been produced.

a useful industrial feedstock in the manufacture of detergents and adipic acid-based polymers. The production of adipic acid from petroselinic acid via an environmentally clean process represents an attractive alternative to the present manufacturing techniques, whereby adipic acid is produced from non-renewable petroleum feedstocks in a process which gives rise to substantial emissions of the ozone-depleting greenhouse gas, N_2O (Draths and Frost, 1994).

Epoxy fatty acids. Epoxy fatty acids can be used in the manufacture of resins and coatings, such as paints. Several plants contain large amounts of epoxy fatty acids in their seed oils, most notably *Vernonia stokesia* and *Euphorbia* species. In *V. stokesia*, the epoxidases responsible for synthesis of vernolic acid (*cis*-12,13-epoxy-*cis*-9-octadecenoic acid) from linoleic acid, appear to resemble fatty acid desaturases in some respects and may be structurally related to them (Bafor *et al.*, 1993). This suggests that PCR-based strategies may be used to clone epoxidase genes from such species for insertion into rapeseed in order to produce transgenic rapeseed varieties capable of accumulating commercially useful epoxy fatty acids.

Non-oilseed lipids

Wax esters. Wax esters are relatively uncommon storage compounds in plants, but do make up the major seed reserve in the desert shrub jojoba. Jojoba wax consists of a C_{20} or C_{22} fatty acid esterified to a fatty alcohol of similar chain length. This liquid wax has numerous industrial uses, such as in lubricants and cosmetics. Very-long-chain fatty alcohols are produced from fatty acids in jojoba by a reductase and then esterified to very-long-chain fatty acids by a ligase. The jojoba reductase gene has recently been cloned and it is anticipated that the ligase gene will also be cloned in the future (Anon., 1992). The aim is then to transfer these two genes into high erucic rapeseed in order to create a transgenic variety capable of accumulating commercially valuable waxes, rather than triacylglycerols, as its seed reserve.

Polyhydroxybutyrate. Polyhydroxybutyrate (PHB) and polyhydroxybutyrate hydroxyvalerate (PHB/V) are polymers produced commercially via fermentation, using the bacterium *Alcaligenes eutrophus*. The product is marketed by Zeneca plc as BIOPOL and is the only thermoplastic derived from renewable resources that is highly durable and yet completely biodegradable. More economic production of PHB may be achieved if it could be produced in plants. To this end, genes encoding the three enzymes of PHB synthesis have been transferred from the bacterium to rapeseed (Smith *et al.*, 1993). Preliminary experiments suggest that very low levels of PHB formation are possible in transgenic plants. Since PHB synthesis uses the same precursor as oil synthesis, i.e. acetyl-CoA, the two pathways will compete with each other. This can be resolved by using antisense methods for the partial suppression of oil synthesis from acetyl-CoA, hence allowing higher levels of PHB accumulation (Elborough *et al.*, 1994). The ultimate aim is to produce a novel transgenic rapeseed variety which contains up to 50% of its seed lipid as PHB but which still contains sufficient storage oil to allow normal seed germination and growth.

Seed meal modification

Glucosinolates

The presence of glucosinolates in *Brassica* seed meals significantly reduces their nutritional quality and potential use in animal feeds. Reductions in the glucosinolate content of rapeseed have been a major goal of oilseed *Brassica* breeders. Conventional breeding programmes have achieved large reductions in total aliphatic glucosinolate content through the use of the cultivar Bronowski. Further reduction in both aliphatic glucosinolates (derived from methionine) and indolyl glucosinolates (derived from tryptophan) are desired. Synthetic *B. napus* lines have been developed which have significantly lower levels of indolyl glucosinolates compared to high glucosinolate cultures (Kräling *et al.*, 1990). A reduction of 97% of indolyl glucosinolates in rapeseed has been achieved by the introduction of a gene which encodes for tryptophan decarboxylase. This gene resulted in tryptophan being directed into tryptamine, rather than indolyl glucosinolate biosynthesis (Chavedej *et al.*, 1994). Biotechnological approaches are also being adopted to reduce aliphatic glucosinolate content. For example, the glucosinolate biosynthetic enzyme, thiohydroxyimate *S*-glucosyltransferase has recently been purified (Reed *et al.*, 1993). This enzyme could be used to obtain clones of the corresponding gene for use in antisense-mediated down-regulation of aliphatic glucosinolate biosynthesis.

In addition to modifying total glucosinolate content in seeds, it may be desirable to modify the levels of individual aliphatic glucosinolates in seeds. This could enhance meal quality and pest and disease resistance. It has been shown that the types of aliphatic glucosinolates which are present in *B. napus* are under simple genetic control and that they can be manipulated through the introgression of genes from synthetic *B. napus* lines (Magrath and Mithen, 1993; Magrath *et al.*, 1994). For example, the hydroxylation of alkenyl glucosinolates has been shown to be dependent upon two genes in *B. napus* which occur on homologous linkage groups in the A and C genome (Parkin *et al.*, 1994). RFLP markers closely linked to these genes can be used to introgress null alleles into breeding lines. Eliminating hydroxyalkenyl glucosinolates may be expected to enhance meal quality, these being the major goitrogenic compounds in

rapeseed. Their elimination would also decrease the palatability of the leaf tissue to non-specific vertebrate pests such as pigeons and rabbits (Mithen, 1992) due to the concomitant increase in alkenyl glucosinolates. Other types of modifications are also possible, both through introgression of genes from exotic sources and through map-based cloning of genes from *Arabidopsis*.

Essential amino acids

Brassica oilseeds typically contain 20–25% (w/w) of a highly nutritious protein meal comparable in quality to soy meal. Nevertheless, most plant seed meals are deficient in certain essential amino acids, such as lysine and methionine. Efforts are now underway to transfer genes encoding methionine-rich storage proteins from species such as the Brazil nut, *Bertholletia excelsa*, into rapeseed to produce a balanced protein meal which would compare favourably with animal-derived proteins such as meat and dairy products.

Molecular farming

This involves the engineering of oilseeds to supply products such as pharmaceuticals where the tonnages required may be relatively low but the value of the end product is extremely high. Recent developments make molecular farming highly relevant in considering future options for the development of novel *Brassica* oilseed crops. It has been recognized for some time that plants, and in particular, seeds could potentially be used as bioreactors to produce relatively large amounts of high-value products, including pharmaceutical peptides and industrial enzymes. The problem has always been that it is both difficult and expensive to extract such peptides from the transgenic seeds (Krebbers *et al.*, 1993).

Recent developments in Canada offer an ingenious way to circumvent this problem (Moloney, 1993). The high-value peptides or proteins are synthesized as fusion products attached to the oleosin proteins which coat seed oil bodies. This means that the fusion products can be purified away from all other soluble proteins by a simple flotation process. The novel peptides or proteins can then be removed from the oil droplets by a straightforward enzymatic cleavage reaction and recovered in virtually pure form. Much developmental work needs to be done but rapeseed molecular farming may offer a genuine alternative to conventional microbial fermentation for the production of pure high-value peptides or proteins for a variety of end uses, ranging from pharmaceutical peptides, such as hirudin or interleukin, to industrial enzymes, such as cellulases, lipases or proteases.

the genetics and molecular biology of *Arabidopsis thaliana* will be of almost immediate practical benefit to *Brassica* oilseed improvement. This has already been demonstrated by the cloning of most of the major fatty acid desaturase genes from *Brassica* species using *Arabidopsis* molecular genetics. The availability of effective *Brassica* hybrid systems, some of which are based on biotechnology, will enable seed companies to protect their often considerable investment in the development of novel cultivars. In the broader agricultural context, the application of biotechnology will lead to new crops for farmers to grow, the production of industrial oils from a renewable resource, cheaper production of high-value pharmaceuticals and more environmentally sustainable agricultural systems.

CONCLUSIONS

In conclusion, the application of biotechnology to *Brassica* oilseed development will have manifold benefits ranging from the enhancing of the speed and efficiency of conventional programmes to the introduction of genes from unrelated species and the production of novel products, such as industrial oils and pharmaceuticals. The considerable investment in research into

ACKNOWLEDGEMENTS

We are grateful to all of our colleagues in the Brassica and Oilseeds Research Department for their assistance in the preparation of this chapter and particularly to Ray Mathias, Steve Rawsthorne and Joanne Ross for their comments on the manuscript.

REFERENCES

Agnihorti, A., Gupta, V., Lakshmikumaran, M.S., Shivanna, K.R., Prakash, S. and Jagannathan, V. (1990) Production of *Eruca–Brassica* hybrid by embryo rescue. *Plant Breeding* 104, 281–289.

Akbar, M.D. (1987) Artificial *Brassica napus* flowering in Bangladesh. *Theoretical and Applied Genetics* 73, 465–468.

Anon. (1992) Firm reports cloning of key reductase gene. *Inform* 3, 1220.

Arondel, V., Lemieux, B., Hwang, I., Gibson, S., Goodman, H.M. and Somerville, C.R. (1992) Map-based cloning of a gene controlling omega-3 fatty acid desaturation in *Arabidopsis*. *Science* 258, 1353–1355.

Bafor, M., Smith, M.A., Jonsson, L., Stobart, K. and Stymne, S. (1993) Biosynthesis of vernoleate (*cis*-12-epoxyoctadeca-*cis*-9-enoate) in microsomal preparations from developing endosperm of *Euphoribia lagascae*. *Archives of Biochemistry and Biophysiology* 303, 145–151.

Banga, S.S. (1993) Heterosis and its utilisation. In: Labanda, K.S., Banga, S.S. and Banga, S.K. (eds) *Breeding Oilseed Brassicas*. Springer Verlag, Berlin, pp. 21–43.

Batra, V., Prakash, S. and Shivanna, K.R. (1990) Intergeneric hybridisation between *Diplotaxis siifolia*, a wild species, and crop brassicas. *Theoretical and Applied Genetics* 80, 537–541.

Bell, C.J. and Ecker, J.R. (1994) Assignment of 30 microsatellite loci to the linkage map of *Arabidopsis. Genomics* 19, 137–144.

Brandle, J.E. and McVetty, P.B.E. (1989) Heterosis and combining ability in hybrids derived from oilseed rape cultivars and inbred lines. *Crop Science* 29, 1191–1195.

Brogile, K., Chet, I., Holliday, M., Cressman, R., Biddle, P., Knowlton, S., Mauvais, C.J. and Brogile, R. (1991) Transgenic plants with enhanced resistance to the fungal pathogen *Rhizoctonia solani. Science* 254, 1194–1197.

Cahoon, E.B., Shanklin, J. and Ohlrogge, J.B. (1992) Expression of a coriander desaturase results in petroselinic acid production in transgenic tobacco. *Proceedings of the National Academy of Sciences of the United States of America* 89, 11,184–11,188.

Chavedej, S., Brisson, N., McNeil, J. and DeLuca, V. (1994) Redirection of tryptophan leads to production of low indole glucosinolate canola. *Proceedings of the National Academy of Sciences of the United States of America* 91, 2166–2170.

Comai, L., Moran, P. and Maslyar, D. (1990) Novel and useful properties of a chimeric plant promoter combining CaMV 35S and MAS elements. *Plant Molecular Biology* 15, 373–381.

Crouch, J., Lewis, B. and Mithen, R. (1994) The effect of A genome substitution on the resistance of *Brassica napus* to *Leptosphaeria maculans. Plant Breeding* 112, 265–278.

Darvasi, A. and Soller, M. (1992) Selective genotyping for determination of linkage between a marker locus and a quantitative trait locus. *Theoretical and Applied Genetics* 85, 353–359.

Delourme, R., Eber, F. and Cherve, A.M. (1989) Intergeneric hybridization of *Diplotaxis erucoides* with *Brassica napus*. I Cytogenetic analysis of F1 and BC1 progeny. *Euphytica* 41, 123–128.

Draths, K.M. and Frost, J. (1994) Environmentally compatible synthesis of adipic acid from D-glucose. *Journal of American Oil Chemists' Society* 116, 399–400.

Elborough, K.M., Farnsworth, L., Winz, R.A., Simon, J.W., Swinhoe, R. and Slabas, A.R. (1994) Regulation of primary storage products of oilseeds by manipulating the level of genes involved in lipid metabolism on plant acetyl-CoA carboxylase. *Journal of Cellular Biochemistry* 18A, 113.

Fahleson, J., Eriksson, I., Landgren, M., Stymne, S. and Glimelius, K. (1994) Intertribal somatic hybrids between *Brassica napus* and *Thlaspi perfoliatum* with high content of the *T. perfoliatum*-specific nervonic acid. *Theoretical and Applied Genetics* 87, 795–804.

Green, A.G. (1986) Genetic control of polyunsaturated fatty acid biosynthesis in flax (*Linum usitatissium*) seed oil. *Theoretical and Applied Genetics* 72, 654–661.

Inomata, N. (1977) Production of interspecific hybrids between *Brassica campestris* and *Brassica oleracea* by culture of *in vitro* excised ovaries. I. Effect of yeast extract and casein hydrolysate on the development of excised ovaries. *Japanese Journal of Breeding* 27, 295–304.

Kennard, W.C., Slocum, M.K., Figdore, S.S. and Osborn, T.C. (1994) Genetic analysis of morphological variation in *Brassica oleracea* using molecular markers. *Theoretical and Applied Genetics* 87, 721–732.

Knutzon, D.S., Thompson, G.A., Radke, S.E., Johnson, W.B., Knauf, V.C. and Kridl, J.C. (1992) Modification of *Brassica* seed oil by antisense expression of a stearoyl-acyl carrier protein desaturase gene. *Proceedings of the National Academy of Sciences of the United States of America* 89, 2624–2628.

Kräling, K., Robellen, G., Theis, W., Herrmann, M. and Ahmadi, M.R. (1990) Variation of seed glucosinolates in lines of *Brassica napus. Plant Breeding* 105, 33–39.

Krebbers, E., Bosch, D. and Vanderkerckhore, J. (1993) Production of foreign proteins and peptides in transgenic plants. In: Van Beck, T.A. and Breteler, H. (eds) *Phytochemistry and Agriculture.* Clarendon Press, Oxford, pp. 346–355.

Lander, E.S. and Botstein, S. (1989) Mapping mendelian factors underlying quantitative traits using RFLP linkage maps. *Genetics* 121, 185–199.

Landry, B.S., Hubert, N., Etoh, T., Harada, J.J. and Lincoln, S.E. (1991) A genetic map for *Brassica napus* based on restriction fragment length polymorphisms detected with expressed DNA sequences. *Genome* 34, 543–552.

Lee, M. and Phillips, R.L. (1988) The chromosomal basis of somaclonal variation. *Annual Review of Plant Physiology and Plant Molecular Biology* 39, 413–437.

Lelivet, C.L.C., Lange, W. and Dolstra, O. (1993) Intergeneric crosses for the transfer of resistance to the beet cyst nematode from *Raphanus sativus* to *Brassica napus. Euphytica* 68, 111–120.

Lydiate, D., Sharpe, A., Lagercrantz, U. and Parkin, I. (1993) Mapping the *Brassica* genome. *Outlook on Agriculture* 22, 85–92.

Magrath, R. and Mithen, R. (1993) Maternal effects on the expression of individual aliphatic glucosinolates in seeds and seedlings of *Brassica napus*. *Plant Breeding* 111, 249–252.

Magrath, R., Bano, F., Morgner, M., Parkin, I., Sharpe, A., Lister, C., Dean, C., Turner, J., Lydiate, D. and Mithen, R. (1994) Genetics of aliphatic glucosinolates. I. Side chain elongation in *Brassica napus* and *Arabidopsis thaliana*. *Heredity* 72, 290–299.

Mariani, C., Gossele, V., De Beuckeleer, M., De Block, M., Goldberg, R.B., De Greef, W. and Leemans, J. (1992) A chimeric ribonuclease-inhibitor gene restores fertility to male sterile plants. *Nature (London)* 357, 384–387.

Mithen, R. (1992) Leaf glucosinolate profiles and their relationship to pest and disease resistance in oilseed rape. *Euphytica* 63, 71–83.

Mithen, R. and Herron, C. (1991) Transfer of disease resistance to oilseed rape from wild *Brassica* species. In: McGregor, D.I. (ed.) *Proceedings of the Eighth International Rapeseed Congress*, Saskatoon, Canada. Organizing Committee, Saskatoon, pp. 244–250.

Moloney, M.M. (1993) Oil-body proteins as carriers of high-value peptides in plants. International Patent Publication Number WO 93/21320.

Moloney, M.M. and Holbrook, L.A. (1993) Transformation and foreign gene expression. In: Labanda, K.S., Banga, S.S. and Banga, S.K. (eds.) *Breeding Oilseed Brassicas*. Springer-Verlag, Berlin, pp. 148–167.

Moloney, M.M., Walker, J. and Sharma, K. (1989) High efficiency transformation of *Brassica napus* using *Agrobacterium* vectors. *Plant Cell Reports* 8, 238–242.

Murphy, D.J. (1993) Structure, function and biogenesis of storage lipid bodies and oleosins in plants. *Progress in Lipid Research* 32, 247–280.

Murphy, D.J. (1994) Biotechnology of oil crops. In: Murphy, D.J. (ed.) *Designer Oil Crops*. VCH Press, Weinheim, Germany, pp. 219–251.

Okuley, J., Lightner, J., Feldmann, K., Yadav, N., Lark, E. and Browse, J. (1994) *Arabidopsis* FAD2 gene encodes the enzyme that is essential for polyunsaturated lipid synthesis. *Plant Cell* 6, 147–158.

Olsson, G. (1986) Allopolyploids in *Brassica*. In: Olsson, G. (ed.) *Svalöf, 1886–1986. Research and Results in Plant Breeding*. Svalöf AB, Svalöf, Sweden, pp. 114–119.

Ozminkowski, R.H. Jr and Jourdan, P.S. (1993). Expression of self incompatibility and fertility of *Brassica napus* L. resynthesised by interspecific somatic hybridisation. *Euphytica* 65, 153–160.

Parker, J.E., Szabo, V., Staskawicz, B.J., Lister, C., Dean, C., Daniels, M.J. and Jones, J.D.G. (1993) Phenotypic characterisation and molecular mapping of the *Arabidopsis thaliana* locus RPP5, determining resistance to *Peronospora parasitica*. *Plant Journal* 4, 821–832.

Parkin, I., Magrath, R., Keith, D., Sharpe, A., Mithen, R. and Lydiate, D. (1994) Genetics of aliphatic glucosinolates. II. Hydroxylation of alkenyl glucosinolates in *Brassica napus*. *Heredity* 72, 594–598.

Plant, A.L., van Rooijen, G.J.H., Anderson, C.P. and Moloney, M.M. (1994) Regulation of an *Arabidopsis* oleosin gene promoter in transgenic *Brassica napus*. *Plant Molecular Biology* 25, 193–205.

Rasmussen, U., Giese, H., Mikkelsen, J.D. (1992) Induction and purification of chitinase in *Brassica napus* L. ssp. *oleifera* infected with *Phoma lingam*. *Planta* 187, 328–334.

Reed, D.W., Davin, L., Jain, J.C., DeLuca, V., Nelson, L. and Underhill, E.W. (1993) Purification and properties of UDP-glucose:thiohydroximate glucosyltransferase from *Brassica napus* seedlings. *Archives of Biochemistry and Biophysics* 305, 526–532.

Ripley, V.L. and Arnison, P.G. (1990) Hybridisation of *Sinapis alba* L. and *Brassica napus* L. via embryo rescue. *Plant Breeding* 104, 26–33.

Sacristan, M.D. (1982) Resistance response to *Phoma lingam* of plants regenerated from selected cell and embryogenic cultures of haploid *Brassica napus*. *Theoretical and Applied Genetics* 62, 193–200.

Schenck, H.R. and Röbbelen, G. (1982) Somatic fusion of protoplasts from *Brassica oleracea* and *B. campestris*. *Zeitschrift für Pflanzenzücht* 89, 278–288.

Sjodin, C. and Glimelius, K. (1989) Transfer of resistance to *Phoma lingam* to *Brassica napus* by asymmetrical somatic hybridisation combined with toxin selection. *Theoretical and Applied Genetics* 78, 513–520.

Slocombe, S.P., Cummins, I., Jarvis, P. and Murphy, D.J. (1992) Nucleotide sequence and temporal regulation of a seed-specific *Brassica napus* cDNA encoding a stearoyl-acyl carrier protein (ACP) desaturase. *Plant Molecular Biology* 20, 151–156.

Slocum, M.K., Figdore, S.S., Kenard, W.C., Suzuki, J. and Osborn, T. (1990) Linkage arrangement of restriction fragment length polymorphism loci in *Brassica oleracea*. *Theoretical and Applied Genetics* 80, 57–64.

Smith, E., White, K.A., Aves, V.A., Holt, D.C., Fentem, A.P. and Bright, S.W.J. (1993) The production of poly-β-hydroxybutyrate in transgenic oilseed rape plants. In: *Proceedings of the Second European Symposium on Industrial Crops and Products*, Pisa, Italy.

Stayton, M., Harpster, M., Brosio, P. and Dunsmuir, P. (1991) High-level, seed-specific expression of foreign coding sequences in *Brassica napus*. *Austrian Journal of Plant Physiology* 18, 507–517.

Sunberg, E. and Glimelius, K. (1986) A method for the production of interspecific hybrids within the Brassiceae via somatic hybridisation, using resynthesis of *Brassica napus* as a model. *Plant Science* 43, 155–162.

Takahata, T. and Takeda, T. (1990) Intergeneric (intersubtribe) hybridisation between *Moricandia arvensis* (L.) DC and *Brassica* A and B genome species by ovary culture. *Theoretical and Applied Genetics* 80, 38–42.

Taylor, D.C., MacKenzie, S.L., McCurdy, A.R., McVetty, P.B.E., Giblin, E.M., Pass, E.W., Stone, S.J., Scarth, R., Rimmer, S.R. and Pickard, M.D. (1994) Stereospecific analyses of seed triacylglycerols from high-erucic acid Brassicaceae: detection of erucic acid at the sn-2 position in *Brassica oleracea* L. genotypes. *Journal of American Oil Chemists' Society* 71, 163–167.

Thompson, C., Dunwell, J.M., Johnstone, C.E., Lay, V., Ray, J., Schmitt, M., Watson, H. and Nisbet, G. (1993) Degradation of oxalic acid by transgenic canola plants expressing oxalate oxidase. In: *Proceedings of the Eighth Crucifer Genetics Workshop*, Saskatoon, Canada, p. 31.

Voelker, T.A., Worrell, A.C., Anderson, L., Bleibaum, J., Fan, C. and Hawkins, D.J. (1992) Fatty acid biosynthesis redirected to medium chains in transgenic oilseed plants. *Science* 257, 72–73.

⑨ Environmental Impact of Rapeseed Production

R. Marquard[1] and K.C. Walker[2]

[1]*Institut für Planzenbau und Pflanzenzüchtung, Giessen;*
[2]*Scottish College of Agriculture, Aberdeen*

INTRODUCTION

Under intensive farm management environmental influences of individual crops such as oilseed rape may be difficult to differentiate from those of other crops in a rotation. Nevertheless speculation about these influences arouses controversy and some environmental aspects of rapeseed production are considered in this chapter.

NITROGEN

Nitrates in soils and groundwater

In the European Union (EU), compliance with the mandatory $50 \text{ mg} \text{l}^{-1}$ nitrate limit in water is often a problem in regions of intensive agriculture (Schneider and Haider, 1992), where considerable amounts of nitrate may enter surface and groundwater (Maidl *et al.*, 1991). The level of nitrate leaching is primarily related to groundwater regeneration, i.e. the amount and distribution of rainfall (Scheffer and Schachtschabel, 1989). However, the level of nitrogen fertilizer application, soil type and crop rotation also have a strong

influence. Furthermore, the effect of crop production on nitrate leaching varies depending on the extent of crop canopy development and root depth.

Rapeseed is considered to have a beneficial effect in delaying nitrate leaching after sowing, but increased leaching may occur after harvest unless control measures are adopted. Rapeseed crops have an autumn nitrogen requirement of 40–80 kg $N \text{ ha}^{-1}$ and, therefore, most of the leachable nitrogen from autumn mineralization is taken up and fixed in the biomass (Sauermann, 1993). Inserting a winter-sown rapeseed crop in a cereal–sugarbeet rotation can decrease the amount of nitrate leached from sandy soils from 30 to 15 kg $N \text{ ha}^{-1}$ (Scheffer and Schachtschabel, 1989). However, at high levels of nitrogen fertilization, rapeseed is rated unfavourably as a rotation crop because rape straw has a relatively high nitrogen content and considerable amounts of nitrogen will therefore remain on the field after harvest to be subsequently mineralized (Sauermann, 1993).

An experiment carried out over several years involving rotation of winter wheat and winter rape on heavy soil (flood plain

Table 9.1. Nitrate-N contents in soil and drainage water.
(a) Nitrate-N content (0 to 90 cm)

| Year | Culture | Soil nitrate-N content (kg nitrate-N ha^{-1}) | |
		Without N application	With N application[1]
1984	Winter wheat	–	13
1985	Winter rape	–	193
1986	Winter wheat	37	109
1987	Winter wheat	25	74
1988	Winter wheat	11	20
1989	Winter wheat	66	116
1990	Winter wheat	8	27
1991	Winter wheat	4	15
1992	Winter rape	60	110

(b) Nitrate content in drainage water under rapeseed

| Year | Culture | Drainage water Nitrate content (mg nitrate l^{-1}) | |
		Without N application	With N application[1]
1984/85	Winter rape	–	18.8
1985/85	Winter wheat	25.2	28.7
1986/87	Winter wheat	6.2	18.4
1987/88	Winter wheat	3.2	20.1
1988/89	Winter rape	0.1	12.8
1989/90	Winter wheat	9.3	24.8
1990/91	Winter wheat	5.4	11.2
1991/92	Winter rape	2.0	6.5
1992/93	Winter wheat	12.5	28.9

Source: Scheffer (1993).
[1] 160–200 kg N ha^{-1} depending on N_{min} values.

loam) showed that an increased concentration of soil nitrogen appeared in the drainage water of the wheat crop following rapeseed (Table 9.1). Accordingly, Scheffer (1993) recommended that rapeseed should not be grown in water conservation areas, since a high degree of nitrogen leaching is likely.

Uptake of nitrogen fertilizer not only depends on the amount applied, it is also influenced by productivity. This was illustrated in investigations by Cramer (1990) who showed that, while only 56% of nitrogen fertilizer was utilized when applied at a rate of 195 kg N ha^{-1} to a rape crop yielding 3.44 t ha^{-1}, this rose to 79% when the nitrogen treatment was the same but the yield increased to 4.45 t ha^{-1}. The nitrogen balance remained clearly positive throughout the year with less nitrogen removed than supplied.

Nitrogen fertilization was generally reduced in the EU when the market price for rapeseed fell following the withdrawal of direct crop subsidies in favour of support on an area basis, resulting in a considerable reduction in the optimum economic nitrogen level (Makowski *et al.*, 1993). In another series of nitrogen response trials the optimum economic nitrogen application in 15 experiments was between 80 and 200 kg N ha^{-1} (Baumgärtel, 1993). The nitrogen removed by rapeseed at harvest was often more than the level of nitrogen applied, especially at the lower nitrogen levels, thus resulting in a negative nitrogen balance (Cramer, 1993). This indicates that high nitrate concentrations in groundwater of rapeseed-growing areas do not reflect a characteristic specific to rapeseed. The problem is more likely to arise where growers are striving for maximum yield by applying nitrogen at rates which the crop cannot fully utilize (Maidl *et al.*, 1991).

Changes in the EU rapeseed price regime have not only stimulated reduction in nitrogen fertilizer use, they have also resulted in modified crop protection practices (Krostitz, 1993). Recent trials have shown that expenditure on herbicides, for example, can be cut by 40% by using reduced rates of active ingredient. Previously, high prices encouraged high application rates in seeking crops free from weeds. Consequently, the changed price structure is likely to be beneficial environmentally by reducing nitrate leaching and pesticide use.

Emission of nitrogen compounds

The release of nitrous oxide by rapeseed crops, as reported by the German Federal Environment Office in a study of biodiesel production (UBA-Studie, 1993), has attracted much attention. The gas emission arises from the denitrification of nitrogen compounds in the soil, which then escape into the atmosphere (Benckiser and Syring, 1992). The rate of emission is positively correlated with nitrogen levels whether derived from fertilizer or soil organic matter (von Rheinbaben, 1990).

Denitrification losses vary widely ranging from the equivalent of 10% to >40% of nitrogen fertilizer (Schneider and Haider, 1992). The Intergovernment Panel on Climate Change (Houghton *et al.*, 1990) gives the emission rate of nitrous oxide from nitrogen fertilizer as 0.4 to 3.2%, without giving details of the level of nitrogen application or the crop being fertilized.

In a study commissioned by the Union zur Förderung von Oel- und Proteinpflanzen (UFOP) (Scharmer *et al.*, 1993), calculations from bibliographical data showed that, when 100 kg N ha^{-1} is applied to rapeseed crops, 30% is utilized, 15% is denitrified and only 3–10% nitrous oxide (0.31 to 1.05 kg N ha^{-1}) escapes into the atmosphere.

These figures concur with calculations by Bouwmann (1990), who showed that nitrous oxide emissions from rapeseed crops account for only about 0.45 kg ha^{-1} of the nitrogen fertilizer applied. Rapeseed crops should not therefore be classified as 'ozone killers' (Makowski, 1993) since nitrous oxide released is correlated with the level of nitrogen applied rather than crop type. In practice, nitrogen fertilizer applications to rapeseed crops in the European Union have been sharply curtailed for economic reasons and they do not threaten greater nitrous oxide emissions than other crops.

Oilseed rape has a particularly high sulphur requirement, twice that of cereals, with removal in the harvested seed alone accounting for 20–30 kg ha^{-1} of sulphur in a 3 t ha^{-1} crop (Klessa and Sinclair, 1989). Booth *et al.* (1991) showed that, where sulphur was limiting, increasing the level of nitrogen applied could depress yield with consequent risk of nitrogen leaching. They also pointed out that, where interveinal chlorosis associated with sulphur deficiency is mistaken for nitrogen deficiency, growers might aggravate these problems by applying more nitrogen.

Anthropogenic sulphur deposition has been decreasing over Europe for many years and this trend is continuing (Anon., 1990b). As a consequence, sulphur deficiency in all crops, but particularly oilseed rape, is becoming increasingly common. Unless growers adapt their sulphur fertilizer policy there is a progressive risk of increased nitrogen leaching (Booth and Walker, 1993).

ALLELOPATHY

Definition of allelopathy

There is some variation in the definition of allelopathy used by different authors. The word literally means 'mutual suffering' but was originally used by Motisch in 1937 to describe both stimulatory and inhibitory chemical interaction among plants and microbes (Putnam and Tang, 1986). Some authors, however, use the term to describe only inhibitory effects between higher plants, and discussion of the topic sometimes encompasses interaction between plants and invertebrates.

Inhibitory allelopathic effects are distinct from competition as they are mediated by the addition of toxins to the environment. The role of allelopathy in natural ecosystems is reviewed by Rice (1984) and Putnam and Tang (1986) describe some agricultural implications.

Allelopathic potential of oilseed rape

Aspects of allelopathy which have been considered with respect to oilseed rape include the effect of rapeseed residues on following crops and the allelopathic effect of the oilseed rape crop on weeds.

Although domestication of rapeseed may have reduced allelopathic potential relative to the original wild type (Mason-Sedun *et al.*, 1986; Oleszek, 1987), studies have indicated that oilseed rape residues may adversely affect subsequent crops. Effects reported include reduction of plant dry weight, plant height, mean tiller number per plant and seed yield (Horricks, 1969; Mason-Sedun *et al.*, 1986). Horricks (1969) suggested that poor seed placement at drilling and lower soil temperatures in the vicinity of heavy residues could have contributed to stunting of plant growth, while Mason-Sedun and Jessop (1989) working with aqueous extracts from oilseed rape residues found an indication of the action of toxins.

Toxicity of residues seems to be related to environmental and seasonal factors during crop growth, such as daylength, temperature and nutrient supply (Mason-Sedun and Jessop, 1989). In addition the persistence of the residue toxins in the soil is dependent on weather factors (Mason-Sedun and Jessop, 1988) – oilseed rape residues lose their toxicity with time, but low temperatures over winter may delay the process of microbial breakdown. Thus the allelopathic effect from rapeseed residue on following crops may vary from year to year and geographically.

Recorded responses to rapeseed residues have not all been negative. Waddington and Bowren (1978) observed yield increases of subsequent crops in response to rape straw and proposed that yield reduction in crops following rapeseed is due to nitrogen deficiency rather than toxins.

Attempts have been made to exploit allelopathy as a weed management strategy (Putnam and Tang, 1986), thereby reducing the need for herbicides. The allelopathic potential of *Brassica* species has been attributed to the isothiocyanates (Oleszek, 1987; Vera *et al.*, 1987) breakdown products of glucosinolates (see Chapter 11). However, Choesin and Boerner (1991) found no evidence of an allelopathic effect linked to allyl isothiocyanate release by *Brassica napus* and Bialy *et al.* (1990) concluded that, as the most potent of these compounds, 2-phenyl isothiocyanate, occurs in only small amounts in oilseed rape, it has limited value as a natural toxicant.

Recent work by Halbrendt and Jing (1994) has demonstrated the nematicidal effect of glucosinolates in rapeseed residues and indicated the potential financial and environmental benefits this may have over chemical nematicides for perennial crops.

WILDLIFE

Roe deer and other deer

In December 1986, farmers and hunters in rapeseed-growing areas of Austria and Germany observed roe deer displaying distinctive abnormal behaviour (Onderscheka et al., 1987; Schneller, 1987). These symptoms, although less pronounced, were observed again in December 1987 (Tataruch et al., 1990). Observations of roe deer were also made in Scotland from mid-December to mid-January 1990 (Boag et al., 1990), where they were seen grazing on rapeseed crops but it was clear that their first preference was winter barley in adjacent fields. Post-mortem analysis of the rumen content of three roe deer shot showed that barley was indeed the main constituent of their diet. Results of these investigations gave no evidence to support poisoning of roe deer as apparently observed in Austria and Germany.

Studies of grazing intensity of roe deer, as well as red and fallow deer, were carried out in different parts of Germany. The traditional high glucosinolate forage cultivar, Akela, was grown alongside the low glucosinolate cultivars, Lirabon, Liratop, Rubin and Santana. The results showed that roe deer, but not red deer, had a strong preference for low glucosinolate cultivars but there was no evidence of poisoning (Ueckermann et al., 1988, 1990).

In the absence of such evidence it is only possible to speculate about the cause of problems associated with deer grazing rapeseed in 1986. Glucosinolates, especially the alkenyl fraction from which the sharp-tasting aglucone is derived, have been largely bred out of double-low cultivars. Consequently they taste better and, because of the limited availability of other plant food during the 1986/87 winter, large amounts were ingested by roe deer. This may have led to their poisoning.

Illness in roe deer grazing rapeseed crops was recorded before double-low cultivars were developed (Ströse, 1912; Rauchfuss et al., 1956). Symptoms of poisoning in domestic ruminants have also been confirmed after feeding forage rape. These were described by Völker (1950) and Rosenberger (1970) and summarized by Schmid and Schmid (1992) as having adverse effects on lungs, kidneys, liver and other organs.

The breakdown products of glucosinolates and S-methylcysteine sulphoxide (SMCO), also known as the kale anaemia factor, are considered toxic substances in rape that cause the above symptoms (Whittle et al., 1976; Smith, 1980). S-methylcysteine sulphoxide is a free amino acid which is transformed into dimethyldisulphide, pyruvate and ammonia in the first stomach of the ruminant:

$$2 \ H_3C-S(O)-CH_2-CH(NH_2)-COOH + H_2 \rightarrow$$
$$CH_3-S-S-CH_3 + 2 \ H_3C-CO-COOH + 2NH_3$$

The toxic substance is the dimethyldisulphide, which is absorbed from the digestive tract and gives rise to extrusion of haemoglobin from red corpuscles, forming Heinz bodies (Maxwell, 1981).

Findings on roe deer in the wild

Comprehensive examinations carried out on affected roe deer have been reported (Onderscheka et al., 1987; Tataruch et al., 1990) and summarized as below.

Clinical examinations.   Their physical condition was generally poor with apathetic

Table 9.2. Blood parameters from roe deer.

Parameters	1987 (n = 17)	1988 (n = 18)	Min.	Max.
Blood sedimentation	30.0	23.4	1.0	157
Haematocrit	0.31	0.34	0.13	0.48
Erythrocytes (10^{12} l^{-1})	6.26	6.95	3.5	10.8
Haemoglobin (mmol l^{-1})	9.14	9.37	5.06	13.4
Haemoglobin (%)	19.7	20.6	4.0	36.9
Alkaline phosphatase (U l^{-1})	28.2	30.7	16.5	42.9
Creatinine (μmol l^{-1})	104.3	–	–	–
Cholesterol (mmol l^{-1})	1.16	1.43	0.24	2.57
Total lipids (g l^{-1})	1.59	–	–	–
Triglycerols (mmol l^{-1})	0.15	0.25	0.01	0.54
Copper (μmol l^{-1})	11.6	9.6	3.2	17.3
Zinc (μmol l^{-1})	12.7	11.3	4.6	19.9

Source: Tataruch *et al.* (1990).

to somnolent behaviour and no response when touched.

Likewise, there was no response to optical stimuli, although some reacted to sound. Heart rate and body temperature were slightly lower and frequency of breathing was lower than normal.

Rumen activity was reduced and ruminal stasis found in several cases. The eyes showed no pathological changes, implying that the 'blindness' observed was due to disturbance of the central nervous system. The animals voluntarily avoided bright sunshine, showing that their re-action to light was apparently normal. The pupillary reflex was reduced but always present.

Some animals showed convulsions shortly before death. There were no signs of lameness. Food intake was reduced and sometimes animals ate abnormal plants or substances.

Sick animals drank much more water than usual and ate a lot of snow when it was present.

Blood analyses. Among the results obtained by Tataruch *et al.* (1990) are some (Table 9.2) showing significantly different reference values for roe deer.

The remarkable reductions in erythro-cytes, haemoglobin and haematocrit con-firm the severe degree of anaemia. This was classified as haemolytic due to the presence of more than 25% Heinz bodies. The reduced values for haemoglobin are also consistent with a diagnosis of haemolysis.

Pathological and histological examina-tions. Roe deer found dead in the field at the beginning of winter mainly showed signs of acute indigestion and bloat. Those which survived some days in captivity did not exhibit symptoms of acute indigestion but did show signs of a severe jaundice as a consequence of haemolytic anaemia.

Several authors (Gräfner, 1986; Kress, 1963) suspected fungi as the cause of rape poisoning but mycological tests by Tataruch *et al.* (1990) provided no such evidence.

Results of feeding experiments with roe deer

Fehlberg *et al.* (1989) carried out a feeding experiment with roe deer over a period of 12 weeks involving three groups, O, A and B, each of four animals. These received the following:

Table 9.3. Changes in liveweight of roe deer after feeding experiment.

Group A[1]			Group B[2]		
Weight at beginning (kg)	Length of experiment (days)	Weight at end (kg)	Weight at beginning (kg)	Length of experiment (days)	Weight at end (kg)
20.0	94	16.5	17.0	94	15.0
14.5	30	11.0	25.5	79	23.5
23.0	89	15.0	21.0	94	19.0
16.0	2	16.0	17.0	35	16.7

Source: Fehlberg *et al.* (1989).
[1] Group A animals received ad lib fresh green double-low rapeseed fodder and water.
[2] Group B animals received fresh green double-low rapeseed fooder, plus hay and an assortment of twigs from willow, Norway spruce and hornbeam.

Group O – no rape feed; provided the control for the determination of haematological standards.
Group A – ad lib fresh green double-low rapeseed fodder and water.
Group B – fresh green double-low rapeseed fodder, plus hay and an assortment of twigs from willow, Norway spruce and hornbeam.

The first disease symptoms appeared in group A after 10 days, largely supporting the observations of Tataruch *et al.* (1990). The animals in group B also became sick. The sickness in both groups led to progressive apathy and balance disorders, and finally to death. During the feeding period and when sick, the animals lost considerable weight (see Table 9.3).

Haematological investigations confirmed that the animals fed on rapeseed showed a clear deviation from the standard for almost all measurements (see Table 9.4). According to Fehlberg *et al.* (1989) the measurements for group A were substantially less favourable than those for group B. Pathological investigations of the rape-fed deer gave further support to the observations of Tataruch *et al.* (1990) but the symptoms were more severe.

The dry matter, protein and raw fibre contents of the rumen showed considerable deviation from those of free-living roe deer (Fehlberg *et al.*, 1989). The results are summarized in Table 9.5.

Traditionally, it is accepted that fresh green rape fodder should not be fed exclusively to ruminants since the nutritive content is unsuitable: the crude protein content is too high and crude fibre content too low (Marquard, 1990a). Consequently, the results of Fehlberg *et al.* (1989) may confirm that feeding green rape fodder exclusively could cause the death of roe deer but this does not explain the losses of roe deer in the wild, since they are not forced to eat rape alone.

Both groups of workers, Oderscheka in Vienna and Fehlberg in Hanover respectively, were convinced that the roe deer deaths were caused by ingestion of rape and both proved that substances in the rape were responsible. However, only rare incidences of deaths possibly linked to rape ingestion have been reported since the winter of 1986/87 when concern was aroused. This is contrary to expectations given that rapeseed cultivation has expanded and that rape green matter has an unequivocal toxic potential

Table 9.4. Effect on blood parameters of feeding roe deer on rapeseed.

Parameters	Normal values group O[1]	Fed on rapeseed Mean A and B	n
Haematocrit (vol %)	60.8	7.1	23
Erythrocytes (10^{12} l^{-1})	11.2	1.6	14
Haemoglobin (g %)	21.5	1.5	14
Heinz bodies (per 1000 erythrocytes)	18.0	5.4	14
Leukocytes (no. μl^{-1})	4135	1970	21
Urea (mmol l^{-1})	7.33	1.77	14
Glucose (mmol l^{-1})	175	46	21
Total protein (g l^{-1})	63.2	9.7	13
Alkaline phosphatase (U l^{-1})	48.3	18.3	15
Creatinine (mmol l^{-1})	89.6	19.2	13
Cholesterol (mg %)	53.3	6.8	13
Calcium (mmol l^{-1})	2.51	0.25	13
Inorganic phosphate (mmol l^{-1})	2.23	0.89	13
Sodium (mmol l^{-1})	152.4	3.6	14
Potassium (mmol l^{-1})	6.0	1.1	14
Lactate dehydrogenase (U l^{-1})	408.5	78.3	14
AST (GOT) (U l^{-1})	55.8	26.9	15
Bilirubin total (mg %)	0.48	0.24	13
Bilirubin direct (mg %)	0.41	0.25	13
Bilirubin indirect (mg %)	0.13	0.08	13

Source: Fehlberg *et al.* (1989).
[1] No rapeseed.

Table 9.5. Composition of the rumen content of roe deer.

	Components (%)		
	Dry matter	Crude protein	Crude fibre
Group A, rape feed only	9.0	34.7	7.4
Group B, mainly rape feed	10.7	29.6	11.9
Average of free-living roe deer	16	12	19

for ruminants, including roe and red deer.

Hares and rabbits

In the autumn of 1986 deaths among hares were reported throughout Germany for which at first there was no explanation (Goldhorn, 1987; Schneller, 1987; Eskens, 1990). Acute liver dystrophy was diagnosed generally as the cause of death (Eskens *et al.*, 1987). This condition is now termed EBHS (European brown hare syndrome) but at the time was almost unknown. In searching for the cause, double-low rape cultivars came under suspicion: conversion from single- to double-low rapeseed cultivars was underway in Germany at the time and hare corpses were often found in or adjacent to rapeseed crops.

Two hypotheses were advanced.

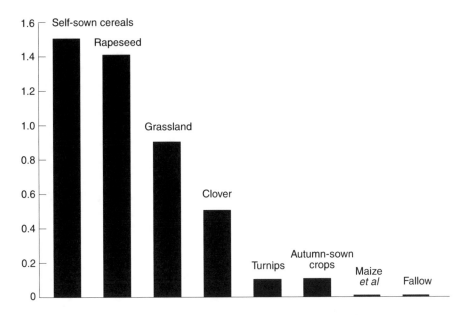

Fig. 9.1. Preference shown in the wild by brown hares grazing in the autumn.

Marquard (1990b) suggested the 'rape hypothesis': the improved taste of rapeseed plants, following the removal of glucosinolates in double-low cultivars, led to increased ingestion by wild animals and resultant poisoning and death. It had been known for some time, following observations in breeding nurseries, that double-low rapeseed, i.e. low in both glucosinolate and erucic acid content, was more heavily grazed by wild animals than single-low rapeseed (low in erucic acid content only). A preference for cultivars with reduced glucosinolate content was also demonstrated in grazing experiments with sheep (Zobelt *et al.*, 1986). Furthermore, Petrak *et al.* (1988) showed that rapeseed crops are preferred by brown hares in autumn.

The breakdown products of glucosinolates in rapeseed vegetation are connected with liver disease in farm animals: particularly, the nitriles and epithionitriles which are produced under acidic conditions (Papas *et al.*, 1979). Liver damage from nitrosamine was not ruled out in

formulating this hypothesis (Preussmann, 1982). The prerequisites for production of nitrosamines could be provided by the green rapeseed fodder arising from high protein and nitrate contents, and the presence of free thiocyanates. These appear as glucosinolate breakdown products and are among the most effective catalysts for the formation of nitrosamine (Röper, 1982).

The second '*Clostridium* hypothesis' considered that mortality was caused by clostridial toxins formed as a result of dietary changes. These would lead to so-called enterotoxaemia in the affected animals and double-low rape could be a cause (Boehnel, 1990).

Different species of *Clostridia* (*C. perfringens, C. difficilis*, etc.) live and multiply in the intestines. Abrupt changes in diet, such as high intake of double-low rapeseed, could encourage excessive multiplication of these bacteria to the detriment of other intestinal flora, thus leading to formation of different toxins. When feed returns to normal after a few days, multiplication stops, sporulation may begin and some toxins will again be formed. The clinical symptoms are diarrhoea, sometimes fever, metabolic disorders and nervous disorders, such as dullness or aggression. Histological changes of varying intensity may be found in the liver, kidneys, muscles or brain (Smith, 1975).

Several workers (Douville de Franssu, 1990; Pegel, 1990; Richter *et al.*, 1990) carried out feeding experiments with brown hares (*Lepus europaeus*, syn. *L. capensis*) and rabbits (*Oryctolagus cuniculus*).

Plants of both single- and double-low cultivars were grown in a phytotron in order to simulate the effect of unfavourable climatic conditions (Marquard, 1990a). Using available weather data, climatic conditions from two sites in Germany were simulated in the phytotron, which enabled their effects on growth constituents to be investigated. Green matter produced in the phytotron proved to have an extremely unfavourable composition for animal nutrition: protein content was >40%, nitrate content was up to 5% and leaf content of glucosinolates was very low, only 4 to 6 μmol g^{-1}.

The 'rape hypothesis', tested in feeding experiments with hares, was not confirmed (Marquard, 1990b). Nor could the disease symptoms, described by Eskens *et al.* (1987), be reproduced in trials where rape was fed alone: there were no mortalities either. Infection due to bacteria, such as *Clostridium* species, could not be detected. These results and those obtained by other workers exclude direct poisoning of hares and rabbits by rapeseed. Research at Giessen led to the conclusion that the 'hare epidemic of 1986' may have been caused by viral infection (Eskens and Volmer, 1989). Various working groups subsequently agreed and dismissed double-low cultivars as a direct cause of hare mortality.

Bees

Intensive agriculture of cereal crops has tended to reduce the availability of flowering plants to bees: this has led to reduced population numbers of some species, such as bumblebees (*Bombus* species). In contrast, the increased area of oilseed rape has been beneficial for bees; this crop is attractive to bees because it produces large quantities of nectar and pollen. At the end of a foraging trip a bee's pollen load weighs about 20 mg. One bee larva consumes approximately 200 mg of pollen during its development.

Honeybees and oilseed rape

Beekeepers often move hives to rapeseed crops during flowering and the potential honey yield from oilseed rape can be 50 kg ha^{-1} (Anon., 1986). However, oilseed rape honey crystallizes very quickly in the combs, making it difficult to extract.

Bumblebees and oilseed rape

Free (1982) states that winter and spring rapeseed crops can provide sufficient forage for bumblebees to provide for colony development from initiation to the production of sexual forms. Although in the short term rapeseed crops may attract bumblebees from other crops (e.g. clover, field bean, runner bean, tree fruit), more rapeseed crops should increase populations and be of general benefit.

Oilseed rape pesticides and bees

Although rapeseed is beneficial to bees, the application of pesticides is potentially detrimental and leads to a conflict of interests as the most effective time for spraying often coincides with peak foraging times. The risk of bee poisoning from pesticide application to rapeseed may be greater in Europe than in countries where production is less intensive and spraying less common.

In some countries farmers are advised

to warn local beekeepers at least 48 h prior to spraying and to use insecticides with least toxicity to bees where possible (Anon., 1990a; Coll, 1991). Control is stricter in Sweden, where no pesticides registered as dangerous to bees may be used on flowering crops, but, despite this, bee poisonings still occur (Fries and Wibran, 1987).

Contaminated pollen

Bees may carry pollen contaminated with pesticides into hives which, if toxic to bees, may adversely affect brood rearing as nurse bees or the larvae. The effect of contaminated pollen varies greatly depending on the pesticide used. Thus, although pyrethroids tend to repel bees and therefore substantially reduce foraging on treated crops, some foraging still occurs and bees carry contaminated pollen into hives. In a study of the toxicity of the pyrethroid insecticide, PP321, low-dose applications had no significant effect on brood rearing (Fries and Wibran, 1987). In another study of the long-term effects of systemic pesticides on honeybees, levels in pollen collected by bees were similar to those obtained on pollen traps after field treatment at recommended application rates and these levels were equivalent to those which killed larvae when fed in hives (Ferguson, 1987).

Direct spraying of bees

Bees foraging in a field when pesticide is applied may be sprayed directly. In estimating the effects, Fries and Wibran (1987) found that 80% of bees receiving a high dose of cypermethrin did not recover from the knock-down effect. Bees receiving the PP321 treatment were paralysed within 30 min of treatment but began to recover within 2 to 4 h and showed no difference from the untreated controls after 24 h. Bees that recovered from both cypermethrin and PP321 had a mortality rate only slightly higher than those untreated.

Foraging on crops after spraying

Foraging on crops treated with pesticide may have a detrimental effect on bees. Pankiw and Jay (1992) studied the effect of ultra-low-volume malathion sprays (at mosquito control rate) on the honeybee, *Apis mellifera*. This has been used extensively in North America to control adult mosquitoes and grasshoppers. The study involved spraying oilseed rape before and after the peak foraging times for bees. Colonies sprayed with ultra-low-volume malathion exhibited significantly less weight gain for up to 28 days after spraying. This may be attributed to bee losses, but the number in dead bee traps was not sufficient to account for the lack of weight gain and the majority must have been lost in the field. The population in sprayed colonies remained significantly lower than those unsprayed throughout the study period. Dead bee counts at the hive were more than in the controls for 3–4 days after spraying and the foraging activity was reduced for 2–3 days after spraying after which pre-spray levels were reached or exceeded.

It can therefore be concluded that oilseed rape is advantageous to honeybees and other bee species, although care is required in the use of pesticides and rapid crystallization can cause problems in extracting honey from combs (C. Coll, Aberdeen, 1994, personal communication).

ALLERGY PROBLEMS

Human response

Expansion of oilseed rape production in the UK from the early 1970s has been accompanied by increasing numbers of people complaining of allergic symptoms such as headaches, respiratory problems and eye irritation. Mayer (1989) found that the acreage in one region (Essex) closely paralleled figures for asthma admissions to a local hospital. Such findings are not conclusive evidence of a cause-and-effect relationship but many people attribute symptoms to the proximity of the crop.

Although rapeseed-related problems have been identified in some parts of Scandinavia, Germany and France, complaints of this type seem to be less frequent in other rapeseed-growing countries than in the UK, where the association has aroused much media coverage, resulting in adverse publicity for farmers, poor public perception of the crop and the development of a lobby seeking to ban the crop from high-population areas.

Some people in the vicinity of flowering rapeseed, particularly those working with the crop in growth cabinets and glasshouses, may develop allergic symptoms. In assessing the prevalence and severity of rapeseed allergy, a distinction should be made between those who may be affected by proximity to the volatile chemicals released by the flowering crop (see below) and those who may be affected by other causes.

Defining and testing for allergic reactions

An allergy was originally defined as 'changed reactivity of the host when meeting an agent on a second or subsequent occasion'. Advances in immunology have led to a better understanding. The term hypersensitivity is now used to describe beneficial immune responses acting inappropriately and sometimes causing inflammatory reaction and tissue damage. Four levels of hypersensitivity are differentiated in terms of immune response and it is the first level (Type 1), immediate hypersensitivity, which most people associate with the term allergy. Some people are atopic, or genetically predisposed to an allergic reaction, but not all atopics necessarily react and not all those who react are atopic. Genetic studies suggest that atopy is inherited independently from predisposi-

tion to other specific allergic reactions but atopy can enhance the likelihood of allergies being experienced.

Tests for allergy include provocation tests, e.g. nasal or bronchial provocation, skin tests, and the radioallergosorbent test (RAST), which is a serum test for antigen-specific immunoglobulin. A positive skin test is usually correlated with a positive RAST and the relevant provocation test (e.g. nasal provocation) with the allergen. However, some people with a clear history of, for example, allergic rhinitis may give negative results for skin or RAST tests. This group of people form a local mucosal antibody response.

History of complaints

Little documented information was available when concern was first expressed about a link between oilseed rape and allergic symptoms. An isolated case, believed to be the first published description of hypersensitivity to pollen of rapeseed or any *Cruciferae*, was reported by Colldahl (1954) but no further examples were published until 1978 (McSharry, 1992), when a study (Bucur and Arner, 1978) in southern Sweden showed that 23% of 366 patients with asthma and other allergies had positive intradermal responses to rapeseed pollens; and 44 of these patients showed positive conjunctival provocation with rapeseed pollens. Since then a small number of studies have examined the occurrence of sensitivity to oilseed rape pollen, monitored oilseed rape pollen counts and considered other factors which may account for the occurrence of allergic symptoms.

Occurrence of sensitivity to oilseed rape pollen

Fell *et al.* (1992) undertook a study of three populations of patients who were exposed to flowering oilseed rape in order to ascertain the prevalence of rapeseed pollen allergy in naturally and occupationally exposed subjects. The populations included a complete village surrounded by the crop, a group of patients attending a health centre when rapeseed pollen counts were high and who complained of conjunctivitis and rhinitis, and a group of scientists who were occupationally exposed to the pollen.

Subjects were skin-tested and blood samples were taken for RAST to determine whether they were allergic to rapeseed pollen. Fell *et al.* (1992) found that rapeseed pollen was not a major problem as the sole inducer of allergic reactions as, out of 1478 naturally exposed subjects, 24 thought they had symptoms due to rapeseed pollen of whom only 3 were clinically proven to be allergic, or 0.2% of the population studied. In contrast the incidence of rapeseed sensitivity in subjects occupationally exposed was much higher: 9 of 29 were skin-test positive to rapeseed pollen, but only 1 of the 9 was sensitive to rapeseed pollen alone. It was concluded that immediate hypersensitivity to the oilseed rape pollen allergen is not common and in general only occurs in atopics who are allergic to other pollens. With one exception this was also true for the occupationally exposed group.

Another study carried out in the Grampian region of Scotland (Harker *et al.*, 1992, 1993; Soutar *et al.*, 1994), between October 1989 and September 1992, took a different approach, but gave complementary results. While Fell *et al.* (1992) studied three different populations exposed to rapeseed Soutar *et al.* (1994) compared symptoms among inhabitants in oilseed-rape-growing areas with those where little or no rape was grown. There was a clear increase in the prevalence of headache, itchy skin, wheeze, rhinitis and itchy eyes in both areas from May to August. People tended to attribute these symptoms to oilseed rape in the oilseed rape areas and to other plants in the non-oilseed rape area. Harker *et al.* (1992) found no major differences in the prevalence of symptoms in

spring and summer among people in oilseed-rape- and non-oilseed-rape-growing rural communities. The symptoms were common, occurring in over 20% of the population in both areas, and it was concluded that oilseed rape may be one of several contributing factors.

A third study, also carried out in the Grampian area, tested 45 male and 36 female atopic children aged 3–14 years for oilseed rape pollen allergy. Asthma, hayfever or allergic rhinitis was diagnosed clinically in each child. Although allergy to oilseed rape pollen was suspected clinically in all the children examined, Ninan *et al.* (1990) found that oilseed rape pollen caused infrequent and only mild sensitization and could find no evidence to support concern that rape pollen is a potent source of sensitization.

The three studies described above (Ninan *et al.*, 1990; Fell *et al.*, 1992; Harker *et al.*, 1992, 1993; Soutar *et al.*, 1994) all conclude that, while rapeseed pollen can cause allergic symptoms, the allergy is not common. However, a serological study by Parratt *et al.* (1990) indicated that oilseed rape pollen is one of the most potent allergens described to date, confirming public concern. Parratt *et al.* (1990) and Parratt (1990) compared the incidence of sensitization to oilseed rape pollen in sera that had been sent to the allergy testing laboratories serving Glasgow and more rural Tayside. The frequency of RAST positivity to oilseed rape pollen was higher than in other studies: 26.3% of samples from Tayside and 7.8% of the samples from Glasgow, implying that sensitization is higher where there is greater exposure to oilseed rape.

Although sensitivity to rapeseed pollen was very high in this study, this does not necessarily contradict the work of Fell *et al.* (1992), who concluded that rapeseed allergy was not common in the population at large and only occurred in atopics allergic to other pollens. Some 85% of the Tayside samples and 82% of the Glasgow serum samples used by Parratt *et al.* (1990) had come from atopic donors. These results (Parratt, 1990; Parratt *et al.*, 1990) seem, however, to contradict those of Ninan *et al.* (1990), who found only mild and infrequent allergy to oilseed rape pollen amongst atopic children. Variation between studies in the response to oilseed rape pollen, together with the difficulty so far in identifying a single factor in the crop to explain so-called allergic reactions, may indicate that other factors, for example weather, may have some effect.

Oilseed rape pollen counts

As part of the study described above, Fell *et al.* (1992) monitored oilseed rape pollen counts in a rural community, 300 m from the nearest rapeseed field. They found that, although the peak oilseed rape pollen season was mid-May, rapeseed was not the predominant pollen at this time. Silver birch (*Betula* sp.), plane (*Platanus* sp.), pine (*Pinus* sp.), mugwort (*Artemisia vulgaris*) and spores of the fungus *Alternaria* were as much, if not more, prevalent. At their peak rapeseed pollen levels reached 90 grains m^{-3} of air, only one-fifth of those recorded for pine pollen or *Alternaria* spores.

Soutar *et al.* (1994) compared pollen counts 300 m from a rapeseed crop, directly adjacent to the crop, and in the centre of a village surrounded by the crop. Levels adjacent to the field were generally low, reaching 100 grains m^{-3} or more on only 6 days, while levels 300 m downwind from the field were extremely low, with weekly averages between 1 and 6 grains m^{-3} in only 5 weeks of the 12-week season. Like Fell *et al.* (1992) they found that birch and pine pollen was predominant and mould spores were more prevalent than rapeseed pollen on the high rapeseed pollen-count days. If these data are compared with the data on occurrence of sensitivity to rapeseed recorded by Ninan *et al.* (1990), Fell *et al.* (1992) and Harker *et al.* (1992) it might be

concluded that symptoms resulting from exposure to other pollens are being mistakenly attributed to the more conspicuous rapeseed crop. However, the correlation between pollen counts and allergic symptoms reported by Parratt (1990) does not support this conclusion. He carried out a clinical survey in which individuals were asked to record their symptoms and pollen counts were monitored at the same time. A strong correlation was found between mean scores for eye irritation, cough, wheeze and the daily rapeseed pollen count, but it should be noted that this study did not include a control group. Parratt (1990) states that birch was the only other significant pollen during the study, and no correlation was found between the recorded symptoms and birch pollen levels. Parratt (1990) concludes that, although his results support the hypothesis that allergy is caused by rapeseed pollen, sensitization to moulds and chemicals released by the plant cannot be excluded.

Fungal spores

Harker *et al.* (1993) concluded that, although large numbers of fungal spores were detected, the ubiquitous nature of such spores during much of the year makes this an unlikely explanation for those affected by proximity of the crop. Fell *et al.* (1992) also considered the possibility that fungal spores associated with oilseed rape caused allergic reaction and carried out nasal challenge tests with extracts of *Alternaria alternatum* but only report one positive result.

Chemicals released by the plant

Soutar *et al.* (1994) sampled air over a field of rape for chemicals, and detected terpenes, predominantly pinene, sabine and limonene. Other terpenes and aromatics present were not identified. A study by Tollsten and Bergstrom (1988) identified

34 volatile compounds released by six *Cruciferae* species. The terpene, sesquiterpene farnesene, was the major compound released by *B. napus*. Tollsten and Bergstrom (1988) state that this terpene is a flower fragrance component. Soutar *et al.* (1994) state that terpenes, which are also produced by pine trees and are present in cut grass, are locally irritant to mucous membranes and may cause skin sensitization, as when allergy to turpentine occurs. They conclude that it is plausible that these substances cause nasal and eye symptoms in people close to oilseed rape and, since they are also produced in pine trees, in other areas as well.

Soutar *et al.* (1994) also suggest that other irritant molecules including aldehydes and ultimately ozone will be formed downwind of oilseed rape crops due to reactivity of the compounds released by the plant. They argue that, while unlikely to cause symptoms adjacent to fields, they will undoubtedly contribute to the overall irritancy of air in country districts.

No evidence of hypersensitivity

Research on this topic is at an early stage, but overall the studies described suggest that classical Type 1 hypersensitivity to oilseed rape pollen is not sufficiently common to explain the widespread complaints from the general population attributed to oilseed rape. This does not eliminate the possibility that the oilseed rape plant causes these symptoms, although it seems likely that symptoms are often mistakenly attributed to oilseed rape because of its conspicuousness. It seems unlikely that symptoms attributed to oilseed rape are caused by fungal spores associated with it. It is more likely that symptoms associated with the crop are irritant rather than allergic, caused by volatile organic compounds released by the

plant. If terpenes are confirmed as a cause, the wide variation in terpene emission from different cultivars suggests that plant breeding could reduce the problem (Harker *et al.*, 1993).

SUMMARY

Rapeseed can influence the environment in many different ways. The growing of the crop may aggravate nitrate levels in groundwater but this is as much an impact of the overall cropping system as the actual oilseed crop and changes in oilseed price support no longer justify high-input growing techniques. Similarly, emission of nitrous oxide from the soil is more related to level of nitrogen applied rather than crop type. However, efficiency of utilization of nitrogen by crops is a complex process and changing circumstances, e.g. reduced sulphur depositions in Europe, may require continuous review of fertilizer policy.

The impact of oilseed rape on wildlife and man is much less clear but research appears to discount any particular intrinsic problems associated with the rapeseed crop. Whilst death of wild deer in mid-Europe initially caused much concern among the rapeseed industry this now appears to be related to S-methylcysteine sulphoxide content – a danger associated with most *Brassica* crops, not just oilseed rape. Similarly the cause of the European brown hare syndrome (EBHS) is now attributed to a virus rather than the rapeseed crop. Whilst complaints of allergy in man to the rapeseed crop have not been completely answered, the evidence suggests that the numbers of people affected are relatively low and elimination of the crop from farm systems would have no effect on overall health levels within the community.

REFERENCES

Anon. (1986) *The Birds and the Bees*. May and Baker, Essex, UK.

Anon. (1990a) *The Protection of Honey Bees*. MAFF Information Leaflet, London.

Anon. (1990b) *Third Report of the United Kingdom Review Group on Acid Rain*. Warren Spring Laboratory, London.

Baumgärtel, G. (1993) Stickstoffdüngung und Nitratreste nach der Ernte im Rapsanbau. *Raps* 11, 90–92.

Benckiser, G. and Syring, K.-M. (1992) Denitrifikation in Agrarstandorten (Bedeutung, Quantifizierung und Modellierung). *Bio-Engineering* 3, 46–52.

Bialy, Z., Oleszek, W., Lewis, J. and Fenwick, G.R. (1990) Allelopathic potential of glucosinolates (mustard oil glycosides) and their degradation products against wheat. *Plant and Soil* 129, 277–281.

Boag, B., Macfarlane Smith, W.H. and Griffiths D.W. (1990) Observations on the grazing of double low oilseed rape and other crops by roe deer. *Applied Animal Behaviour Science* 28, 213–220.

Boehnel, H. (1990) Preliminary results of investigations in clostridial enterotoxaemia in wildlife and its alleged connection with increased oilseed rape cultivation in Germany. In: Askew, M.F. (ed.) *Report EUR11771 en. – Rapeseed OO and Intoxication of Wild Animals*. Commission of the European Communities, Luxembourg, pp. 221–232.

Booth, E.J. and Walker, K.C. (1993) The effect of site and foliar sulphur on oilseed rape: comparison of sulphur responsive and non-responsive. *Phyton* 32(3), 9–13.

Booth, E.J. Walker, K.C. and Schnug, E. (1991) The effect of site, foliar sulphur and nitrogen application on glucosinolate content and yield of oilseed rape (*Brassica napus* L). In: McGregor, D.I. (ed.) *Proceedings of the Eighth International Rapeseed Congress*, Saskatoon, Canada. Organizing Committee, Saskatoon, pp. 567–572.

Bouwmann, A.F. (1990) *Soils and the Greenhouse Effect*. John Wiley & Sons, New York.

Bucur, I. and Arner, B. (1978) Rape pollen allergy. *Scandinavian Journal of Respiratory Diseases* 59, 222–227.

Choesin, D.N. and Boerner, R.E.J. (1991) Allyl isothiocyanate release and the allelopathic potential of *Brassica napus* (Brassicaceae). *American Journal of Botany* 78, 1083–1090.

Coll, C. (1991) *Insect Pests of Oilseed Rape in Scotland*. Technical Note 284, The Scottish Agricultural College, Aberdeen.

Colldahl, H. (1954). Rape pollen allergy. *Acta Allergologica* 7, 367–369.

Cramer, N. (1990) *Raps – Anbau und Verwertung*. Eugen Ulmer Verlag, Stuttgart.

Cramer, N. (1993). Umweltverträgliche N-Versorgung des Rapses. *Raps* 11, 4–7.

Douville de Franssu, P. (1990) Controlled feeding trials on captive hares with rapeseed. In: Askew, M.F. (ed.) *Report EUR 11771 en. – Rapeseed OO and Intoxication of Wild Animals*. Commission of the European Communities, Luxembourg, pp. 203–210.

Eskens, U. (1990) Pathology and epidemiology of the so-called 'hare death' in 1986–87 in central Hesse. In: Askew, M.F. (ed.) *Report EUR 11771 en. – Rapeseed OO and Intoxication of Wild Animals*. Commission of the European Communities, Luxembourg, pp. 211–220.

Eskens, U. and Volmer K. (1989) Untersuchungen zur Ätiologie der Leberdystrophie des Feldhasen (*Lepus europaeus* Pallas). *Deutsche Tierärztliche Wochenschrift* 96, 464–466.

Eskens, U., Klima, H., Nilz, J. and Wiegand, D. (1987) Leberdystrophie bei Feldhasen (*Lepus europaeus* Pallas). *Tierärztliche Praxis* 15, 229–235.

Fehlberg, U., Schoon, H.-A., Kamphues, J., Kikovic, G. and Sodeikat, G. (1989) Auswirkung der Fütterung von erucasäurefreiem und glukosinolatarmem Ölraps auf Rehwild im Gehege. *Zeitschrift Jagdwiss.* 35, 50–63.

Fell, P.J., Soulsby, S., Blight, M.M. and Brostoff, J. (1992) Oilseed rape – a new allergen? *Clinical and Experimental Allergy* 22, 501–505.

Ferguson, F. (1987) Long term effects of systemic pesticides on honey bees. *Australasian Beekeeper* 89, 49–54.

Free, J.B. (1982) *Bees and Mankind*. George Allen and Unwin, London.

Fries, I. and Wibran, K. (1987) Effects on honey-bee colonies following application of the pyrethroids cypermethrin and PP321 in flowering oilseed rape. *American Bee Journal* 127, 266–269.

Goldhorn, W. (1987) Das 'Hasensterben 1986'. *Der prakt. Tierarzt* 68, 42–43.

Gräfner, G. (1986) *Wildkrankheiten, 3. Auflage*. Fischer Verlag, Jena.

Halbrendt, J.M. and Jing, G.N. (1994) Cruciferous green manure as an alternative to nematicide: the effect of glucosinolate content. In: *Proceedings of the Third International Conference on New Industrial Crops and Products*, September 1994, Catamurea, Argentina (in press).

Harker, C., Soutar, A. and Seaton, A. (1992) An epidemiological study of seasonal symptoms in a rural population. In: *British Thoracic Society Winter Meeting*, 9–11 December, Kensington Town Hall, London.

Harker, C.G, Soutar, A., Walker, K.C. and Booth E.J. (1993) Oilseed rape allergy: fact or fiction? *GCIRC Bulletin* 9, 88–89.

Horricks, J.S. (1969) Influence of rape residue on cereal production. *Canadian Journal of Plant Science* 49, 632–634.

Houghton, J.T., Jenkins, G.I. and Ephraums, J.J. (eds) (1990) *Climate Change. The IPCC Scientific Assessment*. Cambridge University Press, Cambridge.

Klessa, D.A. and Sinclair, A.H. (1989) *Sulphur in Soils, Fertilisers and Crops*. Technical Note T160, The Scottish Agricultural College, Perth.

Kress, F. (1963) Beobachtungen über das Vorkommen blinder Rehe. *Österreichs Waidwerk* 5, 186–191.

Krostitz, J. (1993) Die Anpassung des Pflanzenschutzes bei Raps an neue Preis-Kostenverhältnisse. *Raps* 11, 58–65 and 125–128.

McSharry, C. (1992) New aeroallergens in agricultural and related practice. *Clinical and Experimental Allergy* 22, 423–426.

Maidl, F.X., Funk, R., Müller, R. and Fischbeck, G. (1991) Ein Tiefbohrgerät zur Ermittlung des Einflusses verschiedener Formen der Landbewirtschaftung auf den Nitrateintrag in tiefere Bodenschichten. *Zeitschrift für Pflanzenernaehrung und Bodenkunde* 154, 259–263.

Makowski, N. (1993) Raps, ein Ozonkiller? *Raps* 11, 164–165.

Makowski, N., Schulz, R.-R. and Michel, H.-J. (1993) Welche Stickstoffdüngung lohnt bei den heutigen Rapspreisen? *Raps* 11, 182–183.

Marquard, R. (1990a) Glucosinolate, nitrate and protein contents in green matter of oilseed rape (*Brassica napus*) and their possible effects on wild life. In: Askew, M.F. (ed.) *Report EUR11771 en. - Rapseed OO and Intoxication of Wild Animals*. Commission of the European Communities, Luxembourg, pp. 143–149.

Marquard, R. (1990b) Untersuchungen von Inhaltstoffen in der Raps-Grünmasse im Zusammenhang mit dem beobachteten 'Hasensterben 1986'. *GCIRC Bulletin* 6, 56–62.

Mason-Sedun, W. and Jessop, R.S. (1988) Differential phytotoxicity among species and cultivars of the genus *Brassica* to wheat. II Activity and persistence of water soluble phytotoxins from residues of the genus *Brassica*. *Plant and Soil* 107, 69–80.

Mason-Sedun, W. and Jessop, R.S. (1989) Differential phytotoxicity among species and cultivars of the genus *Brassica* to wheat. III Effects of environmental factors during growth on the phytotoxicity of residue extracts. *Plant and Soil* 117, 93–101.

Mason-Sedun, W., Jessop, R.S. and Lovett, J.V. (1986) Differential phytotoxicity among species and cultivars of the genus *Brassica* to wheat. I. Laboratory and field screening of species. *Plant and Soil* 93, 3–16.

Maxwell, M. (1981) Production of a Heinz body anaemia in the domestic fowl after ingestion of dimethyl disulphide: a haematological and ultrastructural study. *Review of Veterinary Science* 30, 233–238.

Mayer, T. C. (1989) Oilseed rape and asthma. *Journal of the Royal College of General Practitioners* 39, 168.

Ninan, T.K., Milne, V. and Russel, G. (1990) Oilseed rape not a potent antigen. *The Lancet* 336, 808.

Oleszek, W. (1987) Allelopathic effects of volatiles from some *Cruciferae* species on lettuce, barnyard grass and wheat growth. *Plant and Soil* 102, 271–273.

Onderscheka, K., Tataruch, F., Steineck, Th., Klansek, E., Vodnansky, H., and Wagner, J. (1987) Gehäufte Rehwildversluste nach Aufnahme von OO-Raps. *Zeitschrift Jagdwiss.* 33, 139–142 and 191–205.

Pankiw, T. and Jay, S.C. (1992). Aerially applied ultra-low-volume malathion effects on colonies of honey bees (Hymenoptera: Apidae). *Journal of Economic Entomology* 85, 692–699.

Papas, A., Campbell, L.D. and Cransfield, P.A. (1979) A study of the association of glucosinolates to rapeseed meal-induced haemorrhagic liver in poultry and influence of vitamin K. *Canadian Jornal of Animal Science* 59, 133–144.

Parratt, D. (1990) Background and current research on medical aspects of the oilseed rape/allergy problem. In: *Oilseed Rape/Allergy Meeting*, Scottish Crop Research Institute, Invergowrie, Dundee, 4 July 1990.

Parratt, D., Thomson, G., Saunders, C., McSharry, C. and Cobb, S. (1990) Oilseed rape as a potent antigen. *The Lancet* 335, 121–122.

Pegel, M. (1990) Controlled feeding studies on European hares in FRG. In: Askew, M.F. (ed.) *Report EUR11771 en. - Rapseed OO and Intoxication of Wild Animals*. Commission of the European Communities, Luxembourg, pp. 193–202.

Petrak, M., Uhl, H.G. and Pegel, M. (1988) Untersuchungen über Zusammenhänge zwischen dem Anbau von OO-Raps und Todesfällen bei Hasen. 3. Mitteilung: Felduntersuchungen. Nieders. *Jäger* 88(22), 1328–1330.

Preussmann, R. (1982) Biologische Wirkungen, Metabolismus, Dosis-Wirkungs-Beziehung und Risikobetrachtungen. In: Preussmann, R. (ed.) *Das Nitrat-Problem*. Verlag Chemie, Weinheim, pp. 255–266.

Putnam, A.R. and Tang, C. (1986) Allelopathy: state of the science. In: Putnam, A.R. and Chung-Shih Tang (eds) *The Science of Allelopathy*. John Wiley and Sons, New York.

Rauchfuss, H., Lohs, H., Braune, E., Gemmer, B., Pauslien, K. and Graf Bernstorff, B. (1956) Rapskrankheit des Rehwildes. *Wild und Hund* 59, 10.

Rice, E.L. (1984) *Allelopathy*, 2nd edn. Academic Press, Orlando, Florida, USA.

Richter, W.I.F., Klein, F.W. and Hofmann (1990) Feeding studies with hares and domestic rabbits on different rape varieties in Bavaria. In: Askew, M.F. (ed.) *Report EUR11771 en. - Rapseed OO and Intoxication of Wild Animals*. Commission of the European Communities, Luxembourg, pp. 94–101.

Röper, H. (1982) Chemie und Bildung von N-Nitroso-Verbindungen. In: Preussmann, R. (ed.) *Das Nitrat-Problem*. Verlag Chemie, Weinheim, pp. 189–211.

Rosenberger, G. (ed.) (1970) *Krankenheiten des Rindes*. Parey Verlag, Berlin.

Sauermann, W. (1993) Vorfruchteffekte von Winterraps. *Raps* 11, 118–120.

Scharmer, K., Golbs, G., and Muschalek, I. (1993) *Pflanzenölkraftstoffe und ihre Umweltwirkungen - Argumente und Zahlen zur Umweltbilanz*. Studie im Auftrag der UFOP. GET mbH, Aldenhofen, Germany.

Scheffer, B. (1993) Hohe Nitratgehalte im Dränwasser. *Landpost* 93, 46–47.

Scheffer, B. and Schachtschabel, P. (1989) *Lehrbuch der Bodenkunde, 11. Auflage*. Enke Verlag, Stuttgart.

Schmid, A. and Schmid, H. (1992) Rapsvergiftung wildlebender Pflanzenfresser. *Landw. Jahrbuch* 69, 87–94.

Schneider, U. and Haider, K. (1992) Denitrification and nitrate leaching losses in an intensively cropped watershed. *Zeitschrift für Pflanzenernaehrung und Bodenkunde* 155, 135–141.

Schneller, H.P. (1987) Raps alsy mögliche Ursache für Hasen- und Rehsterben. *Tierärztliche Umschau* 42, 902–904.

Smith, L.D. (1975) *The Pathogenic Anaerobic Bacteria*, 2nd edn. Thomas, Springfield, Illinois.

Smith, R. (1980) Kale poisoning: the brassica anaemia factor. *Veterinary Record* 107, 12–15.

Soutar, A., Harker, C., Seaton, A., Brooke, M. and Marr, I. (1994) Oilseed rape and seasonal symptoms. *Epidemiological and Environmental Studies* (in press).

Ströse, A. (1912) Die Rapskrankheit der Rehe. *Deutsche Jäger-Zeitung* 59, 625–626.

Tataruch, F., Onderscheke, K. and Steineck, Th. (1990) Studies on roe deer and rapeseed in Austria. In: Askew, M.F. (ed.) *Report EUR11771 en. - Rapseed OO and Intoxication of Wild Animals*. Commission of the European Communities, Luxembourg, pp. 73–86.

Tollsten, L. and Bergstrom, G. (1988) Headspace volatiles of whole plants and macerated plant parts of *Brassica* and *Synapsis*. *Phytochemistry* 27, 4013–4018.

UBA-Studie (1993) *Ökologische Bilanz von Rapsölmethylester als Ersatz für Dieselkraftstoffe (Ökobilanz Rapsöl)*. Umweltbundesamt, Berlin, Germany.

Ueckermann, E., Lutz, W. and Marquard, R. (1988) Versuche zur Beäsung von OO-Raps durch Schalenwild. *Zeitschrift Jagdwiss*. 34, 55–62.

Ueckermann, E., Lutz, W. and Marquard, R. (1990) Controlled feeding studies on roe, red and fallow deer. In: Askew, M.F. (ed.) *Report EUR11771 en. - Rapseed OO and Intoxication of Wild Animals*. Commission of the European Communities, Luxembourg, pp. 234–241.

Vera, C.L., McGregor, D.L. and Downey, R.K. (1987) Detrimental effects of volunteer *Brassica* on production of certain cereal and oilseed crops. *Canadian Journal of Plant Science* 67, 983–995.

Völker, R. (1950) *Lehrbuch der Toxikologie für Tierärzte, 6. Auflage*. Enke Verlag, Stuttgart.

von Rheinbaben, W. (1990) Nitrogen losses from agricultural soils through denitrification – a critical evaluation. *Zeitschrift für Pflanzenernaehrung und Bodenkunde* 153, 157–166.

Waddington, J. and Bowren, K.E. (1978) Effects of crop residues on production of barley, ryegrass and alfalfa in the greenhouse and of barley in the field. *Canadian Journal of Plant Science* 58, 249–255.

Whittle, P., Smith, R. and Mcintosh, A. (1976) Estimation of S-methyl-cysteine sulphoxide (kale anaemia factor) and its distribution among brassica forage and root crops. *Journal of Science and Food in Agriculture* 27, 633–642.

Zobelt, U., Marquard, R. and Daniel, P. (1986) Vergleichende Untersuchungen über Glucosinolatgehalte in der Grünmasse von Raps und Rübsen und ihre Auswirkung auf das Fressverhalten von Schafen. *VDLUFA-Kongressband* 20, 561–571.

Part II

Processing and Utilization

10 Seed Chemistry

B. Uppström
Svalöf Weibull AB, Sweden

INTRODUCTION

Brassica oilseeds are produced not only for
their oil, which is valued as both an edible
and industrial oil, but also for their meal,
which is a good source of protein for both
animal and human consumption. Like
other vegetable oils, the oil is composed
predominantly of fatty acid containing triacylglycerols with lesser amounts of phospholipids and glycolipids and usually only
trace amounts of monoacylglycerols, diacylglycerols and free fatty acids. During
oil extraction compounds of polyisoprenoid
origin extract into the oil. These include
sterols, tocopherols, carotenoids and chlorophyll, most of which have to be removed
due to the undesirable characteristics they
impart to the crude oil.

The meal remaining after oil extraction is high in protein as a result of accumulation during seed development of
napin and cruciferin storage proteins and
oleosin, a structural protein associated with
the oil bodies. Amino acid composition in
part dictates the protein quality. Next to
protein, carbohydrates are the largest component of the meal. Compared to the polysaccharides, the soluble carbohydrates,

monosaccharides and oligosaccharides are
a relatively small component of the mature
seed. Polysaccharides which have been
isolated and at least partially characterized
include amyloids, arabinans, arabinogalactans, pectins, starch and lignins. Of interest, particularly to animal nutritionists, are
seed and meal fibre content which consists
to a large extent of carbohydrates of little
nutritional value to non-ruminants. Nutritional quality of the seed and meal is also
influenced by phytates, phenolic acids,
including free, esterified and insoluble
bound phenolics, and glucosinolates. Each
of these components is discussed in turn.

OIL COMPOSITION

Press or solvent extracts of vegetable seeds
are called oil, or fat if hard at room temperature, by industry and commerce. Bio-
chemists, on the other hand, refer to such
extracts as lipids due to their solubility in
certain organic solvents, or their deriva-
tion from long-chain fatty acids. Compounds of polyisoprenoid origin, which are
lipid soluble, are included in this fraction.
Triacylglycerols constitute more than 90%

Table 10.1. Systematic name, trivial name and symbol for the fatty acids commonly found in *Brassica* and related seed oils.

Systematic name	Trivial name	Symbol
Hexadecanoic	Palmitic	$16^1 : 0^2$
cis-9^3-Hexadecenoic	Palmitoleic	$16 : 1 (n-7)^4$
Octadecanoic	Stearic	$18 : 0$
cis-9-Octadecenoic	Oleic	$18 : 1 (n-9)$
cis-cis-9,12-Octadecadienoic	Linoleic	$18 : 2 (n-6)$
all-*cis*-9,12,15-Octadecatrienoic	α-Linolenic	$18 : 3 (n-3)$
Eicosanoic	Arachidic	$20 : 0$
cis-11-Eicosenoic	Gondoic	$20 : 1 (n-9)$
cis-13-Eicosenoic[5]		$20 : 1 (n-7)$
cis-11,14-Eicosadienoic[5]		$20 : 2 (n-6)$
Docosanoic	Behenic	$22 : 0$
cis-13-Docosenoic	Erucic	$22 : 1 (n-9)$
cis-15-Docosenoic[5]		$22 : 1 (n-7)$
cis-13,16-Docosadienoic[5]		$22 : 2 (n-6)$
Tetracosanoic	Lignoceric	$24 : 0$
cis-15-Tetracosenoic	Nervonic	$24 : 1 (n-9)$

[1] Number of carbon atoms.
[2] Number of double bonds.
[3] Position of the double bond, numbering from the carboxyl end.
[4] Position of the double bond, numbering from the methyl end.
[5] Present in small amount in high erucic acid oils.

of the oil. Lesser amounts of partial glyceride (about 1%), mono- and diacylglycerol and free fatty acids (<0.5%) may also be present, although these usually only occur in significant amounts in damaged or immature seed. Polar lipids, phospho- and galactolipids constitute about 4 to 5% of the seed oil, when extracted with a polar solvent such as chloroform–methanol (Zadernowski and Sosulski, 1978), but only about 2 to 3% in crude oils extracted commercially with less polar solvents (*n*-hexane or light petroleum distillate, Skellysolve F).

Triacylglycerols

Triacylglycerols, sometimes referred to as triglycerides in industry and commerce, have three fatty acids esterified to the hydroxyl positions of glycerol. Since triacylglycerols are the dominant oil com-

ponent, the fatty acid composition of the oil reflects the overall composition of the triacylglycerols. Fatty acid composition is an important quality parameter dictating the nutritional or industrial value of the oil.

As with other vegetable oils it is customary to refer to the fatty acids by their traditional or trivial names. However, not all fatty acids have trivial names, in which case the systematic name or symbol is used (Table 10.1). The systematic names are of Greek origin. Traditional or trivial names derive from the original source of isolation. The first figure of the symbol designates the number of carbon atoms and the second, following the colon, the number of double bonds. In order to define the position of the double bonds in unsaturated fatty acids, the carbon atoms are numbered from the carboxyl end of the fatty acid and the number of the first unsaturated carbon is affixed as a prefix to the systematic name. A *cis* or *trans* designation preceding the numbering identifies the configuration around the double bond. Most unsaturated fatty acids are of the *cis* configuration, although the *trans* configuration can result from hydrogenation. In the medical profession numbering from the methyl end of the fatty acid is preferred. This scheme is only applied to *cis* double bonds and the position of the double bond is recorded as a suffix following the number of carbon atoms and double bonds of the symbol designation (Table 10.1). Additional methods of nomenclature not shown include numbering from the carboxyl or methyl end of the fatty acid and using, e.g. Δ-9 or ω-9, prefixes to the systematic name respectively.

Traditional *Brassica* seed oils differed from other vegetable oils in containing a significant proportion of the long-chain monoenoic fatty acids, eicosenoic and erucic (Table 10.2). Palmitic, palmitoleic, stearic, oleic, linoleic, linolenic, arichidic, *cis*-11-eicosenoic, behenic and lignoceric constitute the remainder of the fatty acid profile. Although genetically determined, fatty acid composition can vary with environment, being more unsaturated with cooler climate and higher latitude.

In the 1950s and again in the early 1970s feeding experiments with laboratory animals indicated that the nutritional value of rapeseed oil would be substantially improved if the erucic acid content could be reduced to <5% of the total fatty acid content (Kramer *et al.*, 1983). The identification of plants with essentially no erucic acid in their seed oil in *Brassica napus* (Stefansson *et al.*, 1961) and *Brassica rapa* (Dorrell and Downey, 1964) resulted in the development of nutritionally superior canola cultivars (Table 10.3). The isolation of zero erucic acid in *Brassica juncea* plants (Kirk and Oram, 1981; Olsson, 1984) has initiated the development of this species as an edible oil crop. Low erucic

acid lines of *Sinapis alba* have also been identified (Krzymanski *et al.*, 1991).

An additional breeding objective has been to reduce the percentage of linolenic acid in rapeseed from the 8 to 10% to less than 3% while maintaining or increasing the level of linoleic acid. Lower linolenic acid is desired to improve the storage characteristics of the oil while a higher linoleic acid (vitamin F) content may be nutritionally desirable. Rakow (1973) developed mutants with half the normal amount of linolenic acid. This initial work by Rakow led to the development of the *B. napus* cultivar Stellar containing less than 3% linolenic acid and as much as 20% linoleic acid (Scarth *et al.*, 1988). Rakow *et al.* (1987) have made selections for increased linoleic acid content, an essential fatty acid in the human diet.

Increased content of fatty acids with shorter chain lengths is also of interest. Swedish researchers have selected *B. rapa* lines with 10 to 12% of palmitic plus palmitoleic acids compared with 4 to 5% in the unselected population (Persson, 1985). Knutzon *et al.* (1992), using a seed-specific antisense gene construct to inhibit stearoyl acyl carrier protein (ACP) desaturase activity, increased the content of stearic acid from the normal level of less than 2% to over 40%. Levels of lauric acid (12 : 0) have also been increased through genetic engineering to over 40%, accompanied by a small (4%) increase in myristic acid (14 : 0) (Voelker *et al.*, 1992; Ohlrogge, 1994).

The oil from high erucic (>50%) cultivars enters the industrial oil market where there are many applications (see Chapter 15). Through conventional breeding, cultivars of the summer form of *B. napus* have been developed with over 50% erucic acid in their oils (Scarth *et al.*, 1991). Plants normally preferentially esterify C_{18} unsaturated fatty acids in the 2-position, or middle hydroxyl, of the triacylglycerol and esterify saturated fatty acids and long-chain

Table 10.2. Fatty acid composition of seed oils from traditional *Brassica*, *Sinapis* and *Crambe* species.

Genus, species, cultivar or type	Fatty acid composition (%)												Reference
	16:0	16:1	18:0	18:1	18:2	18:3	20:0	20:1	22:0	22:1	24:0	24:1	
Brassica napus													
Victor (winter)	3.0	0.3	0.8	9.9	13.5	9.8	0.6	6.3	0.7	52.3	0.2	1.0	Appelqvist (1969)
Target (summer)	3.0	0.5	1.5	20.9	13.9	9.1	0.5	12.2	tr	38.6	–	0.0	Downey (1971)
Brassica rapa													
Duro (winter)	2.0	0.2	1.0	12.9	13.4	9.1	0.7	8.9	0.2	49.0	0.0	1.1	Appelqvist (1969)
Echo (summer)	4.5	0.3	1.3	33.3	20.4	7.6	0.5	9.4	tr	23.0	–	0.0	Downey (1971)
Yellow sarson	1.8	0.2	0.9	13.1	12.0	8.2	0.9	6.2	0.0	55.5	0.0	1.2	Downey (1983)
Brassica juncea													
Indian origin	2.5	0.3	1.2	8.0	16.4	11.4	1.2	6.4	1.2	46.2	0.7	1.9	Appelqvist (1972)
Cutlass	3.3	0.3	1.2	17.2	21.4	14.1	0.7	11.4	0.4	25.8	0.2	1.7	Downey and Rimmer (1993)
Brassica carinata													
Ethiopian mustard	3.2	0.2	0.9	9.8	16.2	13.9	0.7	7.5	0.7	41.6	0.6	2.0	Downey and Rimmer (1993)
Sinapis alba													
Seco	2.7	0.3	0.9	22.8	8.6	10.5	0.6	8.4	0.5	41.1	0.2	2.7	Appelqvist (1972)
Crambe abyssinica	1.7	0.3	1.0	16.7	7.8	6.9	1.3	2.9	2.7	55.7	–	2.9	McGregor et al. (1961)

Table 10.3. Fatty acid composition of *Brassica napus* rapeseed oils modified by conventional plant breeding or genetic engineering.

Oil type and cultivar	Fatty acid composition (%)								Reference
	16:0	16:1	18:0	18:1	18:2	18:3	20:1	22:1	
Low 18:3 (Stellar)	4	<1	1	59	29	3	1	<1	Scarth *et al.* (1988)
Low 22:1[1] (Westar)	4	<1	2	62	20	9	2	<1	Downey (1990)
High 16:0	10	4	1	51	19	13	1	<1	C. Persson (personal communication)
High 18:0	4	–	28	15	18	22	7[2]	3[3]	Knutzon *et al.* (1992)
High 18:1	4	<1	2	79	7	5	2	–	Wong and Swanson (1991)
High 18:2	5	<1	2	57	28	6	1	<1	Rakow *et al.* (1987)

[1] Canola.
[2] Includes 6% 20:0.
[3] Includes 3% 22:0.

mono-unsaturated fatty acids in the 1- and 3-positions (Appelqvist, 1989). Because of the inability of acyl transferases to insert erucic acid in the 2-position of triacylglycerols, there has been thought to be an apparent upper limit of 66% erucic acid obtainable in *Brassica* species (Taylor *et al.*, 1992). However, recent work has shown that triacylglycerol with erucic acid in the middle position is produced in one line of *Brassica oleracea* (Taylor *et al.*, 1994).

In addition to fatty acid composition, the quality of a vegetable oil may be determined by the structure of the individual triacylglycerols. For low erucic acid oils, triacylglycerols high in oleic acid were observed to be dominant (Prévôt *et al.*, 1990) (Table 10.4). However, the composition of triacylglycerols can vary widely, genetic and environmental differences in fatty acid composition having a substantial effect on triacylglycerol composition (Prévôt *et al.*, 1990) (Table 10.4).

Monoacylglycerol, diacylglycerol and free fatty acids

Monoacylglycerol, diacylglycerol and free fatty acids are metabolites of triacylglycerol synthesis. However, in mature seed they are usually present only in trace amounts

(McKillican, 1966). In immature, damaged or germinated seed amounts may be higher. Some environmental conditions also appear to predispose the seed to higher levels of free fatty acids. In certain years, for example, high free fatty acid levels are a problem with seed produced in the southern Ontario region of Canada while free fatty acids are rarely a problem with seed produced in the continental climate of the Canadian prairies. Zadernowski and Sosulski (1978) reported free fatty acid levels of between 0.5 and 0.8% for low and high erucic acid cultivars, respectively. Daun *et al.* (1986) reported free fatty acid values ranging from 0.1 to 1.4% with a mean of 0.3% for producer samples from the Canadian prairies.

Phospholipids and glycolipids

Phospholipids and glycolipids, collectively referred to as polar lipids, comprise approximately 4 to 5% of *Brassica* seed oil (Zadernowski and Sosulski, 1978). Phospholipids have fatty acids attached to the 2- and 3-positions of glycerol and a phosphate group attached to the 1-position of the glycerol to which is esterified a variety of compounds including choline, ethanolamine and inositol. Sosulski *et al.* (1981)

Table 10.4. Triacylglycerol species of rapeseed oils with varying linolenic acid content.

Triacylglycerol species	Westar (11.3% 18 : 3)	Bienvenu (7.3% 18 : 3)	Low linolenic (3.1% 18 : 3)
18 : 2/18 : 3/18 : 2	0.7	–	–
18 : 2/18 : 2/18 : 3	1.3	0.8	0.6
18 : 1/18 : 3/18 : 3	2.9	0.7	–
18 : 2/18 : 2/18 : 2	1.1	2.2	3.1
18 : 1/18 : 2/18 : 3	9.2	4.7	3.0
16 : 0/18 : 2/18 : 3	0.9	1.0	0.3
18 : 1/18 : 2/18 : 2	9.2	9.1	15.6
18 : 1/18 : 1/18 : 3	11.5	8.2	3.6
16 : 0/18 : 2/18 : 2	1.4	2.0	1.8
16 : 0/18 : 1/18 : 3	1.7	1.6	0.5
18 : 1/18 : 1/18 : 2	19.8	22.9	27.6
16 : 0/18 : 1/18 : 2	4.6	6.7	5.4
18 : 1/18 : 2/20 : 1	1.2	1.7	1.8
18 : 1/18 : 1/18 : 1	23.9	26.1	24.3
18 : 0/18 : 1/18 : 2	1.8	2.1	2.2
16 : 0/18 : 1/18 : 1	4.2	6.3	4.6
16 : 0/16 : 0/18 : 1	1.3	1.7	1.6
18 : 0/18 : 1/18 : 1	2.0	2.1	2.1

Source: Prévôt *et al.* (1990).

reported phosphatidylcholine (cephalin), phosphatidylinositol and phosphatidylethanolamine (lecithin) to represent 48, 18 and 8% of rapeseed oil, respectively. Mixtures of crude phospholipids, including other polar lipids obtained as a by-product of oil refining, are referred to as lecithin by industry. However, biochemically lecithin is phosphatidylcholine.

Brassica seeds develop green chlorophyll-containing chloroplasts during maturation and thus accumulate appreciable amounts of monogalactodiacylglycerol and digalactodiacylglycerol, glycolipids of the chloroplast which extract into the oil and are included to some extent in the crude lecithin. Sosulski *et al.* (1981) reported levels of 0.9% for glycolipids in the oil of rapeseed.

Both phospholipids and glycolipids are usually rich in polyunsaturated fatty acids. Goraj-Moszora and Drozdowski (1987) found that the phospholipid fraction of the low erucic *B. napus* cultivar Jantar was more unsaturated than the triacylglycerol fraction, containing more linoleic acids (32.6 versus 22.8%) but less linolenic acid (6.7% versus 9.9%). Sosulski *et al.* (1981) found, in addition to high levels of saturated fatty acids in the phospholipids from medium and low erucic cultivars, relatively high levels of linoleic acid. Linoleic and linolenic acids were also elevated in the diacylglycerol fraction.

Sterols

Sterols and sterol derivatives together with tocopherols, carotenoids and chlorophylls make up the polyisoprenoid lipids derived from isoprene.

Appelqvist *et al.* (1981) analysed the free and esterified sterol content of a number of *Brassica* and *Sinapis* seed oils. Free sterols represented 0.27 to 0.36% of the oils while esterified sterols made up 0.4 to 1.2% of the oils. Sitosterol followed by campesterol predominated in both free and esterified form while Δ^5-avenasterol and Δ^7-stigmasterol were present in trace amounts in all but the *S. alba* oil (Table 10.5). The *S. alba* oil was unique in that it had measurable amounts of cholesterol. The free sterols were richer in brassica-sterol, a sterol characteristic of *Brassica* and other *Cruciferae* oils and used for identification (Ackman, 1983).

Tocopherols

Tocopherols, although present in small amounts, are an important group of poly-isoprenoids because they retard the oxidation of unsaturated fatty acids and thus lessen the production of off-flavours and rancidity. Reported contents of tocopherol isomers have been variable, probably in part due to differences in methods of analysis (Table 10.6). Environmental effects have also been noted. Marquard (1985) using growth chambers investigated the influence of environment on tocopherol content of *B. napus, B. juncea* and *S. alba* oils. Cool temperature (16.5°C) and either short or long daylength resulted in about 50 mg tocopherol 100 g^{-1} oil, while warm temperature (24.5°C) and either short or long daylength elevated the tocopherol content to about 70 mg 100 g^{-1} oil. Tocopherols can also be either destroyed or removed during processing, especially during deodorization (Müller-Mulot, 1976; Sleeter, 1981) (Table 10.6).

Carotenoids

Embryos of mature sound *Brassica* seed have a yellow pigmentatione due to the presence of carotenoids, which are also derivatives of isoprene. Denecke (1978) reported 100 μg g^{-1} total carotenoids in oil from *B. napus* cultivar Lesira. Box and Boekenoogen (1967) reported 13.9 μg g^{-1} lutein (xanthophyll, 3,3′-dihydroxy-α-carotene), 3.9 μg g^{-1} neo-lutein A and 5.0 μg g^{-1} neo-lutein B for a total of

22.8 μg g^{-1} in a rapeseed oil of unknown origin while Froehling *et al.* (1972) from the same laboratory reported 39.5 μg g^{-1} lutein and 4.7 μg g^{-1} lutein diester in a rapeseed oil of Canadian origin.

Chlorophylls

Chlorophylls are green pigments with a hydrophilic porphyrin ring containing magnesium and a hydrophobic polyisoprenoid side chain (phytol) which makes them lipid soluble. The chlorophylls extract into the oil upon processing, from which they are then difficult to remove (Appelqvist, 1989). In addition to imparting an undesirable colour to the oil, chlorophyll can inhibit hydrogenation catalysts used to harden the oil in the manufacture of margarines and shortening (Abraham and de Man, 1986). Oils from seed with elevated chlorophyll content are less stable, their oxidation resulting in off-flavours and rancidity (Dahlén, 1973). Phytol-deficient chlorophylls such as chlorophyllides and pheophorbides may contribute to photosensitive dermatitis (Clare, 1955).

During seed development chlorophyll accumulates to levels of 500 to 800 μg g^{-1} fresh matter. As the seed matures chlorophyll is reduced to less than 20 μg g^{-1} in sound mature seed. Higher residual chlorophyll contents, resulting from cool environmental conditions during seed ripening or frost (Johnson-Flanagan *et al.*, 1990), can severely downgrade a seed lot (Daun, 1982, 1987).

Analysis of canola seed, crude oil and degummed oil from Canadian processing plants showed that chlorophyll *a* and *b*, in a ratio of 3 : 1, were the major chlorophyll pigments present in seed, with pheophytin *a* present in minor amounts and pheophorbide *a* and its methyl ester, methylpheophorbide *a*, present in trace amounts (Endo *et al.*, 1992) (Table 10.7). The major pigments present in crude oil were pheophytin *a* with lesser amounts of pheophytin

Table 10.5. Free and esterified sterol content of *Brassica* and *Sinapis* seed oils.

Species and form	Cholesterol		Brassicasterol		Campesterol		Sitosterol		Δ^5-Avenasterol		Δ^7-Stigmasterol	
	Free	Esterified	Free	Esterified	Free	Esterified	Free	Esterified	Free	Esterified	Free	Esterified
Brassica napus												
Summer	tr		14.1	6.5	31.0	37.9	54.1	52.2	0.8	2.5	tr	0.9
Brassica rapa												
Winter	tr		13.8	6.1	28.8	37.1	55.8	52.0	1.6	4.7	tr	tr
Summer	tr		10.4	4.8	29.0	38.8	60.6	53.4	tr	2.9	tr	tr
Brassica juncea	tr		19.2	9.1	23.6	34.0	57.2	55.2		1.7		
Sinapis alba	2.0	2.3	10.2	2.6	24.5	32.7	52.0	41.7	11.3	20.7		

Source: Appelqvist *et al.* (1981).

Table 10.6. Tocopherol content of rapeseed oils.

Type of cultivar	Tocopherol isomer content (mg 100 g^{-1})				Reference
	α	β	γ	δ	
High erucic	26.0		42.6		Mordret and Helme (1974)
Canbra	19.2		42.8	4.2	Mordret and Helme (1974)
Primor	26.1		61.4		Mordret and Helme (1974)
Unknown	30.0		48.1	2.9	Balz et al. (1992)
Refined oil	7.0	1.6	17.8	0.7	Müller-Mulot (1976)

b and pyropheophytin *a* and trace amounts of pheophorbide *a* and methylpheophorbide *a* present. On the other hand, in degummed oil pyrophyeophytin *a* was the major pigment present, phyeophytin *a* was present in lesser amounts and methylpheophorbide in trace amounts. Thus it would appear that chlorophylls *a* and *b* in the seed lose magnesium and are converted to their respective pheophytins during the extraction steps of processing and that some loss of phytol occurs, converting the pheophytins to pheophorbides and their methyl esters. Further alteration occurred during degumming, the methoxycarboxyl group of the pheophytins probably being removed during the heat treatment in the presence of citric or malic acid and the pheophorbides being removed with the gums during centrifugation.

Protein

Brassica oilseeds contain 20 to 30% protein on a whole-seed basis which adds to the value of the seed. The meal by-product of oil extraction contains between 36 and 44% protein and is generally used as an animal feed, although some work has explored the preparation of protein isolates and concentrates for human consumption (Jones, 1979; Fenwick, 1982; Rubin *et al.*, 1990). As with oil, protein content is under both genetic and environmental control. Protein content has generally shown an inverse relationship to oil content, protein content being higher and oil content lower when the seed is grown under warm dry conditions, and vice versa. However, successful attempts to increase protein content while maintaining oil content have been reported (Grami and Stefansson, 1977; Bengtsson, 1985). Bengtsson was able to increase protein content to 55% on a oil-extracted meal basis.

Seed protein content is determined by measuring the nitrogen released either by the Kjeldahl digestion technique, or more recently by the Dumas combustion technique (Bicsak, 1993; Daun and DeClercq, 1994) and multiplying by a factor of 6.25. Although this conversion factor is widely used it has been recognized that a factor of 5.53 would be more appropriate for *Brassica* oilseeds (Tkachuk, 1969). This factor takes into account the amino acid compositions and exclusion of non-protein nitrogen from nucleic acids, purine and pyrimidine bases, nitrogen-containing lipids and glucosinolates.

Three major classes of proteins have been identified in *Brassica* oilseeds: albumins, globulins and oleosins. The albumins are water-soluble proteins largely responsible for the metabolic activity of the seed. Globulins are salt-soluble proteins which constitute 70% of the protein of mature seed. Oleosins are structural proteins of the oil bodies which can constitute as much as 20% of the total seed protein (Mieth *et al.*, 1983; Murphy and Cummins, 1989).

Brassica oilseeds contain two classes of

Table 10.7. Chlorophyll pigments in canola seed and processed oils.

Sample type	Chlorophyll *a*	Chlorophyll *b*	Chlorophyll pigments (mg kg^{-1})					
			Pheophytin *a*	Pheophytin *b*	Pheophorbide *a*	Methylpheophorbide *b*	Pyropheophytin *a*
Seed	17.6	5.8	0.3				
Crude oil			22.0	2.0	0.1	0.1	3.2
Degummed oil			9.0	0.2	0.6	0.8	6.1
						0.3	

Source: Endo *et al.* (1992).

globulin storage proteins, napin which is a 1.7S globulin and cruciferin which is a 12S globulin. Napin accounts for about 20% of the total seed protein while cruciferin accounts for about 50%.

Napin

The napin class of storage globulins are small (M_r of 12,000 to 14,000), basic proteins (Lönnerdal and Jansson, 1972; Crouch *et al.*, 1983). Lönnerdal and Jansson (1972) isolated four napin proteins from rapeseed, two of which were further characterized with respect to polypeptide content and amino acid composition. Each napin consisted of two polypeptides (M_r 4000 and 9000) covalently linked by disulphide bonds. Both the napins characterized were found to have identical amino acid compositions, with approximately 25% of their residues as glutamine and, in addition, high amounts of proline.

Cruciferin

Studies have shown that the cruciferin of rapeseed is similar to 12S globulins from other seeds (Schwenke *et al.*, 1981, 1983; Dalgalarrondo *et al.*, 1986). It is a large (M_r 300,000 to 350,000), neutral, oligomeric protein composed of six subunit pairs. Each pair consists of one heavy α chain (30 kDa) and one light β chain (20 kDa). Four different subunit pairs exist with the majority of α and β chains of each subunit disulphide-linked (Rödin and Rask, 1990). Analysis of subunit composition of cruciferin hexamers by ion-exchange chromatography has suggested that a large array of hexamers exist, composed of mixed combinations of the four subunits (Rödin and Rask, 1990). Amino acid composition resembles other 12S globulins (Bhatty *et al.*, 1968; Schwenke *et al.*, 1981).

Oleosin

Oleosins are not nitrogen-rich, are insoluble in water regardless of the salt concentration or pH, and are synthesized somewhat later than the napins or cruciferins (Murphy and Cummins, 1989; Murphy *et al.*, 1989; Murphy, 1990). Oleosins are believed to serve a structural role as part of the outer membrane of the oil bodies and to provide binding sites for lipase during germination (Huang, 1992). In a strict sense they are not storage proteins as they are not found in seed protein bodies. Nevertheless, they make up an appreciable portion of the protein of the mature seed. Oleosin content appears to be correlated with oil content (Murphy, 1990).

Amino acid composition

The quality of *Brassica* oilseed proteins is determined to a large degree by the amino acid composition. Studies with rapeseed indicate that both *B. napus* and *B. rapa* are rich in lysine and have substantial amounts of the sulphur amino acids, methionine and cystine, which are limiting in most cereal and oilseed proteins (Table 10.8). Rapeseed also contains substantial amounts of threonine. However, amino acid composition can vary appreciably due to both genetic and environmental differences (Table 10.8). Breeding for increased protein content does not seem to have any deleterious effect on the content of essential amino acids (Uppström and Johansson, 1991).

CARBOHYDRATES

The carbohydrate content of *Brassica* oilseeds can be conveniently divided into soluble carbohydrates (monosaccharides and oligosaccharides), insoluble carbohydrates (polysaccharides) and fibre. The

Table 10.8. Amino acid content of rapeseed.

Amino acid	Brassica napus					Brassica rapa		
	Winter[1]	Summer[1]	Summer high protein[1]	Tower[2]	Oro hull[3]	Winter[1]	Summer[1]	Summer high C$_{16}$[1]
Alanine	4.6	4.6	4.6	5.4	4.4	4.6	4.7	4.7
Arginine	6.6	6.8	6.7	6.2	3.5	6.3	6.6	6.6
Aspartic acid	7.7	8.0	7.4	8.8	10.3	6.9	7.8	8.0
Cystine	2.8	2.5	3.0	1.4	1.3	3.1	2.5	2.5
Glutamic acid	18.7	18.3	19.7	17.4	13.5	19.6	18.6	18.2
Glycine	5.2	5.5	5.4	5.7	5.2	5.2	5.5	5.4
Histidine	4.2	4.5	4.6	3.0	2.2	4.2	4.0	4.1
Isoleucine	4.5	4.5	4.5	4.2	4.1	4.5	4.8	4.7
Leucine	7.4	7.4	7.4	7.8	6.5	7.3	7.5	7.4
Lysine	6.3	5.9	6.2	5.8	7.7	6.5	6.1	6.3
Methionine	2.3	2.2	2.3	2.3	1.9	2.5	2.3	2.3
Phenylalanine	4.2	4.2	4.0	4.3	4.4	4.2	4.2	4.3
Proline	6.1	6.0	6.5	6.9	9.7	6.6	6.1	5.9
Serine	4.8	5.0	4.9	5.5	6.5	4.7	4.8	4.6
Threonine	4.8	4.9	4.7	5.9	6.9	4.9	4.9	5.2
Tryptophan	nd	nd	nd	nd	0.4	nd	nd	nd
Tyrosine	3.3	3.1	2.9	3.1	4.0	2.8	3.1	2.9
Valine	5.5	5.5	5.4	4.9	6.3	5.4	5.6	5.6

Amino acid content (g amino acid 16 g^{-1} N)

nd = Not determined.
[1] Analysed at the author's laboratory, Svalöf Weibull, Svalöv, Sweden.
[2] Lajolo et al. (1991).
[3] Sarwar et al. (1981).

soluble carbohydrate content of mature seed is somewhat variable but is composed predominantly of sucrose with lesser amounts of raffinose and stachyose (Naczk and Shahidi, 1990). Siddiqui and Wood (1977) reported the polysaccharide content of oil-extracted dehulled rapeseed meal to be 29%, the relative proportions of which were pectins (50.0%), cellulose (24.1%), amyloids (15.5%), arabinans (6.9%) and arabinogalactan (3.5%). Small amounts of starch (0.1 to 0.16%) have also been reported to be present in samples of rapeseed cotyledons physically separated from the seed coat (Blair and Reichert, 1984). Fibre, which is predominantly associated with the seed coat, ranges from 12 to 30% on a meal basis, depending to a significant degree on the method of analysis.

Soluble carbohydrates

Variable contents of soluble carbohydrates have been reported for Brassica seed meals. Theander and Åman (1974) reported soluble carbohydrate contents of 10 to 12%, and Blair and Scougall (1975) 10.5 to 16%, for oil-extracted rapeseed meal. More recently Shahidi et al. (1990) reported the soluble carbohydrate content, determined by high-performance liquid chromatography, to vary from 5.93 to 7.58% for both canola and mustard seed meal.

Soluble carbohydrate composition has also been reported to vary (Hrdlička et al., 1965; Siddiqui et al., 1973), possibly due to differences in the method of extraction and/or analysis (Naczk and Shahidi, 1990). Shahidi et al. (1990) reported that rapeseed and mustard meals contained 3.93 to 5.73% sucrose, 0.27 to 0.62% raffinose, and 0.83 to 1.61% stachyose as major soluble carbohydrates, while glucose and fructose were present only in trace amounts. Soluble carbohydrates, particularly raffinose and stachyose, accumulate late in seed development, in part as a result of starch remobilization, and are believed to impart desiccation tolerance (Leprince *et al.*, 1990). Thus variations in climatic conditions, particularly towards the end of seed maturation, may contribute to the variability in content and composition of soluble carbohydrates observed in mature seed.

The seed coat fraction (hulls) of rapeseed has been reported to contain approximately three times less soluble carbohydrate than the corresponding dehulled meal fraction (Theander and Åman, 1974; Theander *et al.*, 1977). Separate analysis of rapeseed hulls showed the soluble carbohydrates to be present in proportions similar to those in the meal: sucrose > stachyose > raffinose > glucose and fructose (Theander *et al.*, 1977).

Insoluble carbohydrates

Compared to the monosaccharides and oligosaccharides, polysaccharides have a different distribution in the embryo and seed coat of *Brassica* seed. While the soluble carbohydrates are more abundant in the embryo, polysaccharides are present in larger amounts in the seed coat (Naczk and Shahidi, 1990). Polysaccharides isolated and identified in *Brassica* oilseeds include amyloids, arabinans, arabinogalactans, pectins and starch.

Two subclasses of amyloids have been found in *Brassica* oilseeds, amylose, composed of glucose, galactose and xylose, and fucoamylose, which also contains fucose (Naczk and Shahidi, 1990). Amyloids have been isolated and identified from the seed of *B. rapa* (Siddiqui and Wood, 1971), while fucoamyloids have been isolated and identified from the seed of mustard (Gould *et al.*, 1971) and rapeseed (Siddiqui and Wood, 1977).

Arabinans have been isolated from the seed of rapeseed where they constituted approximately 7% of polysaccharides (Siddiqui and Wood, 1974, 1977). Arabinogalactans isolated from rapeseed embryos constituted 3.4% of the polysaccharides (Siddiqui and Wood, 1972, 1977).

Pectic substances of differing structure are present in both the embryo and seed coats of *Brassica* oilseeds (Siddiqui and Wood, 1977). In the embryo of rapeseed the pectins contained only 30% galacturonic acid, while pectic substances in the seed coat contained 80 to 90% galacturonic acid.

While Blair and Reichert (1984) detected only trace amounts of starch (0.1 to 0.16%) in rapeseed and canola cotyledons, Slominski and Campbell (1991) found canola meal to contain as much as 2.6% starch. During seed development substantially more accumulates. However, accumulation is only transient and most dissipates by the time the seed is mature (Norton and Harris, 1975; Fischer *et al.*, 1988).

Fibre

Fibre is not a homogeneous chemical entity. Nevertheless, it is an important part of the seed as the amount and chemical composition affects the digestibility of the seed, or meal after oil extraction. It is usually measured in *Brassica* seed as crude fibre, which is determined as the loss in mass upon ignition of the dried residue remaining after sample digestion with dilute sulphuric acid and dilute sodium hydroxide under specific conditions (American Oil Chemists' Society, 1993, method Ba6 84). Included in crude fibre are the cellulose, pentosans and lignin of the cell walls. Most of the fibre is present in the hulls with lesser amounts present in the embryo. This has led to attempts to reduce fibre content by removing the hulls (Naczk and Shahidi, 1990; Thakor, 1993). Analysis of acid detergent and neutral detergent fibre are believed to correspond more closely to the total fibre content in a feed (Van Soest and Moore, 1965; Van Soest, 1966) and have been used to analyse fibre content in *Brassica* seed. Fibre analyses of rapeseed meal have yield values of 12 to 13% for crude fibre, 17 to 23% for acid detergent fibre and 20 to 30% for neutral detergent fibre (see Chapter 14, Table 14.1).

PHYTATES

Phytic acid (myoinositol 1,2,3,4,5,6-hexakis-dihydrogen phosphate) in rapeseed and canola were recently reviewed by Thompson (1990). Phytic acid typically exists in rapeseed as mixed salts (phytates) of Ca, Mg and K (Mills and Chong, 1977; Yiu *et al.*, 1982). Phytates are of interest because these antinutritional properties limit the use of *Brassica* seed protein for animal feed and human consumption. Phytates have long been known to reduce mineral availability, to bind to protein, reducing digestibility and amino acid availability, and to reduce the activity of amylase, thus reducing starch hydrolysis (Thompson, 1990). Phytates are believed to be a primary reserve of phosphorus and myoinositol in most seeds. They are also thought to be an energy reserve and a reserve of cations (Greenwood, 1990) and to act as an antioxidant (Graf *et al.*, 1987). Phytates are found in the crystalline globoids (0.5 to 2.8 µm in diameter) inside protein bodies in the cells of both the radicle and the cotyledon (Yiu *et al.*, 1983).

Up to 10% of the dry weight of *Brassica* oilseeds may be accounted for by the phytate globoids (Hofsten, 1973; Yiu *et al.*, 1983). Phytate levels of 2.0 to 4.0% have been reported for the whole rapeseed, 2.0 to 5.0% for the defatted meal, 5.0 to 7.5% for the protein concentrates, and <1.0 to

9.8% for the protein isolates depending on the method of protein isolation (Uppström and Svensson, 1980; Thompson, 1990). Uppström and Svensson (1980) noted that the content of phytate can be influenced appreciably by the environment, particularly by availability of phosphorus in the soil.

PHENOLICS

The content of phenolics in *Brassica* oilseeds is higher than in other oilseeds. For example, on a dry-matter flour (dehulled oil-extracted meal) basis, rapeseed (6.399 g kg^{-1}) was reported to contain ten times the phenolic content of cotton (567 mg kg^{-1}) and peanut (636 mg kg^{-1}) and 30 times the phenolic content of soyabean (234 mg kg^{-1}) (Kozlowska *et al.*, 1990). Phenolics limit utilization of the seed and meal not only as an animal feed but also a source of food-grade protein. Phenolic compounds may contribute to the dark colour, bitter taste and astringency of *Brassica* seed and meal. They, and/or their oxidized products, can also form complexes with essential amino acids, enzymes and other substances (Shahidi and Naczk, 1992).

Phenolics are predominantly located in the embryo with smaller amounts found in the hull. Rapeseed cultivars of diverse origin have been reported to have similar phenolic contents (Krygier *et al.*, 1982; Kozlowska *et al.*, 1983; Dabrowski and Sosulski, 1984; Naczk *et al.*, 1986; Naczk and Shahidi, 1989). In rapeseed, free phenolic acids constituted 15% and phenolic esters 80% of the total phenolics with insoluble bound phenolics making up the remainder (Shahidi and Naczk, 1992).

Free phenolic acids

Analysis of rapeseed and mustard has yielded free phenolics in the range of

Table 10.9. Free, esterified and insoluble bound phenolic acid content of rapeseed and mustard seed.

Fraction, species and cultivar	Phenolic acid content (g kg^{-1})			Reference
	Free	Esterified	Insoluble bound	
Flour				
Brassica napus				
Tower	0.982	9.820	–	Krygier *et al.* (1982)
Brassica rapa				
Candle	0.845	11.964	–	Krygier *et al.* (1982)
Meal				
Brassica napus				
Tower	2.440	12.020	0.960	Naczk and Shahidi (1989)
Altex	2.480	14.580	1.010	Naczk and Shahidi (1989)
Midas	1.445	15.240	0.687	Shahidi and Naczk (1992)
Triton	0.615	12.120	0.513	Shahidi and Naczk (1992)
Mustard	1.081	15.380	0.224	Shahidi and Naczk (1992)

0.615 to 2.480 g kg^{-1} for flours and meals (Shahidi and Naczk, 1992) (Table 10.9). Sinapine, the choline ester of 3,5-dimethoxy-4-hydroxycinnamic acid (sinapic acid), has been reported to be the predominant free phenolic acid, constituting approximately 73% of the free phenolic acids (Krygier *et al.*, 1982). Minor phenolic acids present were *p*-hydroxybenzoic, vanillic, gentisic, protocatechuic, syringic, *p*-coumaric, *cis*- and *trans*-ferulic and caffeic acids. Trace amounts of chlorogenic acid have also been reported (Lo and Hill, 1972; Kozlowska *et al.*, 1975, 1983; Krygier *et al.*, 1982).

Esterified phenolic acids

Phenolic esters constitute 9.820 to 15.380 g kg^{-1} flour or meal of rapeseed and mustard (Shahidi and Naczk, 1992) (Table 10.9). Sinapine constitutes approximately 70 to 97% of the phenolic acids released from esters and glycosides of rapeseed. Small amount of *p*-hydroxybenzoic, vanillic, protocatechuic, syringic, *p*-coumaric, *cis*- and *trans*-ferulic and caffeic acids have been reported in hydrolysates of soluble

esters extracted from *B. napus*, cultivar Tower, and *B. rapa*, cultivar Candle (Krygier *et al.*, 1982).

Insoluble bound phenolic acids

Insoluble bound phenolics of rapeseed and mustard range from 0.960 to 1.010 g kg^{-1} oil-extracted air-dried meal (Shahidi and Naczk, 1992) (Table 10.9). In total nine phenolic acids have been identified from the insoluble bound fraction of rapeseed, sinapic being the predominant phenolic followed by *p*-coumaric and *trans*-ferulic (Kozlowska *et al.*, 1983).

Tannins

Tannins are complex phenolic compounds which can form soluble and insoluble complexes with protein and as a result diminish the availability of the protein to animals. They are classified either as condensed or hydrolysable based on their reactivity toward hydrolytic agents, particularly acids. Condensed tannins, formed by polymerization of flavan-3-ols or flavan-3,4-diols, were first reported to be present in rapeseed

Table 10.10. Systematic and trivial name of the prevalent glucosinolates found in *Brassica* and related seeds.

Systematic name	Trivial name	Oilseed[1]
Aliphatic		
2-Propenyl (allyl)[2]	Sinigrin	Bj
3-Butenyl	Gluconapin	Bj[3] Bn Br
4-Pentenyl	Glucobrassicanapin	Bn Br
S-2-hydroxy-3-butenyl	Progoitrin	Bn Br
R-2-hydroxy-3-butenyl	Epiprogoitrin	Ca
2-Hydroxy-4-pentenyl	Napoleiferin	Bn Br
Cyclic		
2-Phenylethyl	Gluconasturtin	Bj Bn Br[4]
4-Hydroxybenzyl	Sinalbin	Sa
Heterocyclic (indole)		
3-Indolylmethyl	Glucobrassicin	Bj Bn Br Sa[5]
4-Hydroxy-3-indolylmethyl	4-Hydroxyglucobrassicin	Bj Bn Br Sa
1-Methoxy-3-indolylmethyl	Neoglucobrassicin	Bj Bn Br Sa[4]
4-Methoxy-3-indolylmethyl	4-Methoxyglucobrassicin	Bj Bn Br Sa[4]

[1] Bj = *Brassica juncea*, Bn = *Brassica napus*, Br = *Brassica rapa*, Ca = *Crambe abyssinica*, Sa = *Sinapis alba*.
[2] Common systematic name.
[3] Found in substantial amounts in seed of *B. juncea* of Indian origin.
[4] Found predominantly in the roots.
[5] Found predominantly in the vegetative tissues.

hulls by Bate-Smith and Ribéreau-Gayon (1959). Durkee (1971) subsequently identified cyanidin and pelargonidin among the hydrolytic products of rapeseed hulls. Reports of condensed tannin contents vary widely due to differences in the isolation and quantification methodology (Shahidi and Nazck, 1992). Rapeseed hulls have been reported to contain as much as 0.22% extractable tannins (Mitaru *et al.*, 1982) while rapeseed meal has been reported to contain as much as 2.7% (Fenwick *et al.*, 1984).

Glucosinolates

Over 90 glucosinolates are known to exist in dicotyledonous plants including all members of the genus *Brassica* (Fenwick *et al.*, 1983). Despite the large number of identified glucosinolates, most species con-

tain only a few, and usually only one or two predominate (Table 10.10). Considerable differences in relative abundance also occur between individual plants and plant parts. Even within the same plant part, glucosinolate content can vary with development, being usually highest during the period of most active growth. The seed can accumulate substantial amounts of glucosinolates synthesized in the pod (De March *et al.*, 1989).

Glucosinolates are sulphonated oxime thioesters, predominantly of glucose (Fig. 10.1(1)). The structure of the R group, which is derived from amino acids, may be aliphatic, cyclic or heterocyclic. Glucosinolates coexist in plants but are not in contact with the hydrolytic enzyme myrosinase (thioglucoside glucohydrolase, EC 3.2.3.1). When the plant tissue is crushed in the presence of adequate moisture,

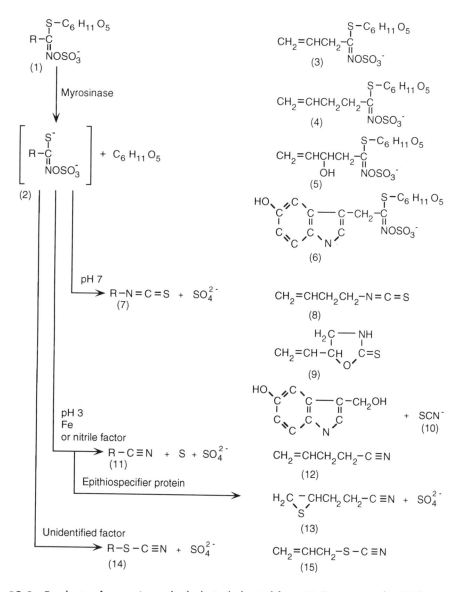

Fig. 10.1. Products of myrosinase hydrolysis (adapted from McGregor *et al.*, 1983).

myrosinase rapidly hydrolyses the gluco-sinolate to yield glucose and an unstable aglucone (McGregor *et al.*, 1983) (Fig. 10.1). The aglucone then undergoes a Lossen-type rearrangement to yield sulphate and a variety of products the nature of which is dependent upon a number of factors, including the structure of the gluco-

sinolate side chain, reaction conditions, such as pH, temperature and duration, the age and condition of the plant tissue, and the presence of cofactors such as metal ions, ascorbic acid and specific proteins.

Many glucosinolates give rise to stable isothiocyanates (Fig. 10.1(7)) particularly under neutral and alkaline conditions.

For example, 3-butenyl glucosinolate (Fig. 10.1(4)), a major glucosinolate in the seed of rapeseed and crambe (*Crambe abyssinica* Hochst. Ex. R.E. Fries), gives rise to 3-butenyl isothiocyanate (Fig. 10.1(8)). However, glucosinolates which possess a β-hydroxyl group in their side chain, such as 2-hydroxy-3-butenyl glucosinolate (Fig. 10.1(5)), also present in the seed of rapeseed and crambe, give rise to isothiocyanates that spontaneously cyclize to form oxazolidinethiones, i.e. 5-vinyloxazolidine-2-thione (Fig. 10.1(9)). Some cyclic and heterocyclic glucosinolates, notably 4-hydroxybenzyl glucosinolate present in yellow mustard seed, and the indole glucosinolates, such as 4-hydroxy-3-indolylmethyl glucosinolate (Fig. 10.1(6)) prevalent in *Brassica* seed, give rise to isothiocyanates which are unstable at neutral and alkaline pH and break down to release an alcohol and free thiocyanate ion (Fenwick *et al.*, 1989) (Fig. 10.1(10)).

In addition to isothiocyanates, nitriles (Fig. 10.1(11)) are formed as autolysis products, their production resulting from the liberation of elemental sulphur. The relative proportion of isothiocyanate to nitrile can vary widely depending upon the conditions of autolysis. In general weakly acidic conditions favour nitrile formation. But some plant species are predominantly isothiocyanate producing while others are predominantly nitrile producing. In rapeseed under acidic conditions 3-butenyl glucosinolate (Fig. 10.1(4)) gives rise to 3-butenyl nitrile (Fig. 10.1(12)). In crambe seed the presence of an as yet unidentified cofactor, believed to be a protein, has been observed to favour nitrile production. An additional cofactor, originally identified in crambe and referred to as an epispecifier protein, in combination with ferrous ion can convert glucosinolates into epithionitriles. For example, 3-butenyl glucosinolate (Fig. 10.1(4)) is converted to 1-cyano-3,4-epithiobutane (Fig. 10.1(13)). Isothiocyanates and nitriles can also arise

non-enzymatically at higher temperatures.

The production of thiocyanates (Fig. 10.1(14)) is less common but prevalent in certain species. Allyl glucosinolate (Fig. 10.1(3)) in stinkweed (*Thlaspi arvense* L.), a common crucifer weed on the Canadian prairies, is hydrolysed predominantly to allyl thiocyanate (Fig. 10.1(15)). Formation is believed to involve a cofactor which may also be a protein, since it has been shown to be labile to both heat and polar organic solvents.

Intact glucosinolates have generally been considered to be innocuous. Hydrolysis products, on the other hand, produce several physiological effects when they are present in large quantities in animal feeds. These include depressed growth related to the goitrogenicity of several hydrolysis products, haemorrhagic livers in poultry possibly related to the presence of epithionitriles, and skeletal abnormalities in poultry (see Chapter 14).

The presence of glucosinolates in *Brassica* oilseeds has hindered the use of the meal in animal feed. As a result, in the 1970s plant breeders developed cultivars which, in addition to having a lower erucic acid content, have a lower glucosinolate content. Nutritional studies established that meals from these low glucosinolate cultivars could be used to a much greater extent in animal feed rations. In Canada, where low glucosinolate cultivars were first introduced, the effect on the oilseed industry was so impressive that this new type of oilseed was designated 'canola'. By definition, canola is the seed of *B. napus* or *B. rapa*, the oil component of which contains less than 2% erucic acid (22 : 1) and the meal (solid) component of which contains less than $30 \mu mol\ g^{-1}$ oil-extracted, air-dried meal of any one or any mixture of the aliphatic glucosinolates, 3-butenyl, 4-pentenyl, 2-hydroxy-3-butenyl and 2-hydroxy-4-pentenyl glucosinolate (Canola Council of Canada, 1990). In 1997 a new canola definition is scheduled to come into effect in Canada which will include seed from the whole of the *Brassica* genus, if it contains less than 18 µmol of total glucosinolates g^{-1} whole seed at a moisture content of 8.5% (Canola Council of Canada, personal communication). Meal derived from such seed must contain less than 30 µmol of total glucosinolates g^{-1} meal at a moisture content of 8.5%.

Rapeseed and canola seeds contain predominantly aliphatic glucosinolates derived from methionine and heterocyclic indole glucosinolates derived from tryptophan (Larsen, 1981; Daun and McGregor, 1991). The glucosinolate content of *Brassica* oilseeds and meals varies widely due to differences between species and cultivars, environmental effects and the effects of processing. Even the method of analysis can affect reported contents.

Seeds of traditional cultivars of *B. napus* have glucosinolate contents between 61 and 150 µmol g^{-1} oil-extracted, air-dried meal while *B. rapa* varies between 77 and 167 µmol g^{-1} oil-extracted, air-dried meal (Sang and Salisbury, 1988) (Table 10.11). Canola cultivars of both species have been bred to have 30 µmol g^{-1} aliphatic glucosinolate on an oil-extracted, air-dried meal basis. Recently it has been shown that it is possible by genetic engineering to reduce the content of indole glucosinolates to only 3% of the original value (Chavadej *et al.*, 1994).

Environment, in particular the presence of sulphur (Mailer and Wratten, 1985; Mailer, 1989), can have a large effect on glucosinolate content. Where availability of sulphur is limited, glucosinolate levels are substantially reduced due to its preferential incorporation into protein (Mailer and Wratten, 1985).

Processing conditions commonly used to extract oil from rapeseed and canola can result in a 40 to 60% reduction in glucosinolate content (Campbell and Cansfield, 1983). This reduction has generally been assumed to be a thermal decomposition which takes place during 'desolventizing' and toasting of the extracted meal. The composition of residual decomposition products in the meal is not well understood but the major components found have been nitriles and thiocyanate ions (Campbell and Cansfield, 1983; Gardrat *et al.*, 1988).

Crambe abyssinica seed has been grown on an experimental basis in the United States as a potential source of erucic acid for industrial and chemical application. Crambe seed contains substantial amounts of (*S*)-2-hydroxy-3-butenyl glucosinolate (epiprogoitrin) which limits the use of crambe meal as a high-protein animal feed ingredient. In a commercial-scale processing study the amount of intact glucosinolates in crambe seed (105 to 164 µmol g^{-1}, whole seed, defatted) was found to decrease substantially during desolventization (20 to 80 µmol g^{-1}) (Carlson *et al.*, 1985). The nitrile formed during hydrolysis of the glucosinolate, 1-cyano-2-hydroxy-3-butene was found to be present in the desolventized meal in amounts of 15 to 50 µmol g^{-1}.

In the Indian subcontinent, *B. juncea* mustard is grown as a source of edible oil. In North America and Europe, however, *B. juncea* and *S. alba* mustards are grown for use as a condiment, the desirable flavour of which is due to hydrolysis products of the glucosinolates present in the seed. *Sinapis alba* contains upwards of 200 µmol g^{-1} oil-extracted, air-dried meal of 4-hydroxybenzyl glucosinolate (sinalbin). *Brassica juncea* of European or North American origin contains 150 to 200 µmol g^{-1} oil-extracted, air-dried meal of allyl glucosinolate (sinigrin). *Brassica juncea* from the Indian subcontinent contains variable amounts of allyl glucosinolate and 3-butenyl glucosinolate. Recently, plant breeders have developed, as sources of edible oil, lines of *B. juncea* which are low in both erucic acid and glucosinolates (Love *et al.*, 1990a, b).

The most common cruciferous weed

Table 10.11. Glucosinolate composition of the seed of *Brassica* and *Sinapis* oilseeds.

Genus, species and cultivar	Glucosinolate content (μmol g^{-1} oil-extract meal)[1]								
	Allyl	3-Butenyl	4-Pentenyl	2-Hydroxy-3-butenyl	2-Hydroxy-4-pentenyl	4-Hydroxy-benzyl	3-Indolylmethyl	4-Hydroxy-3-indolylmethyl	Total
Brassica napus									
High glucosinolate type		37	7	93	3			5	145
Canola type[2]		3		9				5	17
Brassica rapa									
High glucosinolate type		39	30	17	4			5	95
Canola type[2]		4	2	7	1			7	21
Brassica juncea									
High glucosinolate type	200	1						4	205
Low glucosinolate type[2]	5	20	3				1	1	30
Sinapis alba									
High glucosinolate type				5		210			215
Low glucosinolate type[2]				5		1			6

[1] Analysed at the author's laboratory, Svalöf Weibull, Svalöv, Sweden.
[2] Genetically altered to reduce glucosinolate content.

seeds associated with rapeseed or canola in Canada are wild mustard (*Sinapis arvensis* L.) and stinkweed (*Thlaspi arvense* L.). Analyses of wild mustard seed have shown up to 200 μmol g^{-1} oil-extracted, air-dried

meal of 4-hydroxybenzyl glucosinolate while the seed of stinkweed was found to have similar contents of allyl glucosinolate (Daun and McGregor, 1991).

REFERENCES

Abraham, V. and de Man, J.M. (1986) Hydrogenation of canola oil as affected by chlorophyll. *Journal of the American Oil Chemists' Society* 63, 1185–1188.

Ackman, R.G. (1983) Chemical composition of rapeseed oil. In: Kramer, J.K.G., Sauer, F.D. and Pigden, W.J. (eds) *High and Low Erucic Acid Rapeseed Oils.* Academic Press, Toronto, pp. 85–129.

American Oil Chemists' Society (1993) *Official Methods and Recommended Practices of the American Oil Chemists' Society*, 5th edn, Ba 6–84. The American Oil Chemists' Society, Champaign.

Appelqvist, L.-Å. (1969) Lipids in Cruciferae IV. Fatty acid patterns in single seeds and seed populations of various Cruciferae and in different tissues of *Brassica napus* L. *Hereditas* 61, 9–44.

Appelqvist, L.-Å. (1972) Chemical constituents of rapeseed. In: Appelqvist, L.-Å. and Ohlson, R. (eds) *Rapeseed: Cultivation, Composition, Processing and Utilization.* Elsevier Publishing Company, Amsterdam, pp. 123–173.

Appelqvist, L.-Å. (1989) The chemical nature of vegetable oils. In: Röbbelen, G., Downey, R.K. and Ashri, A. (eds) *Oil Crops of the World. Their Breeding and Utilization.* McGraw-Hill, Toronto, pp. 22–37.

Appelqvist, L.-Å., Kornfeldt, A.K. and Wennerholm, J.E. (1981) Sterols and steryl esters in some *Brassica* and *Sinapis* seeds. *Phytochemistry* 20, 207–210.

Balz, M., Schulte, E. and Thier, H.-P. (1992) Trennung von Tocopherolen und Tocotrienolen durch HPLC. *Fat Science Technology* 94, 209–213.

Bate-Smith, E.C. and Ribéreau-Gayon, P. (1959) Leuco-anthocyanins in seeds. *Qualitas Plantarum et Materiae Vegetabiles* 5, 189–198.

Bengtsson, L. (1985) Improvement of rapeseed meal quality through breeding for high protein content. PhD Thesis, Swedish Agricultural University, Svalöv, Sweden.

Bhatty, R.S., McKenzie, S.L. and Finlayson, A.J. (1968) The proteins of rapeseed (*Brassica napus* L.) soluble in salt solutions. *Canadian Journal of Biochemistry* 46, 1191–1197.

Bicsak, R.C. (1993) Comparison of Kjeldahl method for determination of crude protein in cereal grains and oilseeds with generic combustion method: collaborative study. *Journal of the Association of Official Analytical Chemists* 76, 780–786.

Blair, R. and Reichert, R.D. (1984) Carbohydrate and phenolic constituents in a comprehensive range of rapeseed and canola fractions: nutritional significance for animals. *Journal of the Science of Food and Agriculture* 35, 29–35.

Blair, R. and Scougall, R.K. (1975) Chemical composition, nutritive values of rapeseed meals. *Feedstuffs* 47, 26–27.

Box, J.A.G. and Boekenoogen, H.A. (1967) Die Farbstoffe der Pflanzenöle: Carotinoide und Phäophytine im Soja-, Raps- und Leinöl. *Fete Seifen Anstrichmittel* 69, 724–729.

Campbell, L.D. and Cansfield, P.E. (1983) Influence of seed processing on intact glucosinolates and implications regarding the nutritive quality of canola meal. In: McGregor, E.E. (ed.) *Research on Canola Seed, Oil and Meal, Seventh Progress Report*, Canola Council of Canada, Winnipeg, pp. 50–54.

Canola Council of Canada (1990) *Canola Oil and Meal: Standards and Regulations.* Canola Council of Canada Publication, Winnipeg, Canada, 4 pp.

Carlson, K.D., Baker, E.C. and Mustakas, G.C. (1985) Processing of *Crambe abyssinica* seed in commercial extraction facilities. *Journal of the American Oil Chemists' Society* 62, 897–905.

Chavadej, S., Brisson, N., McNeil, J.N. and De Luca, V. (1994) Redirection of tryptophan leads to production of low indole glucosinolate canola. *Proceedings of the National Academy of Sciences of the United States of America* 91, 2166–2170.

Clare, N.T. (1955) Photosensitization in animals. *Advances in Veterinary Science* 2, 182–211.

Crouch, M.L., Tenbarge, K.M., Simon, A.E. and Ferl, R. (1983) cDNA clones for *Brassica napus* seed storage proteins: evidence from nucleotide sequence analysis that both subunits of napin are cleaved from a precursor polypeptide. *Journal of Molecular and Applied Genetics* 2, 273–283.

Dabrowski, K.J. and Sosulski, F.W. (1984) Composition of free and hydrolyzable phenolic acids in defatted flours of ten oilseeds. *Journal of Agriculture and Food Chemistry* 32, 128–130.

Dahlén, J.A.H. (1973) Chlorophyll content monitoring of Swedish rapeseed and its significance in oil quality. *Journal of the American Oil Chemists' Society* 50, 312A–327A.

Dalgalarrondo, M., Robin, J.-M. and Azanza, J.-L. (1986) Subunit composition of the globulin fraction of rapeseed (*Brassica napus*). *Plant Science* 43, 115–124.

Daun, J.K. (1982) The relationship between rapeseed chlorophyll, rapeseed oil chlorophyll and percentage green seeds. *Journal of the American Oil Chemists' Society* 59, 15–18.

Daun, J.K. (1987) Chlorophyll in Canadian canola and rapeseed and its role in grading. In: *Proceedings of the Seventh International Rapeseed Congress*, Poznan, Poland. The Plant Breeding and Acclimatization Institute, Poznan, pp. 1451–1456.

Daun, J.K. and DeClercq, D.R. (1994) Comparison of combustion and Kjeldahl methods for determination of nitrogen in oilseeds. *Journal of the American Oil Chemists' Society* 71, 1069–1072.

Daun, J.K. and McGregor, D.I. (1991) Glucosinolates in seed and residues. In: Rossell, J.B. and Pritchard, J.L.R. (eds) *Analysis of Oilseeds, Fats and Fatty Foods*. Elsevier Applied Science, London, pp. 185–225.

Daun, J.K., Cooke, L.A. and Clear, R.M. (1986) Quality, morphology and storability of Canola and rapeseed harvested after overwintering in northern Alberta. *Journal of the American Oil Chemists' Society* 63, 1333–1340.

De March, G., McGregor, D.I. and Séguin-Shwartz, G. (1989) Glucosinolate content of maturing pods and seeds of high and low glucosinolate summer rape. *Canadian Journal of Plant Science* 69, 929–932.

Denecke, P. (1978) Erfahrungen aus der Verarbeitung von erucasäurearmen Rapssorten (II). *Lebensmittelindustrie* 25, 450–454.

Dorrell, D.G. and Downey, R.K. (1964) The inheritance of erucic acid content in rapeseed (*Brassica campestris*). *Canadian Journal of Plant Science* 44, 499–504.

Downey, R.K. (1971) Agricultural and genetic potentials of Cruciferous oilseed crops. *Journal of the American Oil Chemists' Society* 48, 718–722.

Downey, R.K. (1983) The origin and description of the *Brassica* oilseed crops. In: Kramer, J.K.G., Sauer, F.D. and Pidgen, W.J. (eds) *High and Low Erucic Acid Rapeseed Oils: Production, Usage, Chemistry and Toxicological Evaluation*. Academic Press, Toronto, pp. 1–20.

Downey, R.K. (1990) Breeding canola for yield and quality. In: *Proceedings of the International Canola Conference*, Atlanta, Georgia, USA, 2–6 April. Potash and Phosphate Institute, Atlanta, Georgia, pp. 41–50.

Downey, R.K. and Rimmer, S.R. (1993) Agronomic improvement in oilseed Brassicas. *Advances in Agronomy* 50, 1–66.

Durkee, A.B. (1971) The nature of tannin in rapeseed (*Brassica campestris*). *Phytochemistry* 10, 1583–1585.

Endo, Y., Thorsteinson, C.T. and Daun, J.K. (1992) Characterization of chlorophyll pigments present in canola seed, meal and oil. *Journal of the American Oil Chemists' Society* 69, 564–568.

Fenwick, G.R. (1982) The assessment of a new protein source – rapeseed. *Proceedings of the Nutrition Society* 41, 277–288.

Fenwick, G.R., Heaney, R.K. and Mullin, W.J. (1983) Glucosinolates and their breakdown products in food and food plants. *CRC Critical Reviews in Food Sciences and Nutrition* 18, 123–201.

Fenwick, G.R., Curl, C.L., Butler, E.J., Greenwood, N.M. and Pearson, A.W. (1984) Rapeseed meal and egg-taint: effects of low glucosinolate *Brassica napus* meal, dehulled meal, and hulls, and of neomycin. *Journal of the Science of Food and Agriculture* 35, 749–756.

Fenwick, G.R., Heaney, R.K. and Mawson, R. (1989) Glucosinolates. In: Cheeke, P.R. (ed.) *Toxicants of Plant Origin, Vol. II: Glycosides*. CRC Press, Boca Raton, Florida, pp. 1–41.

Fischer, W., Bergfeld, R., Plachy, C., Schfer, R. and Schopfer, P. (1988) Accumulation of storage materials, precocious germination and development of desiccation tolerance during seed maturation in mustard (*Sinapis alba* L.). *Botanica Acta* 101, 344–354.

Froehling, P.E., van den Bosch, G. and Boekenoogen, H.A. (1972) Fatty acid composition of carotenoid esters in soybean and rapeseed oils. *Lipids* 7, 447–449.

Gardrat, C., Coustille, J.L., Gauchet, C. and Prévôt, A. (1988) Dégradation des glucosinolates au cours du processus technologique de traitement des tourteaux de colza. Identification et dosage de nitriles volatils. Mémoirs scientifique. *Revue Française des Crops Gras* 35, 99–104.

Goraj-Moszora, I.E. and Drozdowski, B. (1987) Composition of phospholipids of double low rapeseed. In: *Proceedings of the Seventh International Rapeseed Congress*, Poznan, Poland. The Plant Breeding and Acclimatization Institute, Poznan, pp. 1463–1468.

Gould, S.E.B., Rees, D.A. and Wight, N.J. (1971) Polysaccharides in germination: xyloglucans ('amyloids') from the cotyledons of white mustard. *Biochemical Journal* 124, 47–53.

Graf, E., Empson, K.L. and Eaton, J.W. (1987) Phytic acid: a natural antioxidant. *Journal of Biological Chemistry* 262, 11647–11650.

Grami, B. and Stefansson, B.R. (1977) Gene action for protein and oil content in summer rape. *Canadian Journal of Plant Science* 57, 625–631.

Greenwood, J.S. (1990) Phytin synthesis and deposition. In: Taylorson, R.B. (ed.) *Recent Advances in the Development and Germination of Seeds*. Plenum Press, New York, pp. 109–125.

Hofsten, A.V. (1973) X-ray analysis of microelements in seeds of *Crambe abyssinica*. *Physiologia Plantarum* 29, 76–81.

Hrdlička, J., Kozlowska, H., Pokorny, J. and Rutkowski, A. (1965) Über Rapsschrote. 7. Mitt. Saccharide in Extraktionsschroten. *Die Nahrung* 9, 71–76.

Huang, A.H.C. (1992) Oil bodies and oleosins in seeds. *Annual Reviews of Plant Physiology and Plant Molecular Biology* 43, 177–200.

Johnson-Flanagan, A.M., Singh, J. and Thiagarajah, M.R. (1990) The impact of sublethal freezing during maturation on pigment content in seeds of *Brassica napus*. *Journal of Plant Physiology* 136, 385–390.

Jones, J.D. (1979) Rapeseed protein concentrate preparation and evaluation. *Journal of the American Oil Chemists' Society* 56, 716–721.

Kirk, J.T.O. and Oram, R.N. (1981) Isolation of erucic acid-free lines of *Brassica juncea*: Indian mustard now a potential oilseed crop in Australia. *Journal of the Australian Institute of Agricultural Sciences* 47, 51–52.

Knutzon, D.S., Thompson, G.A., Radke, E.R., Johnson, W.B., Knauf, V.C. and Kridl, J.C. (1992) Modification of *Brassica* seed oil by antisense expression of a stearoyl-acyl carrier protein desaturase gene. *Proceedings of the National Academy of Sciences of the United States of America* 89, 2624–2628.

Kozlowska, H., Sabir, M.A., Sosulski, F.W. and Coxworth, E. (1975) Phenolic constituents of rapeseed flour. *Canadian Institute of Food Science and Technology Journal* 8, 160–163.

Kozlowska, H., Rotkiewicz, D.A., Zadernowski, R. and Sosulski, F.W. (1983) Phenolic acids in rapeseed and mustard. *Journal of the American Oil Chemists' Society* 60, 1119–1123.

Kozlowska, H., Naczk, M., Shahidi, F. and Zadernowski, R. (1990) Phenolic acids and tannins in rapeseed and canola. In: Shahidi, F. (ed.) *Canola and Rapeseed. Production, Chemistry, Nutrition and Processing Technology*. Van Nostrand Reinhold, New York, pp. 193–210.

Kramer, J.K.G., Sauer, F.D. and Pidgen, W.J. (1983) *High and Low Erucic Acid Rapeseed Oils: Production, Usage, Chemistry and Toxicological Evaluation*. Academic Press, Toronto, 582 pp.

Krygier, K., Sosulski, F. and Hogge, L. (1982) Free, esterified, and insoluble-bound phenolic acids. 2. Composition of phenolic acids in rapeseed flour and hulls. *Journal of Agriculture and Food Chemistry* 30, 334–336.

Krzymanski, J., Peitka, T., Ratajska, I., Byczynska, B. and Krotka, K. (1991) Development of low glucosinolate white mustard (*Sinapis alba* syn. *Brassica hirta*). In: McGregor, D.I. (ed.) *Proceedings of the Eighth International Rapeseed Congress*, Saskatoon, Canada. Organizing Committee, Saskatoon, pp. 1545–1547.

Lajolo, F.M., Marquez, U.M.L., Filiseffi-Cozzi, T.M.C.C. and McGregor, D.I. (1991) Chemical composition and toxic compounds in rapeseed (*Brassica napus*, L.) cultivars grown in Brazil. *Journal of Agricultural and Food Chemistry* 39, 1933–1937.

Larsen, P.O. (1981) Glucosinolates. In: Conn, E.E. (ed.) *The Biochemistry of Plants, a Comprehensive Treatise*, Vol. 7. Academic Press, New York, pp. 501–525.

Leprince, O., Bronchart, R. and Deltour, R. (1990) Changes in starch and soluble sugars in relation to acquisition of desiccation tolerance during maturation of *Brassica campestris* seed. *Plant Cell and Environment* 13, 539–546.

Lo, M.T. and Hill, D.C. (1972) Composition of the aqueous extracts of rapeseed meals. *Journal of the Science of Food and Agriculture* 23, 823–830.

Lönnerdal, B. and Janson, J.-C. (1972) Studies on *Brassica* seed proteins. 1. The low molecular weight proteins in rapeseed. Isolation and characterization. *Biochimica et Biophysica Acta* 278, 175–183.

Love, H.K., Rakow, G., Raney, J.P. and Downey, R.K. (1990a) Development of low glucosinolate mustard. *Canadian Journal of Plant Science* 70, 419–424.

Love, H.K., Rakow, G., Raney, J.P. and Downey, R.K. (1990b) Genetic control of 2-propenyl and 3-butenyl glucosinolate synthesis in mustard. *Canadian Journal of Plant Science* 70, 425–429.

McGregor, D.I., Mullin, W.J. and Fenwick, G.R. (1983) Review of analysis of glucosinolates: analytical methodology for determining glucosinolate composition and content. *Journal of the Association of Official Analytical Chemists* 66, 825–849.

McGregor, W.G., Plessers, A.G. and Craig, B.M. (1961) Species trials with oil plants. I. Crambe. *Canadian Journal of Plant Science* 41, 716–719.

McKillican, M.E. (1966) Lipid changes in maturing oil-bearing plants. IV. Changes in lipid classes in rape and crambe oils. *Journal of the American Oil Chemists' Society* 43, 461–464.

Mailer, R.J. (1989) Effects of applied sulfur on glucosinolate and oil concentrations in the seeds of rape (*Brassica napus* L.) and turnip rape (*Brassica rapa* L. var. *silvestris* (Lam.) Briggs). *Australian Journal of Agricultural Research* 40, 617–624.

Mailer, R.J. and Wratten, N. (1985) Comparison and estimation of glucosinolate levels in Australian rapeseed cultivars. *Australian Journal of Experimental Agriculture* 25, 932–938.

Marquard, R. (1985) The influence of temperature and photoperiod on fat content, fatty acid composition, and tocopherols of rapeseed (*Brassica napus*) and mustard species (*Sinapis alba, Brassica juncea,* and *Brassica nigra*). *Agrochimica* 29, 145–153.

Mieth, G., Schwenke, K.-D., Raab, B. and Brückner, J. (1983) Rapeseed: constituents and protein products. Part 1. Composition and properties of proteins and glucosinolates. *Die Nahrung* 27, 675–697.

Mills, J.T. and Chong, J. (1977) Ultrastructure and mineral distribution in heat-damaged rapeseed. *Canadian Journal of Plant Science* 57, 21–34.

Mitaru, B.N., Blair, R., Bell, J.M. and Reichert, R.D. (1982) Tannin and fiber contents of rapeseed and canola hulls. *Canadian Journal of Animal Science* 62, 661–663.

Mordret, F. and Helme, J.P. (1974) Caractéristiques analytiques de l'huile de colza primor. In: *Proceedings of the Fourth International Rapeseed Congress,* Giessen, Germany. Deutsche Gesellschaft für Fettwissenschaft, Münster pp. 283–289.

Müller-Mulot, W. (1976) Rapid method for the quantitative determination of individual tocopherols in oils and fats. *Journal of the American Oil Chemists' Society* 53, 732–736.

Murphy, D.J. (1990) Storage lipid bodies in plants and other organisms. *Progress in Lipid Research* 29, 299–324.

Murphy, D.J. and Cummins, I. (1989) Biosynthesis of seed storage products during embryogenesis in rapeseed, *Brassica napus. Journal of Plant Physioliology* 135, 63–69.

Murphy, D.J., Cummins, I. and Ryan, A.J. (1989) Immunocytochemical and biochemical study of the biosynthesis and mobilisation of the major seed storage proteins of *Brassica napus. Plant Physiology and Biochemistry* 27, 647–657.

Naczk, M. and Shahidi, F. (1989) The effect of methanol–ammonia–water treatment on the content of phenolic acids of canola. *Food Chemistry* 31, 159–164.

Naczk, M. and Shahidi, F. (1990) Carbohydrates of canola and rapeseed. In: Shahidi, F. (ed.) *Canola and Rapeseed. Production, Chemistry, Nutrition and Processing Technology.* Van Nostrand Reinhold, New York, pp. 211–220.

Naczk, M., Diosady, L.L. and Rubin, L.J. (1986) The phytate and complex phenol content of meals produced by alkanol-ammonia/hexane extraction of canola. *Lebensmittel-Wissenschaft und Technologie* 19, 13–16.

Norton, G. and Harris J.F. (1975) Compositional changes in developing rape seed (*Brassica napus* L.). *Planta* 123, 163–174.

Ohlrogge, J.B. (1994) Design of new plant products: engineering of fatty acid metabolism. *Plant Physiology* 104, 821–826.

Olsson, G. (1984) Selection for low erucic acid in *Brassica juncea. Sveriges Utsädesförenings Tidskrift* 94, 187–190.

Persson, C. (1985) High palmitic acid content in summer turnip rape (*Brassica campestris* var. *annua* L.). *Eucarpia Cruciferae Newsletter* 10, 137.

Prévôt, A., Perrin, J.L., Laclaverie, G., Auge, Ph. and Coustille, J.L. (1990) A new variety of low-linolenic rapeseed oil: characteristics and room-odor tests. *Journal of the American Oil Chemists' Society* 67, 161–164.

Rakow, G. (1973) Selektion auf Linol- ind Linolensäuregehalt in Rapssamen nach mutagener Behandlung. *Zeitschrift für Pflanzenzüchtung* 69, 62–82.

Rakow, G., Stringam, G.R. and McGregor, D.I. (1987) Breeding *Brassica napus* L. canola with improved fatty acid composition, high oil content and high seed yield. In: *Proceedings of the Seventh International Rapeseed Congress*, Poznan, Poland. The Plant Breeding and Acclimatization Institute, Poznan, pp. 27–32.

Rödin, J. and Rask, L. (1990) Characterization of the 12S storage protein of *Brassica napus* (cruciferin): disulfide bonding between subunits. *Physiologia Plantarum* 79, 421–426.

Rubin, L.J., Diosady, L.L. and Tzeng, Y.-M. (1990) Ultrafiltration in rapeseed processing. In: Shahidi, F. (ed.) *Canola and Rapeseed: Production, Chemistry, Nutrition and Processing Technology.* Van Nostrand Reinhold, New York, pp. 307–330.

Sang, J.P. and Salisbury, P.A. (1988) Glucosinolate profiles of international rapeseed lines (*Brassica napus* and *Brassica campestris*). *Journal of the Science of Food Agriculture* 45, 255–261.

Sarwar, G., Bell, J.M., Sharby, T.F. and Jones, J.D. (1981) Nutritional evaluation of meals and meal fractions derived from rape and mustard seed. *Canadian Journal of Animal Science* 61, 719–733.

Scarth, R., McVetty, P.B.E., Rimmer, S.R. and Stefansson, B.R. (1988) Stellar low linolenic–high linoleic acid summer rape. *Canadian Journal of Plant Science* 68, 509–511.

Scarth, R., McVetty, P.B.E., Rimmer, S.R. and Stefansson, B.R. (1991) Hero summer rape. *Canadian Journal of Plant Science* 71, 865–866.

Schwenke, K.D., Raab, B., Linow, K.-J., Pähtz, W. and Uhlig, J. (1981) Isolation of the 12S globulin from rapeseed (*Brassica napus* L.) and characterizatioin as a 'natural' protein. On seed proteins, Part 13. *Die Nahrung* 25, 271–280.

Schwenke, K.D., Raab, B., Plietz, P. and Damaschun, G. (1983) The structure of the 12S globulin from rapeseed *Brassica napus* L. *Die Nahrung* 27, 165–175.

Shahidi, F. and Naczk, M. (1992) An overview of the phenolics of canola and rapeseed: chemicals, sensory and nutritional significance. *Journal of the American Oil Chemists' Society* 69, 917–924.

Shahidi, F., Naczk, M. and Myhara, R.M. (1990) Effect of processing on the soluble sugars in *Brassica* seeds. *Journal of Food Science* 55, 1470–1471.

Siddiqui, I.R. and Wood, P.J. (1971) Structural investigation of water-soluble rape-seed (*Brassica campestris*) polysaccharides. Part I. Rapeseed amyloid. *Carbohydrate Research* 17, 97–108.

Siddiqui, I.R. and Wood, P.J. (1972) Structural investigation of water-soluble rapeseed (*B. campestris*) polysaccharides. Part II. An acidic arabinogalactan. *Carbohydrate Research* 24, 1–9.

Siddiqui, I.R. and Wood, P.J. (1974) Structural investigation of oxalate-soluble rapeseed (*B. campestris*) polysaccharides. Part III. An arabinan. *Carbohydrate Research* 36, 35–44.

Siddiqui, I.R. and Wood, P.J. (1977) Carbohydrates of rapeseed: a review. *Journal of the Science of Food and Agriculture* 28, 530–538.

Siddiqui, I.R., Wood, P.J. and Khanzada, G. (1973) Low molecular weight carbohydrates from rapeseed (*B. campestris*) meal. *Journal of the Science of Food and Agriculture* 24, 1427–1435.

Sleeter, R.T. (1981) Effects of processing on quality of soybean oil. *Journal of the American Oil Chemists' Society* 58, 239–247.

Slominski, B.A. and Campbell, L.D. (1991) The carbohydrate content of yellow-seeded canola. In: McGregor, D.I. (ed.) *Proceedings of the Eighth International Rapeseed Congress*, Saskatoon, Canada. Organizing Committee, Saskatoon, pp. 1402–1407.

Sosulski, F., Zadernowski, R. and Babuchowski, K. (1981) Composition of polar lipids in rapeseed. *Journal of the American Oil Chemists' Society* 58, 561–564.

Stefansson, B.R., Hougen, F.W. and Downey, R.K. (1961) Note on the isolation of rape plants with seed oil free from erucic acid. *Canadian Journal of Plant Science* 41, 218–219.

Taylor, D.C., Weber, N., Hogge, L.R., Underhill, E.W. and Pomeroy, M.K. (1992) Formation of trierucoylglycerol (trierucin) from 1,2-dierucoylglycerol by a homogenate of microspore-derived embryos of *Brassica napus* L. *Journal of the American Oil Chemists' Society* 69, 355–358.

Taylor, D.C., MacKenzie, S.L., McCurdy, A.R., McVetty, P.B.E., Giblin, E.M., Pass, E.W., Stone, S.J., Scarth, R., Rimmer, S.R. and Pickard, M.D. (1994) Stereospecific analyses of seed triacylglycerols from high-erucuc acid Brassicaceae: detection of erucic acid at the sn-2 position in *Brassica oleracea* L. genotype. *Journal of American Oil Chemists* 71, 163–167.

Thakor, N.S. (1993) Dehulling of canola by hydrothermal treatment. PhD Thesis, University of Saskatchewan, Saskatoon, Canada.

Theander, O. and Åman, P. (1974) Carbohydrates in rapeseed and turnip rapeseed meals. In: *Proceedings of the Fourth International Rapeseed Congress*, Giessen, Germany. Deutsche Gesellschaft für Fettwissenschaft, Münster, pp. 429–438.

Theander, O., Åman, P., Miksche, G.E. and Yasuda, S. (1977) Carbohydrates, polyphenols, and lignin in seed hulls of different colors from turnip rapeseed. *Journal of Agriculture and Food Chemistry* 25, 270–273.

Thompson, L.U. (1990) Phytates in canola/rapeseed. In: Shahidi, F. (ed.) *Canola and Rapeseed. Production, Chemistry, Nutrition and Processing Technology*. Van Nostrand Reinhold, New York, pp. 173–192.

Tkachuk, R. (1969) Nitrogen-to-protein conversion factors for cereals and oilseed meals. *Cereal Chemistry* 46, 419–423.

Uppström, B. and Johansson, M. (1991) Amino acid composition of protein fractions in rapeseed. In: McGregor, D.I. (ed.) *Proceedings of the Eighth International Rapeseed Congress*, Saskatoon, Canada. Organizing Committee, Saskatoon, pp. 1377–1379.

Uppström, B. and Svensson, R. (1980) Determination of phytic acid in rapeseed meal. *Journal of the Science of Food and Agriculture* 31, 651–656.

Van Soest, P.J. (1966) Nonnutritive residues: a system of analysis for the replacement of crude fiber. *Journal of the Association of Official Analytical Chemists* 49, 546–551.

Van Soest, P.J. and Moore, L.A. (1965) New chemical methods for the analysis of forages for the purpose of predicting nutritive value. In: *Proceedings of the Ninth International Grasslands Congress*, São Paulo, Brazil. Departmento da Produção Animal da Secretaria da Agricultura do Estrado de São Paulo, pp. 784–789.

Voelker, T.A., Worrell, A.C., Anderson, L., Bleibaum, J., Fan., C., Hawkins, D.J. and Davies, H.M. (1992) Engineering laurate production in oilseeds. In: *Proceedings of the Miami Biotechnology Winter Symposium*, Vol. 2. IRL Press, Oxford, p. 102.

Wong, R.S.-C. and Swanson, E. (1991) Genetic modification of canola oil: high oleic acid canola. *Advances in Applied Biotechmnology, No. 12, Fat Cholesterol Reduced Foods*, 153–164.

Yiu, S.H., Poon, H., Fulcher, R.G. and Altosaar, I. (1982) The microscopic structure and chemistry of rapeseed and its products. *Food Microstructure* 1, 135–143.

Yiu, S.H., Altosaar, I. and Fulcher, R.G. (1983) The effects of commercial processing on the structure and microchemical organization of rapeseed. *Food Microstructure* 2, 165–173.

Zadernowski, R. and Sosulski, F. (1978) Composition of total lipids in rapeseed. *Journal of the American Oil Chemists' Society* 55, 870–872.

11 Seed Analysis

J.K. Daun
Canadian Grain Commission, Winnipeg

INTRODUCTION

Many analytical methods used to evaluate *Brassica* oilseeds are applicable to all oilseeds. Some were recently reviewed by Rossell and Pritchard (1991). A second treatise on the subject is available from the Centre Technique Interprofessionnel des Oléagineux Métropolitains (CETIOM) (Burghart *et al.*, 1987a). The well-known text on rapeseed (Appelqvist and Ohlson, 1972) reviews seed composition (Appelqvist, 1972) and oil analysis (Persmark, 1972). More recently, Niewiadomski (1990) has reviewed the analysis of both rapeseed and rapeseed oil.

This chapter is primarily concerned with seed constituents frequently analysed for purposes of trade. However, analytical methods for other constituents are included at the end of the chapter as they form an appreciable proportion of the seed and have been the subject of considerable chemical research.

Analytical methods for two enzymes of particular interest in *Brassica* seed are also mentioned along with methodology for cultivar identification, an area of growing interest. It has not been the intention to give detailed descriptions of long-established analytical methods but rather to refer to the methods deemed most appropriate for characterization of *Brassica* oilseeds. Much of the methodology has been developed, adapted or applied to the analysis of *Brassica napus* and *Brassica rapa*, the two *Brassica* species most commonly used as a source of vegetable oil.

Analytical methods

Analytical methods may be divided into two classes. The first are 'reference methods' designed to provide accurate and precise analytical information on a component, which are normally approved by an organization or agency qualified to do so. Reference methods will have been tested in inter-laboratory collaborative trials to establish levels of analytical variation in case of legal dispute. They often utilize basic principles and may sacrifice time for accuracy and precision.

The methods of the second class, referred to in this chapter as 'alternative methods', are usually more rapid but, in many cases, are as accurate and precise as the reference methods. They often involve specialized and expensive equipment, such as a near-infrared spectrometer, nuclear magnetic analyser or X-ray fluorescence spectrometer. While alternative methods may be sanctioned by an official body, they usually require calibration using samples or standards with an appropriate value determined by a reference method. Also included under alternative methods are those which do not have reference status either because they are too new or because there has been no call for it.

Official organizations

International organizations providing reference methods suitable for *Brassica* oilseeds include the following.

- American Oil Chemists' Society (AOCS): an international body with nearly half the membership non-American. It has developed commodity-specific methods but these are now being consolidated and harmonized by its Unified Methods Committee.
- International Organization for Standardization (ISO): develops standards and standard methods in many areas. Oilseed standards, principally for sampling, impurities and chemical components, are developed and maintained under Technical Committee 34 (Agricultural Food Products) Subcommittee 2 (Oleaginous Seeds and Fruits) (TC 34/SC 2).
- International Union of Pure and Applied Chemistry (IUPAC): develops standards for many chemical constituents. A subgroup, the Commission on Oils, Fats and Derivatives, has developed a number of analytical methods for determining basic chemical components of oilseeds.
- Federation of Oils Seeds and Fats Organizations (FOSFA). This organization develops contractual methods and procedures used in international trading.
- International Seed Testing Association

(ISTA): develops methods for assessing seeds for propagation.
- Association of Official Analytical Chemists (AOAC) (AOAC International): develops methods used mainly for characterization of food products. While these are not generally applicable to oilseeds, several are useful for characterization of oilseed components with important food functions.

In addition to these and other organizations (Table 11.1), many national organizations (Table 11.2) also publish official methods, many of which may be applicable to *Brassica* oilseeds.

Attempts have been made recently to harmonize methods, particularly between AOCS, FOSFA, IUPAC and ISO. For example, AOCS, FOSFA and ISO have agreed to a method for oil content determination, and ISO and AOCS have agreed to methods for determining glucosinolates and chlorophyll content. Particular attention will be drawn in this chapter to methods of the AOAC, AOCS, ISO and FOSFA applicable to *Brassica* seeds (Table 11.3).

SEED ASSESSMENT

Reference methods

Purity
Brassica oilseed samples are often impure or are admixtures with other seeds. For analytical purposes, the ISO (ISO method 658:1988) specifies a procedure for separating impurities using sieves and hand-picking. A similar method, used by the Canadian Grain Commission (1993), includes procedures for cleaning as well as determining admixture with weed seeds. Seed analysis methods have been developed by the International Seed Testing Association to ensure the high level of purity required for the maintenance of varietal purity in seed multiplication and

Table 11.1. International organizations sponsoring methods of analysis applicable to *Brassica* seeds.

Organization	Address	Methods for:
ISO (International Organization for Standardization)[1]	Case Postale 56, CH-12111 Genève 20, Switzerland	Sampling, impurities, oil, nitrogen, moisture, acidity, ash, crude fibre, glucosinolates, chlorophyll, fatty acid composition (oil)
International Union of Pure and Applied Chemistry (IUPAC)[2]	IUPAC Secretariat, Bank Court Chambers, 2–3 Pound Way, Cowley Centre, Oxford OX4 3YF, UK	Sampling, impurities, moisture, oil, acidity
Federation of Oils Seeds and Fats Associations (FOSFA)[3]	10 St Dunstans's Hill, London EC3R 8HL, UK	Nitrogen, oil content, volatile oil (glucosinolates)
American Oil Chemist's Society (AOCS)[4]	PO Box 3489 Champaign, Illinois 61826-3489, USA	Nitrogen, oil content, glucosinolates, chlorophyll
International Association of Seed Crushers (IASC)[5]	Salisbury Square House, 8 Salisbury Square, London EC4P 4AN, UK	Sampling, moisture, oil content, acidity
Association of Official Analytical Chemists (AOAC)[6]	AOAC International, Suite 400, 2200 Wilson Blvd, Arlington, Virginia 22201-3301, USA	Protein, fat acidity, nitrogen by combustion, volatile oil in mustard seed

[1] Methods published individually and available from ISO or a national standards organization.
[2] Methods published in book form (Paquot and Hautfenne, 1987), updates every 2 years.
[3] Methods published in book form (Federation of Oils Seeds and Fats Associations, 1986).
[4] Methods published in book form (Firestone, 1990), updates annually.
[5] Methods published in book form (International Association of Seed Crushers, 1980).
[6] Methods published in book form (Helrich, 1990), updates annually.

to provide the basis for good crop establishment (Anon., 1985).

Sampling

Seed sampling procedures for *Brassica* oilseeds are described in ISO method 542: 1990. For accurate analysis, the minimum laboratory sample from commercial seed lots should be 1–2 kg. These samples may be split to produce test samples of a minimum 200 g, which may be further subdivided to provide the analytical samples, the size of which will vary with the method. Larger samples are required for determining mean content of a component which

varies widely between different seeds, such as chlorophyll (Daun, 1994).

MOISTURE

Brassica oilseeds may contain variable amounts of water depending on harvest conditions. Seed storage for extended periods is safe only if the water content is less than 10% with most recommendations calling for 8%. In Canada, seed is usually harvested at water contents of 11% or less and average levels in the seed are usually about 7.5–8.5%. In Europe, seed is

Table 11.2. Some national and industrial organizations publishing methods which may be applicable to *Brassica* seeds.

Organization	Address
Association français de normalization (AFNOR)	Tour Europe – Cedex 7, 92080 Paris, La Défense, France
Australian Oilseeds Federation (AOF)	Unit 9, 2–6 Hunter Street, Narramatta, NSW 2150, Australia
British Standards Institution (BSI)	2 Park St, London W1A 2BS, UK
Canadian Grain Commission (CGC)	303 Main St, Winnipeg, Manitoba, Canada R3C 3G8
German Society for Fat Science (DGF)	Soester Stra., D-4400 Munster, Germany
Japan Oil Chemists' Society (JOCS)	Yushi Kogyo Kaikan Bldg, 13-11, 3-chome, Nihonbashi, Chuo-ku, Tokyo, Japan
Leatherhead Food Research Association	Randalls Road, Leatherhead, Surrey KT22 7RY, UK

occasionally harvested at water contents of 15–20% when wet conditions prevail at harvest, and the seed must be dried before storage.

Reference methods

Gravimetric

Moisture, commonly including volatile matter, is usually determined as the loss of mass after heating at a given temperature for a given period of time. ISO method 665 specifies heating at just over the boiling point of water (usually 103°C) for 3 h and then for further 1-h periods until constant mass is achieved. *Brassica* oilseeds are not ground before analysis as this can be a major source of error. The ISO method is most often specified in international trading contracts under the FOSFA. The AOCS has not published a method specifically for moisture in *Brassica* oilseeds but AOCS methods for other seeds call for heating at 130°C for 3 h.

Efforts are being made to reconcile the differences in methodology.

Alternative methods

Microwave heating

Microwave heating followed by gravimetric measurement when applied to linseed, rapeseed and yellow mustard seed gave results similar to the AOAC vacuum oven method for cereal grains, but required less time for completion of the analysis (Oomah and Mazza, 1992).

Electrical conductance/capacitance

Moisture meters, which rely on the electrical properties of the seed, have been used to rapidly determine moisture in *Brassica* samples. There are two general types, conductance meters which measure free water only and add empirically a constant amount for bound water, and capacitance meters which measure both free and bound water. Capacitance meters have gained

Table 11.3. Reference methods applicable to *Brassica* seeds.

Method number	Method title
AOAC methods[1] (Helrich, 1990)	
925.09	Solids (Total) and Moisture in Flour – Vacuum Oven Method
965.49	Fatty Acids in Oils and Fats – Preparation of Methyl Esters
963.22	Methyl Esters of Fatty Acids in Oils and Fats – Gas Chromatographic Method
985.20	Erucic Acid in Oils and Fats – Thin Layer and Gas Chromatographic Method
975.39	Docosenoic Acid in Oils and Fats – Gas Chromatographic Method
970.55	Volatile Oil in Mustard Seed
973.18	Fiber (Acid Detergent) and Lignin in Animal Feed
985.29	Total Dietary Fiber in Foods – Enzymatic and Gravimetric Method
992.23	Crude Protein in Cereals and Oilseeds
AOCS methods (Firestone, 1990)	
Ak 1-92	Determination of Glucosinolate Content in Rapeseed (Colza) by HPLC
Ak 2-92	Determination of Chlorophyll Content in Rapeseed (Colza) by Spectrophotometry
Am 2-92	Oil Content in Oilseeds
Ba 4b-87	Nitrogen–Ammonia–Protein – Modified Kjeldahl Method – Copper Sulfate Catalyst
Ba 4c-87	Nitrogen–Ammonia–Protein – Modified Kjeldahl Method – Kjel-Foss Automatic Method
Ba-4d-90	Nitrogen–Ammonia–Protein – Modified Kjeldahl Method – Titanium Dioxide + Copper Sulfate Catalyst
Ba 6-84	Crude Fiber in Oilseeds
Am 1-92	Determination of Oil, Moisture and Volatile Matter, and Protein by Infrared Reflectance
Ba 4e-93	Combustion Method for the Determination of Crude Protein
Ce 2-66	Preparation of Methyl Esters of Long-Chain Fatty Acids
Ce 1-62	Fatty Acid Composition by Gas Chromatography
ISO methods[2]	
ISO 542:1990	Oilseeds – Sampling
ISO 658:1988	Oilseeds – Determination of Impurities Content
ISO 659:1988	Oilseeds – Determination of Hexane Extract (or Light Petroleum Ether Extract) Called 'Oil Content'
ISO 664:1990	Oilseeds – Reduction of Laboratory Sample to Test Sample
ISO 665:1977	Oilseeds – Determination of Moisture and Volatile Matter Content
ISO 729:1988	Oilseeds – Determination of Acidity of Oils
ISO 5511:1992	Oilseeds – Determination of Oil Content – Method Using Continuous-Wave Low-Resolution Nuclear Magnetic Resonance Spectrometry (Rapid Method)
ISO 9167-1:1992	Rapeseeds – Determination of Glucosinolates Content Part 2: Method using High Performance Liquid Chromatography
ISO 9167-2:1993	Rapeseeds – Determination of Glucosinolates Content Part 2: Method using X-ray Fluorescence Spectrometry
ISO 5498:1981	Agricultural Food Products Determination of Crude Fibre Content – General Method

Table 11.3. (contd).

Method number	Method title
ISO 5508:1990	Animal and Vegetable Fats and Oils – Analysis by Gas Chromatography of Methyl Esters of Fatty Acids
ISO 5509:1978	Animal and Vegetable Fats and Oils – Preparation of Methyl Esters of Fatty Acids
ISO 7700-2:1987	Check of the Calibration of Moisture Meters Part 2: Moisture Meters for Oilseeds
ISO 10519:1992	Rapeseed – Determination of Chlorophyll Content – Spectrometric Method
ISO 10565:1993	Oilseeds – Simultaneous Determination of Oil and Water – Pulsed NMR Method

FOSFA methods (FOSFA, 1986)
  Determination of the Oil Content of Rapeseed by Nuclear Magnetic Resonance, pp. 71–72
  Determination of Oil Content in Oilseeds, pp. 59–64
  Determination of Nitrogen Content by Kjel-Foss Automatic, pp. 55–58

[1] AOAC method numbers are for the 15th edn of the methods manual. Numbers may differ in earlier editions.
[2] Published by the International Organization for Standardization, Case Postale 56, Ch-1211 Geneva 20, Switzerland.

considerable acceptance in Canada, the United States and Great Britain (Pritchard, 1991). ISO method 7700-2:1987 gives a protocol for the calibration of moisture meters.

In Canada, moisture meters have been calibrated using a single-stage vacuum oven method for oilseeds (International Union of Pure and Applied Chemists, 1948). Comparison of this method with the ISO gravimetric method (ISO method 665) has shown good agreement for rapeseed. Analysis of 44 samples with moisture ranging from 6.9 to 12.8% (typical for Canadian conditions) gave mean contents of 8.97% and 8.99% for the IUPAC and ISO methods, respectively (R. Wallis, Canadian Grain Research Laboratory, Winnipeg, 1994, personal communication).

Near-infrared reflectance spectrophotometry

Moisture content may also be determined by near-infrared (NIR) spectroscopy. The AOCS procedure (AOCS method Am 1-92) has a repeatability of 0.3%. Grinding of the seed is to be avoided as it results in a loss of moisture content which will affect calibration and analysis. Riaillier and Maviel (1984) compared three NIR reflectance instruments and obtained standard errors of performance of 0.21 to 0.35 for ground rapeseed samples calibrated against a forced-air oven method. Panford et al. (1988) reported a standard error of performance of 0.064% for 40 samples of rapeseed ranging in moisture content from 5.5 to 14%. Similarly, Hartwig and Hurburgh (1990) obtained 0.26% for 60 samples of crambe seed ranging from 5 to 8.2% in moisture content. This followed drying of ground samples for 1 h at 130°C as a reference method and analysis using a filter instrument. Williams and Sobering (1993), working with whole seed on a scanning reflectance instrument, reported a standard error of performance for moisture of 0.22%.

OIL CONTENT

Oil content is defined as the whole of the substances extractable by n-hexane (or light petroleum ether, Skellysolve F) under specified conditions. This should not be confused with total lipid content, which may involve extraction with a variety of solvents, such as a mixture of chloroform and methanol, in order to obtain a more complete extraction, particularly of polar lipids. Inappropriate application of methods such as the 'Bligh and Dyer' extraction system, commonly used to extract lipids from tissues (Kates, 1986), can result in incomplete extraction of the neutral lipids of oilseeds. A good method for obtaining total lipids from *Brassica* oilseeds, involving homogenization in the presence of methanol and chloroform, is described by Hammond (1993).

Brassica oilseeds are produced primarily for their oil and thus oil content is an important determinant of their value in both national and international markets. In the case of mustard seed grown for culinary use, solvent-extractable oil is often referred to as 'fixed oil content' in order to differentiate it from 'volatile oil content' or 'mustard oil', the latter being the distillable hydrolysis products of the glucosinolates. Oil content is commonly expressed as a percentage by mass of the sample as received, but some methods make allowance for expressing results on the basis of cleaned seed, or on a dry matter basis. A number of methods for determining oil content were reviewed by Daun and Snyder (1989).

Reference methods

Gravimetric

Reference methods for assessing oil content involve gravimetric determination after extraction. These methods are being harmonized by AOCS, FOSFA and ISO (AOCS method Am 1-92; Federation of Oils Seeds and Fats Associations, 1986; ISO method 659:1988). Each of these methods involves analytical extraction approximating the industrial extraction process except that it is repeated several times to achieve an exhaustive extraction for precision purposes. Exhaustive extraction of *Brassica* oilseeds requires regrinding of the partially extracted flour after each extraction. Regrinding is performed to reduce mean particle size to less than 100 μm. Experiments in the Canadian Grain Research Laboratory have shown that as much as 0.5% more oil may be extracted in the third step which is comparable to the amount of oil left in meal after commercial extraction (prepress solvent). Furthermore the percentage of $C_{18:1}$ (n-7) fatty acids (10% of total fatty acids) in the last extract is comparable to that in the oil left in commercial meal. The (n-7) isomer content in the last extract is much higher than in the earlier extracts ($C_{18:1}$ (n-7) = 3% total fatty acids). This suggests that the triacylglycerol material removed in the last stages of oil extraction in the analytical method originates from the seed coat lipids. These contain up to 25% of $C_{18:1}$ (n-7) fatty acids (Hu *et al.*, 1994) in the triacylglycerol portion.

Alternative methods

Rapid extraction gravimetric procedures

A rapid method for extraction of oil from *Brassica* seeds in a ball mill was first described by Troëng (1955) and elaborated by Appelqvist (1967). This is known as the 'Swedish tube method' in Canada. Oil is extracted by grinding seed in steel centrifuge tubes containing steel ball-bearings and solvent, so that the mean seed particle size is reduced to less than 100 μm. Oil extraction is complete within 45 min and, after settling, a known aliquot is withdrawn and evaporated for gravimetric determination of the extracted oil. The method gives results only slightly lower than regrind extraction procedures.

A modification of this method, developed by Burghart *et al.* (1987b), utilizes a Dangoumou ball mill which, due to its higher rate of pulverization, completes the extraction within 5 min. Canessa and Snyder (1991) used a homogenizer to grind rapeseed in the presence of solvent and showed that results equivalent to a two-stage regrind gravimetric method could be obtained.

A more promising method for rapid determination of total oil content is analytical supercritical fluid extraction. Initial studies showed that this may provide a quantitative extraction without the use of organic solvents (Taylor *et al.*, 1993). It is already used for analytical extraction of components such as pesticide residues. The ability to carry out analytical extraction rapidly without the need to dispose of waste solvent could make this method a promising alternative.

Densitometry

A commercial apparatus which measures oil content as a function of change in density when the seed is ground in the presence of a dense halogenated solvent is available (Burghart *et al.*, 1987a). While it is suitable for rapid analysis of oil content in a variety of matrices, the use of halogenated solvents has limited its acceptance.

Nuclear magnetic resonance

Both continuous-wave and pulsed nuclear magnetic resonance (NMR) methods have been developed for rapid determination of oil content in oilseeds (Gambhir, 1992). The 1984 version of the ISO continuous wave method (ISO method 5511:1984) required a single-point calibration against an oil sample typical of the sample for analysis with a correction made for the matrix effect from the meal. This has recently been modified to include a multiple-point calibration (ISO method 5511:1992). A multi-point seed calibration method, using an oil/silica matrix for an instrumental set point developed by FOSFA (Federation of Oils Seed and Fats Associations, 1986), does not require correction for the effect of non-oil seed components on the NMR signal.

Continuous-wave NMR requires predrying of the sample for accurate determination of oil content. It is important that reference samples used for calibration are treated in the same manner as the samples for analysis. Nuclear magnetic resonance without drying has been used by some plant breeders for screening seed of lines with a low (about 6%) and constant moisture content. This has the advantage that the seed remains viable and may be grown after analysis. A pulsed NMR method developed by ISO (ISO method 10565:1993) uses a multi-point calibration and allows simultaneous determination of oil and moisture for samples with less than 10% moisture.

Near-infrared reflectance spectrophotometry

A general standard for NIR spectrophotometric determinations of oil, protein and moisture in ground samples has been published by AOCS (AOCS method Am 1-92) and this standard suggests that a repeatability of 0.4% should be achieved for oil content. While some studies of ground seeds (Starr *et al.*, 1985; Hartwig and Hurburgh, 1990; McGregor, 1990b) noted relatively high standard errors of performance for oil content (about 1%), Ribaillier and Maviel (1984) recorded errors in the order of 0.5% for routine operations with three different instruments. More recently, Panford *et al.* (1988) obtained a standard error of performance of 0.2% for analysis of ground rapeseed using a scanning spectrophotometer.

Near-infrared reflectance analysis of whole seeds promises to provide a more rapid method for determination of oil content than analysis of ground seed. Although initial studies (McGregor, 1990a; Williams and Sobering, 1993) found standard errors of performance for oil content to be about 1%, recent work in the Canadian Grain Research Laboratory (Daun *et al.*, 1994) obtained standard errors of performance in the order of 0.5%. The Laboratory has also been studying the use of NIR spectroscopy in Canadian harvest surveys and calibrations for 1992 and 1993 found standard errors of performance of about 0.5%.

FATTY ACID COMPOSITION

The fatty acid composition of *Brassica* oilseeds can be quite variable, particularly when the changes made by plant breeding and biotechnology are taken into account. As with other oilseeds, the value of the oil is determined largely by the fatty acid composition, and content of an individual fatty acid is usually expressed as a percentage of the total fatty acids released upon esterification of the oil.

Reference methods

Gas-liquid chromatography

Analysis of fatty acid composition is usually carried out by gas–liquid chromatography (GLC) of methyl esters prepared from oil after extraction from the seed. AOAC, AOCS and ISO have protocols for the preparation of methyl esters (AOAC method 965.49; AOCS method Ce 2-66; ISO method 5509:1978) and for gas-liquid chromatographic determination of the fatty acid composition (AOAC method 985.20; AOCS method Ce 1-62; ISO method 5508:1990). The AOCS method for methyl ester preparation and the gas-liquid chromatographic methods of the AOCS and ISO have been recently updated. ISO is in the process of updating

its method for methyl ester preparation. AOAC methods have become slightly out of date but may be brought up to date as methods are harmonized.

Erucic acid (22 : 1 (n-9)) is of particular interest in *Brassica* oilseeds. Regulations in some countries stipulate that oil destined for human consumption must contain less than 2% erucic acid, while oil destined for many industrial applications should ideally have a content in excess of 50%. The AOAC has published two methods for determining erucic acid (AOAC method 985.20 and AOAC method 975.39). The first (AOAC method 985.20) combines thin-layer chromatography and GLC and is specific for erucic acid. Similar methods published by AOCS and ISO have been withdrawn as they are time-consuming and have been little used. The second method (AOAC method 975.39) uses an internal standard in gas chromatographic analysis and is intended for determination of all docosenoic acids including erucic acid (*cis*-13-docosenoic acid).

Alternative methods

Gas-liquid chromatography

Reviews on the preparation of fatty acid methyl esters have recently been published (Christie, 1990a, b, 1992). While simple base-catalysed *trans*-esterification should be possible for most work with *Brassica* oilseeds, it is important to note that free fatty acids are not esterified by base-catalysed esterification. In addition, long-chain acids (i.e. C_{22}) react more slowly than shorter-chained fatty acids (Craske, 1993) and thus care must be taken that reactions proceed to stoichiometric end-point. A fairly wide range of gas chromatographic equipment and columns have given acceptable results for fatty acid analysis (Christie, 1991; Craske, 1993). In the Canadian Grain Research Laboratory, good results have been obtained with open tubular columns 15 m × 0.32 mm containing a 0.5 µm coating of Supelcowax 10. With an optimal flow rate and temperature programming from 220°C to 245°C, resolution of methyl esters from 12 : 0 to 24 : 1 were obtained in about 6 min.

Analyses of high erucic acid oils should include eicosadienoic (20 : 2) and docosadienoic (22 : 2) which may be present in significant amounts. *Brassica* oilseeds may also have significant amounts of both n-7 and n-9 positional isomers of the unsaturated fatty acids (Laakso *et al.*, 1983) which can be resolved even on packed columns. Failure to include these peaks in the calculations can cause substantial errors in the analysis.

Screening methods

Fatty acid composition can be determined with oil extracted from a single cotyledon leaving the remainder of the seed for growing (Downey and Harvey, 1963). To facilitate analyses of large numbers of seeds, oil from a single cotyledon may be extracted and saponified in one step and the fatty acids separated by paper chromatography (Thies, 1971, 1974). A photometric procedure has also been developed to determine the ratio of linoleic acid to linolenic acid (Rakow and Thies, 1972; Thies, 1974).

Recent publications have demonstrated potential for fatty acid analysis using NIR spectrophotometry (Reinhardt *et al.*, 1992; Salgó *et al.*, 1992; Daun *et al.*, 1994). Reinhardt *et al.* (1992) used it to screen for oleic acid and erucic acid. Daun *et al.* (1994) stressed the importance of considering variation of both oil content and fatty acid composition in the calibration procedure. Variation in different samples changes the amount of each fatty acid in the seed even if the relative percentage is unchanged. By accounting for this they were able to obtain good calibrations for linolenic acid and for total saturated fatty acids.

ACIDITY (FREE FATTY ACIDS)

Seed acidity may be expressed as 'acid value' (number of milligrams potassium hydroxide required to neutralize 1 g of sample) or as 'free fatty acids' (the percentage, by weight of fatty acid of specified M_r – usually oleic, M_r 282). Acid value can be converted to percentage free fatty acids (as oleic acid) by the formula:

$$\text{Per cent free fatty acids (as oleic acid)} = 0.503 \times \text{acid value}$$

Reference methods

Titration

The ISO method 729:1988) is the only one for acid value designed for oilseeds. Acidity is determined by titration of the sample with ethanolic potassium hydroxide in a mixture of diethyl ether and ethanol, with phenolphthalein or alkali blue 6b as an indicator. Other organizations (AOCS, IUPAC) have methods for determination of oil acidity but do not specify the method of extraction to be applied to the seed. The ISO method specifies that the oil should be extracted according to ISO method 659:1988. Free fatty acid analyses, performed on samples which have been obtained by rapid or incomplete extraction, have been shown to give low results (May and Hume, 1993).

Alternative methods

Titration

It is the experience of the author that titrations using ISO method 729:1988 may become turbid requiring addition of further solvent. However, this may be overcome using a ternary solvent system of methanol : isopropanol : chloroform (1 : 2 : 2, v/v/v) (Ke and Woyewoda, 1978). With this system and the alkali blue 6b indicator, samples of dark-coloured oil may be titrated to an easily detected endpoint with no cloudiness.

Gas-liquid chromatography

Free fatty acids may also be determined in rapeseed by GLC (May and Hume, 1993). A ground seed sample is extracted with a mixture of heptane and acidic isopropanol. The free fatty acid content of the extract is then determined by direct GLC using heptadecanoic acid as an internal standard. Results were slightly lower than the ISO titration method, possibly due to incomplete extraction.

High-performance liquid chromatography

The Canadian Grain Research Laboratory has recently developed a high-performance liquid chromatography (HPLC) method which allows rapid determination of free fatty acids extracted into isopropanol using the 'Swedish tube' extraction apparatus mentioned above (Troëng, 1955). The samples are extracted in the presence of an internal standard (heptadecanoic acid) and the free fatty acids are determined by reversed-phase (C_{18}) HPLC with a solvent system consisting of methanol : acetonitrile : water : trifluoracetic acid (67.5 : 25 : 7.5 : 1, v/v/v/v) and the fatty acids detected using a light-scattering detector. Results are comparable to the titration method but considerable time is saved and sample size is reduced.

CHLOROPHYLL

High levels of chlorophyll result from incomplete maturation of the seed before harvest. This is predominantly a problem in spring rapeseed from northern growing areas, such as Scandinavia and Canada, but will occasionally be a problem with winter rapeseed. The spice trade also has low tolerances for green seed in mustard.

Reference methods

Spectrophotometric determination after solvent extraction

The ISO and AOCS have a harmonized method (ISO method 10519:1992 and AOCS method Ak 2-92) for determining chlorophyll in rapeseed, which is based on an official method used for many years in Scandinavia (Appelqvist and Johansson, 1968). Chlorophyll is extracted into a solvent mixture of ethanol and isooctane (3 : 10, v/v) and then determined by spectrophotometry.

Green seed count

For grading purposes, chlorophyll may be determined indirectly as 'green seed counts' (Canadian Grain Commission, 1993). Distinctly green seeds are counted in sample of 1000 seeds, which are then individually plated using a plastic template on sticky paper, and crushed with a roller. A level of 2% distinctly green seeds (the maximum for the top grade) corresponds to about $25 \, mg \, kg^{-1}$ chlorophyll, while 6% (the maximum for the second grade) corresponds to about $50 \, mg \, kg^{-1}$ chlorophyll (Daun, 1993). The error on green seed analysis is relatively high, being governed mainly by the binomial distribution (Daun, 1994).

Alternative methods

Near-infrared reflectance spectrophotometry

Near-infrared reflectance spectrophotometers can be used to facilitate chlorophyll determination in *Brassica* oilseeds (Tkachuk et al., 1988). Strictly speaking, this is not NIR spectrophotometry as measurements are made only in the visible range of the spectrum but the spectrophotometer provides a convenient means of scanning a sample with reflected light. Using a Technicon 500 spectrophotometer a standard error of performance of 1.08 was obtained for rapeseed (McGregor, 1990a). Similar results have been obtained by Williams and Sobering (1993) and Daun *et al.* (1994) for whole seeds.

High-performance liquid chromatography

Johnson-Flanagan and Thiagarajah (1990) determined seed chlorophyll pigments extracted into acetone using HPLC with fluorescence detection. Endo *et al.* (1992) carried out a similar separation with chlorophyll and related pigments extracted from seed by the reference method (ISO method 10519:1992). Ward *et al.* (1994a) showed that chlorophyll pigments could be detected on a diode array detector using the absorbance bands near 665 nm. The pigments were quantified using response factors derived from the relationships between their differential absorptivitiy. Comparisons between spectrophotometry and HPLC measurements (Ward *et al.*, 1994b) showed that relationships between the absorptivity are complex. While time-consuming, HPLC is capable of generating quantitative data for individual chlorophylls, or their breakdown products which predominate in processed oils (Endo *et al.*, 1992).

NITROGEN (PROTEIN) CONTENT

Total nitrogen is measured to obtain an estimate of protein content. It should be noted that values based on seed analysis, expressed on an oil-free basis, will not be the same as those for meal prepared from the same seed by commercial extraction. Commercial meals usually contain more lipid, especially in the form of gums extracted during processing and added back to the meal. Measurements of total nitrogen are usually converted to protein estimates by multiplying by a factor of 6.25, which is based on a protein of 16% nitrogen. However, it has been shown for

Brassica oilseeds that this factor yields high results since a portion of the nitrogen (10–15%) does not originate from protein. Tkachuk (1969) estimated that a more appropriate factor would be 5.53 for rapeseed meal and 5.40 for mustard meal. More recently, Moussé and Pernollet (1982) have reported a value of 5.67 for rapeseed. A recent study in the Canadian Grain Research Laboratory found N : P ratios for 14 varieties of *B. napus* and *B. rapa* varied from 5.4 to 5.6 with no apparent difference in factor between species or between high and low glucosinolate cultivars within a species.

Reference methods

Kjeldahl

Until recently, the Kjeldahl method has been most commonly used to determine total nitrogen. Ground seed is digested in acid, nitrogen in the form of ammonia recovered by distillation and measured by titration. Catalysts and modifiers are added to the digestion mixture to ensure that the nitrogen is quantitatively converted to ammonia. The most popular catalyst for use with agricultural products was mercury. However, environmental and safety concerns during the 1970s led to the prohibition of mercury catalysts in many countries. Problems with the analysis of oilseeds ensued, laboratories reporting results with alternative catalysts that were lower than with mercury (Davidson and Daun, 1980; Daun, 1994).

Automated versions of the Kjeldahl method have been introduced, in particular the Kjel-Foss method in which hydrogen peroxide is added to speed up the digestion process. The Kjel-Foss method is an official method of the FOSFA (Federation of Oils Seeds and Fats Associations, 1986) and AOCS (AOCS method Ba 4c-87) for the determination of nitrogen. In 1987, the AOCS introduced a revised Kjeldahl method using copper sulphate rather than mercury (AOCS method Ba 4b-87), which is similar to the FOSFA method. A later modification (AOCS method Ba 4d-90) allows the use of titanium dioxide and copper sulphate as catalyst. However, Berner (1992) noted that the prescribed method using titanium dioxide is based on studies with soyabean meal, cotton seed meal and whole cotton seed, and that modifications are necessary in order to obtain optimal results for other oilseeds.

Dumas combustion

Sample combustion, followed by analysis of the combustion gases, was an approach originally developed by Dumas in 1835 more than 50 years before Kjeldahl developed the acid digestion method. It has recently gained reference method status in AOAC (AOAC 992.23) and is a recommended practice of AOCS (AOCS Ba 4e-93). Environmental concerns with the Kjeldahl method and improved instrumentation for Dumas combustion are encouraging conversion. Instruments have been developed which allow automated analysis of ground seed samples of 0.5 g or less. Measurements are completed within only a few minutes. Comparative studies (Berner, 1992; Bicsak, 1993; Daun and DeClercq, 1994) have reported similar results for the Kjeldahl and combustion methods when applied to rapeseed, although the latter gave slightly higher values, possibly due to more complete recovery of nitrogen.

Alternative methods

Near-infrared reflectance spectrophotometry

Near-infrared reflectance spectrophotometry has been used to estimate protein content of *Brassica* oilseeds. The AOCS procedure (AOCS method Am 1-92) presents guidelines for repeatability of 0.3% for duplicate results for protein (%N × 6.25). While most studies on ground seed

(Ribaillier and Maviel, 1984; Starr *et al.*, 1985; Hartwig and Hurburgh, 1990; McGregor, 1990a) found standard errors of performance greater than 0.4% and often near 1%, Panford *et al.* (1988) recorded about 0.09, which would produce the required level of repeatability recommended by the AOCS. Studies on wholeseed analysis using both transmittance and reflectance techniques have also found standard errors of performance in the order of 0.4% to 0.5% (McGregor, 1990a; Williams and Sobering, 1993; Daun *et al.*, 1994). Williams also calculated a 'standard error per test' of 0.13% for protein. The standard error per test is a measure of the variability of a single sample (possibly a control sample) analysed many times during the course of an experiment. The standard error of performance is the measure of variability between the true value and the measured value for a set of samples having a range of values for the constituent of interest.

GLUCOSINOLATES

Glucosinolates are present in all *Brassica* seeds where they are either important antinutritional factors, such as in rapeseed, or valuable flavour components, notably in mustard. No other single constituent has received more attention from analysts than the glucosinolates. Between 1987 and 1993 *Chemical Abstracts* listed 91 citations with 'analysis' or 'determination' and 'glucosinolate' in the title. Several reviews are available on the analysis of glucosinolates in rapeseed (McGregor *et al.*, 1983; Wathelet, 1987; Daun and McGregor, 1991). Early methods relied on measurement of myrosinase hydrolysis products, sulphate, glucose or the isothiocyanate. Many methods which measure isothiocyanates suffer from lack of stringent conditions to ensure quantitative hydrolysis to the isothiocyanate, and from the fact that not all glucosinolates yield stable iso-thiocyanates. Over the years methods have been developed to measure the glucosino-lates, or desulphoglucosinolates, directly. These methods have gained wide accep-tance because they tend to be more accu-rate and precise as they avoid the problems associated with myrosinase hydrolysis. They also provide quantitative information on individual glucosinolates.

Reference methods

High-performance liquid chromatography

High-performance liquid chromatographic determination of desulphated glucosino-lates extracted from whole seed as devel-oped by the ISO (ISO method 9167-1: 1992) has recently been harmonized as an AOCS method (AOCS method Ak 1-92) and is being considered by the AOAC. A method for determination of intact gluco-sinolates in residues (meals) is also cur-rently under study by ISO. Avoidance of elevated temperatures, particularly during extraction and purification steps, make HPLC particularly suited to the analysis of heat-labile indole glucosinolates.

Isothiocyanate distillation

At the time of writing the ISO has main-tained for rapeseed the old method of myrosinase hydrolysis of the glucosinolates to volatile isothiocyanates followed by dis-tillation and titration but this method will probably be withdrawn at the next review. The method includes only those glucosino-lates which form volatile isothiocyanates and thus excludes 2-hydroxy-3-butenyl and 2-hydroxy-4-pentenyl glucosinolates, which form non-volatile oxazolidine-thiones, and the indole glucosinolates, which form free thiocyanate ions. The AOAC has a distillation method for deter-mination of the glucosinolate or 'volatile oil' content of mustard seed which is similar (AOAC method 970.55) but a gas–liquid

chromatographic method for determina-tion of allyl isothiocyanate may be sub-stituted for the titration. Both methods are applicable only to mustard seed used as a condiment, as they contain almost exclu-sively allyl glucosinolate in the case of *Brassica juncea* and *p*-hydroxybenzyl gluco-sinolate in the case of *Sinapis alba*, both of which form volatile isothiocyanates. The AOAC method also suffers from lack of defined hydrolysis conditions which would ensure quantitative conversion of the glucosinolate to the volatile isothiocyanate.

Alternative methods

X-ray fluorescence spectrophotometry

The ISO has developed a method for gluco-sinolate determination based on X-ray fluorescence determination of the sulphur content of rapeseed (ISO method 9167-2: 1993). The method is rapid and simple to perform. However, it must be calibrated with seed with known glucosinolate and sulphur content and corrections are neces-sary for protein content when the latter deviates from a given range.

High-performance liquid chromatography

An isocratic HPLC procedure, recently proposed for rapeseed analysis (Quinsac *et al.*, 1991), would allow the use of simpler high-performance liquid chromatographic systems since only one pump is involved. Also, the analysis time is more rapid. A method for HPLC of intact glucosinolates has also been developed (Helboe *et al.*, 1980). This avoids the necessity of enzy-matic desulphation of glucosinolates before analysis but still requires the use of ion-exchange chromatographic purification prior to injection as well as a relatively expensive counter ion in the chromatog-raphy solvent.

Gas-liquid chromatography

Many plant breeding programmes have relied on gas–liquid chromatographic determination of trimethylsilyl derivatives of desulphoglucosinolates. This approach can be rapid when a capillary column is used, and yield information on glucosinolate composition as well as estimates of total glucosinolate content (Sosulski and Dabrowski, 1984; Landerouin *et al.*, 1987). Gas–liquid chromatography of trimethylsilyl derivatives of aliphatic glucosinolates is currently specified in the definition of canola (Daun and McGregor, 1981) although the method underestimates indole glucosinolate content because of the heat applied to inactivate myrosinase, extract the glucosinolates and form the trimethylsilyl derivatives. Quantitative analyses comparable with HPLC can be obtained by extracting with aqueous methanol and derivatizing at room temperature (McGregor, 1990b).

Glucose determination

Total glucosinolate content may be estimated by determining the glucose released by myrosinase hydrolysis. This is the basis for a flow injection analysis method utilized in a British plant breeding programme (Smith and Barber, 1991). Methods for measurement of glucose utilizing immobilized myrosinase have also been reported (Leoni *et al.*, 1991; Wang and McGregor, 1991). Glucose-sensitive test papers have long been used to screen for glucosinolate content in breeding programmes (Lein, 1970; McGregor and Downey, 1975; Ribaillier and Quinsac, 1988). A modification by Truscott *et al.* (1991) introduced a photometer to read the colour change on the paper, thereby making the method more objective.

Sulphate determination

Elemental analysis has been used to estimate simultaneously glucosinolate (S), protein (N) and dry matter (C) in small (3–5 mg) seed samples (Johansson and Engwall, 1991). Correlating the carbon content with dry matter facilitated analysis by eliminating the need to weigh the sample accurately. Determination of sulphate by ion chromatography coupled with HPLC of intact glucosinolates has been used to estimate the amounts of glucosinolates destroyed during processing or through seed damage (Fiebig *et al.*, 1990).

Near-infrared reflectance spectrophotometry

Near-infrared reflectance spectrophotometry has been applied to estimate glucosinolate content. Lila and Furstos (1986) established that wavelengths in the region of 1630 to 1680 nm were associated with the presence of glucosinolates. Biston *et al.* (1988), using whole-rapeseed samples ranging in glucosinolate content from 2 to 100 $\mu M\,g^{-1}$, reported a standard error of performance of 2.15 $\mu M\,g^{-1}$. This is close to the values reported by McGregor (1990b) who, using both whole and ground seed samples with a glucosinolate range of 5 to 72 $\mu M\,g^{-1}$, reported standard errors of performance of 4 $\mu M\,g^{-1}$, and of 1.5 $\mu M\,g^{-1}$ for samples with a glucosinolate range of 5 to 20 $\mu M\,g^{-1}$. Salgó *et al.* (1992) obtained somewhat poorer results apparently due to the low precision of the palladium method used as the reference method. Work in the Canadian Grain Research Laboratory, using samples from widely dispersed sites in Canada over 5 years, showed that glucosinolates could be determined with a standard error of performance of 3 $\mu M\,g^{-1}$ (Daun *et al.*, 1994). Similarly, 5 $\mu M\,g^{-1}$ was obtained for both *B. juncea* and *S. alba* mustard seed with a range of glucosinolate content of 70 to 150 $\mu M\,g^{-1}$. This was considered sufficiently precise for survey work and for assisting industry in segregating seed lots.

Capillary electrophoresis

Recently, capillary electrophoresis has shown promise for the determination of intact glucosinolates (Bjergegaard et al., 1991). This method requires only small amounts of solvent and is extremely rapid. The requirement for injection volumes in the nanolitre range necessitates sample concentration before analysis. The concentration step results in a considerable increase in time and requires specialized equipment to be carried out without risk of damaging the analysis.

FIBRE

Fibre, although not a homogeneous entity, is of considerable interest because it is not readily digested and can devalue seed and meal as a component of animal feeds. Fibre is derived from the cell walls, particularly of the hull (seed coat). Its analytical value is dependent on the method of analysis (crude, neutral detergent and acid detergent), which extracts different wall components.

Reference methods

Crude fibre

The fibre content of seed and seed products is usually measured as 'crude fibre' using the so-called Weende procedure (Van Soest and Robertson, 1985). A sample is ground, defatted if necessary, digested with 0.255 M sulphuric acid, the residue digested with 0.313 M sodium hydroxide and, following washing and drying, the insoluble residue weighed. Subsequent loss in mass during ashing is taken as a measure of the 'crude fibre'. Both the AOCS and the ISO have methods based on this approach (AOCS method Ba 6-84; ISO method 5498:1991). Included in crude fibre are the cellulose, pentosan and lignin components of the cell wall.

Neutral and acid detergent fibre

In recent years acid detergent fibre (AOAC method 973.18) and neutral detergent fibre (Van Soest and Robertson, 1985) have been used to analyse the fibre content in *Brassica* oilseeds as they are believed to correspond more closely than crude fibre to total fibre content (Van Soest and Robertson, 1985). Neutral detergent fibre represents the total insoluble matrix of the cell wall, unlike crude fibre which does not contain the lignin and hemicelluloses, while acid detergent fibre does not contain the hemicelluloses. However, the correlation with digestibility of neutral detergent fibre is inferior to that of acid detergent fibre for ruminant animals (Van Soest and Robertson, 1985). Neutral and acid detergent fibre are determined by boiling with a neutral or acid detergent solution, respectively, and, following washing, drying and weighing of the insoluble residue, ashing to determine the loss in mass.

Enzymatic analysis

Assays for fibre content have also been developed based on enzymatic degradation of starch and protein from the defatted flour (AOAC method 985.29). These methods have not had widespread acceptance in the oilseeds industry but may eventually replace those based on chemical degradation as they more accurately simulate the digestive process.

Alternative methods

A semi-micro method for crude fibre

A semi-micro method for determining crude fibre is based on the digestion of small (approximately 50 mg) samples with chromic acid (Stringam et al., 1974). The residual chromic acid is determined by titration thus facilitating the analysis for crude fibre by replacing the ashing stage.

Near-infrared reflectance spectrophotometry

Michalski *et al.* (1992) analysed rapeseed for both neutral detergent fibre and acid detergent fibre using an NIR reflectance analyser in transmittance mode. Standard errors of calibration of 1.73%, for samples with a neutral detergent fibre ranging from 22.1% to 31.4%, and of 1.62%, for samples with an acid detergent fibre ranging from 15.7% to 25.1%, were reported. The crude fibre content of rapeseed has been determined using NIR reflectance spectrophotometry of ground seed. Ribaillier and Maviel (1984), studying three commercially available NIR reflectance instruments, reported standard errors of performance ranging from 0.77% to 1.45%. Panford *et al.* (1988) working with a scanning instrument obtained a standard error of performance of 0.03%.

ADDITIONAL CONSTITUENTS OF IMPORTANCE

Many of the components of *Brassica* seed do not have specific reference methods designated. Nevertheless, they form an appreciable proportion of the seed and have been the subject of extensive chemical studies.

Carbohydrates

After hull removal and defatting, rapeseed contain 52% carbohydrate (Niewiadomski, 1990). Slominski and Campbell (1991) studied the carbohydrate content of yellow-seeded canola. They extracted low M_r carbohydrates in 80% ethanol and determined the composition by GLC of the trimethylsilyl derivatives. Starch was determined by hydrolysis with an enzyme mixture (α-amylase/Termamyl/amyloglucosidase) followed by enzymatic determination of glucose. Non-starch polysaccharides were determined by GLC (component neutral sugars) and by colorimetry (uronic acids).

Slominski and Campbell also describe a scheme to subdivide various water-soluble and non-water-soluble polysaccharides. Oligosaccharides, with a degree of polymerization greater than stachyose, were determined by fractionation on Sephadex LH-20 followed by hydrolysis and determination of component sugars by GLC as alditol acetates. While Slominski and Campbell found about 2.5% starch in oil-free flour from canola seeds, Blair and Reichert (1984) found only traces of starch using another enzymatic method. Studies in the Canadian Grain Research Laboratory have confirmed the presence of about 2.5% starch in defatted canola flour. It is possible that the earlier study did not fully extract the starch from the flour.

Kanya and Urs (1983) determined carbohydrates in *B. juncea* flour by paper chromatography of an 80% ethanol extract. They also present a fractionation scheme for determining free sugars, soluble polysaccharides, pectic substances, hemicelluloses, celluloses and lignin. The carbohydrate fraction of *B. juncea* was compared to that of Dijon mustard by Fournier and Vangheesdaele (1980). These authors used column chromatography to characterize the different fractions prepared from their separation scheme and liquid chromatography to determine different carbohydrates. In this case, starch was also determined enzymatically.

Mucilage

Brassica seeds are known to contain mucilaginous materials which consist of various proteinaceous and carbohydrate-containing substances. Mucilage has been estimated by its release from seeds immersed in water (Van Caeseele and Mills, 1983). In a study of *S. alba* mustard cultivars, Woods and Downey (1980) used an aqueous extraction followed by precipitation with acidic acetone and gravimetric determination to estimate the mucilage content. A slightly more complicated method of extraction was described by Eskin (1992) in a study of the effect of variety and geographical location on the incidence of mucilage in canola seeds. Colman's of Norwich, UK, has used a method for estimation of mucilage in mustard based on the viscosity of aqueous suspensions (J. Hemingway, Norwich, 1994, personal communication).

Phenolic compounds

Brassica seeds contain various phenolic acids, in particular sinapic acid and its choline ester sinapine, which have been associated with palatability problems. Free phenolic acids and phenolic acid esters have been determined by gas chromatography of their trimethylsilyl derivatives after extraction with alcohol, hydrolysis of the esters and further fractionation (Zadernoski *et al.*, 1981). Sinapine and sinapic acid have been estimated spectrophotometrically (Legueut *et al.*, 1981; Naczk *et al.*, 1992).

Phytic acid

Phytic acid in *Brassica* seed meals has been associated with decreased utilization of zinc and other minerals when included in animal rations. Lemmieux *et al.* (1985) compared three methods for determining phytic acid in rapeseed flour and rapeseed protein concentrates. They noted that the ion-exchange spectrophotometric method of Latta and Eskin (1980) was cheaper and quicker than the two other methods investigated and gave good reproducibility.

Seed enzymes

Analytical methods have been developed for two enzymes in *Brassica* seeds, myrosinase and lipoxygenase. A spectrophotometric coupled-enzyme assay for myrosinase was found to be more sensitive than two other assays (pH-stat and spectrophotometric disappearance of substrate) (Wilkinson *et al.*, 1984). This assay involves measurement of the glucose released from sinigrin using the hexokinase/glucose-6-phosphate dehydrogenase assay. Lipoxygenase has been measured by incubating ground seeds with buffer and linoleate (Pokorny *et al.*, 1990).

CULTIVAR IDENTIFICATION

With plant breeders' rights, and the ability to patent genetic modifications to plants in some countries, there is a growing need for methods to identify genetic material. Hitherto, *Brassica* cultivar identification has depended on agronomic characteristics, the assessment of which can be highly subjective. Identification by electrophoresis of proteins has been attempted for *Brassica* species with limited success (Cooke, 1988). White and Law (1991) reported some success in differentiating between cultivars on the basis of fatty acid profiles. However, Davik and Heneen (1993) found that fatty acid and glucosinolate profiles were too similar to allow reliable discrimination of cultivars. High-performance liquid chromatography of ethanol extracts of seed was reported to allow differentiation of 29 *Brassica* cultivars (Mailer *et al.*, 1993). The same group have also investigated the use of polymerase chain reaction (PCR) technology using restriction fragment length polymorphisms (RFLPs) as a means of discriminating between 40 different cultivars (Mailer *et al.*, 1994).

REFERENCES

Anon. (1985) International rules for seed testing. *Seed Science and Technology* 13, 307–519.

Appelqvist, L.-Å, (1967) Further studies on a multisequential method for determination of oil content in oilseeds. *Journal of the American Oil Chemists' Society* 44, 209–214.

Appelqvist, L.-Å. (1972) Chemical constituents of rapeseed. In: Appelqvist, L.-Å. and Ohlson, R. (eds) *Rapeseed Cultivation, Composition, Processing and Utilization*. Elsevier Publishing Company, Amsterdam, pp. 123–173.

Appelqvist, L.-Å. and Johansson, S.-Å. (1968) Fettkvaliteten hos svenskt oljeväxtfrö. II. Bestämning av klorofyllhalt i raps-, ribs- och vitsenapsfrö. *Sveriges Utsädesförenings Tidskrift* 78, 415–431.

Appelqvist, L.-Å. and Ohlson, R. (eds) (1972) *Rapeseed Cultivation, Composition, Processing and Utilization*. Elsevier Publishing Company, Amsterdam, 391 pp.

Berner, D. (1992) Protein in soybeans. *INFORM* 3, 1150.

Bicsak, R. (1993) Comparison of Kjeldahl method for determination of crude protein in cereal grains and oilseeds with generic combustion method: collaborative study. *Journal of the Association of Official Analytical Chemists* 76, 780–786.

Biston, R., Dardenne, P., Cwikowski, M., Marlier, M., Severin, M. and Wathelet, J.-P. (1988) Fast analysis of rapeseed glucosinolates by near infrared reflectance spectroscopy. *Journal of the American Oil Chemists' Society* 65, 1599–1600.

Bjergegaard, C., Michaelsen, S., Møller, P. and Sørenson, H. (1991) High performance capillary electrophoresis: determination of individual anions, carboxylates, intact-desulfoglucosinolates. In: McGregor, D.I. (ed.) *Proceedings of the Eighth International Rapeseed Congress*, Saskatoon, Canada, 9-11 July. Organizing Committee, Saskatoon, pp. 822–827.

Blair, R. and Reichert, R.D. (1984) Carbohydrate and phenolic constituents in a comprehensive range of rapeseed and canola fractions: nutritional significance for animals. *Journal of the Science of Food and Agriculture* 35, 29–35.

Burghart, P., Lagarde, F., Merrien, A., Nouat, E., Prevot, A., Ribaillier, D. and Wolff, J.-P. (1987a) *Guide pratique de l'analyse des graines oléagineuses*. CETIOM/AFNOR, Paris, 225 pp.

Burghart, P., Lagarde, F., Merrien, A., Nouat, E., Prevot, A., Ribaillier, D. and Wolff, J.-P., (1987b) Détermination de la teneur en huile: méthode rapide par extraction à l'hexane. *Guide pratique de l'analyse des graines oléagineuses*. CETIOM/AFNOR, Paris, pp. 97–100.

Canadian Grain Commission (1993) Canola rapeseed. In: *Official Grain Grading Guide*. Canadian Grain Commission, Industrial Services Division, Winnipeg, Canada, Section 07/15.

Canessa, C.E. and Snyder, H.E. (1991) Total oil analysis of soybeans by simultaneous grinding and solvent extraction. *Journal of the American Oil Chemists' Society* 68, 675–677.

Christie, W.W. (1990a) Analysis: preparation of methyl esters – Part 1. *Lipid Technology* 2, 48–49.

Christie, W.W. (1990b) Analysis: preparation of methyl esters – Part 2. *Lipid Technology* 2, 79–80.

Christie, W.W. (1991) Gas chromatographic analysis of fatty acid methyl esters with high precision. *Lipid Technology* 3, 97–98.

Christie, W.W. (1992) Preparation of fatty acid methyl esters. *INFORM* 3, 1032–1034.

Cooke, R.J. (1988) Electrophoresis in plant testing and breeding. *Advances in Electrophoresis* 2, 171–261.

Craske, J.D. (1993) Separation of instrumental and chemical errors in the analysis of oils by gas chromatography – a collaborative evaluation. *Journal of the American Oil Chemists' Society* 70, 325–334.

Daun, J.K. (1993) Oilseeds processing. In: Bass, E. (ed.) *Grains and Oilseeds Handling, Marketing, Processing*. Canadian International Grains Institute, Winnipeg, Canada, pp. 883–926.

Daun, J.K. (1994) Sampling as a source of error in the estimation of green seeds and chlorophyll in canola and rapeseed. *Journal of the American Oil Chemists' Society* 5, 535.

Daun, J.K. and DeClercq, D.R. (1994) Comparison of combustion and Kjeldahl methods for determination of nitrogen in oilseeds. *Journal of the American Oil Chemists' Society* 71, 1069–1072.

Daun, J.K. and McGregor, D.I. (1981) *Glucosinolate Analysis of Rapeseed (Canola). Method of the Canadian Grain Commission Grain Research Laboratory*. Canadian Grain Commission Publication, Winnipeg, Canada, 32 pp.

Daun, J.K. and McGregor, D.I. (1991) Glucosinolates in seeds and residues. In: Rossell, J.B. and Pritchard, J.L.R. (eds) *Analysis of Oilseeds, Fats and Fatty Foods*. Elsevier Applied Science, London, pp. 185–226.

Daun, J.K. and Snyder, H. (1989) Total oil analysis of oilseeds. *Journal of the American Oil Chemists' Society* 66, 1074–1076.

Daun, J.K., Clear, K.M. and Williams, P.C. (1994) Comparison of three whole-seed near infrared analyzers for measuring quality components of canola seed. *Journal of the American Oil Chemists' Society* 71, 1063–1068.

Davidson, L.D. and Daun, J.K. (1980) Determining optimum conditions for Kjeldahl analysis of oilseeds. In: Daun, J.K., McGregor, D.I. and McGregor, E.E. (eds) *Analytical Chemistry of Rapeseed and Its Products – a Symposium.* Canola Council of Canada, Winnipeg, Canada, pp. 181–188.

Davik, J. and Heneen, W.K. (1993) Identification of oilseed turnip (*Brassica rapa* L. var *oleifera*) cultivar groups by their fatty acid and glucosinolate profiles. *Journal of the Science of Food and Agriculture* 63, 385–390.

Downey, R.K. and Harvey, B. (1963) Methods for breeding for oil quality in rape. *Canadian Journal of Plant Science* 43, 271–275.

Endo, Y., Daun, J.K. and Thorsteinson, C.T. (1992) Characterization of chlorophyll pigments present in canola seed, meal and oils. *Journal of the American Oil Chemists' Society* 69, 564–568.

Eskin, N.M.A. (1992) Effect of variety and geographical location on the incidence of mucilage in canola seeds. *Canadian Journal of Plant Science* 72, 1223–1225.

Federation of Oils Seeds and Fats Associations (FOSFA) (1986) *Standard Contractual Methods List.* Federation of Oils Seeds and Fats Associations, London, 78 pp.

Fiebig, H.-J., Jörden, M. and Aitzetmüller, K. (1990) Gesamtkonzept zur Glucosinolatbestimmung – Mööglichkeiten zur Erstellung von Bilanzen in Rapssamen und -schroten. *Fat Science and Technology* 92, 173–178.

Firestone, D. (ed.) (1990) *Official Methods and Recommended Practices of the American Oil Chemists' Society*, 4th edn, 2 vols. American Oil Chemists' Society, Champaign.

Fournier, N. and Vangheesdaele, G. (1980) Composition de la fraction glucidique de la *Brassica juncea* et de la moutarde de Dijon. *Revue Français de Corps Gras* 27, 513–519.

Gambhir, P. (1992) Applications of low resolution pulsed NMR to the determination of oil and moisture in oilseeds. *Trends in Food Science and Techology* 3, 191–196.

Hammond, E.W. (1993) *Chromatography for the Analysis of Lipids.* CRC Press, Boca Raton, Florida, p. 14.

Hartwig, R.A. and Hurburgh, C.R. (1990) Near-infrared reflectance measurement of moisture, protein and oil content of ground crambe seed. *Journal of the American Oil Chemists' Society* 67, 435–437.

Helboe, P., Olsen, O. and Sørenson, H. (1980) Separation of glucosinolates by high-performance liquid chromatography. *Journal of Chromatography* 197, 199–205.

Helrich, K. (ed.) (1990) *Official Methods of Analysis of the Association of Official Analytical Chemists.* Association of Official Analytical Chemists, Arlington, USA, 1298 pp.

Hu, X., Daun, J.K. and Scarth, R. (1994) Proportions of 18 : 1 (n-7) and 18 : 1 (n-9) fatty acids in canola seedcoat surface and internal lipids. *Journal of the American Oil Chemists' Society* 71, 221–222.

International Association of Seed Crushers (IASC) (1980) *Oilseeds Oils and Fats.* International Association of Seed Crushers, London, 100 pp.

International Union of Pure and Applied Chemists (1948) Méthodes unifiées pour l'analyse des matières grasses. *Troisième rapport de la commision internationale pour l'étude des matières grasses.* International Union of Pure and Applied Chemists Secretariat, Oxford, UK, p. 1b.

Johansson, S.-Å. and Engwall, E. (1991) Simultaneous determination of protein and glucosinolates in farmers' oilseed samples by elementary analysis. In: McGregor, D.I. (ed.) *Proceedings of the Eighth International Rapeseed Congress*, Saskatoon, Canada, 9–11 July. Organizing Committee, Saskatoon, pp. 1314–1318.

Johnson-Flanagan, A.M. and Thiagarajah, M.R. (1990) Degreening in canola (*Brassica napus*, cv. Westar) embryos under optimum conditions. *Journal of Plant Science* 136, 180–186.

Kanya, T.C.S. and Urs, M.K. (1983) Carbohydrate composition of mustard (*Brassica juncea*) seed meal. *Journal of Food Science and Techology* 20, 125–126.

Kates, M. (1986) *Techniques of Lipidology. Isolation, Analysis and Identificaion of Lipids.* Elsevier, Amsterdam, 106 pp.

Ke, P.J. and Woyewoda, A.D. (1978) A titrimetric method for determination of free fatty acids in tissues and lipids with ternary solvents and *m*-cresol purple indicator. *Analytica Chimica Acta* 99, 387–391.

Laakso, I., Hiltunen, R., Seppänen, T. and von Schantz, M. (1983) Relationships between some fatty acid isomers in rapeseed oil. *Acta Pharmaceutica Fennica* 92, 127–135.

Landerouin, A., Quinsac, A. and Ribailler, D. (1987) Optimization of silation reactions of desulfoglucosinolates before gas chromatography. In: Wathelet, J.-P. (ed.) *Glucosinolates in Rapeseed: Analytical Aspects.* Martinus Nijhoff Publishers, Dordrecht, pp. 26–37.

Latta, M. and Eskin, N.A.M. (1980) A simple and rapid colorimetric method for phytate determination. *Journal of Agricultural and Food Chemistry* 28, 1313–1315.

Legueut, C., Hocquemiller, R. and Cav, A. (1981) Dosage de la sinapine. *Annales Pharmaceutiques Français* 39, 557–561.

Lein, K.A. (1970) Methods for quantitative determination of seed glucosinolates of *Brassica* species and their application in plant breeding of rape low in glucosinolate content. *Zeitschrift für Pflanzenzüchtung* 63, 137–154.

Lemmieux, I., Amiot, J. and Brisson, G.J. (1985) Determinaton of phytic acid in rapeseed flour and protein concentrate by different methods. *Canadian Institute of Food Science and Technology Journal* 18, 29–33.

Leoni, O., Iori, R. and Palmieri, S. (1991) Immobilization of myrosinase on membrane for determining the glucosinolate content of cruciferous material. *Journal of Agriculture and Food Chemistry* 39, 2322–2326.

Lila, M. and Furstos, V. (1986) Détermination de longueurs d'onde spécifique pour la mesure des glucosinolates du colza par spectrophotométrie de réflexion dans le proche infrarouge. *Agronomie* 6, 703–707.

Mailer, R.J., Daun, J. and Scarth, R. (1993) Cultivar identification in *Brassica napus* L. by reversed-phase high-performance liquid chromatography of ethanol extracts. *Journal of the American Oil Chemists' Society* 70, 863–866.

Mailer R.J., Scarth, R. and Fristensky, B. (1994) Discrimination among cultivars of rapeseed (*Brassica napus* L.) using DNA polymorphisms amplified from arbitrary primers. *Theoretical and Applied Genetics* 87, 697–704.

May, W.E. and Hume, D.J. (1993) An automated gas–liquid chromatographic method of measuring free fatty acids in canola. *Journal of the American Oil Chemists' Society* 70, 229–233.

McGregor, D.I. (1990a) Selected methods for glucosinolate analysis. In: *Proceedings of the Oil Crops Network, Brassica Sub-Network Workshop,* Shanghai, China, 21–23 April.

McGregor, D.I. (1990b) Application of near infrared to the analysis of oil, protein, chlorophyll, and glucosinolates. In: Shahidi, F. (ed.) *Canola/Rapeseed in Canola and Rapeseed Production, Chemistry, Nutrition and Processing Technology.* Van Nostrand Reinhold, New York, pp. 221–231.

McGregor, D.I. and Downey, R.K. (1975) A rapid and simple assay for identifying low glucosinolate rapeseed. *Canadian Journal of Plant Science* 55, 191–196.

McGregor, D.I., Mullin, W.J. and Fenwick, G.R. (1983) Review of analysis of glucosinolates: analytical methodology for determining glucosinolate composition and content. *Journal of the Association of Official Analytical Chemists* 66, 825–849.

Michalski, K., Ochodski, P. and Cicha, B. (1992) Determination of fibre, sulfur amino acids and lysine in oilseed rape by NIT. In: Murray, I. and Cowe, I.A. (eds) *Making Light Work: Advances in Near Infrared Spectroscopy.* VCH, Weinheim, pp. 333–335.

Moussé, J. and Pernollet, J.C. (1982) Storage proteins of legume seeds. In: Arora, S.K. (ed.) *Chemistry and Biochemistry of Legumes.* Edward Arnold, London, pp. 111–193.

Naczk, M., Wanasundara, P.K.J.P.D. and Shahidi, F. (1992) Facile spectrophotometric quantification method of sinapic acid in hexane-extracted and methanol–ammonia–water-treated mustard and rapeseed meals. *Journal of Agriculture and Food Chemistry* 40, 444–448.

Niewiadomski, H. (1990) Methods of analysis and characterization of properties. In: Niewiadomski, H. (ed.) *Rapeseed Chemistry and Technology.* Elsevier, Amsterdam, pp. 90–121.

Oomah, D. and Mazza, G. (1992) Microwave drying for moisture determination in flax, canola and yellow mustard seeds. *Lebensmittel- Wissenschaft und Technologie* 25, 523–526.

Panford, J.A., Williams, P.C. and deMan, J.M. (1988) Analysis of oilseeds for protein, oil and moisture by near-infrared reflectance spectroscopy. *Journal of the American Oil Chemists' Society* 65, 1627–1633.

Paquot, C. and Hautfenne, A. (eds) (1987) *Standard Methods for the Analysis of Oils, Fats and Derivatives,* 7th edn. Blackwell Scientific Publications, Oxford, 347 pp.

Persmark, U. (1972) Analysis of rapeseed oil. In: Appelqvist, L.-Å., and Ohlson, R. (eds) *Rapeseed Cultivation, Composition, Processing and Utilization.* Elsevier Publishing Company, Amsterdam, pp. 174–197.

Pokorny, J., Meshehdani, T., Panek, J. and Davidek, J. (1990) Determination of the lipoxygenase activity of rapeseed. *Bulletin CETIOM* 6, 133–135.

Pritchard, J.L.R. (1991) Analysis and properties of oilseeds. In: Rossell, J.B. and Pritchard, J.L.R. (eds) *Analysis of Oilseeds, Fats and Fatty Foods*. Elsevier Applied Science, London, pp. 39–102.

Quinsac, A., Ribaillier, D., Elfakir, C., Lafosse, M. and Dreux, M. (1991) A new approach to the study of glucosinolates by isocratic liquid chromatography. Part I. Rapid determination of desulfated derivatives of rapeseed glucosinolates. *Journal of the Association of Official Analytical Chemists* 74, 932–939.

Rakow, G. and Thies, W. (1972) Schnelle und einfache Analysen Fettsäurezusammenstzung in einzelnen Rapskotyledonen. II Photometrie der Polyenfettsäuren. *Zeitschrift für Pflanzenzüchtung* 67, 257–266.

Reinhardt, T.C., Paul, C. and Röbbelen, G. (1992) Quantitative analysis of fatty acids in intact rapeseed by NIRS. In: Murray, I. and Cowe, I.A. (eds) *Making Light Work: Advances in Near Infrared Spectroscopy*. VCH, Weinheim, pp. 323–327.

Ribaillier, D. and Maviel, M.F. (1984) L'analyse des graines oléagineuses par spectroscopie de réflexion dans la proce infra-rouge. *Revue Français Corps Gras* 31, 181–189.

Ribaillier, D. and Quinsac, A. (1988). Le dosage des glucosinolates à la réception des graines de colza. *Bulletin CETIOM* 97, 11–13.

Rossell, J.B. and Pritchard, J.L.R. (eds) (1991) *Analysis of Oilseeds, Fats and Fatty Foods*. Elsevier Applied Science, London, 558 pp.

Salgó, A., Fábián, Z., Ungár, E. and Weinbrenner Varga, Z. (1992) Determination of anti-nutritive factors by near infrared techniques in rapeseed. In: Murray, I. and Cowe, I.A. (eds) *Making Light Work: Advances in Near Infrared Spectroscopy*. VCH, Weinheim, pp. 336–341.

Slominski, B.A. and Campbell, L.D. (1991) The carbohydrate content of yellow-seeded canola. In: McGregor, D.I. (ed.) *Proceedings of the Eighth International Rapeseed Congress*, Saskatoon, Canada, 9–11 July. Organizing Committee, Saskatoon, pp. 1402–1405.

Smith, C.B. and Barber, M.G. (1991) Measuring glucosinolates by flow injection analysis. In: McGregor, D.I. (ed.) *Proceedings of the Eighth International Rapeseed Congress*, Saskatoon, Canada, 9–11 July. Organizing Committee, Saskatoon, pp. 845–849.

Sosulski, F.W. and Dabrowski, K.J. (1984) Determination of glucosinolates in canola meal and protein products by desulfation and capillary gas–liquid chromatography. *Journal of Agriculture and Food Chemistry* 32, 1172–1175.

Starr, C., Suttle, J., Morgan, A.G. and Smith, D.B. (1985) A comparision of sample preparation and calibration techniques for the estimation of nitrogen, oil and glucosinolate content of rapeseed by near infrared spectroscopy. *Journal of Agricultural Science, Cambridge* 104, 317–323.

Stringam, G.R., McGregor, D.I. and Pawlowski, S.H. (1974) *Proceedings of the Fourth International Rapeseed Congress*, Giessen, Germany, 4–8 June. Deutsche Gesellschaft für Fettwissenschaft, Münster, pp. 99–108.

Taylor, S.L., King, J.W. and List, G.R. (1993) Determination of oil content in oilseeds by analytical supercritical fluid extraction. *Journal of the American Oil Chemists' Society* 70, 437–439.

Thies, W. (1971) Schnelle und einfache Analysen Fettsäurezusammenstzung in einzelnen Rapskotyledonen. I. Gaschromatographische und papierschromatographische Methoden. *Zeitschrift für Pflanzenzüchtung* 65, 181–202.

Thies, W. (1974) New methods for the analysis of rapeseed constituents. In: *Proceedings of the Fourth International Rapeseed Congress*, Giessen, Germany, 4–8 June. Deutsche Gesellschaft für Fettwissenschaft, Münster, pp. 275–282.

Tkachuk, R. (1969) Nitrogen to protein conversion factors for cereals and oilseed meals. *Cereal Chemistry* 46, 419–423.

Tkachuk, R., Mellish, V.J., Daun, J.K. and Macri, L.J. (1988) Determination of chlorophyll in ground rapeseed using a modified near infrared reflectance spectrophotometer. *Journal of the American Oil Chemists' Society* 65, 281–385.

Troëng, S. (1955) Oil determination of oilseed, gravimetric routine method. *Journal of the American Oil Chemists' Society* 32, 124–126.

Truscott, R.J.W., Tholen, J.T., Buzza, G. and McGregor, D.I. (1991) Glucosinolate measurement in rapeseed using reflectance. The Trubluglu meter. In: McGregor, D.I. (ed.) *Proceedings of the Eighth International Rapeseed Congress*, Saskatoon, Canada, 9–11 July. Organizing Committee, Saskatoon, pp. 1425–1429.

Van Caeseele, L. and Mills, J.T. (1983) Mucilage in canola seeds: rapid detection and interaction with storage fungi. In: McGregor, E.E. (ed.) *Seventh Progress Report: Research on Canola Seed Oil,*

Meal and Meal Fractions. Canola Council of Canada Publication No. 61, Winnipeg, Canada, pp. 170–173.

Van Soest, P.J. and Robertson, J.B. (1985) *Analysis of Forages and Fibrous Foods.* Cornell University Publication, Ithaca, New York, 165 pp.

Wang, X., and McGregor, D.I. (1991) Application of enzyme immobilization to achieve rapid low cost glucosinolate analysis of seed and meal of canola/rapeseed. In: McGregor, D.I. (ed.) *Proceedings of the Eighth International Rapeseed Congress*, Saskatoon, Canada, 9–11 July. Organizing Committee, Saskatoon, pp. 850–855.

Ward, K., Scarth, R., Daun, J.K. and Thorsteinson, C.T. (1994a) Effects of processing and storage on chlorophyll derivatives in commercially extracted canola oil. *Journal of the American Oil Chemists' Society* 71, 811–815.

Ward, K., Scarth, R., Daun, J.K. and Thorsteinson, C.T. (1994b) A comparison of HPLC and spectrophotometry to measure chlorophyll in canola seed and oil. *Journal of the American Oil Chemists' Society* 71, 931–934.

Wathelet, J.-P. (ed.) (1987) *Glucosinolates in Rapeseed: Analytical Aspects.* Martinus Nijhoff Publishers, Dordrecht, 198 pp.

White, J. and Law, J.R. (1991) Differentiation between varieties of oilseed rape (*Brassica napus* L.) on the basis of the fatty acid composition of the oil. *Plant Varieties and Seeds* 4, 125–132.

Wilkinson, A.P., Rhodes, M.J.C. and Fenwick, G.R. (1984) Determination of myrosinase (thioglucoside glucohydrolase) activity by a spectrophotometric coupled enzyme assay. *Analytical Biochemistry* 139, 284–289.

Williams, P.C. and Sobering, D.C. (1993) Comparison of commercial near infrared transmittance and reflectance instruments for analysis of whole grains and seeds. *Journal of Near Infrared Spectroscopy* 1, 25–32.

Woods, D.L. and Downey, R.K. (1980) Mucilage from yellow mustard. *Canadian Journal of Plant Science* 60, 1031–1033.

Zadernoski, R., Rotkiewicz, D., Kozlowska, H. and Sosulski, F. (1981) The procedure of determination of different forms of phenolic acid in rapeseed flours. *Acta Alimentaria Polonica* 7, 147–156.

12 Processing the Seed and Oil

R.A. Carr
POS Pilot Plant Corp., Saskatoon, Canada

INTRODUCTION

Over a long period, improvements have been made to the technology used to separate *Brassica* seeds into oil and solid components, both of which have food and non-food uses. Motivating an increasingly rapid evolution has been the race to satisfy the expanding needs of the marketplace.

The common objectives of all these evolving processes have been to:

1.   maximize yield of oil from oil-bearing material,
2.   minimize damage to the oil and solid fractions,
3.   produce components as free as possible from undesirable impurities,
4.   make quality products to meet market demands,
5.   produce a meal of the highest value.

SEED PROCESSING

Traditional *Brassica* seed processing systems are based on 'crushing' seeds to separate oil and meal components. The process involves seed cleaning, flaking, conditioning, mechanical extraction by prepressing and extrusion, followed by solvent extraction (Fig. 12.1). Optional tempering and dehulling treatments may be applied prior to flaking. Solvent is recovered from the meal by desolventizing–toasting.

Pretreatment

Brassica oilseeds which have been carefully harvested, graded for quality, cleaned and stored are ready for pretreatment and processing (Carr, 1993). Well-cleaned seed, such as that produced by the BM & M Partnership rotary screener (Fig. 12.2) (addresses for equipment suppliers are given at the end of this chapter), is essential for the manufacture of high-quality finished products.

Each small *Brassica* seed contains tiny oil bodies within the cells. The crusher's objective is to remove the maximum quantity of oil present, without damaging the quality of the resultant oil and protein components.

A successful extraction operation must be able to:

1.   rupture the walls of the cells,

267

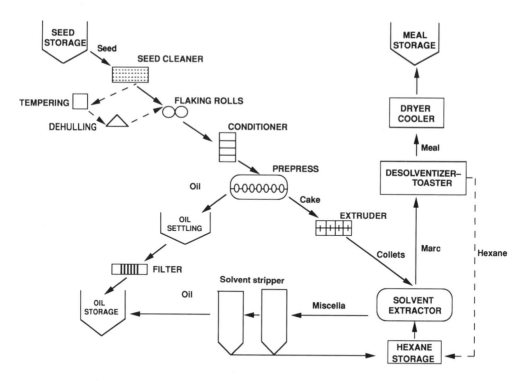

Fig. 12.1. Schematic diagram of *Brassica* seed processing.

2. obtain diffusion and agglomeration of the oil component,

3. obtain a final and complete separation of oil and solids.

Following optional tempering and dehulling, seed is carefully reduced in size by flaking, then conditioned by controlled heating to facilitate separation of the oil and solids components.

Tempering

Some seed processing plants in colder climates temper the cleaned seed prior to processing. Preheating the seed to approximately 35 °C, through units such as grain driers, reduces the amount of pulverizing which occurs when chilled seed from very cold storage enters a flaking unit. Undue pulverization affects the efficiency of the flaker as flakes need to be of an appropriate size for subsequent processing.

Dehulling

Brassica seeds are small and high in oil content. As a result, they are difficult to dehull economically in commercial plants. However, for some limited markets, an optional dehulling system may be used for cleaned whole seeds to obtain *Brassica* meal with reduced fibre and increased protein content. Dehulling combines mechanical or pneumatic impact separation of the hull from the seed with separation of the hulls through air aspiration and/or a fluidized bed sorter.

In 1985 an industrial dehulling system was utilized in France (Chone, 1985). The unit was installed at the COMEXOL oil mill of the Bunge group, located in Chalon/Saône. Dehulling was successful using a combination of a mechanical or pneumatic dehuller with a fluid grader, as developed by the Centre Technique Interprofessional des Oléagineux Métropoli-

Fig. 12.2. Rotary screener for cleaning *Brassica* seed. (Courtesy of BM and M.)

tains (CETIOM) (Burghart and Evrard, 1992). It successfully separated a significant portion of the hulls prior to the crushing operations and increased the protein content of the meal from 35% for nondehulled meal to approximately 42% for dehulled meal.

The process uses a dehulling system patented by CETIOM which involves firing the seeds against a target to fracture them. They are then sorted and separated into kernels, hulls and undehulled seeds, which is accomplished using fluid bed sorters manufactured by TECMACHINE in Saint Etienne. CETIOM observed that their dehulling system reduced the fibre content in the meal by approximately 40%

and increased protein content by 18%. Metabolizable energy content of the dehulled meal was improved by 11% for poultry, and digestible energy by 19% for swine.

The advantages of dehulling are primarily related to improved meal quality. Reduced fibre content and increased protein content make *Brassica* dehulled meal more competitive with meal from dehulled soyabeans. In addition, hull removal prior to processing reduces the carry-over of various impurities, such as hull pigments, to the crushing process. These impurities reduce product quality and increase processing costs during subsequent processing operations. An efficient, economical dehulling process should increase *Brassica*

Fig. 12.3. Flaking mill for converting *Brassica* seed to flakes. (Courtesy of Ferrell-Ross.)

meal utilization in animal feed and may become a major step towards the use of meal products in human food.

European information indicates that an oil content of approximately 3% remains within the hull fraction, which would lead to an increased oil loss of approximately 0.55% after extraction (Niewiadomski, 1990). A higher price for the hulls will be necessary to compensate for reduced oil yield. Large-scale conversion to dehulling by industry will depend upon careful economic analyses. Dehulled meal would have to generate a premium of at least 15% to be viable. This will determine whether or not improved *Brassica* meal value and reduced oil and meal processing costs will justify the additional cost of dehulling.

Unpublished information from the POS Pilot Plant Corporation, Saskatoon, Canada, indicates that fibre content in meal may be reduced from approxi-mately 12 to 6.5% by an experimental dehulling system, and protein content improved to levels close to that of soya-bean meal. A different configuration of equipment was used for the confidential POS Pilot Plant Corporation dehulling process.

Flaking

To facilitate oil extraction, cleaned *Brassica* seed is flaked by rolling to break open the hulls. Flaking mills, such as a Ferrell-Ross unit (Fig. 12.3), are capable of processing up to 450 tonnes day^{-1}. Flaking mills are equipped with one or two pairs of cast-iron smooth rollers, revolving on large swivel suspension roller bearings. The rollers turn at different speeds to shear the seeds into flakes. Vibrating feeders ensure an even seed distribution over the whole width of the flaking rollers. Precise separation dis-tance between the rollers is achieved by micrometric adjustment. The pressure of the rollers can then be adjusted to control the flake thickness. When the two stainless-steel rollers are properly positioned with a narrow clearance, the seed coats and some of the oil cells within the seeds will be ruptured, prior to the extraction process.

The final particle size that leads to most efficient extraction must be determined by experimentation. Generally, smaller-sized pieces are better for oil removal, but, if the pieces are too small, they may contaminate the oil and be difficult to remove during subsequent filtration. Thin flakes will facilitate solvent extraction, both by moderating the disruptive effect of rolling and by reducing the distances that solvent and oil must diffuse in and out of the flake during the extraction process. The thickness of these flakes is very important. *Brassica* seed can be flaked to a thickness of 0.30–0.38 mm with good results. Flakes thinner than 0.20 mm are very fragile. Flakes thicker than 0.40 mm process less satisfactorily and may reduce oil yield. Flaking is done on a continuous basis and should be linked directly to the subsequent conditioning process to ensure that the *Brassica* flakes are processed immediately without storage.

Conditioning

Thermal treatment (conditioning or cooking) is required, in addition to size reduction, to break open the oil bodies. Heat is also necessary to destroy enzymes within the seeds, particularly myrosinase (thioglucoside glucohydrolase EC 3.2.3.1). When the seed is crushed the presence of moisture can break down glucosinolates, releasing sulphur into the oil and compounds toxic to animals in the meal. Controlled heating can also improve protein availability in the resultant meal fraction obtained from the expelled oilseed cake.

Conditioning oil-bearing materials prior to expelling tends to:

1.   rupture cells thermally which have integrally survived the flaking process,
2.   reduce oil viscosity and promote coalescing of minute droplets of oil,
3.   increase the diffusion rate of the prepared oil cake,
4.   denature hydrolytic enzymes,
5.   adjust moisture control of the expeller feedstock to an optimum level of 4–6%.

Failure to inactivate myrosinase can result in severe damage to both oil and meal components. In the intact seed myrosinase is compartmentalized from its glucosinolate substrate. Only when the seed is crushed in the presence of moisture are the glucosinolates hydrolysed to isothiocyanates, oxazolidinethiones and nitriles. These sulphur-containing products are oil soluble and their extraction with the oil can adversely affect subsequent oil processing by, for example, contaminating the nickel catalyst used to hydrogenate the oil. The isothiocyanates and nitriles are also physiologically active compounds which when left in the meal can have deleterious effects on animals consuming it.

Seed should pass through the flaking process rapidly to minimize glucosinolate breakdown before myrosinase is inactivated during conditioning. In the conditioner, the temperature of the seed must be raised rapidly to inactivate the myrosinase early in the process (Youngs and Wetter, 1969; Daun, 1993).

Either drum or stack-type conditioners can be utilized satisfactorily for conditioning of *Brassica* oilseeds (Fig. 12.4). The drum conditioner may offer higher heat transfer rates, but at the expense of the integrity of the fragile flakes. Stack conditioners consist of a series of four to eight vertical, closed, cylindrical steel kettles, with each kettle usually 30 to 50 cm in diameter and 50 to 70 cm high. Conditioner capacities can range from 50 to 1000 tonnes day^{-1}. Flaked seed is mixed and turned by the sweeps, which are blades

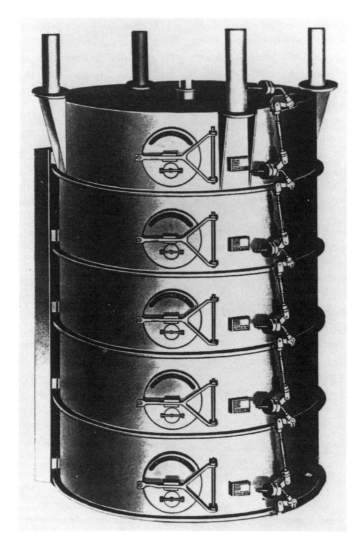

Fig. 12.4. Stacked drier for conditioning *Brassica* flakes. (Courtesy of Anderson International Corp.)

attached to a slowly turning vertical shaft. The flaked seed cascades from kettle to kettle by the mixing and pushing action of the sweeps in each kettle. Bed depth and sweep blade clearance have a profound effect on heat transfer and thus conditioning capacity.

Flaked seed is continuously conveyed into the top of the conditioner and rapidly heated to above 80°C, with the conditioner vents closed. The stacked kettles can be utilized either to add moisture by steam injection or to dehydrate seed. An initial moisture content of 6 to 10% is commonly used in the top kettle of a conditioner, where the myrosinase is destroyed. Careful moisture control is important during conditioning. Moisture contents above 10%

will enhance myrosinase activity whereas moisture contents below 6% will retard heat inactivation.

Temperatures may range from 80 to 105°C during the conditioning cycle. Protein damage and marked thermal decomposition of glucosinolates occur if conditioning is done above 100°C for an extended period of time (Campbell and Slominski, 1990). Extended conditioning time should also be avoided to minimize degradation of the protein. Desirable conditioning conditions are approximately 88°C and 8% moisture for 15–20 min. This amount of time is necessary for good oil release.

Flakes may then be dried in the lower trays at temperatures of 94–105°C. The vents are left open to complete conditioning. During discharge, the sweep blades are allowed to scrape the bottom of the conditioner to prevent scorching and to provide continuous feeding to the extraction process.

Mechanical extraction

The type of extraction system used to separate oil from oilseeds depends primarily on the oil content of the seed. Mechanical pressing with continuous expellers is generally used for initial extraction of *Brassica* seed oil because of the relatively high oil content of 40–45%. Solvent extraction may then be used to extract most of the remaining oil from the resultant press cake, which following mechanical extraction has an oil content below 20%.

Prepressing

The objective of prepressing is to remove as much oil as possible from the flaked seeds while optimizing the expeller and extractor outputs with press cake of acceptable quality. There is general agreement throughout the industry that prepressing followed by solvent extraction in medium- to large-scale plants does produce better

overall oil extraction economy, when high-oil-content seeds (>20%) are processed. This holds true, despite the power requirements and mechanical maintenance costs required.

During processing, a significant amount of protein denaturation can take place. This denaturing is usually considered desirable for feeding purposes because of the resultant improvement in digestibility. However, excessive heat during processing can result in extensive loss of certain amino acids, particularly the essential amino acid, lysine. The extent of such damage will depend on the time and temperature of processing, moisture status and the content of reducing sugar and foreign matter in the flakes.

Conditioned flakes containing approximately 42% oil and 7–9% moisture are fed to a series of low-pressure continuous screw presses or expellers where they receive a moderate press (Fig. 12.5). These units, which consist of a rotating screw shaft within a cylindrical barrel and cage unit, can press up to 350 tonnes of flaked seed day^{-1}. The cage supports a series of hardened-steel, flat bars set edgewise around its periphery, but carefully spaced to retain the cake solids within the barrel, while the oil is allowed to flow out of the barrel through precisely positioned gaps between the bars.

The rotating shaft presses the cake against an adjustable choke, which partially constricts the discharge of the cake from the end of the barrel. A plug of compressed meal is formed by the choke device at the discharge end of the barrel. This acts as a continuous hydraulic presshead, with 'new cake' being formed at the choke as the 'old cake' is continuously discharged from the expeller, past the choke device. This action removes most of the oil while avoiding excessive pressure, power consumption and temperature. Expellers, with a power consumption of approximately 125–300 horsepower, can reduce oil

Fig. 12.5. Mega 200 expeller press for separating conditioned flakes into crude oil and cake. (Courtesy of Anderson International Corp.)

content of well-flaked and cooked rapeseed from 42% to approximately 16% at a throughput of up to 250 tonnes day^{-1} per unit.

Good-quality cake is spongy and permeable, and resists disintegration during conveying to the solvent extractor. It is characterized as having a dry texture, with a preferred moisture level of 4–5% and an oil content of 15–18%. The cake should be between 3.2 and 4.8 mm thick with good physical integrity and durability. Inadequate moisture levels during pressing result in a granular discharge from the press or expeller, while excessive moisture results in a 'sloppy' product. Either type of poor expeller cake will cause problems in the solvent extraction process.

Pressing also consolidates the tiny flakes into larger units of cake fragments. Cake fragments substantially larger than the original flakes are essential in the subsequent extraction operation, to obtain satisfactory rates of solvent gravity percolation through the cake bed. A compromise must be reached between percolation rate, which is a function of size and thickness of the cake fragment, and the mean diffusion dimension, cake thickness. Therefore, cake thickness and durability are the two main variables which must be carefully controlled at the expellers so that fragments of optimum size enter the solvent extractor after delivery through a long conveying system.

Extrusion

Over the last 10 years, an increasing number of *Brassica* crushing plants have included an additional process called mechanical extrusion (Buhr, 1990). Various equipment supply companies, such as Anderson International and Technol, provide units which can be used for both low and high-oil-content seed preparations, ahead of solvent extraction. Extrusion of prepress *Brassica* cake is a recent application. Anderson International reports that an extruder (also known as an enhancer) has the ability to process oilseeds at high capacity and minimum cost by restructuring the cake to increase bulk density, improve extractability, rupture oil cells and fibres efficiently, inactivate enzymes, remove free liquids/oils and fats from their solid components, and cook the protein constituent for satisfactory agglomeration (Crawford, 1993).

Cake material from the prepress operation is conveyed into the extruder where it is moved through the barrel by flights on a centre shaft. It is mixed by a series of breaker screws along the length of the barrel. The temperature within the barrel is derived from friction from the rotating shaft and from injected steam. When the material is propelled to the end of the barrel, it is continuously forced against small die openings in the end plate. During discharge, the sudden release of pressure 'expands' the material, creating rope-like segments of porous collets ready for solvent extraction.

During the extrusion process, the feed material is subjected to a high-shear mastication in the presence of water hot enough to be a vapour, but which is compressed into a liquid. This increases the liberation rate of the oil and causes the solids to adhere together so that collets can be formed at the die openings. Hence, incoming feed material is mixed and kneaded with its natural moisture and additional steam-injected moisture. As the extruded product leaves the unit, pressure is released and water vaporizes instantly. This action causes the collets to swell and liberate the oil with the vaporizing moisture, thus creating pores in the collets. Once the moisture vaporizes, the liberated oil is reabsorbed into the collets, which is readily extractable in a solvent extractor due to the pore structure.

Some preliminary success has been achieved through delivery of flaked *Brassica* directly to the extruder, and eliminating the prepress operation. The flakes need not be as thin as those for the prepress opera-

tion, because the high-temperature short-time cook within the extruder will aid in rupturing a great percentage of the oil cells and liberate the oil efficiently. When processing full-fat flaked *Brassica*, a drainage cage is usually included on the extruder to take away the large quantity of oil liberated during extrusion. With drainage capability, discharging collets contain approximately 25–30% oil, from an inlet flake feed containing 42% oil.

Oil settling and filtering

All expelled oils contain approximately 3% solid matter and should be gravity-settled for approximately 3 h in a screening tank. The tank is usually equipped with a high-level controller which activates an oil discharge pump. The oil is maintained at approximately 66°C. The solids which settle, also called 'foots', develop from pressure within the expeller on the conditioned flakes. The amount of foots developed can be minimized by proper flaking and conditioning.

The foots may be continuously dredged off, drained and recycled back through the conditioner for re-pressing, or they may be re-pressed in a separate foots screw press, a specialized version of the previously described expeller. Screening and centrifugal separation may be used as an alternative. The re-pressed cake is then sent to the solvent extraction operation, along with the main cake stream. Expelled oil from the foots press is recycled back to the screenings tank for resettlement of the suspended 'fines' (small solid particles). The dredges may be sent directly to solvent extraction along with the main cake stream, provided their inclusion does not adversely affect the overall efficiency of the solvent extraction operation. As an alternative, they may be recycled back to the conditioners.

Oil is continuously drawn off from the unfiltered oil tank and the remaining suspended fines in the oil can be removed by either filtration or centrifugation. A totally enclosed multiple-screen filter is commonly used. The enclosed filtration plates consist of double-sided stainless-steel screens, pre-coated with the fines themselves. Many units are powered to open and close automatically for cleaning. The cake from the filter is returned to the system for reprocessing.

Oil from the prepress cage and extruder operations is collected and sent to a settling and filtration system. The expeller cake at 83–94°C with 15–18% residual oil is conveyed to the cake sizer for size reduction, before being sent to the solvent plant.

Solvent extraction

When dealing with relatively low-oil-content materials (less than 20% oil), extraction with a solvent, normally n-hexane, is the most efficient technique for oil recovery (Fig. 12.6). The objective for solvent extraction is to remove as much oil as possible with a minimum of solvent loss, because the oil is usually the most valuable component of the seed. The solvent dissolves the trapped oil and separates it from the solids as a 'miscella' (oil plus solvent). Oil produced by this method is of satisfactory quality, because very little heat treatment is required. In addition, the resultant meal fraction contains protein which has suffered minimal deterioration from heat damage.

There are, however, several disadvantages related to the solvent extraction process including:

1.   more expensive equipment compared to other extraction systems,
2.   the danger of fire and explosion unless non-flammable solvents can be used,
3.   dust explosion hazards since low-oil-content meal tends to be dusty.

Hexane extraction

Satisfactory cake from the expeller, or collets from the extruder, are conveyed to a solvent extractor for treatment with commercial n-hexane. The conveyor must be well insulated, resistant to corrosion by organic acids and moisture, and must convey the cake with minimal damage to its integrity. An inclined drag conveyor equipped with heavy wooden casing is suitable for this operation.

A number of different mechanical designs for solvent extractors are used to move the collets or sized cake and miscella in opposite directions with good inter-mixing, and to obtain a final separation of miscella from the solvent-saturated meal. The majority of extractors are designed to operate countercurrently and continuously. The extracted meal leaves at one end, and the miscella at the other. A basket percolation-type extractor, such as the Rotocell, is widely used, as well as shallow-bed loop extractors, such as the Crown Iron Works unit (Anderson, 1987). Extractors must position cake or collets for full contact with the n-hexane, and allow sufficient solvent : oil retention time for the large volumes of solvent to wash and separate the miscella with minimum solvent carry-over in the discharged solids.

The cake is introduced into the basket extractor through a vapour-seal unit. It is deposited into a basket, which is then flooded with a solvent or miscella through five to eight stages. A series of pumps spray the miscella over the baskets, with each stage containing a higher solvent : oil ratio. In addition to temperature, the extraction operator can, within limits, control bed depth, solvent feed and solvent : miscella distribution within the extractor.

The solvent percolates by gravity through the cake bed, diffusing into and saturating the cake fragments. Oil diffuses into a miscella solution, with a viscosity much lower than that of the oil alone. It then diffuses to the surface of the cake

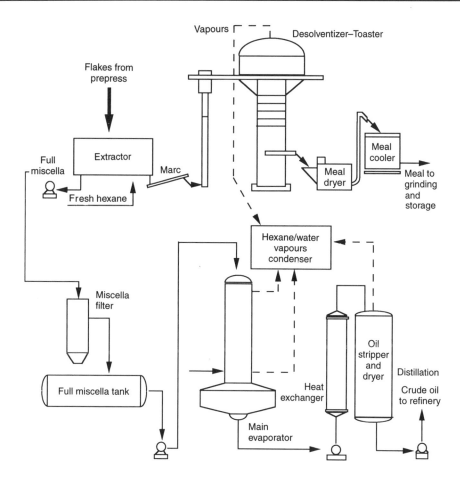

Fig. 12.6. Schematic diagram of n-hexane extraction and recovery.

particle where it is continuously washed away by the percolating flow of miscella. The oil containing miscella flows out through the cake support screen at the bottom of each basket. This miscella is then pumped to the next successive basket of cake. Just before cake discharge, each basket is given a final wash with pure solvent.

Continuous shallow-bed extractors use the same basic principles as the basket extractor. Cake from prepress is transported on a perforated steel belt through the length of the extractor, past a series of stationary solvent and miscella sprays. Fresh solvent is introduced before the cake

discharge section and circulates against the flow of the cake via a series of miscella pumps. It eventually becomes an oil-rich miscella at the cake inlet to the extractor, prior to its discharge from the extractor. The marc (hexane-saturated meal) leaves the extractor after the fresh solvent wash, containing less than 1% oil.

The vapour pressure of n-hexane limits the practical operating temperature of the extractor and its contents to approximately 50–55°C. Higher temperatures increase unduly the quantity of solvent vapour which the recovery systems must capture and recycle, or lose. Furthermore, if the cake temperature is at or near the boiling temperature of the solvent, a vapour phase may occur at the interface between the cake fragments and the solvent, effectively blocking liquid diffusion.

Solvent recovery

The full miscella leaving the extractor contains approximately 25 % oil and must be separated into crude oil for refining and *n*-hexane which is reused in the system. It is sent to a surge tank, from which it is pumped at a steady rate through a series of stills, stripping columns and associated condensers, to free the oil from the miscella. The hexane-free oil is cooled and filtered, before leaving the solvent extraction plant for storage or further treatment. As much heat as possible, such as that from condensation from vapours leaving the desolventizer–toaster, is recovered for use elsewhere as in distillation, etc.

The air and vapours from the solvent condensers and other parts of the extraction plant must be essentially solvent-free before being discharged to the atmosphere. This is frequently achieved by 'scrubbing' the vapours in a column with mineral oil. All the recovered solvent is separated from water in a gravity separation tank and used over and over again in the solvent extraction operation.

Desolventizing-toasting

The meal discharged from the solvent extractor is saturated with solvent (30 to 35%) and must therefore be transported in a closed conveyor system to a desolventizer–toaster. The purpose of the desolventizer–toaster is to remove the solvent by evaporation from the meal, which can then be used directly as a high-protein supplement for animal feed.

The desolventizer–toaster consists of a vertical stack of cylindrical gas-tight pans, each having a steam-heated bottom. The pans are equipped for direct steam injection to flash off the solvent. The meal enters at approximately 57 °C and is heated to 105 °C, before dropping by gravity through the unit from tray to tray via automatic doors, as the solvent is gradually volatilized and recovered for reuse. Steam is added to the meal in the top trays to displace the *n*-hexane absorbed by the protein. If required, the meal can be further heated to assist in the desolventization. By passing over the successive trays, the temperature can be controlled between 103 and 105 °C for drying and crisping the meal. Steam injection into the lower pans removes the last traces of solvent from the meal. The meal is discharged from the desolventizer–toaster to a drier–cooler at approximately 100 °C and with 10–12 % moisture. Meal from the desolventizer-toaster is virtually solvent-free and contains approximately 1 % lipid and 15–18 % moisture. It is dried to between 8 and 10 % moisture, cooled, and milled to a uniform particle size for delivery to feed manufacturers in either bulk or pelletized form.

The combination of temperature and moisture during desolventizing assures complete inactivation of any remaining myrosinase and also lowers the glucosinolate content by thermal decomposition. To obtain significant heat recovery savings, the hot solvent vapour from the desolventizer–toaster can be used to heat evaporating units before being condensed and pumped to the solvent work tank for reuse.

Processes such as cooking, expelling, mechanical extrusion and desolventizing can all damage the meal protein quality. Excessive heating during processing can result in reduced animal digestibility of some amino acids, particularly lysine. Processors must exercise strict process control of time, temperature and moisture variables to ensure amino acid damage is minimized by not overheating flakes in the cooker or meal in the desolventizer.

Table 12.1. Characteristics of combined crude oil from extraction operations and methods for quantifying them.

Characteristic/constituent	Quantity	Method
Colour	15 red, 70 yellow (1″ tube)	AOCS Cc 13b-45
Chlorophyll	35 ppm	AOCS Cc 13d-55
Free fatty acids	0.5%	AOCS Ca 5a-40
Phosphorus	450 ppm	AOCS Ca 12-55
Sulphur	8 ppm	Daun and Hougen (1976)

High-pressure extraction

A high-temperature/pressure batch method is technically feasible for extracting fats and oils from oilseeds (Mchugh and Krukonis, 1986). It utilizes carbon dioxide, heated and compressed above its critical temperature and pressure, to alter its properties. Such supercritical carbon dioxide is an ideal solvent because it is non-toxic, non-explosive, cheap, readily available and easily removed from the extracted products. It is effective in removing triglycerides while yielding a high-quality, gum-free, light-coloured crude oil with low iron content. However, additional development work is required to overcome the high-pressure engineering problems related to an economical, high-volume continuous process.

Despite the evolution of processing technology, techniques have not yet been developed to make *Brassica* meal suitable for human consumption. The presence of unacceptably high contents of glucosinolates, phenolics, phytate and fibre is an obstacle yet to be overcome (McCurdy, 1990).

OIL PROCESSING

Brassica oil recovered from the mechanical and solvent extraction processes consists of approximately 98% triglycerides (neutral oil), esters resulting from the combination of one molecule of glycerol with three molecules of various fatty acids. The remainder of the oil consists of phospholipids (gums), free fatty acids, pigments, sterols, waxes, meal, oxidized materials, moisture and dirt.

The combined crude oil from the extraction operations has approximate characteristics which may be measured by the American Oil Chemists' Association (AOCS) methods, as shown in Table 12.1.

The objective of oil processing is to remove the impurities while maximizing yield of (neutral) oil, and minimizing damage to its quality and the natural tocopherol antioxidant content. Impurities requiring closest attention are the phospholipids, free fatty acids and pigments. Most impurities are readily removed during processing (Fig. 12.7). Phospholipids are removed either by degumming processes or alkali refining, free fatty acids and odour/flavour components by physical refining, and pigments by bleaching. Waxes may then be removed by winterization, and the oil hardened by hydrogenation or interesterification. Deodorization may also be applied as a final step to remove odour and flavour components.

Refining

Phospholipid removal (degumming) is the major oil refining challenge and has been the main focus of research and development in recent years. Phospholipids are

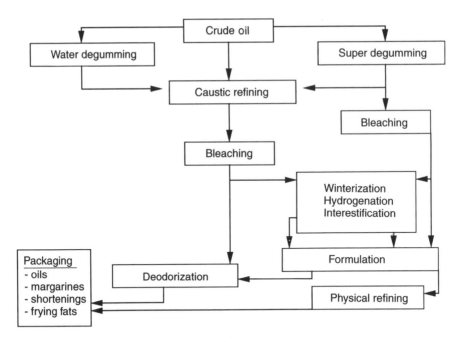

Fig. 12.7. Schematic diagram of *Brassica* oil processing.

esters resulting from the combination of one molecule of glycerol, two molecules of various fatty acids and phosphoric acid. Additional moieties may be esterified to the phosphoric acid. In *Brassica* these are choline, ethanolamine, inositol and a nitrogen base referred to as phytosphingosine. *Brassica* crude oils contain approximately 450 to 500 ppm phospholipid, which must be removed by the selected refining process.

Until recently, alkali refining was the most utilized commercial refining process for phospholipid removal. However, growing interest in water and acid degumming, in combination with physical refining to remove free fatty acids, has resulted in a major emphasis to improve its phospholipid removal capabilities. These advances have created greater assurance in the use of physical refining for quality products.

Degumming

Brassica crude oil containing soluble phospholipids has the propensity to form emulsions. Hence, phospholipid should be removed promptly after the extraction operations and prior to storage. If phospholipids are left in the crude oil, they will cause processing problems through higher than necessary chemical (alkali) refining losses, or by settling out in storage tanks. Some phospholipids are hydratable and can be removed by water degumming while others must be treated with acid first. When phosphatidylcholine (lecithin) from *Brassica* oils has a viable market, water degumming can be used to produce this by-product.

Water degumming. Water degumming exploits the affinity of the hydratable phospholipids for water through their conversion to hydrated gums, which are insoluble in oil and readily separated by centrifugal action. The principal phospho-

Fig. 12.8. PX 90 crude oil degumming separator.

lipids commonly found in seed oils are phosphatidylcholine and phosphatidylethanolamine (cephalin). The remaining phospholipids are not hydratable, cannot be removed by the conventional water degumming operation, and must be removed by acid degumming or alkali refining processes.

For water degumming, crude oil at 80°C is mixed for 10–30 min with a small quantity of water or steam (2–4%), in order to precipitate the hydratable gums.

Precipitated gums and the water are separated from the oil in continuous disc centrifuges, such as a Tetra Laval PX 90 separator (this unit can also be used for separations required in acid degumming, dewaxing and chemical refining operations) (Fig. 12.8). Hermetic-type centrifuges are recommended, to minimize contact with air while the oil is hot.

The discharged oil is dried under vacuum and pumped to oil storage tanks. Drying removes all traces of water which would cause further hydration and gum precipitation during storage. Water-degummed oil, still containing approximately 150–200 ppm non-hydratable phospholipids, should meet quality specifications before being further processed or sold as crude degummed oil. Removal of gums prior to alkali refining often improves the overall yield of refined oil because phospholipids act as an emulsifier in a caustic solution and increase the quantity of neutral oil retained in the soapstock.

Usually, the extracted gums are blended back into the meal, by adding them to the top tray of the desolventizer–toaster in the solvent extraction plant. The gums make the meal agglomerate more readily.

Acid degumming. If the non-hydratable phospholipids are also to be removed, they must be converted to the hydratable form before centrifugation. Most of them can be precipitated from the oil through the application of acid (phosphoric, citric, malic) and water during the degumming process. Phosphoric acid is normally the acid of choice. Acids act as sequestering agents breaking the calcium- and iron-based cations which form a complex with the non-hydratable phospholipids. Hence, the metal/phospholipid complexes are split into oil-insoluble metals, salts and hydratable phospholipid. Once the complex is broken, the phospholipids can be made oil-insoluble by hydration with water.

Various acid treatment systems have been described, but the basic practice is to add such a treatment prior to the water degumming process. After heating the incoming oil to 70°C, it is mixed with approximately 0.1% of 75% phosphoric acid for up to 20 min. Water is then added to a level of 2% with mixing to complete the hydration. The hydrated gums are removed from the neutral oil by centrifugal separation. Oil leaving the centrifuge still contains 0.3–0.5% moisture and, therefore, must be vacuum dried if destined for storage prior to bleaching. Drying is not usually necessary if the oil is bleached immediately. Separated gums from acid degumming are not used for human consumption but can be disposed of in the meal for animal feed.

In 1977, a Unilever group patented a superdegumming process which allowed a guarantee of residual phosphorus content. It utilized the principle that hydratable phospholipids contain polar groups which form liquid crystals in combination with water. Below 40°C they have a laminar structure and are insoluble in oil, but are capable of absorbing other impurities. After a 15-min agitation as acid/oil, the mixture is cooled to 25°C before addition of water. After a 3-h crystallization period, the mixture is heated to 75°C and separated by centrifuge.

A second stage (Uni-degumming) can be used for superdegummed oil (30 ppm maximum gum content) to obtain approximately 8 ppm phosphorus content in the oil. The Uni-degummed process cools the oil again to approximately 40°C, then the oil is mixed with a small quantity of caustic soda for 2 h to agglomerate the crystals, which are then separated by a clarifier centrifuge.

Vandemoortele/Belgium, in cooperation with Westfalia Separator AG, patented a degumming process (known as the TOP process) between 1987 and 1989 (Eickhoff, 1992). In this process crude oil is heated to between 90 and 105°C and mixed with a small amount of phosphoric acid for about 3 min. Sufficient dilute caustic soda is added to partially neutralize the phosphoric acid, but not enough to form soap. After mixing, it is conveyed to the first separator to remove most of the gums. Remaining fine phospholipid particles in oil are removed by a subsequent hot water treatment followed by separation in a high-performance, high-*g*-force centrifuge. The fine stream of phospholipids removed by the centrifuge is fed back into the first centrifuge, to be recombined in the large mass of phospholipids. Phosphorus content in the finished oil ranges from 2.8 to 6.2 ppm, with calcium 1.1 to 2.5 ppm, magnesium 0.3 to 0.7 ppm and iron 0.03 to 0.08 ppm. For physical refining, iron content should be less than 0.2 ppm. Both the Super/Uni-degumming and the TOP processes guarantee a residual phosphorus content below 10 ppm.

Acid degumming of crude oil to approximately 30 ppm phosphorus content, followed by thorough vacuum bleaching, can produce acceptable-quality super-degummed oils. Finished oil has been judged equal in quality to finished oil produced by the conventional alkali-refining process.

Physical refining

The purpose of physical refining is to remove free fatty acids, odour, flavour and some colour from low phospholipid content oils, using steam distillation. Step 1 is to contact acid-degummed oil with 0.05–0.1% of 85% phosphoric acid at about 110°C under vacuum, then mix the oil with approximately 1.0% of acid-activated bleaching clay for 10–15 min. An enclosed filtration unit separates the spent clay and impurities from the oil. This bleached oil is then pumped through a deaerator to a deodorizer–deacidifier for approximately 1 h at 260°C and 1–3 mm mercury vacuum. Injection of steam volatilizes most of the free fatty acids and odour/flavour impurities. The desired steam-distilled oil is cooled and passed through a filter into a storage tank.

The removal of metals and phospholipids by careful degumming is a crucial preparation for the physical refining of quality oils. Physical refining can also lower the loss of neutral oil in by-products, reduce the number of operations in the purification process, and eliminate the acidulation problems associated with the soapstock by-product produced by alkali refining. Hence, capital and operating costs may be lower for physical refining than for chemical refining, and environmental effluent problems are minimized.

Alkali refining

Climate and agronomic practices influence seed quality and can cause considerable variability from year to year in the undesirable impurities which affect the colour, appearance, odour, taste and keeping quality of the finished oil. With variable seed quality, alkali refining has proven better than physical refining for those processors who do not always have top-quality *Brassica* seed to deal with. Hence, most processors utilize alkali refining as their primary process, or as a standby emergency process when physical refining is not able to meet product specifications.

In the basic process, crude or degummed oil is transferred from a storage tank to a 'day tank' in preparation for alkali refining. It is treated with a solution of phosphoric acid, then continuously mixed with dilute sodium hydroxide (caustic soda) solution and heated to obtain a break in the emulsion. Soapstock is continually separated from the oil by centrifuges, such as the PX 90 separator. The resultant refined oil is mixed with hot, soft water and again centrifugally separated, using for example the Tetra Laval SSB 215 centrifuge (Fig. 12.9), to remove small amounts of residual soap. This water-washed refined oil, containing traces of moisture, is then passed through a continuous vacuum-drying stage and on to the refined oil storage tank. There have been few changes in chemical refining over recent years and the process has been documented in various publications (Carr, 1976).

Key factors in determining the success

Fig. 12.9. SSB 215 solid-bowl separator for water-washing refined oil. (Courtesy of Tetra Laval Food.)

of any edible fat and oil refining operation are:

1. uniform feedstock,
2. acid pretreatment of crude oil,
3. proper quantity of refining agents,
4. proper mixing of the oil and refining agents,
5. proper control of residual contact time and temperature,
6. efficient centrifugation,
7. sufficient water wash and drying.

Moderate agitation in the oil tanks and accurate metering pumps are essential for providing a uniform flow of homogeneous material to the alkali refining process. *Brassica* crude oil should be pretreated with 100 to 500 ppm of food-grade 75% phosphoric acid, then carefully mixed prior

Table 12.2. Characteristics of alkali-refined and washed oil.

Characteristic/constituent	Quantity	Method
Colour	15 red, 70 yellow (1″ tube)	AOCS Cc-13b-45
Chlorophyll	30 ppm	AOCS 13d-55
Phosphorus	20 ppm	AOCS 12-55

to addition of sodium hydroxide. The strength of the sodium hydroxide solution is usually measured using a specific gravity hydrometer calibrated in degrees Baumé (Bé). Sodium hydroxide concentrations of 15–18°Bé are usually prescribed for *Brassica* oils. The quantity of sodium hydroxide applied to the crude oil will vary with the free fatty acid and phospholipid content. Usually, a 0.1% excess over the theoretical amount of sodium hydroxide necessary to neutralize the free fatty acid is sufficient to neutralize and remove the gums. In North America, the sodium hydroxide is mixed with oil using the 'long mix' method. After mixing the sodium hydroxide with the oil at 30°C with a high-speed inline mixer, a series of dwell mixers are used to allow approximately 16 min of contact prior to separation. European processors favour a 'short mix' method, which starts with the crude oil heated to between 85 and 105°C. The oil/sodium hydroxide mixture is promptly delivered to the centrifuge at about 90°C, to separate the soapstock impurities from the neutral oil.

Refined oil from the primary centrifuge is reheated to between 85 and 90°C and treated with 10–20% of softened water, by weight of the oil flow. The oil/water/soap mixture passes through a high-speed inline mixer and is delivered to a secondary separator, such as the DeLaval B-214-C centrifuge. Washed oil at approximately 85°C is passed through nozzles into the evacuated section of a continuous vacuum drier operating at 70 mmHg, which dries the oil to a level of less than 0.1% moisture. The dried oil is cooled to

approximately 25°C before being sent to storage or the bleaching process.

A typical alkali-refined and washed oil will approximate characteristics which may be measured by the AOCS methods as shown in Table 12.2.

Bleaching

Water-washed, vacuum-dried oil from the refinery contains carotenoid and chlorophyll pigments, which must be removed to reduce the oil colour to an acceptable level. The oil may also contain small quantities of soap (10–50 ppm), which must be removed to ensure proper hydrogenation and deodorization. A 'medium vacuum' bleaching operation is used to remove most of the pigments from the oil by adsorption on acid-treated bentonite bleaching clay. At the same time, detrimental contaminants, such as soaps, metals and oxidation products, will be retained on the clay and removed from the oil.

Continuous vacuum bleaching is usually employed because it protects the oil from oxidation during periods at elevated bleaching temperatures. Incoming oil is heated with deaeration to approximately 100°C as it enters the bleacher, where it comes in contact under vacuum (50 mmHg) with a metered addition of acid-activated bleaching clay. The ratio of bleaching clay to oil in the flow is controlled to ensure that a constant percentage (0.5–2.0%) of the bleaching clay is in contact with the oil. The flow rates are adjusted to provide a contact time of 10–20 min, depending upon oil quality and the type

Table 12.3. Characteristics of a typical refined, bleached *Brassica* oil.

Characteristic/constituent	Quantity	Method
Colour	1.4 red, 22 yellow ($5\frac{1}{4}$″ tube)	AOCS Cc 13b-45
Free fatty acid	0.04% (as oleic)	AOCS Ca 5a-40
Chlorophyll	50 ppb	AOCS Cc 13d-55
Peroxide value	1 m/equiv. kg^{-1}	AOCS Cd 8-53
Acetone insoluble	0.02%	AOCS Ja 4-46
Phosphorus	<1 ppm	AOCS Ca 12-55
Sulphur	1 ppb	Daun and Hougen (1976)

of equipment and bleaching clay used. The bleached oil is continuously pumped from the bottom of the bleacher into a filter, to separate the spent clay. Filtration usually involves enclosed hermetic leaf filters with stainless-steel mesh elements, followed by a polishing filtration. Spent cake is dried in the filter by blowing with steam. The recovered oil is recycled. At the end of the cycle, the filter is opened and the dried spent clay is detached from the leaves by vibration.

A typical refined, bleached *Brassica* oil will usually have analytical characteristics which may be measured by the American Oil Chemists' Association (AOCS) methods (see Table 12.3).

Winterization

Growing conditions in some seasons result in traces of wax, esters of fatty acids and long-chain alcohols in the oil. Where consumers demand perfectly clear salad and cooking oils, both on store shelves and in the refrigerator, processors will utilize a low-temperature winterization process to precipitate and remove waxes before the oil is bottled. Because these contaminants tend to solidify at cool temperatures, winterization precipitates them as solid particles which can be removed by low-temperature filtration. Bleached oil is usually processed, but the process can be applied after refining or deodorization.

Incoming oil is rapidly cooled, initially

by a heat exchanger. Filter aid can be mixed with the oil to improve the subsequent filtration. Crystallization is initiated by slowly chilling the mixture to between 0 and 19°C and will be complete after about 4 h if this temperature is maintained. Hermetic tank filters in a cool room separate the filter aid and precipitate from the oil. Because of the cool temperature, filters can be blown dry with compressed air. The clear oil is sent to a storage tank or on to the next process, usually deodorization, for use as salad and cooking oils.

Oil modification

Refined, bleached liquid oil can be converted to semi-solid, plastic fats suitable for shortening or margarine manufacture. Change in the melting-point is effected either by hydrogenation or by interesterification.

Hydrogenation

Hydrogenation is used to add hydrogen to the unsaturated fatty acid double bonds in the neutral oil. In addition to increasing the solids content and melting-point of the oil, hydrogenation also enhances the stability and colour of the final product. For efficient hydrogenation the quality and purity of both the hydrogen and the oil are important.

For hydrogenation, the oil, gaseous hydrogen and a nickel catalyst are combined by mechanical agitation, at specific temperatures and pressures in a closed reaction vessel. The nickel catalyst is usually combined with a Kieselgur clay to improve filtration efficiency. Temperature, the nature of the oil being hydrogenated, the activity and concentration of the catalyst, agitation, and the rate at which hydrogen and unsaturated oil molecules are combined at the active points of the catalyst surface all affect the rate at which the hydrogenation reaction proceeds, the final solids curve and melting characteristics of the fat.

In practice, batch dead-end hydrogenators or converters are held under vacuum until approximately 4500–18,000 kg of liquid oil are loaded and the catalyst in oil suspension is charged. During the heat-up period, the vacuum deaerates and dries the oil before it reaches the elevated reaction temperature. When the required hydrogenation temperature is reached, the vacuum line is closed and hydrogen is metered into the vessel until the pressure builds to the desired level. Each company may use different hydrogenation techniques, but conditions would usually be varied within the following ranges to obtain hydrogenated base stocks with suitable sharp melting-point curves for margarine, or flat ones for shortening formulations:

1. catalyst concentration, 0.01–0.1% nickel,
2. reaction temperature, 150–210°C,
3. hydrogen pressure, 35–450 kPa,
4. agitation speed, approximately 150 rpm,
5. reaction time, 10–60 min.

The hydrogenation reaction endpoint is controlled by determining the refractive index of the oil in the converter, which can be used as an indicator of the iodine value. The solids content, iodine value, fatty acid composition and melting-point are evaluated later in the laboratory. At the end of the reaction, hydrogen flow is stopped and the hydrogen in the head space, with accumulated impurities, is evacuated. Vacuum is maintained in the vessel while the hydrogenated oil is cooled to approximately 80–85°C before being discharged.

In order to improve overall heat recovery and increase hydrogenation capacity, external heat exchangers can be used. An exchanger, such as a spiral unit, can heat incoming oil to the charge tank with hot oil leaving the converter. The cooled, hydrogenated oil is then sent through a catalyst filtration unit (black press) and a post-bleaching operation before being sent to storage. Oil from the black press can contain from 5 to 20 ppm nickel. Most is nickel catalyst on Kieselguhr clay, but some can be oil-soluble nickel soap. Using precoat and bag filters, the final nickel content in the oil can be decreased to less than 0.1 ppm (Kokken, 1993). The recovered catalyst may be reused until its activity becomes unsuitable for meeting hydrogenated product specifications.

Concerns over *trans*-isomer formation during hydrogenation will probably reduce the use of hydrogenation in future and more liquid oils and blends for interesterification/fractionation are likely to be used to compensate. In addition, considerable effort is being focused on the development of new custom-tailored oilseeds with high solid contents (stearic and palmitic). These combined efforts could ultimately minimize or eliminate the need for hydrogenation.

Interesterification

The interesterification process is used to modify the triglyceride structure for *Brassica* oil alone, or in blends with other oils and fats of vegetable or animal origin. Major changes in the melting characteristics of the original oil can result. Thus, it has the capability of providing an alternative to hydrogenation for producing base stocks with significant solid content, suitable for preparing margarine and

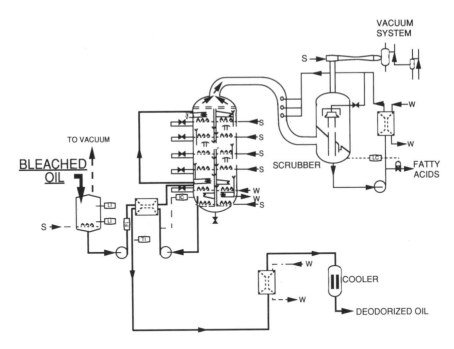

Fig. 12.10. 'Unistock' continuous deodorizer. (Courtesy of De Smet Engineering.)
FI, flow indicator; IC, indicator controller; LC, level controller;
LI, level indicator; S, steam; TI, temperature indicator; W, water.

shortening blends with a minimum content of *trans*-isomers.

To interesterify, the oil or oil blend under vacuum in a reactor is brought into contact with 0.1–0.2% of an alkaline metal catalyst, such as sodium methoxide. After 20–30 min of mixing at 135–150°C, the fatty acids have been separated from the triglycerides, well mixed and then randomly reconnected to the glycerols to form different triglycerides, with altered melting characteristics. To obtain additional flexibility in melting properties, the process can be conducted at various lower temperatures, to direct the higher melting-point triglycerides to crystallize and remove themselves from the reaction mass. As a result, the formation of higher melting-point triglycerides is promoted. Both the quantity and type of these triglycerides can be altered by adjusting the reaction temperature.

Considerable flexibility in preparing margarine and shortening base stocks can be obtained by 'random' and 'directed' interesterification, or blends of the two. This capability will be enhanced with the developing *Brassica* cultivars having a wider range of fatty acid composition.

Deodorization

Large differences between the volatility of less volatile triglycerides and their more volatile natural flavour and odour components makes steam distillation (deodorization) feasible for stripping these impurities from the oil. In addition, the process destroys peroxides in the oil, removes aldehydes and ketones or other volatile products resulting from atmospheric oxidation, and reduces the oil colour by the destruction of the relatively unstable carotenoid pigments. Prior to deodorization, various combinations of liquid oils and solids-containing base stocks are blended to meet specific product melting curve requirements.

Table 12.4. Typical characteristics of a deodorized oil.

Characteristic/constituent	Quantity	Method
Colour	0.4 red, 4 yellow ($5\frac{1}{4}$" tube)	AOCS Cc 13b-45
Free fatty acid	0.02%	AOCS Ca 5a-40
Peroxide value	0.0 mequiv. kg^{-1}	AOCS Cd 8-53
Tocopherol	300 ppm	AOCS H 19-58
Chlorophyll	<0.1 ppm	AOCS Cc 13d-55
Sulphur	<1.0 ppm	Daun and Hougen (1976)

The preferred method of deodorization is a 'continuous' or 'semi-continuous' process carried out in vertical stainless-steel deodorizers fitted with approximately five to seven shallow trays (Fig.12.10). The oil is deaerated under vacuum and then charged to the top steam heating tray. It then cascades down to a second high-temperature heating tray, in which it is heated by Dowtherm heat-transfer liquid to about 240–260°C. The heated oil then flows through two deodorization trays, counter-current to a superheated stripping steam flow. Under normal operating conditions, the stripping steam requirements are approximately 4.5 kg 45 kg^{-1} of oil. Dwell time in each of the trays is usually 20 min. A three-stage steam ejector system maintains a vacuum of 3–6 mmHg to enhance volatilization of the impurities and to protect the quality of the neutral oil. The deodorized oil then exits the second deodorizer tray and is cooled in a water-chilled cooling tray, before being filtered and transferred to a deodorized oil storage tank. Typical characteristics of a deodorized oil are shown in Table 12.4. Bland, deodorized fats and oils from the deodorized oil storage tanks are ready for finishing as packaged cooking oils, salad oils, shortenings, salad dressings and margarines.

EQUIPMENT SUPPLIERS

Anderson International Corporation: 6200 Harvard Avenue, Cleveland, Ohio, USA 44105.

BM and M Partnership: 9377 – 193rd Street, Surrey, British Colombia, Canada V3T 4W2.

Campro-Agra Limited: 2200 Argentia Road, Mississauga, Ontario, Canada L5N 2X7.

Crown Iron Works: PO Box 1364, Minneapolis, Minnesota 55440-1364, USA.

ER and F Turner Ltd.: Knightdale Road, Ipswich IP1 4LE, UK.

Extraction De Smet: 265 Prins Boudewijniaan, 2520 Edegem, Belgium.

Ferrell-Ross: 805 S. Decker Drive, PO Box 256, Bluffton, Indiana 46714, USA.

French Oil Mill Machinery Co.: PO Box 920, Piqua, Ohio 45356, USA.

Industrial Filter and Pump Co.: 5936 Ogden Avenue, Cicero, Illinois 60650-3888, USA.

Tetra Laval Food: Tumba, S-14780, Sweden.

REFERENCES

Anderson, G. (1987) *Solvent Extraction of Canola.* Crown Iron Works Technical Presentation, POS Short Course 1–12.

Buhr, N. (1990) Mechanical pressing. In: Erickson, D.R. (ed.) *Edible Fats and Oils Processing: Basic Principles and Modern Practices.* American Oil Chemists' Society, Champaign, Illinois, pp. 43–48.

Burghart, P. and Evrard, J. (1992) *Dehulling of Rapeseed: The French Experience.* CETIOM, Paris, France, pp. 1-6.

Campbell, L. and Slominski, B. (1990) Extent of thermal decomposition of indole glucosinolate during the processing of canola seed. *Journal of the American Oil Chemists' Society* 67, 73-75.

Carr, R. (1976) Degumming and refining practices in the USA. *Journal of the American Oil Chemists' Society* 53, 347-352.

Carr, R. (1993) Oilseed harvesting, storage and transportation. In: Applewhite, T.H. (ed.) *AOCS Proceedings of the World Conference on Oilseed Technology and Utilization*, Budapest, Hungary. American Oil Chemists' Society, Champaign, Illinois, pp. 118-125.

Chone, E. (1985) Le Epelliculage des graines de colza. *Centre Technique Interprofessional des Oléagineux Métropolitains (CETIOM) Information Technical* 32, 18.

Crawford, J. (1993) The Anderson oilseed expander line. *Technology Update for the Worldwide Oilseed Processing Industry*, 1-15.

Daun, J. (1993) Oilseeds-processing. *Grains and Oilseeds: Handling, Marketing, Processing*, Vol. 2, 4th edn. Canadian International Grains Institute, Winnipeg, Manitoba.

Daun, J.K. and Hougen, F.W. (1976) Sulfur content of rapeseed oils. *Journal of the American Oil Chemists' Society* 53, 169-171.

Eickhoff, K. (1992) *Activities of Westfalia Separator AG in Edible Oil Degumming.* Westfalia Separator, Technical Information 2-28.

Kokken, M. (1993) Hydrogenation in practice. In: Applewhite, T.H. (ed.) *AOCS Proceedings of the World Conference on Oilseed Technology and Utilization*, Budapest, Hungary. American Oil Chemists' Society, Champaign, Illinois.

McCurdy, S. (1990) Effects of processing on the functional properties of canola and rapeseed protein. *Journal of the American Oil Chemists' Society* 67, 281-284.

Mchugh, M. and Krukonis, V. (1986) *Supercritical Fluid Extraction: Principles and Practices.* Butterworth Publishers, Boston, Massachusetts, pp. 6-97.

Niewiadomski, H. (1990) *Rapeseed Chemistry and Technology.* PWN, Warsaw, 6-1-1, 162.

Youngs, C. and Wetter, L. (1969) *Processing of Rapeseed for High Quality Meal, Rapeseed Meal for Livestock and Poultry*, Vol. 3. Rapeseed Association of Canada, Winnipeg, Manitoba, pp. 2-3.

13 Oil Properties of Importance in Human Nutrition

B.E. McDonald
University of Manitoba, Winnipeg, Canada

INTRODUCTION

Dietary fat has a number of nutritional functions. It is an important source of energy; for example, it constitutes 45–50% of the calories for the breast-fed infant. It is also the source of essential fatty acids, namely those of the n-6 and n-3 (also known as omega-6 and omega-3) families of polyunsaturated fatty acids. Members of these families are important constituents of cell membranes and serve as precursors of a variety of biologically active compounds known collectively as eicosanoids (e.g. prostaglandins, thromboxanes, leukotrienes). In addition, dietary fat serves as a carrier for the fat-soluble vitamins and plays an important role in the palatability of foods. Much of the current interest in dietary fat, however, stems from its implication in the aetiology of a number of chronic diseases, such as cardiovascular disease, cancer, obesity and hypertension. As a result, dietary recommendations in many countries call for a reduction in total fat intake to 30% of total energy and in saturated fat intake to 10% of total energy. In some cases, the nutritional recommendations (Health and Welfare Canada,

1990) also specify the recommended levels of n-6 and n-3 fatty acids in the diets of specific age groups. Hence the source of fat in the diet has assumed considerable importance over the past 20 years.

The discussion in this chapter will be restricted to the nutritional properties of low erucic acid rapeseed oil (also known as canola-type rapeseed oil and hereafter referred to simply as canola oil). Very little research has been reported on any *Brassica* oils other than those of rapeseed, namely *Brassica napus* and *Brassica rapa*. Furthermore, very few animal studies on the nutritional properties of the *Brassica* oils have been published since Kramer *et al.* (1983) summarized so eloquently the profusion of publications on high and low erucic acid rapeseed oils that occurred during the 1970s. Hence, the discussion will deal exclusively with the nutritional properties of canola oil for humans.

Appreciable research on the nutritional properties of canola oil has been reported since the reviews by McDonald (1983) and Ackman (1990). Studies with humans in Canada, Finland, Sweden and the United States have confirmed the nutritional benefits of canola oil alluded to by Ackman

(1990). These studies, which have dealt with the effect of canola oil on two of the primary processes predisposing to coronary heart disease, namely atherosclerosis and thrombosis, arose from the finding that dietary monounsaturated fatty acids were as effective as polyunsaturated fatty acids in reducing blood cholesterol levels (Mattson and Grundy, 1985; Mensink and Katan, 1989) and reports that n-3 fatty acids alter platelet activity (Kinsella, 1988). Canola oil, which is characterized by a low level of saturated fatty acids and a relatively high level of monounsaturated fatty acids, also contains an appreciable amount of α-linolenic acid (18 : 3 n-3, hereafter referred to simply as linolenic acid). Canola oil is second only to olive oil, among the common edible fats and oils, in oleic acid (18 : 10 n-9) content and, together with soyabean oil, is the only common edible oil that contains a significant amount of linolenic acid.

EFFECT OF CANOLA OIL ON CHOLESTEROL AND LIPOPROTEIN METABOLISM

The report by Mattson and Grundy (1985) that dietary monounsaturated fatty acids (viz. oleic acid) were just as effective as polyunsaturated fatty acids (viz. linoleic acid) in reducing total plasma and low-density lipoprotein (LDL) cholesterol provided a possible explanation for the observation that canola oil was as effective as soyabean oil in reducing plasma cholesterol in normolipidaemic men (McDonald et al., 1974). Prevailing theory had held that saturated fatty acids raised plasma cholesterol, that polyunsaturated fatty acids reduced plasma cholesterol and that monounsaturated fatty acids had no effect, that is, they neither raised nor reduced plasma cholesterol (Hegsted et al., 1965; Keys et al., 1965).

Several studies have found canola oil just as effective as sunflower oil (McDonald et al., 1989; Valsta et al., 1992), soyabean oil (Chan et al., 1991) or safflower oil (Wardlaw et al., 1991) in reducing total plasma and LDL cholesterol (Table 13.1). All treatment diets resulted in significant decreases in total plasma cholesterol (mean of − 0.47 to − 0.88 mM) from the levels on the baseline diets (diets typical of usual fat intake) which were appreciably higher in total saturated fatty acid content (14–19 % of total dietary energy on the baseline diets versus 7–13 % on the canola and polyunsaturated fatty acid diets). Furthermore, the decrease in total cholesterol was primarily due to a decrease in LDL cholesterol (mean of − 0.43 to − 0.74 mM) which is consistent with the primary objective of intervention programmes aimed at reducing the risk of coronary heart disease. Elevated plasma LDL level is a major risk factor in the development of coronary heart disease (Consensus Conference, 1985; McNamara, 1990). Also, the decreases in total and LDL cholesterol were the same with the canola oil (i.e. monounsaturated fatty acid diets) and the polyunsaturated fatty acid diets, which coincides with the earlier study by Mensink and Katan (1987), who also used diets based on customary foods. Likewise, decreases in the level of apolipoprotein B (mean of − 190 to − 25 %), which is the lipoprotein characteristic of the LDL fraction, were similar for the canola and polyunsaturated fatty acid diets. The changes observed in plasma cholesterol and lipoprotein levels in response to canola oil are consistent with what would be expected when customary dietary fats are replaced by dietary fat that is low in saturated fatty acids and high in monounsaturated fatty acids.

None of the diets used in the studies cited in Table 13.1 had any effect on the level of plasma high-density lipoprotein (HDL) cholesterol even though the levels of polyunsaturated fatty acids in the sun-

Table 13.1. Effect of dietary canola oil versus polyunsaturated fatty acid sources on total plasma and lipoprotein cholesterol levels of human subjects.

Dietary PUFA oil source[1]	Plasma lipid parameter	Baseline (mM)	Canola diet (% change from baseline)	PUFA diet (% change from baseline)	Reference
Sunflower	Total cholesterol	4.42	−20	−15	McDonald et al. (1989)
	LDL cholesterol	2.76	−25	−21	
Soyabean	Total cholesterol	4.40	−18	−16	Chan et al. (1991)
	LDL cholesterol	2.98	−25	−18	
Safflower	Total cholesterol	5.39	−9	−15	Wardlaw et al. (1991)
	LDL cholesterol	3.71	−12	−15	
Sunflower	Total cholesterol	5.35	−15[2]	−12[2]	Valsta et al. (1992)
	LDL cholesterol	3.17	−23[2]	−17[2]	

[1] PUFA = polyunsaturated fatty acid.
[2] Significant difference between diets, $P < 0.01$.

flower, soyabean and safflower oil diets (13–22% of the total energy) were appreciably higher than the levels in the average diet. This finding is consistent with the lack of any adverse effect of a relatively high intake of polyunsaturated fatty acid on plasma HDL levels (Mensink and Katan, 1987). Mattson and Grundy (1985) found that very high intakes of polyunsaturated fatty acids (29% of total energy) reduced total plasma HDL level whereas very high intakes of monounsaturated fatty acids had no effect on HDL level.

Canola oil also was effective in reducing total serum and LDL cholesterol levels in hyperlipidaemic subjects. Miettinen and Vanhanen (1994) found that replacing 50 g of fat in the regular diet with 50 g of canola oil mayonnaise resulted in a 9% decrease in total cholesterol and a 10% decrease in LDL cholesterol in a group of men with a mean total serum cholesterol level of 7.1 mM. Although mean serum total and LDL cholesterol levels rose slightly during an additional 15 weeks on the canola diet, they did not differ significantly ($P > 0.05$) from the levels after the initial 6 weeks on the canola oil diet. Serum triglyceride and very low-density lipoprotein (VLDL) levels were also decreased with the canola oil diet whereas HDL levels increased. Serum apo-

lipoprotein levels followed the expected patterns; apolipoprotein B decreased whereas apolipoprotein A increased on the canola oil diet. Normally, type of fat, at the level of exchange used in this study, has not affected serum HDL cholesterol or apolipoprotein A levels. However, the decrease in total and LDL cholesterol is consistent with other studies on canola oil (Table 13.1). Bierenbaum et al. (1991) also found that substituting canola oil for a portion of the usual oils in the diets of hyperlipidaemic patients resulted in a decrease in plasma LDL cholesterol levels. Replacing 30 g of the usual oils and spreads with canola oil resulted in an average decrease in LDL cholesterol of 0.18 mM ($P < 0.025$). Total cholesterol, however, did not differ from baseline values.

Substitution of canola oil for other fats in the diets, at much lower levels than those used in the studies cited above, also resulted in decreases in plasma cholesterol level. Seppanen-Laakso et al. (1992) substituted canola oil or canola oil margarine for butter in a group of subjects with mildly elevated plasma cholesterol levels (mean baseline of 6.15–6.32 mM). On average, substituted fat represented one fifth of the total fat and 8% of total energy in the diets. Replacement of butter with canola oil resulted in significant decreases in total plasma (– 0.49 mM; – 9%) and LDL (– 0.590 mM; – 13%) cholesterol levels during the first 3 weeks of the study. For the canola oil margarine group the decreases were – 0.39 mM (– 6%) and – 0.33 mM (– 8%), respectively. Similarly, Truswell *et al.* (1992) found that supplying about half of the dietary fat as potato crisps fried in canola oil or palmolein (palm oil fraction: 39% palmitic acid, 45% oleic acid) resulted in a lower total plasma cholesterol level with the canola oil than with the palmolein regimen ($P < 0.001$); total cholesterol decreased an average of 0.39 mM on the canola oil crisps and increased 0.14 mM on the palmolein crisps. In contrast, HDL cholesterol was significantly higher with the palmolein than the canola oil regimen; HDL increased 0.12 mM on average from baseline on the palmolein crisps whereas there was no change on the canola crisps (mean increase of 0.02 mM). Although low-fat diets have been found to decrease plasma HDL levels (Mensink and Katan, 1987; Grundy *et al.*, 1988; Barr *et al.*, 1992), type of fat is generally not believed to affect plasma HDL levels in contrast to the marked effect of type of fat on plasma LDL levels.

The low level of saturated fatty acids and the relative high level of mono-unsaturated fatty acids are undoubtedly major factors in the effectiveness of canola oil in reducing plasma cholesterol level. Like other dietary fats that are rich in monounsaturated fatty acids, its primary effect is on the LDL fraction which is a major risk factor in coronary heart disease. Canola oil has been shown to be equally as effective as sunflower oil (Lasserre *et al.*, 1985; McDonald *et al.*, 1989; Valsta *et al.*, 1992), soyabean oil (Chan *et al.*, 1991) and safflower oil (Wardlaw *et al.*, 1991) in reducing plasma cholesterol and LDL levels when they replace saturated fat sources in the diet.

CANOLA OIL, PLATELET ACTIVITY AND THROMBOGENESIS

There are two major events in cardio-vascular disease: atherosclerosis, the formation of plaques on the intima (innermost lining) of major blood vessels; and thrombosis, the formation of a massive clot in these damaged vessels. Atherosclerosis is a relatively slow process that develops over several years and the risk factors associated with it have been extensively studied over the past half-century. In contrast, clot formation, which occurs in a matter of minutes, and its relationship to coronary heart disease have only recently received the attention of researchers and clinicians. Dietary fat has been implicated in both of these processes (Ulbricht and Southgate, 1991).

The marked difference in the incidence of coronary heart disease among Danes and Greenland Eskimos (Dyerberg, 1986) led to interest in the antithrombogenic effect of the n-3 polyunsaturated fatty acids. The long-chain n-3 polyunsaturated fatty acids of fish oils, namely eicosapentaenoic acid (20 : 5 n-3) and docosahexaenoic acid (22 : 6 n-3), have been shown to inhibit platelet aggregation (Herold and Kinsella, 1986). This effect is associated with changes in the fatty acid composition of platelet phospholipids and an accompanying change in prostanoid metabolism.

These developments have led to an interest in the possible role of linolenic acid (18 : 3 n-3) in this process. However, there is controversy surrounding the importance of dietary linolenic acid to platelet function (Kinsella, 1988). Although humans are able to convert linolenic acid to eicosapentaenoic acid, it appears they do not do so to any great extent (Sanders and Younger, 1981; Lasserre *et al.*, 1985). Nevertheless, Renaud *et al.* (1986a, b) found an inverse relationship between dietary linolenic acid and platelet aggregation in humans.

The ingestion of canola oil has been

shown to alter the fatty acid composition of plasma and platelet phospholipids (Renaud *et al.*, 1986a; Corner *et al.*, 1990; Weaver *et al.*, 1990; Kwon *et al.*, 1991; Mutanen *et al.*, 1992; Chan *et al.*, 1993). In general, phospholipid fatty acid composition reflects the fatty acid composition of the diet; canola oil results in higher levels of oleic acid (18 : 1 n-9) whereas soyabean oil, sunflower oil and safflower oil result in higher levels of linoleic (18 : 2 n-6). Canola and soyabean oil also result in higher levels of linolenic acid (18 : 3 n-3) although very little linolenic acid is incorporated into phospholipids. Even when appreciable levels of linolenic acid were incorporated into the diet from flax (linseed) oil, the level of linolenic acid in plasma and platelet phospholipids did not exceed 1% of the total fatty acids (Sanders and Younger, 1981; Chan *et al.*, 1993). On the other hand, dietary linolenic acid has been found to alter the long-chain polyunsaturated fatty acid composition of phospholipids. Several groups of workers (Renaud *et al.*, 1986a; Weaver *et al.*, 1990; Kwon *et al.*, 1991; Mutanen *et al.*, 1992) reported lower levels of arachidonic acid (20 : 4 n-6) with canola oil diets. In addition, canola oil diets resulted in higher levels of long-chain n-3 polyunsaturated fatty acid, in particular eicosapentaenoic acid (20 : 5 n-3) and docosapentaenoic acid (22 : 5 n-3), than sunflower oil, soyabean oil or customary mixed fat diets (Renaud *et al.*, 1986a; Corner *et al.*, 1990; Weaver *et al.*, 1990; Chan *et al.*, 1993). However, not all studies showed lower levels of arachidonic acid (Corner *et al.*, 1990; Chan *et al.*, 1993) or higher levels of eicosapentaenoic acid or docosapentaenoic acid (Kwon *et al.*, 1991; Mutanen *et al.*, 1992) with canola oil diets. Part of the explanation for apparent differences among studies may relate to the balance among dietary fatty acids. Chan *et al.* (1993) found that changes in fatty acid composition of plasma and platelet phospholipids varied not only

with the level of linolenic acid in the diet but its ratio to linoleic acid (i.e. 18 : 2/ 18 : 3).

Ingestion of canola oil also resulted in a decrease in the *in vitro* aggregation of platelets (Renaud *et al.*, 1986a; Kwon *et al.*, 1991) although similar effects were observed on diets containing sunflower oil and safflower oil (i.e. low levels of linolenic acid and relatively high levels of linoleic acid). Similarly, McDonald *et al.* (1989) found both canola oil and sunflower oil tended to increase clotting time and prostacyclin production (an anti-aggregating substance) and decrease thromboxane production (a pro-aggregating substance). On the other hand, Mutanen *et al.* (1992) found canola oil and sunflower oil had a similar but opposite effect on platelet function; both increased *in vitro* platelet aggregation. There is no obvious explanation for the marked differences among these studies. Possible factors could include diets of subjects prior to the study, the length of the study period, the parameter measured (e.g. aggregation versus thromboxane production), or the type and amount of agonist (aggregating agent) used in the *in vitro* assays. Although the effect of canola oil on platelet activity and clot formation is not as well established as its favourable effect on plasma cholesterol and lipoprotein levels, there is, nonetheless, evidence that it may impede thrombus formation.

DIETARY FAT AND THE OXIDATIVE STABILITY OF LOW-DENSITY LIPOPROTEINS

Fatty streaks, which are characterized by a build-up of lipids in the intima of large vessels, are generally considered an early lesion of atherosclerosis. These lipids are derived primarily from LDL. There is appreciable evidence that the uptake of LDL and formation of fatty streaks is enhanced by the oxidation of the

LDL fraction (Steinberg *et al.*, 1989; Parathasarathy and Rankin, 1992). In addition, there is growing evidence that dietary monounsaturated fatty acids protect against LDL oxidation. Low-density lipoproteins were found to be appreciably more stable to oxidation when subjects were fed diets rich in oleic acid than when fed linoleate-enriched diets (Reaven *et al.*, 1991; Abbey *et al.*, 1993). Thus canola oil might be expected to improve the oxidative stability of LDL because of its relatively high content of oleic acid although the relatively high level of linolenic acid could compromise the protective effect of oleic acid. However, the fact that very little linolenic acid is incorporated into plasma lipids would negate the latter argument. Preliminary results (B.E. McDonald and V.M. Bruce, 1994, unpublished data) showed an appreciably higher rate of conjugated-diene production (a measure of oxidative susceptibility) for LDL from subjects fed a sunflower oil diet than those fed mixed fat, canola oil or low linolenic acid canola oil diets.

CANOLA OIL IN INFANT FORMULAS

Recognition of the functional importance of arachidonic acid (20 : 40 n-6) and docosahexaenoic acid (22 : 6 n-3) in the structural lipids of the central nervous system has generated appreciable interest in the fat composition of infant formulas. Infant formulas differ considerably in fatty acid composition from human breast milk, which is characterized by relatively high levels of palmitic acid (16 : 0) and oleic acid (18 : 1 n-9). In addition, breast milk contains small, but some would claim important (Van Aerde and Clandinin, 1993), levels of long-chain polyunsaturated fatty acids (viz. arachidonic acid and docosahexaenoic acid). Although the infant is capable of desaturating and elongating linoleic acid and linolenic acid to the

longer-chain homologues, the efficiency and effectiveness of this process are only now being clarified. Since the same pathway is involved in the desaturation–elongation of linoleic acid and linolenic acid (Fig. 13.1), there is competition for the enzymes in the pathway. In addition, there are differences in the rate of incorporation of linoleic acid and linolenic acid into tissue lipids and in their rate of utilization for energy. Since the factors regulating the partition among these pathways are not understood (Innis, 1993) and, since breast milk contains small amounts of both n-6 and n-3 long-chain polyunsaturated fatty acids, there is some question of the appropriateness of the n-6 : n-3 ratio of between 4 : 1 and 10 : 1 recommended for infants (Neuringer and Connor, 1986; Health and Welfare Canada, 1990) even though it is based on ratios found in human milk. Thus, there has been interest in the use of canola oil in the formulation of infant formulas because of the balance between linoleic acid and linolenic acid (18 : 2/18 : 3 ratio approximately 2.5 : 1) (Arbuckle and Innis, 1992; Rioux and Innis, 1992; Innis *et al.*, 1993). Although inclusion of canola oil in infant formulas holds potential nutritional benefits, there is hesitation on the part of regulatory agencies to approve its use because of the lingering concern with the myocardial lipidosis observed in weanling rats fed high erucic acid rapeseed oil (the progenitor of canola).

SUMMARY

Canola oil (low erucic acid rapeseed oil) is characterized by a low level of saturated fatty acids (less than 4% palmitic acid) and relatively high levels of oleic acid (55–60%) and linolenic acid (8–10%). Studies in several countries have confirmed the effectiveness of canola oil in reducing plasma cholesterol levels. It has been found equally as effective as sunflower oil, safflower oil

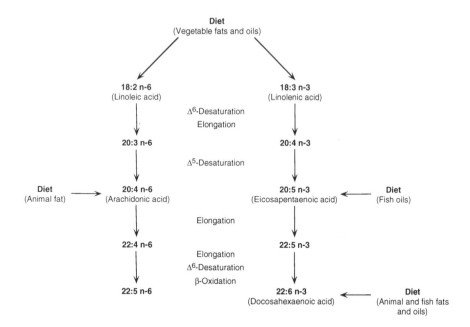

Fig. 13.1. Dietary sources and the desaturation-elongation pathway of n-6 and n-3 polyunsaturated fatty acids.

and soyabean oil in reducing total plasma and LDL cholesterol levels when each replaced saturated fat in the diet. There is also evidence that platelet activity and thrombogenesis are altered on canola oil diets although the research supporting these observations is not as convincing as the hypocholesterolaemic effect. Indications that oxidation of the lipid in the LDL fraction renders it more atherogenic and that high intakes of oleic acid increase the oxidative stability of LDL suggest a nutritional benefit of canola oil over linoleic acid-rich fats and oils. The relatively high level of oleic acid and the low ratio of linoleic/linolenic acid in canola oil have created interest in its use in infant formulas. Interest in canola oil continues to mount as research continues to establish its nutritional properties for humans.

REFERENCES

Abbey, M., Belling, G.B., Noakes, M., Hirata, F. and Nestel, P. (1993) Oxidation of low-density lipoproteins: intraindividual variability and the effect of dietary linoleate supplementation. *American Journal of Clinical Nutrition* 57, 391–398.

Ackman, R.G. (1990) Canola fatty acids – an ideal mixture for health, nutrition, and food use. In: Shahidi, F. (ed.) *Canola and Rapeseed: Production, Chemistry, Nutrition and Processing Technology.* Van Nostrand Reinhold, New York, pp. 81–98.

Arbuckle, L.D. and Innis, S.M. (1992) Docosahexaenoic acid in developing brain and retina of piglets fed high or low α-linolenate formula with and without fish oil. *Lipids* 27, 89–93.

Barr, S.L., Ramakrishnan, R., Johnson, C., Hollerman, S., Dell, R.B. and Ginsberg, H.N. (1992) Reducing total dietary fat without reducing saturated fatty acids does not significantly lower plasma cholesterol concentration in normal males. *American Journal of Clinical Nutrition* 55, 675–681.

Bierenbaum, M.L., Reichstein, R.P., Watkins, T.R., Maginnis, W.P. and Geller, M. (1991) Effects of canola oil on serum lipids in humans. *Journal of the American College of Nutrition* 10, 228–233.

Chan, J.K., Bruce, V.M. and McDonald, B.E. (1991) Dietary α-linolenic acid is as effective as oleic acid and linoleic acid in lowering blood cholesterol in normolipidemic men. *American Journal of Clinical Nutrition* 53, 1230–1240.

Chan, J.K., McDonald, B.E., Gerrard, J.M., Bruce, V.M., Weaver, B.J. and Holub, B.J. (1993) Effect of dietary α-linolenic acid and its ratio to linoleic acid on platelet and plasma fatty acids and thrombogenesis. *Lipids* 28, 811–817.

Consensus Conference (1985) Lowering blood cholesterol to prevent heart disease. *Journal of the American Medical Association* 253, 2080–2086.

Corner, E.J., Bruce, V.M. and McDonald, B.E. (1990) Accumulation of eicosapentaenoic acid in plasma phospholipids of subjects fed canola oil. *Lipids* 25, 598–601.

Dyerberg, J. (1986) Linolenate-derived polyunsaturated fatty acids and prevention of atherosclerosis. *Nutrition Reviews* 44, 125–134.

Grundy, S.M., Florentin, L., Nix, D. and Whelan, M.F. (1988) Comparison of monounsaturated fatty acids and carbohydrates for reducing raised levels of plasma cholesterol in man. *American Journal of Clinical Nutrition* 47, 965–969.

Health and Welfare Canada (1990) Nutrition recommendations. In: *Report of the Scientific Review Committee.* Minister of National Health and Welfare, Ottawa, pp. 40–52.

Hegsted, D.M., McGandy, R.B., Myers, M.L. and Stare, F.J. (1965) Quantitative effects of dietary fat on serum cholesterol in man. *American Journal of Clinical Nutrition* 17, 281–295.

Herold, P.M. and Kinsella, J.E. (1986) Fish oil consumption and decreased risk of cardiovascular disease: a comparison of findings from animal and human feeding trials. *American Journal of Clinical Nutrition* 43, 556–598.

Innis, S.M. (1993) Essential fatty acid requirements in human nutrition. *Canadian Journal of Physiology and Pharmacology* 71, 699–706.

Innis, S.M., Quinlan, P. and Diersen-Schade, D. (1993) Saturated fatty acid chain length and positional distribution in infant formula: effects on growth and plasma lipids and ketones in piglets. *American Journal of Clinical Nutrition* 57, 382–390.

Keys, A., Anderson, J.T. and Grande, F. (1965) Serum cholesterol response to changes in the diet. IV. Particular saturated fatty acids in the diet. *Metabolism* 14, 776–787.

Kinsella, J.E. (1988) Food lipids and fatty acids: importance in food quality, nutrition, and health. *Food Technology* 42(10), 126–142.

Kramer, J.K.G., Sauer, F.D. and Pidgeon, W.J. (1983) *High and Low Erucic Acid Rapeseed Oils: Production, Usage, Chemistry, and Toxicological Evaluation.* Academic Press, New York, 582 pp.

Kwon, J.-S., Snook, J.T., Wardlaw, G.M. and Hwang, D.H. (1991) Effects of diets high in saturated fatty acids, canola oil, or safflower oil on platelet function, thromboxane B_2 formation, and fatty acid composition of platelet phospholipids. *American Journal of Clinical Nutrition* 54, 351–358.

Lasserre, M., Mendy, F., Spielmann, D. and Jacotot, B. (1985) Effects of different dietary intake of essential fatty acids on C20 : 3ωm6 and C20 : 4ω6 serum levels in human adults. *Lipids* 20, 227–233.

McDonald, B.E. (1983) Studies with high and low erucic acid rapeseed oil in man. In: Kramer, J.K.G., Sauer, F.D. and Pidgeon, W.J. (eds) *High and Low Erucic Acid Rapeseed Oils: Production, Usage, Chemistry, and Toxicological Evaluation.* Academic Press, New York, pp. 535–549.

McDonald, B.E., Bruce, V.M., LeBlanc, E.L. and King, D.J. (1974) Effect of rapeseed oil on serum lipid patterns and blood parameters of young men. *Proceedings of the Fourth International Rapeseed Congress*, Giessen, Germany. Deutsche Gesellschaft für Fettwissenschaft, Münster, pp. 693–700.

McDonald, B.E., Gerrard, J.M., Bruce, V.M. and Corner, E.J. (1989) Comparison of the effect of canola oil and sunflower oil on plasma lipids and lipoproteins and on *in vivo* thromboxane A_2 and prostacyclin production in healthy young men. *American Journal of Clinical Nutrition* 50, 1382–1388.

McNamara, D.J. (1990) Coronary heart disease. In: Brown, M.L. (ed.) *Present Knowledge in Nutrition*, 6th edn. International Life Sciences Institute Nutrition Foundation, Washington, pp. 349–354.

Mattson, F.H. and Grundy, S.M. (1985) Comparison of effects of dietary saturated, monounsaturated, and polyunsaturated fatty acids on plasma lipids and lipoproteins in man. *Journal of Lipid Research* 26, 194–202.

Mensink, R.P. and Katan, M.B. (1987) Effect of monounsaturated fatty acids versus complex carbohydrates on high-density lipoproteins in healthy men and women. *Lancet* i, 122–125.

Mensink, R.P. and Katan, M.B. (1989) Effect of a diet enriched with monounsaturated or polyunsaturated fatty acids on level of low-density lipoprotein cholesterol in healthy women and men. *New England Journal of Medicine* 321, 436–441.

Miettinen, T.A. and Vanhanen, H. (1994) Serum concentration and metabolism of cholesterol during rapeseed oil and squalene feeding. *American Journal of Clinical Nutrition* 59, 356–363.

Mutanen, M., Freese, R., Valsta, M.L., Ahola, I. and Ahlström, A. (1992) Rapeseed oil and sunflower oil diets enhance platelet *in vitro* aggregation and thromboxane production in healthy men when compared with milk fat or habitual diets. *Thrombosis and Haemostasis* 67, 352–356.

Neuringer, M. and Connor, W.E. (1986) n-3 Fatty acids in the brain and retina: evidence for their essentiality. *Nutrition Reviews* 44, 285–294.

Parathasarathy, S. and Rankin, S.M. (1992) Role of oxidized low density lipoprotein in atherogenesis. *Progress in Lipid Research* 31, 127–143.

Reaven, P., Parthasarathy, S., Grasse, B.J., Miller, E., Almazan, F., Mattson, F.H., Khoo, J.C., Steinberg, D. and Witzum, J.L. (1991) Feasibility of using an oleate-rich diet to reduce the susceptibility of low-density lipoprotein to oxidative modification in humans. *American Journal of Clinical Nutrition* 54, 701–706.

Renaud, S., Godsey, F., Dumont, E., Thevenon, C., Ortchanian, E. and Martin, J.L. (1986a) Influence of long-term diet modification on platelet function and composition in Moselle farmers. *American Journal of Clinical Nutrition* 43, 136–150.

Renaud, S., Morazain, R., Godsey, F., Dumont, E., Thevenon, C., Martin, J.L. and Mendy, F. (1986b) Nutrients, platelet function and composition in nine groups of French and British farmers. *Atherosclerosis* 60, 37–48.

Rioux, F.M. and Innis, S.M. (1992) Arachidonic acid concentrations in plasma and liver phospholipid and cholesterol esters of piglets raised on formulas with different linoleic and linolenic acid contents. *American Journal of Clinical Nutrition* 56, 106–112.

Sanders, T.A.B. and Younger, K.M. (1981) The effect of dietary supplements of ω3 polyunsaturated fatty acids on the fatty acid composition of platelets and plasma choline phosphoglycerides. *British Journal of Nutrition* 45, 613–616.

Seppanen-Laakso, T., Vanhanen, H., Laakso, I., Kohtamaki, H. and Viikari, J. (1992) Replacement of butter on bread by rapeseed oil and rapeseed oil-containing margarine: effects on plasma fatty composition and serum cholesterol. *British Journal of Nutrition* 68, 639–654.

Steinberg, D., Parthasarathy, S., Carew, T.E., Khoo, J.C. and Witztum, J.L. (1989) Modifications of low-density lipoprotein that increases its atherogenicity. *New England Journal of Medicine* 320, 915–924.

Truswell, A.S., Choudhury, N. and Roberts, D.C.K. (1992) Double blind comparison of plasma lipids in healthy subjects eating potato crisps fried in palmolein or canola oil. *Nutrition Research* 12, S43–S52.

Ulbricht, T.L.V. and Southgate, D.A.T. (1991) Coronary heart disease: seven dietary factors. *Lancet* 338, 985–992.

Valsta, L.M., Jauhianinen, M., Aro, A., Katan, M.B. and Mutanen, M. (1992) Effects of a monounsaturated rapeseed oil and a polyunsaturated sunflower oil diet on lipoprotein levels in humans. *Arteriosclerosis and Thrombosis* 12, 50–57.

Van Aerde, J.E. and Clandinin, M.T. (1993) Controversy in fatty acid balance. *Canadian Journal of Physiology and Pharmacology* 71, 707–712.

Wardlaw, G.M., Snook, J.T., Lin, M.-C., Puangco, M.A. and Kwon, J.S. (1991) Serum lipid and apolipoprotein concentrations in healthy men on diets enriched in either canola oil or safflower oil. *American Journal of Clinical Nutrition* 54, 104–110.

Weaver, B.M., Corner, E.J., Bruce, V.M., McDonald, B.E. and Holub, B.J. (1990) Dietary canola oil: effect on the accumulation of eicosapentaenoic acid in the alkenylacyl fraction of human platelet ethanolamine phosphoglyceride. *American Journal of Clinical Nutrition* 51, 594–598.

14 Meal and By-product Utilization in Animal Nutrition

J.M. Bell
University of Saskatchewan, Saskatoon, Canada

INTRODUCTION

Meal is the major by-product resulting from the extraction of oil from *Brassica* oilseeds. For rapeseed, *Brassica napus* and *Brassica rapa*, it represents about 60% of the original weight of the seed, contains 36 to 44% crude protein and a good balance of essential amino acids. But it is high in fibre content which results in a relatively low level of available energy when used in animal feeds. Since the development, through plant breeding, of low glucosinolate cultivars, rapeseed meal has been widely used as a protein supplement for animals. Such meal, because of its superior nutritional quality, has been designated as 'canola' in Canada. In this chapter research on low glucosinolate rapeseed meal is emphasized.

In addition to the meal, certain other by-products of rapeseed crushing and oil refining are useful in animal feeding and are sometimes blended with the meal at the crushing plant. These include gums (containing lecithin), acidulated fatty acids and spent bleaching clay. Cleaning of the seed prior to crushing yields screenings (small and cracked seed, and foreign matter) which also have feed value, and there is interest in using whole seed and rapeseed oil in animal feed formulations.

Some species of mustard, mainly brown or Oriental mustard (*Brassica juncea*), are processed to obtain their oil and as with rapeseed a meal residue is available for animal feed use. While rapeseed and mustard meals are similar in some respects the latter contain higher levels and different kinds of glucosinolates.

COMPOSITION AND CHARACTERISTICS OF RAPESEED MEAL

Nutrient content

Rapeseed meal consists of two major components, oil-extracted embryos and hulls. Hulls represent about 16% of the original seed and about 30% of the meal weight. Compared with the embryo fraction, hulls contain much less protein and more fibre, which is composed largely of cellulose, pentosans and lignin. The digestibility of both protein and fibre in hulls is very low for simple-stomached animals. Since hulls

Table 14.1. Chemical components of low glucosinolate rapeseed meal.

Component	Percentage composition		
	Combined data 1978–1983[1]	Bourdon and Aumaître (1990)	Bell and Keith (1991)
Moisture	9.7	8.1	8.5
Crude protein (N × 6.25)	39.8	39.1	38.3
Ether extract	4.2	1.9	3.4
Ash	7.3	7.7	nd
Crude fibre	13.1	12.7	12.0
Neutral detergent fibre	30.5	28.2	21.5
Acid detergent fibre	23.3	20.5	17.5

nd, Not determined.
[1] Data from 11 reports summarized by Bourdon and Aumaître (1990).

contain over 10% of the total protein and 25% of the gross energy of rapeseed meal, the presence of hulls in the meal depresses the levels of available energy and protein as well as amino acids and minerals. As a consequence, methods of hull reduction and removal are of interest as a means of improving the nutritional value of the meal.

Rapeseed meal is classed as a protein supplement in the feed trade and, for the most part, competes with soyabean meal. Typical rapeseed meal contains 38–40% crude protein (Table 14.1), 2–4% residual ether extract (depending on the efficiency of oil extraction and whether oil-containing refinery by-products are added back to the meal), 12–13% crude fibre and 7–8% ash. Neutral detergent fibre represents 20–30% of the weight of meal.

Variation in chemical composition of rapeseed meal, as with many other feedstuffs, occurs as a result of species of *Brassica*, cultivar, processing conditions in the crushing plant, soil type, weather and other environmental factors. These factors and their effects remain poorly understood and are often difficult to manage or control but they have significant effects on meal quality. For instance, over a 10-year period in western Canada the mean annual crude

protein levels of oil-free seed varied from 36 to 41%, and within a single year a range of 34–47% has been encountered among regions (DeClerq *et al.*, 1989).

The levels of essential amino acids indicate that rapeseed protein is generally of high nutritional quality (Table 14.2). Tests with rapeseed protein concentrate fed to rats have yielded protein efficiency ratios exceeding those of soyabean meal. However, for more meaningful evaluations of protein quality in commercially processed rapeseed meal, computerized linear programming methods, using levels of *available* amino acids, are preferred to protein efficiency ratios (Bell, 1993a). Another useful approach is to compare the levels of essential amino acids expressed as percentages of the crude protein, since protein supplements are fed primarily to supply amino acids. Rapeseed protein tends to contain less lysine than soyabean protein but more of the sulphur amino acids methionine and cystine (Table 14.2). Lysine and sulphur amino acids, in this order, are the amino acids most often deficient in cereal meals. Rapeseed protein is also relatively high in threonine.

Amino acid levels in rapeseed meal vary in proportion to the level of protein which may in turn differ considerably, as

Table 14.2. Amino acid content of rapeseed meal and rapeseed protein.

Component	Percentage in meal[1]		Percentage amino acid in crude protein		
	Bourdon and Aumaître (1990)	Bell and Keith (1991)	Rapeseed meal Bourdon and Aumaître (1990)	Rapeseed meal Bell and Keith (1991)	Soyabean meal[2]
Arginine	2.27	2.38	6.10	6.21	6.44
Histidine	1.12	1.41	3.01	3.68	2.40
Isoleucine	1.53	1.79	4.11	4.67	4.69
Leucine	2.57	2.78	6.91	7.26	7.49
Lysine	2.12	2.28	5.70	5.95	6.22
Methionine	0.74	0.79	1.99	2.06	1.40
Cystine	1.01	1.10	2.72	2.87	1.44
Phenylalanine	1.49	1.56	4.01	4.07	4.80
Tyrosine	1.12	1.16	3.01	3.03	3.53
Threonine	1.64	1.90	4.41	4.96	3.80
Tryptophan	–	0.48	–	1.25	1.20
Valine	2.00	2.29	5.38	5.98	5.00
Crude protein (N × 6.25)	37.2	38.3	–	–	–

–, Not reported.
[1] Adjusted to 8.5% moisture.
[2] Soyabean meal, International Feed No. 5-04-612. Adapted from Anon. (1971).

indicated previously, but the relationships are not exactly parallel. Levels may be predicted using the following formulas: lysine, % = % crude protein × 0.0402 + 0.546 ($r = 0.57$), methionine + cystine, % = % crude protein × 0.0468 – 0.033 ($r = 0.64$) (Beste *et al.*, 1992). In addition to these relationships between protein and amino acid levels, differences have been found among amino acid profiles of rapeseed protein obtained from different geographical regions (Bell and Keith, 1991), suggesting varying proportions of globulins (storage proteins) and albumins (metabolically active proteins) among regions. The composition of storage proteins in terms of the type of protein and polypeptides is genetically controlled but the proportions may vary due to environmental effects (Norton, 1989). Yet another aspect in describing the nitrogenous components of rapeseed is the factor used to convert Kjeldahl nitrogen content to protein. The

conventional value of 6.25 assumes that all proteins contain 16% N, but proteins of oilseeds and cereals contain more than 16% N. Factors of 5.53 for rapeseed and 5.69 for soyabean give truer values for protein content (Tkachuk, 1969). The difference between soyabean and rapeseed protein factors reflects different amino acid profiles and differing amounts of non-protein nitrogen. The dilemma over choice of a factor is partly resolved by observing that amino acids are central to non-ruminant nutrition and that ruminants are less dependent on specific arrays of amino acids and can utilize non-protein nitrogen.

Hull reduction or removal is seen as a means to improving the nutritional value of rapeseed meal. Dehulled rapeseed meal containing 43–45% crude protein was shown to have a digestibility of 81–86% for pigs compared with about 76% digestibility for regular meal (Anon., 1988; Bell, 1993b). However, hulls contain protein

Table 14.3. Amino acid content of the crude protein fraction from meal, dehulled meal and hulls of rapeseed.

Amino acid	Percentage amino acid in crude protein[1]				
	Bell (1993a)			Bourdon and Aumaître (1990)[2]	
	Meal	Dehulled meal	Hulls	Meal	Dehulled meal
Arginine	6.13	7.10	4.90	6.09	6.29
Histidine	3.09	3.57	3.06	3.00	2.89
Isoleucine	4.06	4.50	4.50	4.10	4.09
Leucine	6.45	7.26	5.71	6.90	7.09
Lysine	5.28	4.96	9.36	5.70	5.29
Methionine	1.94	2.00	1.53	1.99	2.00
Cystine	3.70	3.63	3.10	2.70	2.60
Phenylalanine	4.17	4.40	3.78	4.00	4.00
Tyrosine	3.11	2.95	3.28	3.00	2.89
Threonine	4.22	4.70	5.08	4.39	4.29
Tryptophan	1.20	1.27	2.52	–	–
Valine	4.75	4.98	5.62	5.38	5.29

–, Not reported.
[1]Protein = N × 6.25.
[2]Values converted from amino acid composition of low glucosinolate meal.

that is nearly indigestible by pigs (Bell and Shires, 1982) and their removal alters the amino acid characteristics of the meal (Table 14.3). In particular, the protein of the hull fraction of rapeseed meal contains more lysine than in the protein of the embryo fraction.

As with cereal grains, seasonal and regional effects have been shown to affect mineral content of rapeseed meal. Significant differences were observed among canola meals obtained from seven different crushing plants in western Canada for 11 minerals essential for animal nutrition (Bell and Keith, 1991). Causes of the differences could not be identified but might include regional, varietal and environmental factors. Coefficients of variation ranged from 1.4 to 4.1 with the largest values associated with iron, selenium, sulphur and molybdenum. The mineral content of rapeseed meal (Table14.4) generally exceeds that of soyabean meal, especially with regard to phosphorus, calcium, magnesium, selenium and manganese which are about

twice as great. The sulphur content is also higher in low glucosinolate rapeseed meal than in soyabean meal but this varies with glucosinolate level, 1.72 and 0.77% sulphur having been observed in high and very low glucosinolate cultivars grown in a comparative test (Josefsson, 1971).

Few studies have been reported on the vitamin content of rapeseed meal (Table 14.4) but, as with some of the minerals, most levels exceed those of soyabean meal. Choline content is relatively high, some of it possibly being present as sinapine, the ester of choline and sinapic acid.

Carbohydrate and fibre components

Carbohydrates comprise nearly one half of the dry matter of rapeseed meal, have an important bearing on its nutritional value, and thus have received attention from a number of researchers (Norton and Harris, 1975; Theander and Åman, 1976; Siddiqui and Wood, 1977; Theander *et al.*,

14.4. Mineral and vitamin content of rapeseed and soyabean meal.

Component	Units	Rapeseed meal[1,2]	Soyabean meal[1,3]
Mineral			
Calcium	%	0.64	0.3
Phosphorus	%	1.03	0.65
Potassium	%	1.24	2.11
Sulphur	%	0.86	0.42
Magnesium	%	0.52	0.29
Copper	mg kg^{-1}	5.8	23
Iron	mg kg^{-1}	144	140
Manganese	mg kg^{-1}	50	31
Molybdenum	mg kg^{-1}	1.4	
Zinc	mg kg^{-1}	69	52
Selenium	mg kg^{-1}	1.1	0.1
Vitamin[4]			
Choline	%	0.67	0.26
Biotin	mg kg^{-1}	1.1	0.32
Folic acid	mg kg^{-1}	2.3	0.6
Niacin	mg kg^{-1}	160	28
Pantothenic acid	mg kg^{-1}	9.5	16.3
Riboflavin	mg kg^{-1}	5.8	2.9
Thiamine	mg kg^{-1}	5.2	6
Pyridoxine	mg kg^{-1}	7.2	6
Vitamin E	mg kg^{-1}	14	2.4

[1] 8.5% moisture basis.
[2] Low glucosinolate meal. Bell and Keith (1991).
[3] Soyabean meal, International Feed No. 5-04-612. Anon. (1971).
[4] Bell (1993c).

1977; Vohra, 1989; Naczk and Shahidi, 1990; Slominski and Campbell, 1991). Up to 3% starch, 10% sucrose and 5% cellulose occur in rapeseed meal. Starch occurs at relatively high levels in immature seed but declines as the seed matures. Starch and sucrose are readily digested by simple-stomach animals once released from the cells, otherwise their utilization may depend on microbial action in the lower gut. In this regard chemical methods grossly overestimate the amount of available carbohydrates in rapeseed meal compared with chick assay methods (13–14% compared with 4–6%) (Rao and Clandinin, 1972).

Cellulose at 5% of meal is a minor component of the fibre complex. Neutral detergent fibre in canola meal is about 26% and total dietary fibre about 33% (Bell, 1993a). According to Slominski and Campbell (1990, 1991) over 90% of the non-starch carbohydrates are insoluble (in water) and upon hydrolysis yield fractions dominated by arabinose ($C_5H_{10}O_5$) and uronic acids ($C_6H_{10}O_7$) (Table 14.5). While free pentoses can be metabolized to some degree by animals, these and other products cannot be released in the digestive system from their parent polysaccharides without the action of microflora which transforms them into volatile fatty acids. However Dierick et al. (1987) concluded that volatile fatty acids arising from non-starch polysaccharides make an insignificant net contribution to the metabolizable energy intake of the growing pig. It may be assumed that volatile fatty acids arising from rapeseed meal are also of limited value in poultry nutrition, in contrast to their great importance in ruminant nutrition. However, our understanding of the nature and role of the fibre components of rapeseed meal in the diets of animals remains incomplete.

The soluble and insoluble dietary fibre fractions of seed, hulls and dehulled meal from low glucosinolate rapeseed of partly yellow-seeded cultivars were compared to brown-seeded cultivars with respect to their physiological and antinutritional effects on rats by Bjergegaard et al. (1991). Seed or meal from brown cultivars had higher ratios of insoluble to soluble fibre. Protein digestibility was depressed by fibre-associated nitrogen. Soluble fibre promoted microbial growth resulting in increased microbial nitrogen excretion.

Glucosinolates

Since the discovery over 150 years ago of sinigrin in mustard (*B. juncea*) more than 100 glucosinolates have been identified in the plant kingdom, about six of which

Table 14.5. Carbohydrate components, and component carbohydrates of the non-starch polysaccharide fraction, of moisture-free canola meal.

Component	Percentage
Cellulose	4.9
Oligosaccharides[1]	2.5
Sucrose	7.7
Starch	2.5
Non-starch polysaccharides	17.9
Soluble	1.5
Insoluble	16.4
Non-starch polysaccharides components	
Rhamnose	0.2
Fucose	0.2
Arabinose	4.5
Xylose	1.6
Mannose	0.4
Galactose	1.7
Glucose[2]	5.0
Uronic acids	4.3

Source: Slominski and Campbell (1990, 1991).
[1] Excludes sucrose, includes raffinose and stachyose.
[2] Includes glucose from cellulose.

occur in appreciable amounts in the seed of rape (*B. napus* and *B. rapa*). Four are aliphatic glucosinolates, butenyl, pentenyl and their 2-hydroxy analogues, the other two indole glucosinolates, 3-indolylmethyl and 4-hydroxy-3-indolylmethyl glucosinolate (see Chapter 10). An enzyme, called myrosinase (thioglucosidase, EC 3.2.3.1), exists in separate compartments within the seed. When the seed is crushed in the presence of adequate moisture myrosinase hydrolyses the glucosinolates to release glucose and sulphate and various isothiocyanates and thiocyanates. Cyclization usually occurs with the isothiocyanates from the hydroxylated aliphatic glucosinolates to produce oxazolidinethiones. The latter were identified originally as the goitrogenic factor in rapeseed but other hydrolysis products are also goitrogenic or otherwise have toxic effects on animals fed the seed or seed products. Under some conditions, toxic nitriles may be formed. The chemistry and occurrence of glucosinolates in rapeseed were reviewed by Röbbelen and Thies (1980).

One of the reasons for cooking rapeseed prior to expeller or solvent extraction is to inactivate myrosinase before glucosinolate hydrolysis can occur. The ingestion of intact glucosinolates in diets devoid of myrosinase is relatively innocuous but the inclusion of supplementary myrosinase produces adverse effects in mice (Belzile *et al.*, 1963) and pigs (Bell, 1965). However, myrosinase of bacterial origin can hydrolyse glucosinolates within the gastrointestinal tract of the hen (Oginsky *et al.*, 1965; Slominski *et al.*, 1988) and intact glucosinolates have been detected in the blood and livers of roosters (*Gallus domesticus*, adult, male) administered in high single doses (Freig *et al.*, 1987). Sørensen (1990) doubts the existence of an effective mechanism of absorption of unhydrolysed glucosinolates, but appreciable passive transport across the intestinal wall of the rat occurs (Michaelsen *et al.*, 1994). Aside from the goitrogenic and other antinutritional properties of glucosinolates they are regarded as generally unpalatable and responsible for reduced feed intake with diets containing rapeseed meal (Lee *et al.*, 1984).

Animal species appear to differ in response to ingestion of glucosinolates, partly as a result of intestinal microflora activity. Nugon-Baudon *et al.* (1988) compared the responses of axenic (germ-free), conventional, chimeric rats and chicks fed diets containing low glucosinolate rapeseed meal. The intestinal tracts of chimeric animals (initially germ-free) were colonized by microflora from conventional animals of the other species. The axenic and chimeric rats grew more rapidly than the conventional rats. Liver and kidney weights were increased in conventional rats but not in chickens or chimeric rats. Chicken

Table 14.6. Glucosinolate content of rapeseed meals of different types of origin.

Meal type	Glucosinolate content[1] (μmol g^{-1} oil-extracted meal)	
	Mean	Range
High glucosinolate B. napus[2]	166	125–207
Low glucosinolate B. napus[3]	37	9–69
Low glucosinolate B. napus and B. rapa mixed[2]	16	6–28

[1] Excludes indole glucosinolates.
[2] Winter type produced in France. Bourdon and Aumaître (1990).
[3] Summer type produced in Canada. Bell and Keith (1991).

14.7. Effect of commercial processing on glucosinolate content of low glucosinolate (canola) meal.

Glucosinolate	Glucosinolate content (μmol g^{-1} oil-extracted meal)	
	Seed	Meal
3-Butenyl	7.44	4.97
4-Pentenyl	2.55	1.67
2-Hydroxy-3-butenyl	13.44	8.82
2-Hydroxy-4-pentenyl	0.99	0.74
Total	24.42	16.20
3-Indolylmethyl	0.63	0.38
4-Hydroxy-3-indolylmethyl	13.37	4.48
Total	14.00	4.86
Overall total, canola origin	38.42	21.06
Contaminant glucosinolates		
2-Propenyl (allyl)	1.41	1.05
4-Hydroxybenzyl	2.31	2.25

Source: Bell and Keith (1991).

microflora produced dramatic increases in thyroid weights in both chickens and chimeric rats.

The effects of glucosinolates on thyroid and liver morphology and function in pigs were investigated by Spiegel et al. (1992) who found that thyroid and liver weights increased as glucosinolate intake increased up to 4.6 μmol day^{-1} and decreased thereafter. Thyroxine blood serum levels decreased as glucosinolate intake level increased but changes in triiodothyronine were minor and pig growth rate showed little effect.

Since the advent of low glucosinolate cultivars in which the aliphatic glucosinolates have been reduced to 10–20% of their former levels (Table 14.6), rapeseed meal has been widely accepted for use in animal feeds. Meal containing less than 30 μmol g^{-1} has been designated canola in Canada, a trade name since adopted by several other countries to identify a markedly superior product.

Indole glucosinolates have been of less concern than the aliphatic glucosinolates and are not primarily goitrogenic (Bell, 1993a); nevertheless, in some countries it has been decided to monitor total glucosinolates to promote further reduction of all types. Establishing a maximum desirable level of glucosinolate in the meal for nutritional purposes is a controversial issue because many factors are involved. These include the types of glucosinolates, contaminant glucosinolates (Keith and Bell, 1991), the possibility of myrosinase being present in the diet (from certain weed seeds), possible intestinal microfloral action, animal species, age of animal, and level of meal to be used in the diet. In addition, it has been shown that processing of rapeseed in different crushing plants can result in a 15 to 77% reduction in glucosinolates compared with original seed levels (Table 14.7) (Bell and Keith, 1991). Some may be converted to nitriles (Slominski and Campbell, 1989), which may be more toxic. Most of the reduction in glucosinolate content during processing occurs in the desolventizer–toaster.

While the development of low gluco-sinolate cultivars has been a major improvement, nutritionists generally agree that further reduction in glucosinolate levels is desirable even though for most feed purposes the current levels are acceptable. Bjerg *et al.* (1987), citing research on four animal species, recommended a maximum dietary level of 2.5 $\mu mol\ g^{-1}$ and Campbell and Slominski (1991) suggested a maximum of 1.43 $\mu mol\ g^{-1}$ diet to minimize the liver haemorrhage problem in laying hens. A strain of *B. rapa* developed for its low glucosinolate content produced a meal containing 0.53–1.66 $\mu mol\ g^{-1}$ of total glucosinolates (Bell *et al.*, 1991); thus it is probable that cultivars of the future will approach total elimination of glucosinolates in rapeseed.

Other components

Tannins

Tannins are polyphenolic compounds having leukocyanidin as the basic unit and two types are recognized: hydrolysable and condensed (Leung *et al.*, 1979). Condensed tannins were identified in rapeseed hulls by Bate-Smith and Ribéreau-Gayon (1959). Rapeseed meal contains about 1.5% tannin. Tannin–protein complexes have been demonstrated to reduce the digestibility and nutritive value of tannin-containing feeds for both ruminant and non-ruminant animals. However, condensed tannins isolated from rapeseed hulls had no effect on *in vitro* alpha-amylase enzyme activity (Mitaru *et al.*, 1984) and extracted tannins from rapeseed meal added to soyabean meal-supplemented chick diets had no adverse effect on nitrogen absorption although the metabolizable energy value of the diet was reduced (Yapar and Clandinin, 1972). The crude protein of hulls from dark rapeseed has been shown to be indigestible for pigs but protein in hulls from yellow seed was 20% digestible, a finding that may be attribut-able to lower tannin and lignin levels in yellow-seeded hulls (Bell and Shires, 1982).

As reviewed by Shahidi and Naczk (1992), tannins may also have other antinutritional effects. Tannins may be involved with sinapine in the 'fishy egg' syndrome by inhibiting trimethylamine oxidase which converts trimethylamine to a water-soluble, odourless oxide. There is also evidence that tannins may form complexes with carbohydrates such as starch and they may impair absorption of certain minerals and vitamins in the digestive tract.

Phenolic compounds

Rapeseed meal contains a variety of phenolic compounds that exist in free, esterified and insoluble bound forms (Shahidi and Naczk, 1992). Free phenolic acids comprise about 15% of the total phenolics in meal, little or none of which exists in the hull fraction and most of which is sinapic acid. Esterified phenolic acids comprise about 80% of the total phenolics in meal, mostly in the embryo fraction. Sinapic acid is the major component of this group of compounds which includes sinapine. About 5% of the total phenolics in meal is made up of insoluble bound phenolics and sinapic acid is an important constituent although eight other phenolic acids have been identified in meals.

Sinapine is the major compound among several phenolic choline esters present in rapeseed meal (Larsen *et al.*, 1983). Levels of 0.7–3.0% have been reported in meal, 90% of which exists in the embryo (non-hull) fraction. Sinapine has a bitter flavour which may affect feed intake with diets containing rapeseed meal but Lee *et al.* (1984) found that glucosinolates had a greater adverse effect on palatability or feed intake level of pigs. Perhaps more important is the effect of sinapine on the production of off-flavour or 'fishy eggs' by susceptible hens which have insufficient trimethylamine oxidase. Such hens cannot effectively handle the high yield of choline following the hydrolysis of choline in the gut, resulting in a build-up of trimethylamine and its transfer to the developing egg (Fenwick *et al.*, 1979; Goh *et al.*, 1979). Apparently this phenomenon is restricted to certain strains of hens that lay brownshelled eggs.

The high choline content of rapeseed meal (Table 14.4) may include the choline content of sinapine. March and MacMillan (1980) found 4.65–5.72 g kg^{-1} of total choline and 1.66–2.15 g kg^{-1} of free choline in rapeseed meal, compared with 1.85 and 0.85 g kg^{-1}, respectively, for soyabean meal. Chick assays indicated that rapeseed total choline was about one half as available as choline from soyabean meal.

Phytates

Phytic acid (myoinositol hexaphosphoric acid, $C_6H_{18}O_{24}P_6$) is strongly negatively charged at the usual pH of feeds and therefore is very reactive with positively charged groups such as cations and proteins (Thompson, 1990). Depending on pH, phytic acid may form complexes with several mineral ions to form phytates of varying solubility, composition and stability. At pH 7.4 the order of decreasing stability is copper, zinc, nickel, cobalt, manganese, iron and calcium (Vohra *et al.*, 1965).

Phytic acid may account for 60–90% of the total phosphorus and exists mainly as salts of calcium, magnesium and potassium. Rapeseed meal contains 2–5% phytic acid (Nwokolo and Bragg, 1977), largely, if not entirely, located in the embryo and associated with protein bodies.

Mineral availability is adversely affected by ingestion of phytates by nonruminant animals. Availability of minerals for chicks was found to be: calcium, 60–76%; phosphorus, 65–81%; magnesium, 52–66%; manganese, 45–60%; copper, 62–85% and zinc, 23–58% (Nwokolo and Bragg, 1980). Mineral availability of

soyabean meal was not included in this study but the following literature values were cited: calcium, 85%; phosphorus, 89%; magnesium, 78%; manganese, 76%; copper, 51% and zinc, 66%. It was also found that crude fibre level was inversely related to mineral availability in rapeseed meal, as confirmed by Ward and Reichert (1986).

In studies with pigs and rats fed canola meal or soyabean meal in cereal-based diets zinc was the mineral affected most adversely, followed by iron, copper and manganese (Keith and Bell, 1987a, b). Availability of minerals for ruminants is apparently not impaired by phytates because of microbial phytases in the rumen (Reddy *et al.*, 1982). Phytic acid may form complexes with proteins in the food, with digestive enzymes or with proteins closely associated with starch (Thompson, 1990) but a reduced phytate level in rapeseed flour failed to improve *in vivo* protein digestibility (Thompson and Serraino, 1986).

Despite substantial literature on the nature and nutritional implications of phytates, no clear understanding has emerged. As mentioned, fibre and other diet components may also bind minerals. Phytates also vary in composition. At least two kinds of phytase exist and intestinal phosphatases may effect some phytate hydrolysis. Cereal grains with which rapeseed meal may be fed vary in phytase content and this phytase may hydrolyse some phytate in the upper digestive tract of pigs and chickens (Nelson, 1967; Ranhotra and Loewe, 1975). Some studies have involved feeding isolated phytate which may differ from natural phytate. Levels of dietary supplements of calcium and phosphorus may affect availability of minerals bound in phytate. Rapeseed meal apparently contains more phytate (3 to 5%) than soyabean meal (1.4%) (de Boland *et al.*, 1975; Nwokolo and Bragg, 1977; Uppström and Svensson, 1980). However, rapeseed meal contains higher levels of most minerals (Table 14.4) which may compensate for their somewhat lower availability. Further information on phytates may be obtained from the reviews by Erdman (1979), Maga (1982) and Thompson (1990).

Trypsin inhibitors

Trypsin or proteinase inhibitors are widely distributed in plant material (West and Norton, 1991; Visentin *et al.*, 1992). Levels found in rapeseed are relatively low and comprise two or three isoinhibitors that differ from those in other plants. Seed extracted with petroleum ether contains about 10 trypsin inhibitor units' activity g^{-1}. The component types differ in thermostability and there is some *in vitro* evidence that certain double-zero types of rapeseed (low erucic acid, low glucosinolate) may contain inhibitors that are more thermostable. There is no evidence that trypsin inhibitors adversely affect the feeding value of commercially prepared rapeseed meal but this matter has received little attention.

Saponins

Saponins at levels ranging from 0.62 to 2.85% were found in rapeseed meal but, since a 0.5% dietary level of reagent-grade saponin had no adverse effect on the metabolizable energy content of chick diets, it was concluded that nutrient utilization from rapeseed meal was not affected by its saponin content (March and Sadiq, 1974).

DIGESTIBILITY OF ENERGY AND PROTEIN

Knowledge of the availability of nutrients in a feedstuff is essential to an understanding of its value for animal feeding. Digestibility indicates how much of the ingested material can be absorbed into the bloodstream. Further refinements in the procedure may indicate what fraction of the absorbed nutrients is metabolizable and what enters tissue or product formation, such as meat or milk. The key fractions in rapeseed meal are energy and protein.

While rapeseed meal is classed as a protein supplement, its available energy is a critical factor in feed formulation. As with most feed ingredients no single energy value can be assigned to the meal because this is influenced by variability among samples, by processing of the seed, by the method used to assess energy content, by species and age of animal and possibly other factors. Conventional faecal collection methods may be used with pigs or ruminants to estimate digestible energy in megajoules (MJ) or kcal kg^{-1} of meal. Metabolizable energy may be obtained by also measuring gas and urinary nitrogen losses. Since poultry excrete urine and faeces together, it is customary to determine apparent metabolizable energy (AME) using a total faecal collection method. A more recent procedure yields true metabolizable energy (TME) by using roosters force-fed predetermined amounts of test meal in a single dose. Both apparent and true metabolizable energy may be further refined by correcting for protein stored or lost during the trial, yielding corrected apparent metabolizable energy (AME_n) and corrected true metabolizable energy (TME_n) values. Other variations of energy evaluation include use of ileal and re-entrant fistulas to avoid or measure the complicating effects of microflora in the lower gut and inserting nylon bags containing test samples into the rumen via a fistula and later retrieving the digested sample residues to measure rumen digestion. Various *in vitro* procedures have also been developed.

Methods for predicting digestibility of feeds were compared (Aufrere and Michalet-Doreau, 1988); relationships between bioavailable energy values for pigs and poultry have been developed (Sibbald, 1987; Sibbald *et al.*, 1990), new methods for estimating metabolizable and net

Table 14.8. Reported ranges in metabolizable energy of rapeseed (high glucosinolate) meal, canola (low glucosinolate) meal and dehulled low glucosinolate meal.

Test animal	Glucosinolate content[1]	Units[2]	Metabolizable energy Range (MJ kg^{-1})	Reference[3]
Swine[4]	High	ME	10.41–13.73	1, 2
	Low	ME	11.30–14.63	1, 3, 4, 5, 6, 7, 8
	Very low	ME	14.76	9
	Low dehulled	ME	12.49–16.06	10, 11
Poultry				
Hens	High	ME$_n$	7.50–9.35	11, 12
Roosters	High	TME$_n$	8.31–10.60	13
Chicks	Low	AME$_n$	7.41–10.90	14, 15
Roosters	Low	TME$_n$	8.60–9.51	13, 15, 16
Chicks	Low dehulled	ME$_n$	8.79–9.96	17
Ruminants	High and low	ME	11.7–13.4	18, 19, 20
	Low	NE$_l$	7.2–8.3	21, 22

[1] High: >30 μmol g^{-1}; low and low dehulled: <30 μmol g^{-1}; very low: <6 μmol g^{-1} aliphatic glucosinolates.

[2] ME = digestible energy (DE) minus energy losses in urine and gases; ME$_n$ = ME adjusted for body tissue energy gain or loss; AME$_n$ = gross energy corrected for faecal and urine energy losses; TME$_n$ = ME$_n$ corrected for metabolic faecal energy loss; NE$_l$ = net energy for lactation = MJ kg^{-1} in milk produced from 1 kg meal fed in an adequate diet to a lactating cow.

[3] 1, Rundgren (1983); 2, May and Bell (1971); 3, Just et al. (1983); 4, Thomke (1984); 5, Noblet et al. (1989); 6, Agunbiade et al. (1991); 7, Bell and Keith (1991); 8, Bourdon and Aumaître (1990); 9, Bell et al. (1991); 10, Bell (1993b); 11, Rundgren et al. (1985); 12, Askbrant and Håkannson (1984); 13, Sibbald (1986); 14, Clandinin and Robblee (1983a); 15, March et al. (1973); 16, Muztar et al. (1978); 17, Shires et al. (1983); 18, Jarl (1951); 19, Sharma et al. (1980); 20, Bush et al. (1978); 21, Anon. (1981); 22, National Research Council (1989).

[4] Swine results reported as digestible energy were multiplied by 0.93 to convert to metabolizable energy (Bourdon and Aumaître, 1990). When standard deviations rather than ranges were reported the ranges were estimated to be $\bar{Y} \pm 2$ s.d.

energy values for pigs have been proposed and applied to a variety of feedstuffs including rapeseed meal (Noblet et al., 1989, 1993).

Typical metabolizable energy values for rapeseed meal used in feed formulation are 8.7 for broiler chicks, 9.2 for adult hens, 13.2 for growing pigs and 12.1 MJ kg^{-1} for cattle (Table 14.8) but there is appreciable variation among the reported values. Some of this variability is of animal origin.

As indicated in Table 14.8 the metabolizable energy values for adult hens are about 10% greater than those for chicks. Up to 2.1 percentage units' increase in energy digestibility per 10 kg increase in body weight of pigs starting at 25 kg has been reported (Bell and Keith, 1988). Much of the variability, however, seems to be inherent in the samples, as indicated by the variability observed in true metabolizable energy values reported by Sibbald

(1986) using an assay method found to be highly repeatable (Dale and Fuller, 1986).

Low glucosinolate levels in modern rapeseed meal have apparently resulted in improved metabolizable energy values. The partial replacement of the 3–6% glucosinolates present in older cultivars (Appelqvist, 1972) by sources of metabolizable energy may account for some of the improvement. In addition, glucosinolates may impair digestive functions, thereby reducing the metabolizable energy yield of meal (Bjerg et al., 1987; Kawakishi and Kaneko, 1987). The metabolizable energy value is also influenced by variations in the content of fibre, protein and oil in the meal and these in turn may be influenced by cultivar, seed quality, environmental factors and processing methods. Crude protein content of meal ranged from 36 to 41% over a 10-year period (DeClerq et al., 1989) and this affects gross energy content in two ways. Protein contains more gross energy than does carbohydrate and the relative proportions of carbohydrate and protein determine the gross energy of the meal because ash content is relatively constant and supplies no energy and most of the oil has been extracted. In addition, the protein fraction is more digestible than the carbohydrate fraction, consequently elevated protein content increases the digestible energy and metabolizable energy values. Quality of seed as affected by maturity, moisture or heat stress, frost damage, bin heating and content of foreign matter contributes to variation in chemical composition (Bell and Jeffers, 1976) and in digestible and metabolizable energy values.

Protein is more difficult to evaluate than energy because of differences in the digestive systems of animals and differing requirements for essential amino acids among species and ages of animals. In ruminants dietary protein is first subjected to rumen microorganisms which degrade protein into peptides, amino acids and ammonia. If the ammonia production exceeds the ability of the microorganisms to utilize it, the excess may be lost as urea in the urine. Typical rapeseed meal protein is about 70% degraded in the rumen, compared to about 60% for soyabean meal protein, but this is influenced by various dietary factors such as the levels of protein, fibre and fermentable carbohydrate, especially starch, and also by processing conditions in the rapeseed crushing plant (Hill, 1991). While some rumen degradation of protein is necessary, a higher 'by-pass value' for rapeseed meal, allowing for more crude protein digestion in the small intestine, would provide a better balance of essential amino acids for absorption. This is especially important for high-producing cows and ewes in late pregnancy.

Digestion of protein by pigs and poultry is dependent on proteolytic enzymes secreted by or into the stomach, duodenum and small intestine, with microorganisms playing a relatively insignificant role. Amino acid absorption is essentially complete at the terminal ileum. However, nitrogenous matter entering the large intestine or caeca may be altered by microorganisms, thereby complicating attempts to measure amino acid digestibility when using faecal collection methods. In order to circumvent or measure the effects of post-ileal digestion certain surgical interventions have been applied, such as caecectomy in poultry, ileal cannulas and ileo-rectal anastomoses in pigs.

Another factor complicating the nutritional evaluation of protein is the metabolic faecal nitrogen component of faeces and of digesta. The metabolic faecal nitrogen fraction is largely of endogenous origin but contains a diet-related component (Maynard et al., 1979). The metabolic faecal nitrogen amounts to about 2 mg N g^{-1} dry matter consumed (Eggum, 1973) but is affected by the amount and type of fibre in the diet. When metabolic faecal nitrogen is measured and corrected for, true digestibility is obtained, as compared to apparent digestibility of protein or amino acids. True digestibility indicates the maximum absorption of protein or amino acids from a feedstuff and provides useful comparative information. However, the excreted nitrogenous matter, whether as metabolic faecal nitrogen or undigested protein, represents a loss which is best assessed against dietary protein as it is with apparent digestibility.

The apparent digestibility of rapeseed meal protein by pigs has been studied extensively and ranges from about 70 to 90% with the most common values being 72–76%. The variability may be associated with differences in age or weight of the pigs (Bell and Keith, 1988; Shi and Noblet, 1993), dietary fibre level (Bell and Keith, 1989), dietary protein level (Lloyd and Crampton, 1955) as well as differences in the characteristics of the meals. Partridge *et al.* (1987) obtained higher digestibility values for energy and nitrogenous components of rapeseed meal when fed in a starch-based diet than in a barley diet.

While apparent digestibility of protein is a useful indicator of the total faecal protein losses associated with consumption of a diet or ingredient, the use of surgically modified pigs for protein and amino acid evaluation has increased. Most protein digestion occurs before digesta enters the hindgut. No significant amino acid absorption occurs in the hindgut (Just *et al.*, 1983; Shi and Noblet, 1993). Consequently digestion that occurs prior to the ileo-caecal junction provides a good indication of the availability of protein and essential amino acids and avoids microbial protein breakdown in the hindgut from which the animals derive little benefit. Nevertheless, there is a high correlation ($R^2 = 0.97$) between ileal and faecal estimates of protein digestibility according to the equation:

$$DCP_i = 17 + 0.83 \times DCP_g$$

where DCP is digestible crude protein in $g\,kg^{-1}$ of dry matter fed, i is for ileal and g is for faecal collection from growing pigs (Shi and Noblet, 1993).

The true digestibility of rapeseed protein determined by faecal collection and correction for metabolic faecal nitrogen ranged from 85 to 90% (Bell and Keith, 1988). Partridge *et al.* (1987) obtained 67% for ileal and 72% for faecal estimates of protein digestibility when rapeseed meal was fed in barley diets at 37.5%, and 75 and 87%, respectively, when fed in starch diets. The corresponding values for soyabean meal protein were 72 and 79% when fed with barley. Lysine was also more highly available in soyabean meal than in rapeseed meal (79 versus 72%) but methionine showed similar values (83 and 85%, respectively). Similar differences between these two meals were found by Sauer and Thacker (1986), Green and Kiener (1989) and by others.

With adult chickens apparent digestibility of protein in rapeseed meal is about 71–74% (Lodhi *et al.*, 1970; Tao *et al.*, 1971). The latter reported true digestibility of 78% for protein using colostomized roosters whereas 73% digestibility was obtained by Zuprizal *et al.* (1991a, b) who also found 80% average true digestibility for 14 amino acids using adult roosters. Lysine was 77% digestible compared to 87% for soyabean meal lysine. Salmon (1984a), using roosters to evaluate meals derived from three cultivars, reported true availability of amino acids ranging from 80 to 95% with lysine at 81–85%.

There are some apparent anomalies among these assessments of protein and amino acid digestibility. It would be expected that true protein digestibility would exceed apparent digestibility and that true digestibilities of protein and amino acids would be similar. It is possible that problems associated with measuring metabolic faecal nitrogen and amino acid losses were involved. However, when comparisons were made between rapeseed meal and soyabean meal, the latter was about 4%

Table 14.9. Milk composition of cows fed canola, soyabean or cotton seed meal.

Milk parameter	Units	DePeters and Bath (1986)		McLeod (1991)	
		Canola meal	Cotton seed meal	Canola meal	Soyabean seed meal
Whole milk	kg day^{-1}	41.4	39.8	34.2	34.2
Fat	%	3.5	3.5	3.5	3.2
Protein	%	3.1	3.1	3.2	2.9
Thiocyanate ion	mg l^{-1}				1.12
Iodine	mg l^{-1}				0.18

higher in apparent digestibility of protein and 10% higher in true amino acid availability, with lysine being slightly less available than the other essential amino acids.

USES OF RAPESEED MEAL

While the history of rapeseed cultivation spans more than 3000 years (see Chapter 1), little is known about the early use of the residue or cake from the oil presses. Its earliest known use was as a soil improver or organic fertilizer, a use which persists today in some countries. It is not known when rapeseed cake (meal, if ground) first became of interest in animal feeding but Schneider (1947) cited a German report dated 1872 in which digestibility of rapeseed oil meal was determined with cattle. Increased interest in using and evaluating the meal as a protein supplement occurred during and subsequent to World War II. Glucosinolates were soon recognized as undesirable components which prevented extensive use of the meal. With the advent of low glucosinolate cultivars in the mid-1970s the nutritional value and acceptability of the meal was substantially enhanced. It was also demonstrated that inactivation of myrosinase, the glucosinolate-hydrolysing enzyme in rapeseed, was important since hydrolysis products of glucosinolates (isothiocyanates, oxazolidinethione, nitriles) are more toxic

and antithyroid than intact glucosinolates. Also it was found that excessive heating during processing can lead to interactions between amino acids and carbohydrates which reduce the bioavailability of the amino acids, especially lysine, and may also result in the generation of toxic products known as premelanoidins (Yannai, 1980).

Ruminants

Low glucosinolate rapeseed or canola meal is widely accepted as a protein supplement in ruminant rations. Lactating dairy cows performed equally well when fed diets supplemented with low glucosinolate rapeseed meal or with soyabean meal and the composition of milk was equally satisfactory (Hill, 1991) (Table 14.9). Similar responses were reported by DePeters and Bath (1986), McLeod (1991), Emanuelson *et al.* (1993), Grandegger and Steingass (1990) and by Laarveld and McClean (1991) who also reported no significant effects on milk protein or fat content. Likewise the levels of thiocyanate ion were within the normal range found in milk (Virtanen *et al.*, 1963) and well below the levels found in milk of cows fed high levels of *Brassica* plants.

Dairy calves starting at 6 weeks of age and fed starter diets containing 20–30% canola meal grew as well as those fed diets supplemented with soyabean or cotton seed meal (Fisher, 1980; Claypool *et al.*, 1985).

Various studies have also shown that low glucosinolate rapeseed meal can be used successfully in rations of young bull calves and weaned beef (feeder) calves being fed for slaughter (Sharma *et al.*, 1980; Hill, 1991; McKinnon *et al.*, 1993).

Little information is available on the use of low glucosinolate rapeseed meal for sheep production; however, studies involving high glucosinolate rapeseed meal as a protein and energy supplement to native lowland hay (mainly *Carex* sp.) fed to ewes during pregnancy resulted in good performance with ewe weight changes, lambing percentages and birth weights similar to those obtained with diets supplemented with linseed meal (Bell and Williams, 1953). No enlarged thyroids were observed but ewes fed rapeseed meal consumed their daily allowance of 225 g more slowly. Bezeau *et al.* (1960) fed 10, 20 and 30% rapeseed meal diets to mature-range ewes for 2 years and obtained good responses in terms of body weights of ewes and lambs and weight and quality of fleece. Diets containing 30% rapeseed meal lacked palatability, probably because of glucosinolate levels but there was no gross evidence of thyroid enlargement. Vincent *et al.* (1988) fed high glucosinolate meal at 20% of a diet fed at 2.5 kg day^{-1} with barley straw fed *ad libitum*, compared with a similar diet containing 16.5% soyabean meal, and found no treatment differences in overall health, weight, oestrous activity, conception, number of lambs born alive or birth weights. Rapeseed meal was also an effective supplement to an 80% oat straw diet fed to 40 kg wethers (Coombe, 1985).

Little information exists regarding the use of rapeseed meal for other herbivores. Canola meal was fed at 15.5% of the diet to weanling draught-horses from 6 to 12 months of age (Cymbaluk, 1990) and to yearling colts at 15% of the diet (Sutton and Stredwick, 1979) and responses were similar to those with control diets. Goats fed 10% high glucosinolate meal showed

reduced feed intake which could be overcome by the inclusion of molasses in the diet (Morand-Fehr and Hervieu, 1983). Weanling and fryer rabbits fed diets containing 18.7% canola meal, replacing soyabean meal, in isonitrogenous and isocaloric diets, performed well compared to the controls (Throckmorton *et al.*, 1980).

Pigs

Following the conversion to low glucosinolate rapeseed cultivars the resulting meal was markedly superior as a protein supplement in pig rations. Less thyroid enlargement due to glucosinolates occurred (Rundgren, 1983), palatability improved and higher feeding levels could be used without adverse effects on rate of liveweight gain. Feeding low glucosinolate meal to pigs in the 25 to 100 kg weight range at levels up to about 20% has resulted in performance comparable to that obtained with diets supplemented with soyabean meal (Grandhi *et al.*, 1980; Thomke, 1984; Castaing and Grosjean, 1986; Bourdon and Aumaître, 1990). However, Bell and Keith (1993) reported 3–5% lower average gain with pigs fed canola meal than with soyabean meal when these supplements were compared in diets in which wheat, maize (corn) or hulless barley replaced 0–100% of regular (hulled) barley in the grain mixture.

Variability in pig responses to low glucosinolate rapeseed meal compared to soyabean meal may be attributed to several factors including cleaning and quality of the seed prior to crushing (Bell and Jeffers, 1976); processing, especially in regard to overheating (Bourdon and Aumaître, 1990), residual oil content, inclusion in the meal of rapeseed oil refining by-products such as acidulated fatty acids, soapstocks, gums (McCuaig and Bell, 1981) or spent bleaching clay (Smith, 1984; Keith and Bell, 1986). Glucosinolate content, even within samples considered to be low in glucosinolate, may vary considerably, depending on the level in the original seed and on processing conditions. In a survey of seed, meal and press cake obtained from seven crushing plants, the apparent destruction of glucosinolates, mostly in the desolventizer–toaster, ranged from 15 to 77% (Bell and Keith, 1991). Processing conditions that reduce glucosinolate content in the meal may result in production of thiocyanates and indoleacetonitriles as thermal degradation products (Slominski and Campbell, 1989).

Hydrolysis of glucosinolates during the crushing operation is prevented either by rapid elevation to cooking temperature in the cooker or by ensuring that the seed has a very low moisture content (see Chapter 12); however, some myrosinase activity, perhaps insignificant, has been found in rapeseed meals (Bell *et al.*, 1976) and in press cake (Keith and Bell, 1991). Myrosinase present in the diet increases the toxicity of ingested glucosinolates (Belzile *et al.*, 1963; Bell, 1965) and there is evidence that intestinal microflora may hydrolyse glucosinolates and also that dietary supplements such as copper and antibiotics may influence this type of hydrolysis. Ions of copper and iron in moistened rapeseed meal may reduce glucosinolates (Youngs *et al.*, 1971) by converting them to nitriles, hence method of feeding pigs and trace mineral supplementation may affect responses to glucosinolates.

The responses to rapeseed meal as a protein supplement may be influenced by the digestible energy and the protein and amino contents of the cereal grains of the diet (Castell, 1980; Bell and Keith, 1988, 1989). However the strengths, weaknesses and variability of rapeseed meal are ultimately tested in the marketplace. Opportunity pricing (maximum acceptable ingredient cost) and parametric linear programming (ingredient cost effects on level of inclusion in diet formula) using rapeseed meals of different digestible energy or

metabolizable energy, protein, lysine and methionine contents to formulate least-cost diets in various appropriate market situations provide monetary evaluations reflecting quality of meal (Bell, 1993a) and provides indications of the potential benefits from quality improvement.

The use of canola meal as the main protein supplement in the diets of gilts and sows during gestation and lactation gave no evidence of adverse effects on litter size, piglet birth weight, weaning weight or first post-partum oestrus (Flipot and Dufour, 1977; Lewis *et al.*, 1978; Lee and Hill, 1985). However, Etienne and Dourmad (1987) fed gilts meal from a low glucosinolate cultivar (Tandem) and found increased early embryonic mortality and more evidence of thyroid dysfunction in fetuses than in their dams.

Feeding low glucosinolate meal to weanling or starter pigs has been less satisfactory (Castell, 1977; McKinnon and Bowland, 1977; Mcintosh and Aherne, 1981). While canola meal was more palatable than high glucosinolate meal, its use at more than 7–9% of the diet resulted in reduced performance. It is possible that low digestibility of the meal was a major drawback for pigs of this age (Bell and Keith, 1989).

Poultry

The nutritional value of low glucosinolate rapeseed meal for poultry has been assessed in several countries and the subject has been reviewed (Hill, 1979; Fenwick and Curtis, 1980; Clandinin and Robblee, 1983b; Elwinger, 1983; Thomke *et al.*, 1983; and others). The use of up to 15–20% meal in layer diets was usually shown to have no adverse effects on production, egg size, egg quality or bird mortality when isocaloric diets were used. Up to 33.4% canola meal in corn-based isocaloric diets fed to Single Comb White Leghorn laying hens resulted in no differences compared to a soyabean meal control diet in egg production, efficiency of egg production, final body weight, internal and external egg quality, fertility and hatchability of fertile eggs and faecal dropping scores. However, feed consumption declined at the highest level of canola meal and thyroid enlargement occurred (Nassar *et al.*, 1985). Elwinger (1983) reviewed Swedish experiments in which up to 15% low glucosinolate rapeseed meal was fed to layers and concluded that a maximum of 10% should be used because of a tendency for reduced feed intake with increased levels of meal in the diet, evidence of reduced eggshell quality, evidence of egg taint, thyroid enlargement and increased mortality. A level of 15% low glucosinolate rapeseed meal was recommended by Roth-Maier and Kirchgessner (1988) who reported no differences among six meal levels from 0 (control) to 20% for production traits, egg size and average daily egg mass but at the 20% level mortality increased and organoleptic scores for odour of eggs decreased.

Despite the successful experimental use of 15–20% levels of low glucosinolate rapeseed meal and the successful commercial use in layer hen diets there is continuing interest in resolving certain recognized problems in such meal, as identified by Hill (1979), Elwinger (1983) and others, and as referred to previously in this chapter. Egg taint has been linked to trimethylamine which arises from sinapine in rapeseed meal and which renders the meal unsuitable for certain strains of layers, mainly those that produce brown-shelled eggs. This is partly a hen genetics issue but there is now evidence that an unidentified factor in rapeseed meal reduces the liver trimethylamine oxidase activity, thereby exacerbating an already reduced enzyme activity in susceptible birds.

The incidence of liver haemorrhage syndrome attributed to the use of rapeseed meal has declined since the advent of low glucosinolate rapeseed meal but a low

incidence remains, more so with some layer strains than others. Evidence that intact glucosinolates were more detrimental than glucosinolate aglucones (Campbell, 1987) appears contradictory to the finding that rapeseed meal with glucosinolates removed by hot-water extraction, or destroyed by heating, resulted in the same incidence of liver haemorrhages as the control meal fed at a 30% level (Wight *et al.*, 1987).

Numerous experiments have demonstrated the value of low glucosinolate rapeseed meal in broiler chicken diets. Meal derived from *B. napus* and *B. rapa* cultivars were compared alone and in combination with fish meal as partial or complete replacements for soyabean meal in diets based on maize and wheat (Hulan and Proudfoot, 1981a). Rapeseed meal could replace up to 80% of the soyabean meal without adversely affecting weight or feed efficiency and could replace all of the soyabean meal if energy, protein and amino acid differences were compensated. Salmon *et al.* (1981) fed up to 28.1% canola meal in wheat-based diets to broiler chicks and 12.1% to finishing birds and compared two levels of protein and two levels of metabolizable energy with a soyabean meal control diet. Growth rates were not affected by canola meal but feed efficiency declined when canola meal was used in low nutrient-density diets. An increased incidence of leg weakness has been reported by several authors (Hill, 1979), the specific cause of which remains unresolved. Meat flavour was adversely affected when meal was used at levels exceeding 21 and 9% in the starter and finisher diets, respectively. Mortality was not affected by canola meal. The use of 22% low glucosinolate rapeseed meal in starter and finisher broiler diets compared favourably with control diets (Kiiskinen, 1983) and similar results were obtained with 15% low glucosinolate rapeseed meal by Kozlowski *et al.* (1989).

The level of low glucosinolate rapeseed meal used for commercial broiler production is often lower than that found satisfactory experimentally. Klein *et al.* (1981) performed an economic analysis of the use of canola meal for broiler diets involving feed ingredient price scenarios covering 5 years. For example, when canola meal price was below 60% of that of soyabean meal (46% crude protein) at least 8% canola meal would be used; at 70% at least 4% meal. In computer formulations the cost of available nutrients in meal must be competitive with all other available nutrient sources in that situation.

Canola meal in diets of juvenile and adult chicken genotypes bred for meat production resulted in performance equal to that obtained with soyabean meal-supplemented diets (Proudfoot *et al.*, 1982, 1983). Five per cent canola meal was used in the starter and grower stages and 6.5% in the breeder diets. The diets were isonitrogenous and isocaloric within each stage.

Several studies have demonstrated that up to 20% canola meal can be used satisfactorily in the diet of turkey broilers and that meat quality equals that from birds fed soyabean meal control diets (Moody *et al.*, 1978; Hulan *et al.*, 1980; Robblee and Clandinin, 1983; Waibel *et al.*, 1992). Determining the optimum dietary level of canola meal by linear programming indicated that maximum profitability may be achieved without fully compensating for the low nutrient density (metabolizable energy) of canola meal (Klein *et al.*, 1979).

Fish

Rapeseed protein is a potentially valuable source of amino acids in fish diets (Higgs *et al.*, 1994) replacing part of the high-cost fish meal normally used. About 20% canola meal has been used successfully in diets for carp (*Cyprinus carpio*), tilapia (*Sarotherodon mossambicus*), rainbow trout (*Oncorhynchus mykiss*) and chinook salmon (*Oncorhynchus tshawytscha*). Fish may be adversely affected by high levels of carbohydrate including fibre, by phenolic compounds, phytic acid, glucosinolates and sinapine. The extensive removal of most of these antinutritional components from meal results in a protein concentrate of much superior quality that can supply at least 24% dietary protein for trout and salmon (McCurdy and March, 1992).

IMPROVING THE NUTRITIONAL VALUE OF MEAL

The development of low glucosinolate cultivars of rapeseed resulted in major improvement in the nutritional quality of rapeseed meal. Beyond this, however, most breeding research has been devoted to improvements in oil quality and agronomic performance of the crop, the benefits of which have been appreciable, at least in Canada (Nagy and Furtan, 1977). Further opportunities exist for enhancement of meal quality.

Reduction of the fibre or hull content would improve the digestible and metabolizable energy values and improve the digestibility or availability of protein and amino acids. Seeds of yellow colour have less hull and have hulls of greater digestibility than hulls of dark seed. There may also be less polyphenols, lignin and tannins in hulls of light-coloured seed which would be desirable for improving digestibility. Mechanical dehulling or hull reduction has not been widely adopted, partly because of hull disposal problems.

Increasing the protein level in the seed while maintaining or increasing the oil level should be associated with a reduction in carbohydrate (fibre) level. Marked regional and yearly variations have been observed in protein content of the oil-free meal. A consistently higher ratio of protein to carbohydrate in meal would enhance its value, especially for pigs and poultry.

Depending on the kinds of cereal

grains used in pig and poultry diets, one or more amino acids may be needed as supplements if low glucosinolate rapeseed meal is the major protein supplement in the diet. Enhancement of available lysine, methionine plus cystine, and tryptophan would improve meal under certain conditions. On the other hand, for ruminant feeding, improved 'by-pass' value may be more important than the amino acid profile of the protein.

Further reduction in glucosinolate levels would improve the value of meal especially for laying hens, young pigs and fish. Reduction in sinapine levels would reduce or eliminate the 'fishy-egg' problem associated with certain strains of laying hens and would improve palatability of the meal.

The relatively high phytic acid content of meal reduces the availability of calcium, phosphorus, zinc and certain other essential minerals for pigs and poultry and may depress protein digestibility. The most promising counteraction approach now appears to be dietary phytase supplements.

The various carbohydrates (mono-, di-, oligo-, polysaccharides) found in *Brassica* constitute a major fraction of the oil-free meal. The majority of them are of little nutritional value, at least to poultry and pigs, and in general are detrimental. Better understanding of this aspect may provide new opportunities for improvement.

Removal of glucosinolates and other antinutrients from rapeseed by methanol–ammonia processing (Shahidi and Naczk, 1990), by ultra-filtration (Rubin *et al.*, 1990) and by aqueous enzyme processing (Jensen *et al.*, 1990) has been described (see Chapter 12). Such procedures yield concentrates or isolates of relatively high purity which may find use mainly in human foods and in certain formulations requiring higher levels of meal refinement.

The nutritional value of rapeseed hulls is extremely low for non-ruminants. The metabolizable energy is near zero for poultry and pigs (Jones and Sibbald, 1979; Bell and Shires, 1982; Lessire *et al.*, 1991) although hulls from yellow seed were about 25% digestible by pigs. However, neither feeding broilers 5% canola hulls in iso-nitrogenous, isocaloric diets nor feeding them an equivalent amount of tannin extract resulted in adverse effects, indicating that there were no effects on nutrient availability associated with the hulls (Mitaru *et al.*, 1983).

Dehulled meal yielded 30% more apparent metabolizable energy in broiler chick diets (Seth and Clandinin, 1973) and 9.5% more true metabolizable energy (corrected for body protein stored or lost during the trial) in rooster diets (Jones and Sibbald, 1979). Meal from a low fibre cultivar contained 13% more apparent metabolizable energy for hens than a regular meal (Olsson *et al.*, 1990). However, the favourable effect of dehulling on metabolizable energy has not been matched by improved chick performance. Growth rates declined with increasing levels of dehulled meal (Shires *et al.*, 1981; Bougon *et al.*, 1988). This may in part be attributed to an increased glucosinolate level in dehulled meal and to the relatively high fibre (neutral detergent fibre) level remaining in the embryo fraction.

Dehulled rapeseed meal contains 20–25% more digestible energy and metabolizable energy for market pigs than regular meal and has feeding value approaching that for soyabean meal (Bourdon, 1986; Bell, 1993b) but its energy digestibility was too low (63%) for effective use with weanlings (Christison and Bell, 1993). Meal from a low fibre cultivar showed improved digestibility of organic matter (76 versus 69%) and increased metabolizable energy (corrected for body protein stored or lost during the trial) (12.0 versus 11.1 MJ kg^{-1} dry matter) (Olsson *et al.*, 1990) compared to regular meal.

FEEDING WHOLE, 'FULL-FAT', UNEXTRACTED, LOW GLUCOSINOLATE RAPESEED

Whole seed contains about 40% oil and 22% crude protein. The oil is useful for increasing the energy level in the diet, for reducing feed dust and for modifying the fatty acids in eggs, milk and meat. Similarly the protein level is high and the amino acid profile of the protein is favourable. Consequently there has been substantial interest in utilization of rapeseed as a feed ingredient.

Ruminants

Untreated, heat- and steam-treated, crushed, low glucosinolate rapeseed and tallow have been compared using lactating cows fed 1.5 kg day^{-1} per cow mixed into a hay-concentrate ration (Emanuelson *et al.*, 1991). There were no significant effects on total digestibility of the diet and no differences among treatments on milk yield or content of fat, protein or milk solids but the fatty acid proportions of stearic acid ($C_{18:0}$) increased and palmitic acid ($C_{16:0}$) and palmitoleic acid ($C_{16:1}$) decreased when rapeseed was fed. Subsequent research in which 0.9 kg of heat-treated seed day^{-1} was fed (Emanuelson *et al.*, 1993) over several lactations confirmed previous findings and included examination of the thiocyanate and iodine contents of the milk, neither of which was found to be of concern from a human nutritional standpoint. Cooked, crushed rapeseed is currently produced commercially in Sweden for the feed industry.

Kennelly *et al.* (1993) found that 'Jet-Sploded' (heat exchanger, pressure differential) canola seed had improved rumen by-pass value compared with untreated or extruded seed. Reports from various countries have confirmed the potential role of full-fat rapeseed in lactation diets for high-producing dairy cows (Huhtanen and Poutiainen, 1985; Nalecz, 1986; Murphy *et al.*, 1987; Beaulieu *et al.*, 1990; Ashes *et al.*, 1992; Ferlay *et al.*, 1992).

Crossbred beef steers, initially 300 kg, were fed 1.4 kg day^{-1} of ground, raw canola seed resulting in a small depression in daily gain and feed intake but no change in efficiency of feed utilization compared to steers fed soyabean or canola meal (Rule *et al.*, 1989). The adipose tissue contained increased levels of stearic ($C_{18:0}$), linolenic ($C_{18:3}$), arachidic ($C_{20:0}$) and eicosanoic ($C_{20:1}$) acids with decreased levels of palmitic ($C_{16:0}$) and palmitoleic ($C_{16:1}$) acids. Lambs (24 kg initial weight) fed diets containing 6% whole canola seed did not differ from those fed the soyabean meal control diet in growth rate, feed efficiency or carcass characteristics but changes in fatty acid composition of lean tissue fat were observed (Lough *et al.*, 1991).

Pigs

Weaning and growing–finishing pigs have shown no reduction in either growth rates or efficiency of feed utilization when fed diets containing up to 15% ground canola seed (Castell, 1977; Castell and Falk, 1980; Salo, 1980; Aumaître *et al.*, 1989; Shaw *et al.*, 1990). Several investigators observed reduced daily feed intake when low glucosinolate rapeseed was included in pig diets (Cromwell *et al.*, 1989; Grosjean *et al.*, 1989; Han *et al.*, 1990). Variation in findings may be attributed to variability in seed composition (Leclercq *et al.*, 1989), processing (cooking, grinding, pelleting), pig age (de Souza *et al.*, 1990), glucosinolate content of the seed, etc. Cooking was beneficial for pigs of 6.8 kg initial weight fed diets containing 10, 20 or 24% canola seed (Ochetim *et al.*, 1980). Autoclaving (12°C, 30 min) was effective but dry heat (12°C, 1 h) and extruding (85°C, 30 s) were not effective. However, Shaw *et al.* (1990) obtained no benefit from Jet-Sploded seed (260°C, 60 to 90 s) used in pelleted diets.

Metabolizable energy values of 18.72 and 20.26 MJ kg^{-1} (dry matter) were obtained for two samples of canola seed fed to 40 kg pigs (Salo, 1980). The higher value was associated with higher oil and protein contents.

Frost-damaged canola seed containing 22, 27 and 35% oil and 18.2, 18.8 and 19.8% crude protein had digestibility of 32–38% for energy and 12–20% for protein when fed whole and 60–69% for energy and 62–68% for protein when fed ground. Pelleting of the entire diet containing unground canola seed was as effective as grinding the seed (Bell *et al.*, 1985). Digestible energy values of processed seed ranged from 15.6 to 17.7 MJ kg^{-1} (dry matter). Pigs from 23 to 100 kg performed well with 10 and 20% damaged canola seed in their diets but feed intake and daily gain were depressed at 30% seed (Bell and Keith, 1986).

Canola seed as a means of increasing the energy intakes of sows just prior to and during lactation has received limited study. Spratt and Leeson (1985) found that dietary levels above 10% resulted in feed refusal. Diets containing 5 and 10% canola seed had no effect on milk parameters but the 10% level would not have provided the minimum 7.5% supplemental fat required to increase milk fat content (Moser and Lewis, 1980).

Poultry

Ground, raw, high glucosinolate rapeseed fed at 10–15% of the diet has been shown to depress egg production, efficiency of production and egg weight (Leslie and Summers, 1972) but 10% low glucosinolate rapeseed meal resulted in performance equal to that of hens fed the control diet (Leeson *et al.*, 1978). Good performance was also obtained with broiler chicks fed 10 and 20% raw, unground seed although thyroid enlargement occurred. This enlargement could be avoided by autoclaving the seed, or by extruding (Leclercq *et al.*, 1989). Increased production was observed by Al-Bustany and Elwinger (1988). The n-3 fatty acid content of eggs and of the brain tissue of embryos and chicks from such eggs was increased as a result of feeding diets containing 16% ground canola seeds to laying hens (Cherian and Sim, 1991). Similar fatty acid changes were observed in carcass fat of broiler chickens (Ajuyah *et al.*, 1991). Such modification of fatty acids in meat and eggs was perceived as being beneficial in the human diet.

Rapeseed contains about 28 MJ kg^{-1} of gross energy, varying according to oil and protein contents. Estimates of available energy range from 17.7 to 21.4 MJ kg^{-1} for apparent metabolizable energy and from 18.3 to 23.0 MJ kg^{-1} for true metabolizable energy (Sibbald and Price, 1977; Leclercq *et al.*, 1989). Some of the variability found in metabolizable energy values for full-fat rapeseed may be due to oil composition as well as processing of the seed. Pigs digested 87% of low erucic acid oil (0.2%) compared to 75% for oil containing 50% erucic acid (Bell and Shires, 1982). These oils were only about 40% digestible when fed as unextracted, raw, whole rapeseed. The level of rapeseed in the diet was inversely related to the true metabolizable energy value (Salmon, 1984b). Fed at a 10% level the true metabolizable energy was 26.2 MJ kg^{-1} but at 30% it was 23.4. Similar effects of dietary fat levels on available energy values have been reported (Sibbald, 1982; Noblet *et al.*, 1993). Rolling or grinding of the seed improves availability of nutrients (Salmon *et al.*, 1988, 1991) but heating was of little benefit in studies with turkeys.

Turkeys may be fed up to about 16 or 17% whole or flaked canola seed in pelleted diets without affecting performance (Salmon *et al.*, 1988). Ajuyah *et al.* (1993) demonstrated that 10% canola seed resulted in increased levels of n-3 fatty acids in fat deposited in the turkey.

UTILIZATION OF PROCESSING BY-PRODUCTS FOR ANIMAL FEED

Screenings, dockage, damaged seed

Rapeseed as received from farms contains foreign matter, called 'dockage' if it can be removed by commercial cleaning equipment. If it cannot be removed it is referred to as 'inseparable', e.g. wild mustard seed (*Sinapis arvensis* D.C., Wheeler) and cultivated mustard seed. Dockage, once removed, is called 'screenings' and may be sorted into coarse and fine fractions. The former is mainly fibrous matter including pieces of seed pods, stems, etc. and the latter includes small and broken rapeseed, small weed seeds, etc. Failure to clean the seed effectively results in meal with increased fibre and glucosinolate contents as well as decreased protein and available energy (Bell and Shires, 1980). Seed purity can be evaluated in the meal through feed microscopy (Bell and Jeffers, 1976).

Feeding diets containing up to 4% wild mustard seed (raw, ground) or up to 2% stinkweed (*Thlaspi arvense* L.) seed and containing 20% Tower rapeseed (low glucosinolate) has been shown not to affect feed utilization or weight gain of growing mice (Shires *et al.*, 1982) but, when oil-extracted weed seed meals comprising up to 9% of the supplementary protein were compared, increasing levels resulted in reduced feed intake and gains, with the greatest depression arising from wild mustard meal. Levels of contamination by these weed seeds as found in typical commercial meals would involve little risk but their contributions of allyl and *p*-hydroxybenzyl glucosinolates would increase the total glucosinolate content of the meal.

Fine screenings, stinkweed, lamb's-quarters (*Chenopodium album* L.) and smartweed (*Polygonum lapathfolium* L.) seed meals and oils were evaluated for their effects on reproduction in female mice (Rose and Bell, 1982a, b). No teratogenic effects or other evidence of adverse effects was found using dietary levels exceeding those likely to be encountered in practice. Steam cooking of canola fine screenings reduced glucosinolates from 40 to 15 µmol g^{-1} and improved daily gains of pigs fed 10 and 20% dietary levels of rolled screenings (Keith and Bell, 1983). Cooking with ammonia further reduced glucosinolates to 2 µmol g^{-1} but protein digestibility, lysine availability and daily gain declined. Ammoniation of canola seed containing up to 9% wild mustard seed failed to reduce *p*-hydroxybenzyl glucosinolate but did reduce the canola glucosinolates (Bell *et al.*, 1987). Contamination of canola seed with wild mustard seed did not affect daily gain of pigs but feed efficiency and serum thyroxine levels declined (Bell *et al.*, 1987). Frost-damaged canola seed containing 22 to 35% oil and 18 to 20% protein contained 15.6 to 17.7 MJ digestible energy kg^{-1} (dry matter) when ground or pelleted for pigs (Bell *et al.*, 1985). Up to 20% of frost-damaged seed fed to pigs from 23 to 100 kg supported growth rates and feed efficiency equal to the controls. Glucosinolate levels were inversely related to the percentage of frost-damaged kernels.

Rapeseed may contain sclerotinia resulting from crop infection by the common fungal pathogen *Sclerotinia sclerotiorum* (Lib.) de Bary (see Chapter 5). Sclerotia are removed in the screenings. Morrall *et al.* (1978) fed rats diets containing 1% and 5% sclerotia in an 84-day subacute trial. Rats fed 1% sclerotia maintained normal feed intakes and growth rates but performance declined markedly at 5%. Sclerotia are highly unpalatable (for humans) and have an acrid, musty flavour. Examination of blood, liver and kidneys of the rats revealed no changes that could not be attributed mainly to reduced feed intake and consequent general malnutrition.

Gums, acidulated fatty acids, spent clay

Several by-products of rapeseed oil refining may be used in animal feeds. Degumming removes phospholipids (gums) which may contain 35–40% oil and the metabolizable energy value may approach that of rapeseed oil. From 1.5 to 2.0% gums added to rapeseed meal improves the energy value of the meal and reduces dustiness. Various studies have shown good responses from feeding gums to laying hens, broilers and turkeys (Salmon, 1970; Leeson et al., 1977; March, 1977; Summers and Leeson, 1977) but some studies indicated possible differences among gum samples and among strains of laying birds (Summers et al., 1978; Hulan and Proudfoot, 1981b). Gums fed to pigs at levels up to 4% of the rapeseed meal in the diet were 80% digestible and presented no problems (McCuaig and Bell, 1981). Beef steers likewise responded favourably to dietary gums (Mathison, 1978).

Spent bleaching clay (bentonite) typically contains about 30% oil and therefore is a potential source of feed energy. The bentonite per se provides no nutrients and its inclusion in the diet has shown both positive and negative effects associated with its strong adsorptive properties. Studies with mice and rats fed dietary levels of spent clay up to 4% revealed no adverse effects (Keith and Bell, 1986). Blair et al. (1986) likewise found that up to 4% spent clay could be used in broiler and layer diets, although some decrease in yolk pigmentation occurred, probably because the clay adsorbed pigmenting agents in the digestive tract. Inclusion of spent clay in animal feeds may provide protection against the absorption from the digestive tract of T-2 toxin, a trichothecene mycotoxin produced by several species of fungi that may con-

taminate food of man and animals, according to experiments with rats and pigs (Smith, 1984). Spent clay obtained following hydrogenation of oil with a nickel catalyst may contain highly toxic nickel sulphides and should not be incorporated into feeds.

Soapstock derived from alkali refining to remove free fatty acids may be used as an energy source in feeds. Acidulated fatty acids may also be available. For laying hens these provided about 33.5 MJ kg^{-1} when fed at 10% of the diet, compared with low glucosinolate rapeseed oil at 37.3 MJ kg^{-1} (Lall and Slinger, 1973a). The acidulated fatty acids used contained 22.8% erucic acid ($C_{22:1}$) which is poorly digested so the metabolizable energy value of acidulated fatty acids from low glucosinolate rapeseed oil would probably be greater than 33.5 MJ kg^{-1}. These authors further showed that the metabolizable energy value of acidulated fatty acids was influenced by synergistic interactions with other dietary fats and levels (Lall and Slinger, 1973b).

Hulls

Dehulling of rapeseed generates hulls as a potential feedstuff. Experimental and small-scale dehulling done prior to oil extraction resulted in hulls with about 18–20% crude protein, 24–34% crude fibre, 14–20% oil and about 5% ash (Appelqvist, 1972; Theriez and Brun, 1982). Much lower crude protein values (10.0–11.25%) were obtained for 'purified' hulls by Finlayson (1974) who also found that this protein was resistant to chemical hydrolysis and was low in sulphur amino acids. Bell and Shires (1982) found that pigs digested about 2% of the energy from dark hulls compared with 30% for yellow hulls which contained less lignin and fibre. The corresponding values for crude protein digestibility were 0 and 20%. Hulls are also indigestible by poultry (Lessire et al., 1991). It seems evident therefore that the composition and nutritional value of hulls depends on method of removal, subsequent treatment, cultivar of seed and other factors.

Hulls are not suitable dietary ingredients for poultry or pigs. However, rabbits fed up to 40% rapeseed hulls, partly replacing alfalfa (lucerne), grew well and produced good carcasses (Lebas *et al.*, 1981). The high oil content of the hulls (16%) resulted in dietary oil levels increasing from 2.4 (control) to 7.8% and in digestible energy of the diet increasing from 11.8 to 13.2 MJ kg^{-1}. Bourdillon-Sanders (1988) reported digestible energy values of 13.0 to 13.7 MJ kg^{-1} for hulls fed to adult rabbits. Lambs grew well when fed 15 and 30% hulls and digested 60% of the gross energy of the hulls (Theriez and Brun, 1982). Young bulls (337 kg) fed 25% hulls in the diet, replacing maize silage, showed no reduction in daily liveweight gain (Anon., 1988).

Press cake

Press cake obtained after flaking, cooking, and expelling about one half of the oil in a prepress solvent extraction plant was evaluated using lactating cows fed about 11% ground press cake in their diets (Beaulieu *et al.*, 1990). Digestibility of energy was improved compared to the control diet (66.5 versus 64.6%) and similarly for dietary ether extract 68.7 versus 60.4%. Milk production, per cent fat and per cent crude protein were unaffected.

Twenty-eight samples of press cake from seven crushing plants ranged from 17.5 to 26.7% oil and from 29.1 to 35.6% crude protein (Keith and Bell, 1991). Glucosinolate levels were only slightly lower than those of the original seed and a small residual myrosinase activity was found. Energy and protein were 75% digestible in pigs and the digestible energy ranged from 17.4 to 18.8 MJ kg^{-1} dry matter depending on the oil content.

RAPESEED AS A FORAGE

Rape has long been used as a forage crop, as grazing, silage or soilage (cut and fed fresh). Forage rape cultivars may contain 22% or more of highly digestible protein and 0.7% lysine in its dry matter. Organic matter digestibility was 68% for sows (Livingstone and Jones, 1980). Ruminant bloat may occur due to the high digestibility of fresh green matter and nitrate poisoning occasionally occurs (Buret, 1991). Low glucosinolate cultivars are more palatable to grazing sheep (Zobelt *et al.*, 1991) and glucosinolate levels varied according to planting time. Ensiling reduced glucosinolate levels to one tenth of original levels (Fales *et al.*, 1987) but silage made from high glucosinolate cultivars had high digestibility and supported greater daily gains in steers than silage originating from low glucosinolate cultivars (Lancaster *et al.*, 1990).

Animals (roe deer, hares) grazing winter-type rape in Europe have apparently been affected by toxic levels of *S*-methyl-L-cystine sulphoxide (SMCO) which, upon conversion to dimethyldisulphide by microorganisms in the gastrointestinal tract, causes haemolytic anaemia and central nervous system disorders (Schoon *et al.*, 1989) (see Chapter 9). This toxic amino acid is most abundant in the leaf and stem fraction and is quite variable; 0.3 to 5.7 g kg^{-1} versus 0.1 g kg^{-1} in rapeseeds (Rihs *et al.*, 1991). It occurs at similar levels in high and low glucosinolate cultivars but better palatability of low glucosinolate types may lead to increased forage consumption and greater risk of toxicity. Fales *et al.* (1987) found about 2.7 g SMCO kg^{-1} dry matter in rape at time of ensiling and 2.3 g after 300 days in the silo, consequently SMCO is a potential risk factor in feeding rape as a silage.

Rape pollen, airborne when the crop is in bloom, is believed responsible for an increased incidence of equine respiratory disease in the United Kingdom (Dixon and Mcgorum, 1990). This condition has not been reported elsewhere or pertaining to other animals.

MUSTARD MEAL

Since mustard is used mainly in the condiment trade (see Chapter 17) it has received little attention as a protein supplement for animal feeding. The glucosinolates in the seed are important flavour components for condiments but the levels are equal to or higher than those in traditional rapeseed. Two types of *B. juncea* (brown, Oriental) evolved along different lines (Röbbelen and Thies, 1980). One from India contains about 150 µmol g^{-1} of 2-hydroxy-4-pentenyl glucosinolate and the other from Europe has about 110 µmol g^{-1} of allyl glucosinolate (see Chapter 10). *Sinapis alba* (L.) (yellow mustard) contains about 185 µmol g^{-1} of *p*-hydroxybenzyl glucosinolate. Each mustard type is rich in one glucosinolate with only minor amounts of others.

Feeding 10 or 20% *B. juncea* meal (allyl type) to weanling miced depressed feed intake and growth when dietary myrosinase was included. Treatment with iron to destroy glucosinolates resulted in growth comparable to the soyabean meal control diet (Bell *et al.*, 1971). Feeding similar meal to pigs at 13% of the diet resulted in growth and feed efficiency significantly inferior to the performance of pigs fed soyabean meal control diets and inferior to meal derived from *S. alba* (Bell *et al.*, 1981). Both of the *Brassica* meals contained about 13.7 MJ kg^{-1} of digestible energy. The crude protein levels in dry matter were 47.8 and 45.0% for *B. juncea* and *S. alba*, respectively, and the essential amino acid contents of the protein were similar except that *B. juncea* contained more methionine plus cystine (4.6 versus 3.9%). Upon dehulling *S. alba* the solvent-extracted meal contained 48.3% crude protein (dry matter) and the hulls 14.2% crude protein (Sarwar *et al.*, 1981). The hulls contained more lysine in their protein (8.7 versus 4.9%) and more threonine (7.1 versus 4.1%) than the dehulled meal. The inclusion of 3.3 to 5.4% hulls in semipurified diets for mice resulted in improved growth rates and feed intakes but the inclusion of 7–25% *S. alba* meal resulted in about 30% depression in weight gain associated with about 9% reduced feed intake, and a reduction in apparent digestibility of dietary crude protein (77 versus 85%) compared with the soyabean meal control diet.

Ammonia treatment of *B. juncea* meal reduces the glucosinolate content (McGregor *et al.*, 1983). Treated meal with 80% reduction in glucosinolates also had 20% less lysine and its crude protein increased from 44.6 to 51.1% (dry matter) (Bell *et al.*, 1984). Energy and protein digestibility values for pigs were 72 to 75%, respectively, and the digestible energy (dry matter) was 14.4 MJ kg^{-1}. When this meal was included in the diet at 15% as the only supplemental protein in barley–wheat (2 : 1) diets, growth, feed intake and feed efficiency were inferior to values obtained with canola meal but performance improved significantly with lysine supplementation. Similar results were obtained with broiler chicks (Blair, 1984). Subsequently supplementary lysine and isoleucine were found to be more effective (Keith and Bell, 1985).

An interspecific cross between *B. juncea* Indian-type mustard and *B. rapa* canola contained under 10 µmol g^{-1} of butenyl glucosinolates in the oil-free seed (Love *et al.*, 1990). Meal with 42% crude protein had protein containing lower concentrations of all essential amino acids, except arginine, leucine, phenylalanine and cystine, than found in canola protein (J.M. Bell and G. Rakow, unpublished data). Digestibility of energy and protein was inferior to canola values. However, when this meal was fed to growing pigs, replacing one third, two thirds or all of the soyabean meal in diets supplemented with 0.2% lysine, the pigs performed well, indicating that glucosinolates were not a deterrent and that, when dietary levels of digestible energy and amino acids are adequate, good performance could be expected. Nevertheless it was shown through linear programming that the low digestible energy and low lysine values of such meal would incur appreciable price discounts relative to canola meal.

REFERENCES

Agunbiade, J.A., Wiseman, J. and Cole, D.J.A. (1991) Nutritional evaluation of triple low rapeseed products for growing pigs. *Animal Production* 52, 509–520.

Ajuyah, A.O., Lee, K.H., Hardin, R.T. and Sim, J.S. (1991) Influence of dietary full-fat seeds and oils on total lipid, cholesterol and fatty acid composition of broiler meats. *Canadian Journal of Animal Science* 71, 1011–1019.

Ajuyah, A.O., Hardin, R.T. and Sim, J.S. (1993) Studies on canola seed in turkey grower diet: effects on ω-3 fatty acid composition of breast meat, breast skin and selected organs. *Canadian Journal of Animal Science* 73, 177–181.

Al-Bustany, Z. and Elwinger, K. (1988) Whole grains, unprocessed rapeseed and β-glucanase in diets for laying hens. *Swedish Journal of Agricultural Research* 18, 31–40.

Anon. (1971) *Atlas of Nutritional Data on United States and Canadian Feeds.* National Academy of Sciences, Washington, DC, USA.

Anon. (1981) *Nutritional Energetics of Domestic Animals and Glossary of Energy Terms.* National Academic Press, Washington, DC, USA.

Anon. (1988) *Valorization des Produits du Dépelliculage du Colza 00.* Centre Technique Interprofessionel de Oléagineux Métropolitains, Paris, France.

Appelqvist, L.-Å. (1972) Chemical constituents of rapeseed. In: Appelqvist, L.-Å. and Ohlson, R. (eds) *Rapeseed, Cultivation, Composition, Processing and Utilization.* Elsevier Publishing Company, Amsterdam, The Netherlands, pp. 123–173.

Ashes, J.R., Welch, P.S., Gulati, S.K., Scott, T.W., Brown, G.H. and Blakeley, S. (1992) Manipulation of the fatty acid composition of milk by feeding protected canola seeds. *Journal of Dairy Science* 75, 1090–1096.

Askbrant, S. and Håkannson, J. (1984) The nutritive values of rapeseed meal, soya bean meal and peas for laying hens. *Swedish Journal of Agricultural Research* 14, 107–110.

Aufrere, J. and Michalet-Doreau, B. (1988) Comparison of methods for predicting digestibility of feeds. *Animal Feed Science and Technology* 20, 203–218.

Aumaître, A., Bourdon, D., Peiniau, J. and Bengala Freire, J. (1989) Effects of graded levels of raw and processed rapeseed on feed digestibility and nutrient utilization in young pigs. *Animal Feed Science and Technology* 24, 275–287.

Bate-Smith, E.C. and Ribéreau-Gayon, P. (1959) Leucoanthocyanins in seeds. *Qualitas Plantarum et Materiae Vegetabiles* 5, 189–198.

Beaulieu, A.D., Olubobokun, J.A. and Christensen, D.A. (1990) The utilization of canola and its constituents by lactating dairy cows. *Animal Feed Science and Technology* 30, 289–300.

Bell, J.M. (1965) Growth depressing factors in rapeseed meal. VI. Feeding value for growing–finishing swine of myrosinase-free, solvent-extracted meal. *Journal of Animal Science* 24, 1147–1151.

Bell, J.M. (1993a) Factors affecting the nutritional value of canola meal: a review. *Canadian Journal of Animal Science* 73, 679–697.

Bell, J.M. (1993b) Nutritional evaluation of dehulled canola meal for swine. In: *Tenth Project Report, Research on Canola Seed, Oil and Meal.* Canola Council of Canada, Winnipeg, Canada, pp. 63–72.

Bell, J.M. (1993c) Composition of canola meal. In: *Canola Meal. Feed Industry Guide.* Canola Council of Canada, Winnipeg, Canada, pp. 7–11.

Bell, J.M. and Jeffers, H.F. (1976) Variability in the chemical composition of rapeseed meal. *Canadian Journal of Animal Science* 56, 269–273.

Bell, J.M. and Keith, M.O. (1986) Growth, feed utilization and carcass quality responses of pigs fed frost-damaged canola seed (low-glucosinolate rapeseed) as affected by grinding, pelleting and ammoniation. *Canadian Journal of Animal Science* 66, 181–190.

Bell, J.M. and Keith, M.O. (1988) Effects of barley hulls, dietary protein level and weight of pig on digestibility of canola meal fed to finishing pigs. *Canadian Journal of Animal Science* 68, 493–502.

Bell, J.M. and Keith, M.O. (1989) Factors affecting the digestibility by pigs of energy and protein in wheat, barley and sorghum diets supplemented with canola meal. *Animal Feed Science and Technology* 24, 253–265.

Bell, J.M. and Keith, M.O. (1991) A survey of variation in the chemical composition of commercial canola meal produced in Western Canadian crushing plants. *Canadian Journal of Animal Science* 71, 469–480.

Bell, J.M. and Keith, M.O. (1993) Effects of combinations of wheat, corn or hulless barley with hulled barley supplemented with soybean meal or canola meal on growth rate, effect of feed utilization and carcass quality on market pigs. *Animal Feed Science and Technology* 43, 129–150.

Bell, J.M. and Shires, A. (1980) Effects of rapeseed dockage content on the feeding value of rapeseed meal for swine. *Canadian Journal of Animal Science* 60, 953–960.

Bell, J.M. and Shires, A. (1982) Composition and digestibility by pigs of hull fractions from rapeseed cultivars with yellow or brown seed coats. *Canadian Journal of Animal Science* 62, 557–565.

Bell, J.M. and Williams, K. (1953) Growth depressing factors in rapeseed oilmeal. *Canadian Journal of Agricultural Science* 33, 201–209.

Bell, J.M., Youngs, C.G. and Downey, R.K. (1971) A nutritional comparison of various rapeseed and mustard seed solvent-extracted meals of different glucosinolate composition. *Canadian Journal of Animal Science* 51, 259–269.

Bell, J.M., Sharby, T.F. and Sarwar, G. (1976) Some effects of processing and seed source on the nutritional quality of rapeseed. *Canadian Journal of Animal Science* 56, 809–816.

Bell, J.M., Shires, A., Blake, J.A., Campbell, S. and McGregor, D.I. (1981) Effect of alkali treatment and amino acid supplementation on the nutritive value of yellow and Oriental mustard meal for swine. *Canadian Journal of Animal Science* 61, 783–792.

Bell, J.M., Keith, M.O., Blake, J.A. and McGregor, D.I. (1984) Nutritional evaluation of ammoniated mustard meal for use in swine feeds. *Canadian Journal of Animal Science* 64, 1023–1033.

Bell, J.M., Keith, M.O. and Kowalenko, W.S. (1985) Digestibility and feeding value of frost-damaged canola seed (low glucosinolate rapeseed) for growing pigs. *Canadian Journal of Animal Science* 65, 735–743.

Bell, J.M., Keith, M.O., Darroch, C.S. and McGregor, D.I. (1987) Effects of ammoniation of canola seed contaminated with wild mustard seed on growth, feed utilization and carcass characteristics of pigs. *Canadian Journal of Animal Science* 67, 113–125.

Bell, J.M., Keith, M.O. and Hutcheson, D.S. (1991) Nutritional evaluation of very low glucosinolate canola meal. *Canadian Journal of Animal Science* 71, 497–506.

Belzile, R., Bell, J.M. and Wetter, L.R. (1963) Growth depressing factors in rapeseed oil meal. V. Effects of myrosinase activity on the toxicity of the meal. *Canadian Journal of Animal Science* 43, 169–173.

Beste, R., Fontaine, J. and Heinbeck, W. (1992) Research evaluates amino acid composition of canola meal. *Feedstuffs* 18, 16–17.

Bezeau, L.M., Slen, S.B. and Whiting, F. (1960) The nutritional value of rapeseed oil meal for lamb and wool production in mature range ewes. *Canadian Journal of Animal Science* 40, 37–43.

Bjerg, B., Eggum, B.O., Larsen, L.M. and Sørenson, H. (1987) Acceptable concentrations of glucosinolates in double low oilseed rape and possibilities of further quality improvements by processing and plant breeding. In: *Proceedings of the Seventh International Rapeseed Congress*, Poznan, Poland. The Plant Breeding and Acclimatization Institute, Poznan, pp. 1619–1626.

Bjergegaard, C., Eggum, B.O., Jensen, S.K. and Sørensen, H. (1991) Dietary fibres in oilseed rape: physiological and antinutritional effects in rats of isolated IDF and SDF added to a standard diet. *Journal of Animal Physiology and Animal Nutrition* 66, 69–79.

Blair, R. (1984) Nutritional evaluation of ammoniated mustard meal for chicks. *Poultry Science* 63, 754–759.

Blair, R., Gagnon, J., Salmon, R.E. and Pickard, M.D. (1986) Evaluation of spent bleaching clay as a feed supplement in layer diets. *Poultry Science* 65, 1990–1992.

Bougon, M., Boixel, J.L. and Tomassone, R. (1988) Use of dehusked rapeseed 00 oilcake in chickens. *Bulletin d'Information Station Expérimentale d'Aviculture de Ploufragan*, Ploufragan, France, 28, 155–162.

Bourdillon-Sanders, A. (1988) *Les Pellicules de colza dans l'alimentation du lapin.* Centre Technique Interprofessionel de Oléagineux Métropolitains, Paris, France.

Bourdon, D. (1986) Nutritive value of new oilmeals and whole seeds of rape low in glucosinolates for fattening pigs. *Journées de la Recherche Porcine en France* 18, 13–28.

Bourdon, D. and Aumaître, A. (1990) Low-glucosinolate rapeseeds and rapeseed meals: Effect of technological treatments on chemical composition, digestible energy content and feeding value for growing pigs. *Animal Feed Science and Technology* 30, 175–191.

Buret, Y. (1991) *Clinical Case: Acute Poisoning by Nitrates in a Herd of Milking Cows* [in French]. Centre Vétérinaire de Riaille, France, No. 1, pp. 67–68.

Bush, R.S., Nicholson, J.W.G., MacIntyre, T.M. and McQueen, R.E. (1978) A comparison of Candle and Tower rapeseed meals in lamb, sheep and beef steer rations. *Canadian Journal of Animal Science* 58, 369–376.

Campbell, L.D. (1987) Intact glucosinolates and glucosinolate hydrolysis products as causative agents in liver hemorrhage in laying hens. *Nutrition Reports International* 36, 491–496.

Campbell, L.D. and Slominski, B.A. (1991) Nutritive quality of low-glucosinolate canola meal for laying hens. In: McGregor, D.I. (ed.) *Proceedings of the Eighth International Rapeseed Congress*, Saskatoon, Canada. Organizing Committee, Saskatoon, pp. 442–447.

Castaing, J. and Grosjean, F. (1986) Rapeseed of 00 variety combined with wheat or maize in the diet of bacon pigs [in French]. *Journées de la Recherche Porcine en France* 18, 29–33.

Castell, A.G. (1977) Effects of virginiamycin on the performance of pigs fed barley diets supplemented with soybean meal or low-glucosinolate rapeseed meal. *Canadian Journal of Animal Science* 57, 313–320.

Castell, A.G. (1980) Effects of relative contributions of cereal and canola rapeseed meal to the dietary protein on the performance of growing–finishing pigs. *Canadian Journal of Animal Science* 60, 709–716.

Castell, A.G. and Falk, L. (1980) Effects of dietary canola seed on pig performance and backfat composition. *Canadian Journal of Animal Science* 60, 795–797.

Cherian, G. and Sim, J.S. (1991) Effect of feeding full fat flax and canola seeds to laying hens on the fatty acid composition of eggs, embryos, and newly hatched chicks. *Poultry Science* 70, 917–922.

Christison, G.I. and Bell, J.M. (1993) Dehulled canola meal for weanling pigs. In: *Tenth Project Report, Research on Canola Seed, Oil and Meal.* Canola Council of Canada, Winnipeg, Canada, pp. 73–75.

Clandinin, D.R. and Robblee, A.R. (1983a) Apparent and true metabolizable energy values for low glucosinolate-type rapeseed meal. *Feedstuffs* 55, 5 and 20.

Clandinin, D.R. and Robblee, A.R. (1983b) Canola meal can be good source of high-quality protein for poultry: Canadian researchers. *Feedstuffs* 55, 36–37.

Claypool, D.W., Hoffman, C.H., Oldfield, J.E. and Adams, H.P. (1985) Canola meal, cottonseed, and soybean meals as protein supplements for calves. *Journal of Dairy Science* 68, 67–70.

Coombe, J.B. (1985) Rape and sunflower seed meals as supplements for sheep fed oat straw. *Australian Journal of Agricultural Research* 36, 717–728.

Cromwell, G.L., Stahly, T.S. and Randolph, J.H. (1989) Canola seed as a protein and energy source for growing–finishing pigs. In: *Swine Research Progress Report #321.* University of Kentucky, Lexington, USA, pp. 44–45.

Cymbaluk, N. (1990) Using canola meal in growing draft horse diets. *Equine Practice* 12, 16–19.

Dale, N.M. and Fuller, H.L. (1986) Repeatability of true metabolizable energy versus nitrogen corrected true metabolizable energy values. *Poultry Science* 65, 352–354.

de Boland, A.R., Garner, G.B. and O'Dell, B.L. (1975) Identification and properties of phytate in cereal grains and oilseed products. *Journal of Agricultural Food and Chemistry* 23, 1186–1189.

DeClerq, D.R., Daun, J.K. and Tipples, K.H. (1989) *Quality of Western Canadian Canola.* Grain Research Laboratory, Canadian Grain Commission, Winnipeg, Canada.

DePeters, E.J. and Bath, D.L. (1986) Canola versus cottonseed meal as the protein supplement in dairy diets. *Journal of Dairy Science* 69, 148–154.

de Souza, R., Melcion, J.P., Bourdon, D., Giboulot, G., Peiniau, J. and Aumaître, A. (1990) Raw vs extruded whole rapeseed: a raw source of energy and protein in the diet for weaned piglets. *Journées de la Recherches Porcine en France* 22, 151–158.

Dierick, N.A., Vervaeke, I.J., Demeyer, D.I. and Decuypere, J.A. (1987) Approach to the energetic importance of fiber digestion in pigs. 1. Importance of fermentation in the overall energy supply. In: *Proceedings of Organization of Economic Cooperation and Development Workshop on Dietary Fiber in Monogastric Nutrition*, Cornell University, Ithaca, New York, USA, 20 pp.

Dixon, P.M. and Mcgorum, B. (1990) Oilseed rape and equine respiratory disease. *Veterinary Record* 126, 585.

Eggum, B.O. (1973) *A Study of Certain Factors Influencing Protein Utilization in Rats and Pigs.* Institute of Animal Science, Copenhagen, Denmark, Publication 406, pp. 1–173.

Elwinger, K. (1983) Rapeseed meal of low-glucosinolate type fed to broiler chickens and laying hens. In: *Proceedings of the Fourth European Symposium on Poultry Nutrition*, Tours, France, pp. 199–207.

Emanuelson, M., Murphy, M. and Lindberg, J.-E. (1991) Effects of heat-treated and untreated full-fat rapeseed and tallow on rumen metabolism, digestibility, milk composition and milk yield in lactating cows. *Animal Feed Science and Technology* 34, 291–309.

Emanuelson, M., Ahlin, K.A. and Wiktorsson, H. (1993) Long-term feeding of rapeseed meal and full-fat rapeseed of double low cultivars to dairy cows. *Livestock Production Science* 33, 199–214.

Erdman, J.W. (1979) Oilseed phytates: nutritional implications. *Journal of American Oil Chemists' Society* 56, 736–741.

Etienne, M. and Dourmad, J.Y. (1987) Effets de la consommation de tourteau de colza normal ou à faible teneur en glucosinolates sur la réproduction chez la truie. *Journées de la Recherche Porcine en France* 19, 231–238.

Fales, S.L., Gustine, D.L., Bosworth, S.C. and Hoover, R.J. (1987) Concentrations of glucosinolates and S-methylcysteine sulfoxide in ensiled rape (*Brassica napus* L.). *Journal of Dairy Science* 70, 2402–2405.

Fenwick, G.R. and Curtis, R.F. (1980) Rapeseed meal and its use in poultry diets. A review. *Animal Feed Science and Technology* 5, 255–298.

Fenwick, G.R., Hobson-Frohock, A., Land, D.G. and Curtis, R.F. (1979) Rapeseed meal and egg taint: treatment of rapeseed meal to reduce tainting potential. *British Poultry Science* 20, 323–329.

Ferlay, A., Legay, F., Bauchart, D., Poncet, C. and Doreau, M. (1992) Effect of a supply of raw or extruded rapeseeds on digestion in dairy cows. *Journal of Animal Science* 70, 915–923.

Finlayson, A.J. (1974) The amino acid composition of rapeseed hulls. *Canadian Journal of Animal Science* 54, 495–496.

Fisher, L.J. (1980) A comparison of rapeseed meal and soybean meal as a source of protein and protected lipid as a source of supplemental energy for calf starter diets. *Canadian Journal of Animal Science* 60, 359–366.

Flipot, P. and Dufour, J.J. (1977) Reproductive performance of gilts fed rapeseed meal cv. Tower during gestation and lactation. *Canadian Journal of Animal Science* 57, 567–571.

Freig, A.A.H., Campbell, L.D. and Stanger, N.E. (1987) Fate of ingested glucosinolates in poultry. *Nutrition Reports International* 36, 1337–1345.

Goh, Y.K., Clandinin, D.R., Robblee, A.R. and Darlington, K. (1979) The effect of level of sinapine in a laying ration on the incidence of fishy odor in eggs from brown-shelled egg layers. *Canadian Journal of Animal Science* 59, 313–316.

Grandegger, K.J.T. and Steingass, H. (1990) Effects of 00-rapeseed meal on dry matter intake, milk yield and milk composition of cows [in German]. *Agribiological Research* 43, 82–90.

Grandhi, R.R., Narendran, R., Bowman, G.H. and Slinger, S.J. (1980) A comparison of soybean meal and Tower rapeseed meal as supplements to corn in diets of growing-finishing and heavy weight pigs. *Canadian Journal of Animal Science* 60, 123–130.

Green, S. and Kiener, T. (1989) Digestibilities of nitrogen and amino acids in soya-bean, sunflower, meat and rapeseed meals measured with pigs and poultry. *Animal Production* 48, 157–179.

Grosjean, F., Fekete, J. and Gatel, F. (1989) Utilization of low glucosinolate fullfat rapeseed by weaned piglets and growing finishing pigs. In: *Proceedings of the 40th Meeting of the European Association for Animal Production*, Dublin, Ireland, 4 pp.

Han, M.-S., Froseth, J.A., Peters, D.N., Honeyfield, D.C. and Busboom, J. (1990) Performance of growing pigs fed three different types of full-fat rapeseed (canola) at two dietary levels. *Proceedings of Washington State University Swine Information Day*, Pullman, Washington, USA, 5, 47–51.

Higgs, D.A., Dosanjh, B.S., Prendergast, A.F., Beames, R.M., Hardy, R.W., Riley, W. and Deacon, G. (1994) Use of rapeseed/canola protein products in finfish diets. In: Lessa, D. and Lim, C. (eds) *Nutrition and Utilization Technology in Aquaculture*. American Oil Chemists' Society, Champaigne, Illinois, USA, pp. 130–156.

Hill, R. (1979) A review of the 'toxic' effects of rapeseed meals with observations on meal from improved varieties. *British Veterinary Journal* 135, 3–16.

Hill, R. (1991) Rapeseed meal in the diets of ruminants. *Nutrition Abstracts and Reviews* 61, 139–155.

Huhtanen, P. and Poutiainen, E. (1985) Effect of fullfat rapeseed on digestibility and rumen fermentation in cattle. *Journal of the Scientific Agricultural Society of Finland* 57, 67–73.

Hulan, H.W. and Proudfoot, F.G. (1981a) Replacement of soybean meal in chicken broiler diets by rapeseed meal and fish meal complementary sources of dietary protein. *Canadian Journal of Animal Science* 61, 999–1004.

Hulan, H.W. and Proudfoot, F.G. (1981b) Performance of laying hens fed diets containing soybean gums, rapeseed gums or rapeseed meals with or without gums. *Canadian Journal of Animal Science* 61, 1031–1040.

Hulan, H.W., Proudfoot, F.G. and McRae, K.B. (1980) The nutritional value of Tower and Candle rapeseed meals for turkey broilers housed under different lighting conditions. *Poultry Science* 59, 100–109.

Jarl, F. (1951) Experiments on the feeding of Swedish rape-seed oil meal to dairy cows [in Danish]. In: *Kungl. Lantbrukshögskolan och Statens Lantbruksförsök*. Statens Husdjursförsök, Meddelande Nr 45, Denmark, pp. 1–42.

Jensen, S.K., Olsen, H.S. and Sørensen, H. (1990) Aqueous enzymatic processing of rapeseed for production of high quality products. In: Shahidi, F. (ed.) *Canola and Rapeseed. Production, Chemistry, Nutrition and Processing Technology*. Van Nostrand Reinhold, New York, pp. 331–343.

Jones, J.D. and Sibbald, I.R. (1979) The true metabolizable energy values for poultry of fractions of rapeseed (*Brassica napus* cv. Tower). *Poultry Science* 58, 385–391.

Josefsson, E. (1971) Studies of the biochemical background to differences in glucosinolate content of *Brassica napus* L. 1. Glucosinolate content in relation to general chemical composition. *Physiologia Plantarum* 24, 150–159.

Just, A., Jørgensen, H., Fernandez, J.A., Bech-Andersen, A. and Engaard Hansen, N. (1983) Forskellige Forderstoff ers kemiske sammensætning, fordøjelighed, energi- og proteinvaædi til svin. In: *Beretning fra Statens Husdyrbruqsforsøg*. Copenhagen, Denmark, 99 pp.

Kawakishi, S.K. and Kaneko, T. (1987) Interaction of proteins with allylisothiocyanate. *Journal of Agricultural Food and Chemistry* 35, 85–88.

Keith, M.O. and Bell, J.M. (1983) Effects of ammonia and steam treatments on the composition and nutritional value of canola (low glucosinolate rapeseed) screenings in diets for growing pigs. *Canadian Journal of Animal Science* 63, 429–441.

Keith, M.O. and Bell, J.M. (1985) Amino acid supplementation of ammoniated mustard meal for use in swine feeds. *Canadian Journal of Animal Science* 65, 937–944.

Keith, M.O. and Bell, J.M. (1986) Effects of feeding spent bleaching clay from canola oil refining to growing mice and rats. *Canadian Journal of Animal Science* 66, 191–199.

Keith, M.O. and Bell, J.M. (1987a) Effects of canola meal on absorption and tissue levels of trace minerals in rats. *Canadian Journal of Animal Science* 67, 141–149.

Keith, M.O. and Bell, J.M. (1987b) Effects of pH and canola meal on trace mineral solubility *in vitro*. *Canadian Journal of Animal Science* 67, 587–590.

Keith, M.O. and Bell, J.M. (1991) Composition and digestibility of canola press cake as a feedstuff for use in swine diets. *Canadian Journal of Animal Science* 71, 879–885.

Kennelly, J.J., Khorasani, G.R., Robinson, P.H. and de Boer, G. (1993) Effect of jet-sploding and extrusion on the nutritive value of canola meal and whole canola seed for dairy cattle. In: *Tenth Project Report, Research on Canola Seed, Oil and Meal*. Canola Council of Canada, Winnipeg, Canada, pp. 30–159.

Kiiskinen, T. (1983) The effect of diets supplemented with Regent rapeseed meal on performance of broiler chicks. *Annales Agriculturae Fenniae* 22, 206–213.

Klein, K.K., Salmon, R.E. and Larmond, M.E. (1979) A linear programming model for determining the optimum level of low glucosinolate rapeseed meal in diets of growing turkeys. *Canadian Journal of Agricultural Economics* 27, 61–73.

Klein, K.K., Salmon, R.E. and Gardiner, E.E. (1981) Economic analysis of the use of canola meal in diets for broiler chickens. *Canadian Journal of Agricultural Economics* 29, 327–338.

Kozlowski, M., Faruga, A., Mikulski, D., Bock, H.D., Kozlowska, H., Rotkiewicz, D. and Kozlowski, K. (1989) Fattening and slaughter results in broiler chickens fed diets containing rapeseed meal from double zero varieties. *Nahrung* 33, 617–623.

Laarveld, B. and McClean, C.A. (1991) Use of canola meal as a protein supplement for growth-hormone treated dairy cows. In: *Ninth Project Report, Research on Canola Seed, Oil and Meal.* Canola Council of Canada, Winnipeg, Canada, pp. 260–264.

Lall, S.P. and Slinger, S.J. (1973a) The metabolizable energy content of rapeseed oils and rapeseed oil foots and the effects of blending with other fats. *Poultry Science* 52, 143–151.

Lall, S.P. and Slinger, S.J. (1973b) Nutritional evaluation of rapeseed oils and rapeseed soapstocks for laying hens. *Poultry Science* 52, 1729–1740.

Lancaster, L.L., Hunt, C.W., Miller, J.C., Auld, D.L. and Nelson, M.L. (1990) Effects of rapeseed silage variety and dietary level on digestion and growth performance of beef steers. *Journal of Animal Science* 68, 3812–3820.

Larsen, L.M., Olsen, O., Plöger, A. and Sørensen, H. (1983) Phenolic choline esters in rapeseed: possible factors affecting nutritive value and quality of rapeseed meal. In: *Proceedings of the Sixth International Rapeseed Conference*, Paris, France, pp. 1577–1582.

Lebas, F., Seroux, M. and Franck, Y. (1981) The use of rapeseed hulls for feeding the growing rabbit. Fattening results. *Informations Techniques, Centre Technique Interprofessionel de Oléagineux Metropolitains*, Paris, France, 76 III, 18–23.

Leclercq, B., Lessire, M., Guy, G., Hallouis, J.M. and Conan, J.M. (1989) Utilisation de la graine de colza en aviculture. *Institut National de la Recherche Agronomigue, Productions Animales* 2, 129–136.

Lee, P.A. and Hill, R. (1985) Studies on rapeseed meal from different varieties of rape in the diets of gilts. 1. Effects on attainment of puberty, ovulation rate, conception and embryo survival of the first litter. *British Veterinary Journal* 141, 581–591.

Lee, P.A., Pittam, S. and Hill, R. (1984) The voluntary food intake by growing pigs of diets containing 'treated' rapeseed meals or extracts of rapeseed meal. *British Journal of Nutrition* 52, 159–164.

Leeson, S., Slinger, S.J. and Summers, J.D. (1977) Performance of laying hens fed diets containing gums derived from Tower rapeseed. *Canadian Journal of Animal Science* 57, 479–483.

Leeson , S., Slinger, S.J. and Summers, J.D. (1978) Utilization of whole Tower rapeseed by laying hens and broiler chickens. *Canadian Journal of Animal Science* 58, 55–61.

Leslie, A.J. and Summers, J.D. (1972) Feeding value of rapeseed for laying hens. *Canadian Journal of Animal Science* 52, 563–566.

Lessire, M., Buadet, J.J., Hallouis, J.M. and Conan, L. (1991) Metabolizable energy content and protein digestibility of rapeseed hulls in adult cockerels. In: McGregor, D.I. (ed.) *Proceedings of the Eighth International Rapeseed Congress*, Saskatoon, Canada. Organizing Committee, Saskatoon, pp. 1585–1590.

Leung, J., Fenton, T.W., Mueller, M.M. and Clandinin, D.R. (1979) Condensed tannins of rapeseed meal. *Journal of Food Science* 44, 1313–1316.

Lewis, A.J., Aherne, F.X. and Hardin, R.T. (1978) Reproductive performance of sows fed low glucosinolate (Tower) rapeseed meal. *Canadian Journal of Animal Science* 58, 203–208.

Livingstone, R.M. and Jones, A.S. (1980) The digestibility of a new variety of forage rape (*Brassica napus*) by sows. *Animal Production* 30, 494 [Abstract].

Lloyd, L.E. and Crampton, E.W. (1955) The apparent digestibility of the crude protein of the pig ration as a function of its crude protein and crude fiber content. *Journal of Animal Science* 14, 693–699.

Lodhi, G.N., Renner, R. and Clandinin, D.R. (1970) Factors affecting the metabolizable energy value of rapeseed meal. *Poultry Science* XLIX, 991–999.

Lough, D.S., Solomon, M.B., Rumsey, T.S., Elsasser, T.H., Slyter, L.L., Kahl, S. and Lynch,

G.P. (1991) Effects of dietary canola seed and soy lecithin in high-forage diets on performance, serum lipids, and carcass characteristics of growing ram lambs. *Journal of Animal Science* 69, 3292-3298.

Love, H.K., Rakow, G., Raney, J.P. and Downey, R.K. (1990) Development of low glucosinolate mustard. *Canadian Journal of Animal Science* 70, 419-424.

McCuaig, L.W. and Bell, J.M. (1981) Effects of rapeseed gums on the feeding value of diets for growing–finishing pigs. *Canadian Journal of Animal Science* 61, 463-467.

McCurdy, S.M. and March, B.E. (1992) Processing of canola meal for incorporation in trout and salmon diets. *Journal of American Oil Chemists' Society* 69, 213-220.

McGregor, D.I., Blake, J.A. and Pickard, M.D. (1983) Detoxification of *Brassica juncea* with ammonia. In: *Proceedings of the Sixth International Rapeseed Conference*, Paris, France, pp. 1426-1431.

Mcintosh, M.K. and Aherne, F.X. (1981) An evaluation of canola meal (var: Candle) as a protein supplement for starter pigs (3-8 weeks of age). In: *Agriculture and Forestry Bulletin, 60th Annual Feeder's Day Report*. Department of Animal Science, University of Alberta, Edmonton, Canada, pp. 16-17.

McKinnon, J.J., Cohen, R.D.H., Olubobokun, J.A. and Christensen, D.A. (1993) Supplemental protein sources for feedlot cattle. In: *Tenth Project Report, Research on Canola Seed, Oil and Meal*. Canola Council of Canada, Winnipeg, Canada, pp. 165-171.

McKinnon, P.J. and Bowland, J.P. (1977) Comparison of low glucosinolate–low erucic acid rapeseed meal (cv. Tower), commercial rapeseed meal and soybean meal as sources of protein for starting, growing and finishing pigs and young rats. *Canadian Journal of Animal Science* 57, 663-678.

McLeod, G.K. (1991) Canola meal as a protein supplement in corn-based dairy rations. In: *Ninth Project Report, Research on Canola Seed, Oil and Meal*. Canola Council of Canada, Winnipeg, Canada, pp. 255-259.

Maga, J.A. (1982) Phytate: its chemistry, occurrence, food interactions, nutritional significance, and methods of analysis. *Journal of Agricultural Food and Chemistry* 30, 1-9.

March, B.E. (1977) Response of chicks to the feeding of different rapeseed oils and rapeseed oil fractions. *Canadian Journal of Animal Science* 57, 137-140.

March, B.E. and MacMillan, C. (1980) Choline concentration and availability in rapeseed meal. *Poultry Science* 59, 611-615.

March, B.E. and Sadiq, M. (1974) *In vitro* and *in vivo* studies on the digestibility of rapeseed meal. In: *Third Progress Report, Research on Rapeseed Seed, Oil and Meal*. Canola Council of Canada, Winnipeg, Canada, pp. 70-77.

March, B.E., Smith, T. and El-Lakany, S. (1973) Variation in estimates of the metabolizable energy value of rapeseed meal determined with chickens of different ages. *Poultry Science* 52, 614-618.

Mathison, G.W. (1978) Rapeseed gum in finishing diets for steers. *Canadian Journal of Animal Science* 58, 139-142.

May, R.W. and Bell, J.M. (1971) Digestible and metabolizable energy values of some feeds for the growing pig. *Canadian Journal of Animal Science* 51, 271-278.

Maynard, L.A., Loosli, J.K., Hintz, H.F. and Warner, R.G. (1979) *Animal Nutrition*, 7th edn. McGraw-Hill, New York, pp. 408-409.

Michaelsen, S., Otte, J., Simonsen, L.-O. and Sørensen, H. (1994) Absorption and degradation of individual intact glucosinolates in the digestive tract of rodents. *Acta Agriculturae Scandavica, Section A, Animal Science* 44, 25-37.

Mitaru, B.N., Blair, R., Bell, J.M. and Reichert, R. (1983) Effect of canola hulls on growth, feed efficiency, and protein and energy utilization in broiler chickens. *Canadian Journal of Animal Science* 63, 655-662.

Mitaru, B., Reichert, R.D. and Blair, R.B. (1984) Kinetics of tannin deactivation during anaerobic storage and boiling treatments of high tannin sorghums. *Journal of Food Science* 49, 1566-1568, 1583.

Moody, D.L., Slinger, S.J., Leeson, S. and Summers, J.D. (1978) Utilization of dietary Tower rapeseed products by growing turkeys. *Canadian Journal of Animal Science* 58, 585-592.

Morand-Fehr, P. and Hervieu, J. (1983) Essai d'appréciation de l'acceptabilité du tourteau de colza par tests sur caprins. In: *Proceedings of the Sixth International Rapeseed Conference*, Paris, France, pp. 1637-1642.

Morrall, R.A.A., Loew, F.M. and Hayes, M.A. (1978) Subacute toxicological evaluation of sclerotia of *Sclerotinia sclerotiorum* in rats. *Journal of Comparative Medicine* 42, 473–477.

Moser, D.B. and Lewis, A.J. (1980) Adding fat to sow diets – an update. *Feedstuffs* 52, 36.

Murphy, M., Uden, P., Palmquist, D.L. and Wiktorsson, H. (1987) Rumen and total diet digestibility in lactating cows fed diets containing fullfat rapeseed. *Journal of Dairy Science* 70, 1572–1582.

Muztar, A.J., Likuski, H.J. and Slinger, S.J. (1978) Metabolizable energy content of Tower and Candle rapeseeds and rapeseed meals determined in two laboratories. *Canadian Journal of Animal Science* 58, 485–492.

Naczk, M. and Shahidi, F. (1990) Carbohydrates of canola and rapeseed. In: Shahidi, F. (ed.) *Canola and Rapeseed. Production, Chemistry, Nutrition and Processing Technology.* Van Nostrand Reinhold, New York, pp. 211–231.

Nagy, J.G. and Furtan, W.H. (1977) *The Socioeconomic Costs and Returns from Rapeseed Breeding in Canada.* Technical Bulletin BL:77-1, Department of Agricultural Economics, University of Saskatchewan, Saskatoon, Canada.

Nalecz, T. (1986) Effect of feeding rape seeds on productivity and chemical composition of cow milk 1. Productivity and basic chemical composition of milk [in Polish]. *Prace i Materialy Zootechnicze* 37, 53–65.

Nassar, A.R., Goeger, M.P. and Arscoft, G.H. (1985) Effect of canola meal in laying hen diets. *Nutrition Reports International* 31, 1349–1355.

National Research Council (1989) *Nutrient Requirements of Dairy Cattle,* 6th revised edn. National Academy Press, Washington, DC, USA.

Nelson, T.S. (1967) The utilization of phytate phosphorus by poultry – a review. *Poultry Science* 46, 862–871.

Noblet, J., Fortune, H., Dubois, S. and Henry, Y. (1989) *New Approaches for Estimating Digestible Metabolizable and Net Energy Values in Pig Feeds.* Institut National de la Recherche Agronomigue, Station de Recherches Porcines, St. Gilles, France, 106 pp.

Noblet, J., Fortune, H., Dupire, C. and Dubois, S. (1993) Digestible, metabolizable and net energy values of 13 feedstuffs for growing pigs: effect of energy system. *Animal Feed Science and Technology* 42, 131–149.

Norton, G. (1989) Nature and biosynthesis of storage proteins. In: Röbbelen, G., Downey, R.K. and Ashri, A. (eds) *Oil Crops of the World.* McGraw-Hill, New York, pp. 165–191.

Norton, G. and Harris, J.F. (1975) Compositional changes in developing rape seed (*Brassica napus* L.). *Planta (Berlin)* 123, 163–174.

Nugon-Baudon, L., Szylit, O. and Raibaud, P. (1988) Production of toxic glucosinolate derivatives from rapeseed meal by intestinal microflora of rat and chicken. *Journal of the Science of Food and Agriculture* 43, 299–308.

Nwokolo, E. and Bragg, D.B. (1977) Influence of phytic acid and crude fiber on the availability of minerals from four protein supplements in growing chicks. *Canadian Journal of Animal Science* 57, 475–477.

Nwokolo, E. and Bragg, D.B. (1980) Biological availability of minerals in rapeseed meal. *Poultry Science* 59, 155–158.

Ochetim, S., Bell, J.M., Doige, C.E. and Youngs, C.G. (1980) The feeding value of Tower rapeseed for early-weaned pigs 1. Effect of methods of processing and of dietary levels. *Canadian Journal of Animal Science* 60, 407–421.

Oginsky, E.L., Stein, A.E. and Greer, M.A. (1965) Myrosinase activity in bacteria as demonstrated by the conversion of progoitrin to goitrin. *Proceedings of the Society of Experimental Biological Medicine* 119, 360–367.

Olsson, A.C., Bjorklund, K. and Thomke, S. (1990) Use of triple-low rapeseed meal for pigs and poultry. *Husdour,* 4 pp.

Partridge, I.G., Low, A.G. and Matte, J.J. (1987) Double-low rapeseed meal for pigs: ileal, apparent digestibility of amino acids in diets containing various proportions of rapeseed meal, fish meal and soya-bean meal. *Animal Production* 44, 415–420.

Proudfoot, F.G., Hulan, H.W. and McRae, K.B. (1982) The effect of diets supplemented with Tower and/or Candle rapeseed meals on performance of meat chicken breeders. *Canadian Journal of Animal Science* 62, 239–247.

Proudfoot, F.G., Hulan, H.W. and McRae, K.B. (1983) Effect of feeding poultry diets supplemented with rapeseed meal as a primary protein source to juvenile and adult meat breeder genotypes. *Canadian Journal of Animal Science* 63, 957–965.

Ranhotra, G.S. and Loewe, R.J. (1975) Effect of wheat phytate on dietary phytic acid. *Journal of Food Science* 40, 940–942.

Rao, P.V. and Clandinin, D.R. (1972) Chemical determination of available carbohydrates in rapeseed meal. *Poultry Science* 51, 1474–1475.

Reddy, N.R., Sathe, S.K. and Salunkhe, D.K. (1982) Phytates in legumes and cereals. *Advances in Food Research* 28, 1–92.

Rihs, T., Herzog, W. and Blum, J. (1991) Variation of S-methyl-L-cysteine sulfoxide content in fresh plants of 0 and 00-rape varieties. *GCIRC Bulletin* 7, 88–92.

Röbbelen, G. and Thies, W. (1980) Variation in rapeseed glucosinolates and breeding for improved meal quality. In: Tsunoda, S., Hinata, K. and Gomez-Campo, C. (eds) *Brassica Crops and Its Wild Allies. Biology and Breeding.* Japan Scientific Societies Press, Tokyo, pp. 285–299.

Robblee, A.R. and Clandinin, D.R. (1983) Use of canola meal in rations for turkey broilers. In: *Proceedings of the Sixth International Rapeseed Conference*, Paris, France, pp. 1534–1539.

Rose, S.P. and Bell, J.M. (1982a) Reproduction of mice fed low glucosinolate rapeseed meal contaminated with screenings meals and weed seed meals. *Canadian Journal of Animal Science* 62, 607–616.

Rose, S.P. and Bell, J.M. (1982b) Reproduction of mice fed low erucic acid rapeseed oil contaminated with weed seed oils. *Canadian Journal of Animal Science* 62, 617–624.

Roth-Maier, D.A. and Kirchgessner, M. (1988) Long-term feeding of canola meal to laying hens. *Landwirtschaftliche Forschung* 41, 140–150.

Rubin, L.J., Diosady, L.L. and Tzeng, Y.-M. (1990) Ultrafiltration in rapeseed processing. In: Shahidi, F. (ed.) *Canola and Rapeseed. Production, Chemistry, Nutrition and Processing Technology.* Van Nostrand Reinhold, New York, pp. 307–330.

Rule, D.C., Wu, W.H., Busboom, J.R., Hinds, F.C. and Kercher, C.J. (1989) Dietary canola seeds alter the fatty acid composition of bovine subcutaneous adipose tissue. *Nutritional Reports International* 39, 781–786.

Rundgren, M. (1983) Low glucosinolate rapeseed products for pigs. A review. *Animal Feed Science and Technology* 9, 239–262.

Rundgren, M. Askbrant, S. and Thomke, S. (1985) Nutritional evaluation of low- and high-glucosinolate rapeseed meals with pigs, laying hens and rats. *Swedish Journal of Agricultural Research* 15, 182–190.

Salmon, R.E. (1970) Rapeseed gum in poultry diets. *Canadian Journal of Animal Science* 50, 211–212.

Salmon, R.E. (1984a) True metabolizable energy and total and available amino acids of Candle, Altex, and Regent canola meals. *Poultry Science* 63, 135–138.

Salmon, R.E. (1984b) True metabolizable energy and dry matter contents of some feedstuffs. *Poultry Science* 63, 381–383.

Salmon, R.E., Gardinger, E.E., Klein, K.K. and Larmond, E. (1981) Effect of canola (low glucosinolate rapeseed) meal, protein and nutrient density on performance, carcass grade, and meat yield, and of canola meal on sensory quality of broilers. *Poultry Science* 60, 2519–2528.

Salmon, R.E., Stevens, V.I. and Ladbrooke, B.D. (1988) Full-fat canola seed as a feedstuff for turkeys. *Poultry Science* 67, 1731–1742.

Salmon, R.E., Stevens, V.I. and Ladbrooke, B.D. (1991) True amino acid availability of whole or flaked, raw or cooked full-fat or extracted canola seed. In: McGregor, D.I. (ed.) *Proceedings of the Eighth International Rapeseed Congress*, Saskatoon, Canada. Organizing Committee, Saskatoon, pp. 1591–1594.

Salo, M.-L. (1980) Nutritive value of full fat rapeseed for growing pigs. *Journal of the Scientific Agricultural Society of Finland* 52, 1–6.

Sarwar, G., Bell, J.M., Sharby, T.F. and Jones, J.D. (1981) Nutritional evaluation of meals and meal fractions derived from rape and mustard seed. *Canadian Journal of Animal Science* 61, 719–733.

Sauer, W.C. and Thacker, P.A. (1986) Apparent ileal and faecal digestibility of amino acids in barley-based diets supplemented with soya bean meal or canola meal for growing pigs. *Animal Feed Science and Technology* 14, 183–192.

Schneider, B.H. (1947) *Feeds of the World. Their Digestibility and Composition.* Agricultural Experimental Station, West Virginia University, Morgantown, USA.

Schoon, H.A., Brunckhorst, D. and Fehlberg, U. (1989) 00-Rape intoxication in roe deer [in German]. *Praktische Tierarzt* 70, 5-52.

Seth, P.C.C. and Clandinin, D.R. (1973) Metabolisable energy value and composition of rapeseed meal and of fractions derived therefrom by air-classification. *British Poultry Science* 14, 499-505.

Shahidi, F. and Naczk, M. (1990) Removal of glucosinolates and other antinutrients from canola and rapeseed by methanol/ammonia processing. In: Shahidi, F. (ed.) *Canola and Rapeseed. Production, Chemistry, Nutrition and Processing Technology.* Van Nostrand Reinhold, New York, pp. 291-306.

Shahidi, F. and Naczk, M. (1992) An overview of the phenolics of canola and rapeseed: chemical, sensory and nutritional significance. *Journal of American Oil Chemists' Society* 69, 917-924.

Sharma, H.R., Ingalls, J.R. and Deviin, T.J. (1980) Apparent digestibility of Tower and Candle rapeseed meals by Holstein bull calves. *Canadian Journal of Animal Science* 60, 915-918.

Shaw, J., Baidoo, S.K. and Aherne, F.X. (1990) Nutritive value of canola seed for young pigs. *Animal Feed Science and Technology* 28, 325-331.

Shi, X.S. and Noblet, J. (1993) Contribution of the hindgut to digestion of diets in growing pigs and adult sows: effect of diet composition. *Livestock Production Science* 34, 237-252.

Shires, A., Bell, J.M., Blair, R., Blake, J.A., Fedec, P. and McGregor, D.I. (1981) Nutritional value of unextracted and extracted dehulled canola rapeseed for broiler chickens. *Canadian Journal of Animal Science* 61, 989-998.

Shires, A., Bell, J.M., Keith, M.O. and McGregor, D.I. (1982) Rapeseed dockage. Effects of feeding raw and processed wild mustard and stinkweed seed on growth and feed utilization of mice. *Canadian Journal of Animal Science* 62, 275-285.

Shires, A., Bell, J.M., Laverty, W.H., Fedec, P., Blake, J.A. and McGregor, D.I. (1983) Effect of desolventization conditions and removal of fibrous material by screening on the nutritional value of canola rapeseed meal for broiler chickens. *Poultry Science* 62, 2234-2244.

Sibbald, I.R. (1982) Measurement of bioavailable energy in poultry feedingstuffs: a review. *Canadian Journal of Animal Science* 62, 983-1048.

Sibbald, I.R. (1986) *The TME System of Feed Evaluation: Methodology, Feed Composition Data and Bibliography.* Technical Bulletin 1986-4E, Animal Research Centre, Agriculture Canada, Ottawa, Canada.

Sibbald, I.R. (1987) Estimation of bioavailable amino acids in feedingstuffs for poultry and pigs: a review with emphasis on balance experiments. *Canadian Journal of Animal Science* 67, 221-300.

Sibbald, I.R. and Price, K. (1977) The true metabolizable energy values of the seeds of *Brassica campestris, B. hirta* and *B. napus. Poultry Science* 56, 1329-1331.

Sibbald, I.R., Hall, D.D., Wolynetz, M.S., Fernandez, J.A. and Jorgensen, H. (1990) Relationships between bioavailable energy estimates made with pigs and cockerels. *Animal Feed Science and Technology* 30, 131-142.

Siddiqui, I.R. and Wood, P.J. (1977) Carbohydrates of rapeseed: a review. *Journal of the Science of Food and Agriculture* 28, 530-538.

Slominski, B.A. and Campbell, L.D. (1989) Indoleacetonitriles – thermal degradation products of indole glucosinolates in commercial rapeseed (*Brassica napus*) meal. *Journal of the Science of Food and Agriculture* 47, 75-84.

Slominski, B.A. and Campbell, L.D. (1990) Non-starch polysaccharides of canola meal: quantification, digestibility in poultry and potential benefit of dietary enzyme supplementation. *Journal of the Science of Food and Agriculture* 53, 175-184.

Slominski, B.A. and Campbell, L.D. (1991) The carbohydrate content of yellow-seeded canola. In: McGregor, D.I. (ed.) *Proceedings of the Eighth International Rapeseed Congress*, Saskatoon, Canada. Organizing Committee, Saskatoon, pp. 1402-1407.

Slominski, B.A., Campbell, L.D. and Stanger, N.E. (1988) Extent of hydrolysis in the intestinal tract and potential absorption of intact glucosinolates in laying hens. *Journal of the Science of Food and Agriculture* 42, 305-314.

Smith, T.K. (1984) Spent canola oil bleaching clays: Potential for treatment of T-2 toxicosis in rats and short-term inclusion in diets for immature swine. *Canadian Journal of Animal Science* 64, 725–732.

Sørensen, H. (1990) Glucosinolates: structure–properties–function. In: Shahidi, F. (ed.) *Canola and Rapeseed. Production, Chemistry, Nutrition and Processing Technology.* Van Nostrand Reinhold, New York, pp. 149–172.

Spiegel, C., Bestefti, G., Rossi, G. and Blum, J.W. (1992) Feeding of rapeseed presscake meal to pigs: effects on thyroid morphology and function and on thyroid hormone blood levels, on liver and on growth performance. *Journal of Veterinary Medicine* 39, 1–13.

Spratt, R.S. and Leeson, S. (1985) The effect of raw ground full-fat canola on sow milk composition and piglet growth. *Nutrition Reports International* 31, 825–831.

Summers, J.D. and Leeson, S. (1977) Performance and carcass grading characteristics of broiler chickens fed rapeseed gums. *Canadian Journal of Animal Science* 57, 485–488.

Summers, J.D., Leeson, S. and Slinger, S.J. (1978) Performance of egg-strain birds during their commercial life cycle when continuously fed diets containing Tower rapeseed gums. *Canadian Journal of Animal Science* 58, 183–189.

Sutton, E.I. and Stredwick, R.V. (1979) Acceptance of rapeseed meal (cv. Candle) by horses. *Canadian Journal of Animal Science* 59, 819–820.

Tao, R., Belzile, R.J. and Brisson, G.J. (1971) Amino acid digestibility of rapeseed meal fed to chickens: effects of fat and lysine supplementation. *Canadian Journal of Animal Science* 51, 705–709.

Theander, O. and Åman, P. (1976) Low-molecular weight carbohydrates in rapeseed and turnip rapeseed meals. *Swedish Journal of Agricultural Research* 6, 81–85.

Theander, O., Åman, P., Miksche, G.E. and Yasuda, S. (1977) Carbohydrates, polyphenols, and lignin in seed hulls of different colors from turnip rapeseed. *Journal of Agricultural and Food Chemisty* 25, 270–273.

Theriez, M. and Brun, J.P. (1982) Valeur alimentaire des pellicules de colza. *Informations Techniques*, Centre Technique Interprofessionel de Oleagineux Metropolitans, Paris, France, 80 III, 14–19.

Thomke, S. (1984) Further experiments with RSM [rapeseed meal] of a Swedish low-glucosinolate type fed to growing–finishing pigs. *Swedish Journal of Agricultural Research* 14, 151–157.

Thomke, S., Elwinger, K., Rundgren, M. and Ahlström, B. (1983) Rapeseed meal of Swedish low-glucosinolate type fed to broiler chickens, laying hens and growing–finishing pigs. *Acta Agriculturae Scandinavica* 33, 153–174.

Thompson, L.U. (1990) Phytates in canola/rapeseed. In: Shahidi, F. (ed.) *Canola and Rapeseed. Production, Chemistry, Nutrition and Processing Technology.* Van Nostrand Reinhold, New York, pp. 173–192.

Thompson, L.U. and Serraino, M.R. (1986) Effect of phytic acid reduction on rapeseed protein digestibility and amino acid absorption. *Journal of Agricultural and Food Chemistry* 34, 468–469.

Throckmorton, J.C., Cheeke, P.R. and Patton, N.M. (1980) Tower rapeseed meal as a protein source for weanling rabbits. *Canadian Journal of Animal Science* 60, 1027–1028.

Tkachuk, R. (1969) Nitrogen-to-protein conversion factors for cereals and oilseed meals. *Cereal Chemistry* 46, 419–423.

Uppström, B. and Svensson, R. (1980) Determination of phytic acid in rapeseed meal. *Journal of the Science of Food and Agriculture* 31, 651–656.

Vincent, I.C., Williams, H.L. and Hill, R. (1988) Feeding British rapeseed meals to pregnant and lactating ewes. *Animal Production* 47, 283–289.

Virtanen, A.I., Kreula, M. and Kiesvaara, M. (1963) *Investigations on the Alleged Goitrogenic Properties of Milk.* Biochemical Institute, Helsinki, Finland.

Visentin, M., Lori, R., Valdicelli, L. and Palmieri, S. (1992) Trypsin inhibitory activity in some rapeseed genotypes. *Phytochemistry* 31, 3677–3680.

Vohra, P. (1989) Carbohydrate and fibre content of oilseeds and their nutritional importance. In: Röbbelen, G., Downey, R.K. and Ashri, A. (eds) *Oil Crops of the World. Their Breeding and Utilization.* McGraw-Hill, New York, pp. 208–225.

Vohra, P., Gray, G.A. and Kratzer, F.H. (1965) PA-metal complexes. *Proceedings of the Society for Experimental Biology and Medicine* 120, 447–449.

Waibel, P.E., Noll, S.L., Hoffbeck, S., Vickers, Z.M. and Salmon, R.E. (1992) Canola meal in diets for market turkeys. *Poultry Science* 71, 1059–1066.

Ward, A.T. and Reichert, R.D. (1986) Comparison of the effect of cell wall and hull fiber from canola and soybean on the bioavailability for rats of minerals, protein and lipid. *Journal of Nutrition* 116, 233–241.

West, G. and Norton, G. (1991) Rapeseed proteinase inhibitors. In: McGregor, D.I. (ed.) *Proceedings of the Eighth International Rapeseed Congress*, Saskatoon, Canada. Organizing Committee, Saskatoon, pp. 928–933.

Wight, P.A.L., Scougall, R.K., Shannon, D.W.F., Wells, J.W. and Mawson, R. (1987) Role of glucosinolates in the causation of liver haemorrhages in laying hens fed water-extracted or heat-treated rapeseed cakes. *Research in Veterinary Science* 43, 313–319.

Yannai, S. (1980) Toxic factors induced by processing. In: Liener, I.E. (ed.) *Toxic Constituents of Plant Foodstuffs*, 2nd edn. Academic Press, London, pp. 371–427.

Yapar, Z. and Clandinin, D.R. (1972) Effects of tannins in rapeseed meal on its nutritional value for chicks. *Poultry Science* 51, 222–228.

Youngs, G.G., Sallans, H.R. and Bell, J.M. (1971) Iron or copper catalytic decomposition of thioglucosides in rapeseed. *US Patent Office* 3, 560,217.

Zobelt, U., Gütlich, B., Marquard, R. and Daniel, P. (1991) Investigations of fodder quality of forage rape. In: McGregor, D.I. (ed.) *Proceedings of the Eighth International Rapeseed Congress*, Saskatoon, Canada. Organizing Committee, Saskatoon, pp. 1595–1600.

Zuprizal, M., Larbier, M., Chagneau, A.M. and Lessire, M. (1991a) Bioavailability of lysine in rapeseed and soyabean meals determined by digestibility trial in cockerels and chick growth assay. *Animal Feed Science and Technology* 35, 237–246.

Zuprizal, M., Larbier, M., Chagneau, A.M. and Lessire, M. (1991b) Effect of protein intake on true digestibility of amino acids in rapeseed meals for adult roosters force fed with moistened feed. *Animal Feed Science and Technology* 35, 255–260.

15 Industrial Utilization of Long-chain Fatty Acids and Their Derivatives

N.O.V. Sonntag
306 Shadow Wood Trail, Ovilla, Texas, USA

INTRODUCTION

Over the past three decades two long-chain fatty acids derived from *Brassica* oilseeds, behenic (docosanoic, 22 : 0) and erucic (13-docosenoic, 22 : 1), have seen a substantial application in the oleochemical industry. A third, arachidic (eicosanoic, 20 : 0), has also shown promise. Behenic acid and erucic acid have achieved their growth, and will continue to do so, as relatively pure products: behenic acid, 85 % or higher purity, and erucic acid, 90–95 % purity. Approaching the 21st century these three long-chain fatty acids show considerable promise for expanded use. This chapter reviews the suitability of *Brassica* oilseeds as a source of these fatty acids, current and potential production technology, industrial application of these fatty acids and their derivatives, and the economics of their production and use.

BRASSICA OILSEEDS AS A SOURCE OF LONG-CHAIN FATTY ACIDS

Brassica oilseeds, notably *Brassica napus*, are a major source of behenic and erucic acids.

Since the advent for the edible oil market of cultivars low in erucic acid, cultivars have been specifically bred for high erucic acid content. In Germany the *B. napus* winter cultivar Askari and in Canada the *B. napus* spring cultivars Reston, Hero and Mercury have been specifically bred as sources of high erucic acid seed. The *Brassica* oilseeds provide the most economical and readily available source of erucic acid although cultivars grown for the market contain only 45–55 % of their total fatty acids as erucic. The remainder, which is predominantly oleic, linoleic and linolenic (see Chapter 10), constitutes a low-value by-product which the oleochemical industry can have difficulty disposing.

Alternative sources are available but are less economical. Crambe (*Crambe abyssinica* Hochst. Ex R.E. Fries) contains somewhat higher erucic acid, 55–60 %, but still falls short of being economically acceptable because of the low value of its residual fatty acid by-product. Nasturtium (*Tropaeolum majus* L.) contains 80–85 % erucic acid in its seed oil but the seed contains only 7–10 % oil. Within the past two decades the oleochemical industry has turned to inexpensive, readily available fish oils such as menhaden (*Brovoortia tyrannus*), pilchard (*Sardinops sax*) and herring (*Clupea forengus*) as raw materials for the production, after hydrogenation, of behenic acid. Fish oils are also a preferred source of high-quality arachidic acid. However, erucic acid contents are low and the numerous mono- and polyunsaturated fatty acids in fish oils make separation difficult. Fish oils also display a greater latitude in their fatty acid composition, making *Brassica* oilseeds a more dependable source of erucic acid. Although fish oils generally contain a higher amount of C_{20} unsaturated fatty acids than *Brassica* oils, and the hydrogenated content is usually substantially higher when expressed as arachidic acid, *Brassica* oilseeds are a preferred source of homologue-free behenic acid. With a market developing for large quantities of silver behenate for photographic applications (see below), fractionation of the saturated acids is more efficient when *Brassica* oils are used.

PRODUCTION TECHNOLOGY FOR BEHENIC AND ERUCIC ACIDS

Present technology

To obtain behenic and erucic acids, the oil is first split by some innovation of the Colgate Emery process (Brown, 1949) with recovery of glycerol as a by-product. The fatty acids are fractionally distilled, preferably in specialized stills uniquely suited for handling high-boiling, long-chain acids without subjection to high temperatures for prolonged periods. The Stage equipment is designed for this purpose (Stage, 1973, 1975). A Polish approach to fractional distillation of rapeseed fatty acids has involved the use of two 'Polpak'-packed columns in series (Gradon *et al.*, 1982). Considerable care is necessary as unsaturated C_{22} acids exhibit chain cleavage when heated above ordinary distillation temperatures.

An alternative to the Colgate Emery

process is methanolysis with alkaline catalysts and methanol. Glycerol is recovered as a by-product. The crude, washed methyl esters are dried and fractionally distilled, at a lower temperature than are fatty acids, to afford methyl erucate of the desired purity. The still design fixes the degree of separation of the methyl erucate (methyl 13-docosenoate) from methyl 9- and 11-docosenoate homologues that are present at 2.5–3.5% in most high erucic acid rapeseed oils (J. Beare-Rogers, 1990, personal communication). Methyl esters of the fatty acids are separated more readily than the corresponding fatty acids. If the methyl ester route is chosen, the methyl esters are not hydrolysed back to the fatty acids for sale as such. Thus, the methyl ester route offers the twin advantages of less expensive product costs and for certain derivative manufacture, particularly simple esters and amides, methyl esters function more efficiently than do the fatty acids.

Potential improvements

The relatively low erucic acid contents of *Brassica* oilseeds has led to the investigation of processes to improve the efficiency and economics of extraction. Two processes have been under study. One is a 'silicalite' molecular sieve adsorption process similar to the process patented for tallow fatty acid separation (Cleary *et al.*, 1985a). The other is an aqueous surfactant separation, as used to separate mixtures of liquid oleic acid from solid stearic and palmitic acids in tallow, the so-called Henkel 'rewetting' or hydrophilization process (Stein and Hartmann, 1957; Stein, 1968; Sonntag, 1988).

Silicalite adsorption process
Silicalite, a non-zeolite, hydrophobic, crystalline silica (Grose, 1977, 1978), is uniquely suited for the separation of saturated from unsaturated fatty acids, and also for the separation of certain unsaturated fatty acids from each other (Cleary *et al.*,

1985b). The saturated fatty acids are selectively adsorbed and the unsaturated fatty acids pass through the column and are recovered. The saturated fatty acids are desorbed by a solvent of the proper polarity and are recovered by evaporation of the solvent. The process may be designed to run on a continuous basis.

The silicalite absorption process is potentially more suited for separation of rapeseed fatty acids than it is for hydrogenated rapeseed fatty acids, although eventually the latter separation will probably be feasible and the technique cannot be eliminated at this time for this purpose.

The method has at least two distinct disadvantages. Large volumes of solvent are required for the separation, involving the inconvenience of solvent handling, recovery and storage. Second, it is not certain that a high selective absorption can be maintained with the silicalite through a second pass after the displacement solvent has eluted the fatty acid and initial solvent.

Aqueous surfactant process

Aqueous surfactant separation is based on differences in the melting-points between saturated and unsaturated fatty acids. Development work in Poland on rapeseed fatty acid separation used a separation temperature of 16°C with 5% sodium sulphate and 1% sodium lauryl sulphate and a water to oil ratio of 1 : 1.5 (v/v) to separate erucic acid, melting-point 33°C, from other largely unsaturated fatty acids with lower melting-points (Zwierzykowski and Ledochowska, 1977, 1981). The 31.9% fractionation efficiency and low recovery obtained were probably due to the complexity of the mixture, small amounts of oleic, cis-9-eicosenoic (gadoleic, 20 : 1), cis-11-eicosenoic (gondoic, 20 : 1), cis-11-docosenoic (cetoleic, 22 : 1) and cis-9-docosenoic acids, all having melting-points too close to the 16°C operating temperature to be solidified sharply. The author's experience indicates that operation at

46°C, using magnesium sulphate and sodium decyl sulphate in a water phase modified with a small amount of monoglyceride, affords far superior results (N.O.V. Sonntag, unpublished data). Thus, this technology may have some potential when utilized in behenic acid production. It can conveniently reduce the level of solid saturated C_{16} and C_{18} fatty acids and, in so doing, improve the efficiency of the subsequent fractional distillation. It is anticipated that there will be further process development for water–surfactant separation, for rapeseed fatty acid separation and possibly for the separation of C_{20} and C_{22} fatty acids. The technique is likely to be employed as an adjunct to fractional distillation separation, not as a replacement.

UTILIZATION OF BEHENIC AND ERUCIC ACIDS

Development of applications in the United States

In the United States increased industrial use of C_{22} oleochemicals occurred in the 1950s with the industrial application of behenic acid, behenyl alcohol and erucic acid. By 1955 United States consumption in equivalent fatty acids was approximately 1.8×10^6 kg.

Erucamide, produced first from erucic acid about 1962, and then from methyl erucate in the 1970s, was probably the first C_{22} oleochemical to achieve a reasonably high level of production. It slowly replaced stearamide and oleoamide as polyethylene film anti-slip and anti-block additives during the 1950s, 1960s and by the 1970s had approximately 90% of the polyethylene film market. Estimates of its usage in the 1970s vary, from 1.2×10^6 kg in 1970 (United States Tariff Commission, 1970) to 0.4×10^6 kg in 1975 (Hull and Co., 1975). Erucamide had other uses, such as an additive in printing ink, which may account for this variability. Values as high as 6.9×10^6 kg of total fatty amide additive production (including erucamide, stearamide, oleoamide and behenamide) were reported by 1980–1981 (Modern Plastics, 1981). However, the author has estimated the United States erucamide production in 1982 at only 2.3×10^6 kg, which amounts to 58% of the 3.9×10^6 kg of 90% erucic acid consumed in the United States in that year.

Despite the fact that overall United States production of erucamide still continues to show a small growth, as a result of its use pattern having been extended to other plastic materials, it is apparent that this material is in the early stages of obsolescence as a polyolefin film additive. By 1991 secondary amides of erucamide, structurally related to erucamide, but not produced from it, had captured 25% of the market for polyethylene film applications. Although prices are somewhat higher than erucamide, secondary erucamides with palmityl, stearyl or erucyl groups offer lower volatility, better thermal stability, better migration from the molten polyethylene during film formation, and good lubricity with decreased friction, thus facilitating the film-forming process.

Generally, in the past it has been possible to determine reliably whether a behenic acid was prepared from a hydrogenated *Brassica* seed oil or from a variety of hydrogenated fish oils, hydrogenated *Brassica* seed oils having a $C_{20:22}$ ratio of 0.01–0.25 while hydrogenated fish oils have a ratio of 0.3–0.7. However, with the development of improved fractional distillation techniques the ability to make this differentiation has disappeared. Some behenic acid products, such as Hydrofol 2260, with a composition of 12% arachidic acid and 51% behenic acid and a $C_{20:22}$ ratio of 0.196, have actually been produced from either rapeseed or menhaden oils.

European, Japanese, American and worldwide estimates of behenic acid and erucic acid production for 1991 are shown in Table 15.1. It can be appreciated that Unites States production is only about 15% of worldwide behenic acid production and about 7.6% of worldwide erucic acid production.

Present utilization

The uses made of behenic and erucic acids and their derivatives are numerous. In a comprehensive survey of the literature and the marketplace made between 1981 and 1990 well over 1000 applications were identified, although admittedly about one third of these had minimal industrial potential (Sonntag, 1992). Application may be assigned according to the type of derivative or end use. An estimate of the 1991 use pattern of C_{22} oleochemicals by type of derivative is shown based upon equivalent 85% behenic acid or 90% erucic acid (Table 15.2). An estimate of the 1991 use in the United States, Europe, Japan and worldwide for six end-use application groupings is shown for 85% behenic acid (Table 15.3) and 90% erucic acid (Table 15.4). It is estimated that the annual increase for C_{22} oleochemicals will be in the order of 4% for the remainder of the 20th century.

Future utilization

There are several oleochemicals derived from behenic acid which along with behenic acid will probably show substantial growth during the next two or more decades. These include the following.

• Behenic acid. Used in recording materials (Kawaguchi and Hotta, 1991), as a pearlescent in the cosmetic industry (Yamamoto *et al.*, 1992) and in antidandruff shampoos (Nippon Oils and Fats Co., Ltd *et al.*, 1988). Also, 85% behenic acid shows promise as a cosmetic

Table 15.1. Estimated 1991 behenic and erucic acid production.

	60–95% Behenic acid[1] ($\times 10^6$ kg)	80–95% Erucic acid[2] ($\times 10^6$ kg)
United States	2.0 ± 0.2[3]	1.4 ± 0.2[4]
Europe and Great Britain	9.1 ± 0.5[5]	13.6 ± 0.5[6]
Japan	2.3 ± 0.2[7]	2.7 ± 0.2[8]
World	13.4 ± 1.0	17.7 ± 1.0

[1] From hydrogenated fish oils and hydrogenated *Brassica* oils with fractionation of the fatty acids or methyl esters.

[2] From *Brassica* and *Crambe* oils with fractionation of fatty acids or methyl esters.

[3] Witco Corporation, New York, now comprising Humko, Memphis, Tennessee (acquired in 1980), and Sherex, Mapleton (Peoria), Illinois (acquired in 1992), has *c.* 95% of production. Latent production also from several other companies could amount to 9.1×10^6 kg year^{-1}.

[4] Much of the United States production is 'latent', with Witco Corporation, New York, capable of additional latent production of *c.* 4.5×10^6 kg year^{-1} at Memphis Tennessee and Mapleton (Peoria) Illinois.

[5] 90% of production is from Henkel GmbH, Dusseldorf, Germany, and Unichema, Emmerich, Germany. Pronova Oleochemicals, Standefjord, Norway, and Croda Universal, Inc., North Humberside, Great Britain have a combined volume of *c.* 1×10^6 kg year^{-1}.

[6] 87% of production is from Henkel GmbH, Dusseldorf, Germany, and Unichema, Emmerich, Germany. Croda Universal, Inc., North Humberside, Great Britain, produces *c.* 1.4×10^6 year^{-1}.

[7] Kao Corporation, Wakayama and Tokyo, and Nissan Oil and Fats Co., Tokyo have *c.* 95% of production.

[8] Kao Corporation, Wakayama and Tokyo, is the principal Japanese manufacturer.

Table 15.2. Estimated 1991 world use pattern for C_{22} oleochemicals.

Equivalents of 85% behenic acid ($\times 10^6$ kg)		Equivalent of 90% erucic acid ($\times 10^6$ kg)	
Behenic acid and sodium salts	5.4	Erucic acid and sodium salts	4.5
Behenate salts (excluding quaternary and sodium salts)	2.3	Erucate salts (excluding quaternary and sodium salts)	1.6
Behenate esters	2.4	Erucate esters	1.9
		Erucamide	6.8
Behenic nitrogen derivatives	2.9	Erucate nitrogen derivates (excluding erucamide)	1.0
Other behenic derivatives	0.1	Other erucate derivatives	2.4
Behenyl alcohol	5.4	Erucyl alcohol	6.8
Total	18.5		25.0

Table 15.3. Estimated 1991 behenic acid market for the United States, Europe, Japan and the world.

Group	Application	United States	Europe	Japan	World
		\multicolumn{4}{c}{Equivalent $\times 10^3$ kg of 85% behenic acid}			
1	Surfactants	453.6	226.8	272.2	952.5
	Detergents	499.0	1134.0	748.4	2381.4
	Subtotal	952.6	1360.8	1020.6	3333.9
2	Plastics	2.3	0.9	1.4	4.5
	Plastic additives	997.9	1338.1	816.5	3152.5
	Subtotal	1000.2	1339.0	817.9	3157.0
3	Photographic materials	9.1	4.5	54.4	68.0
	Recording materials	90.7	113.4	136.1	340.2
	Subtotal	99.8	117.9	190.5	408.2
4	Food	9.1	9.1	4.5	22.7
	Food additives	9.1	13.6	9.1	31.8
	Cosmetics	362.9	544.3	657.7	1564.9
	Pharmaceuticals	90.7	113.4	158.8	362.9
	Personal care products	317.5	544.3	839.1	1700.9
	Subtotal	789.3	1224.7	1669.2	3683.2
5	Ink additives	181.4	68.0	317.5	567.0
	Paper	13.6	18.1	9.1	40.8
	Textiles	408.2	499.0	113.4	1020.6
	Lubricants	36.3	40.8	13.6	90.7
	Fuel additives	36.3	45.4	31.8	113.4
	Subtotal	675.8	671.3	485.4	1832.5
6	Adhesives	1.4	0.5	0.5	2.3
	Anti-foamers	0.9	1.4	0.5	2.7
	Antistats	1.8	1.8	0.9	4.5
	Corrosion inhibitors	2.7	5.4	0.9	9.1
	Crayons	–	1.4	0.5	1.8
	Waxes and polishes	3.6	6.4	3.6	13.6
	Insecticides	0.5	–	–	0.5
	Leather	2.7	3.2	1.4	7.3
	Water treatment	–	0.5	–	0.5
	Miscellaneous	589.7	308.4	235.9	1134.0
	Subtotal	603.3	329.0	244.2	1176.3
	Grand total	4121.0±82.4	5042.7±151.2	4427.8±176.8	13591.1±406.8

Takahashi, 1991), pharmaceutical (Tsuda *et al.*, 1990) or personal care product (Fujita and Hayahshi, 1992) additive. High-quality C_{18}- and C_{20}-free >95% behenic acid has some potential for the production of speciality derivatives that require high purity.

- Behenic acid salts. Used in nail lacquer (Shiseido Co. *et al.*, 1986) and in anti-dandruff shampoos (Nippon Oils and Fats Co., Ltd *et al.*, 1988).

- Silver behenate. Used in photographic film (Reeves, 1981; Davis, 1988; Fukui *et al.*, 1989; Masato *et al.*, 1989; Kagami *et al.*, 1990) and in recording materials (Ogawa and Hanashi, 1990; Ogawa *et al.*, 1991; Iwata *et al.*, 1992).

- Saturated quaternary salts. Preferred fabric softeners for fabrics other than cotton, especially in Japan and the Orient (Lion Corp., 1982).

- Behenyl hydrogen sulphate, sodium salt.

Table 15.4. Estimated 1991 erucic acid market for the United States, Europe, Japan and the world.

Group	Application	United States	Europe	Japan	World
		Equivalent $\times 10^3$ kg of 90% erucic acid			
1	Surfactants	408.2	680.4	294.8	1383.5
	Detergents	362.9	340.2	45.4	748.4
	Subtotal	771.1	1020.6	340.2	2131.9
2	Plastics	2.3	2.3	4.5	9.1
	Plastic additives	3855.5	4059.7	40.8	7956.0
	Subtotal	3857.8	4062.0	45.3	7965.1
3	Photographic materials	0.5	0.9	1.4	2.7
	Recording materials	0.5	2.7	1.4	4.5
	Subtotal	1.0	3.6	2.8	7.4
4	Food	22.6	27.2	2.3	52.2
	Food additives	272.2	408.2	90.7	771.1
	Cosmetics	680.4	907.2	172.4	1759.9
	Pharmaceuticals	142.9	408.2	36.3	587.4
	Personal care products	90.7	589.7	158.8	839.1
	Subtotal	1208.8	2340.5	460.5	4009.8
5	Ink additives	113.4	158.8	30.8	303.0
	Paper	9.1	18.1	0.9	28.1
	Textiles	152.0	657.7	127.0	936.7
	Lubricants	133.8	272.2	90.7	496.7
	Fuel additives	45.4	136.1	60.0	240.4
	Ore flotation	–	45.4	0.0	45.4
	Subtotal	453.7	1288.3	309.4	2051.4
6	Adhesives	1.8	3.6	1.4	6.8
	Anti-foamers	0.9	2.7	0.9	4.5
	Antistats	1.4	2.7	0.5	4.5
	Corrosion inhibitors	8.2	9.1	0.9	18.1
	Crayons	0.0	0.5	0.0	0.5
	Waxes and polishes	4.5	22.7	4.5	31.8
	Insecticides	0.0	0.5	0.0	0.5
	Leather	4.5	9.1	4.5	18.1
	Water treatment	0.0	0.0	0.0	0.0
	Miscellaneous	453.6	793.8	90.7	1338.1
	Subtotal	474.9	844.7	103.4	1422.9
	Grand total	6767.3±203.0	9559.7±273.2	1261.6±37.8	17588.6±527.4

Used in a contraceptive (Eli Lilly and Co. *et al.*, 1991).

- Behenic esters. Used in recording materials like photosensitive silicon wafers (Naskatsui, 1992), for magnetic recording materials (Ogawa *et al.*, 1991) and laminated recording laser media (Hashida and Ando, 1991). Ricoh Corporation of Tokyo holds 17 Japanese and European patents on the use of behenic acid and its derivatives in recording materials alone.

- Behenate esters and behenyl alcohol. Likely to be a cosmetic, pharmaceutical or personal care product additive.

- *N*-behenyl-dimethylamine-*N*-oxide. Used as an anti-dandruff and sebhorroea treatment (Shiseido Co. *et al.*, 1990) and skin conditioner (Nishiyama and Tamoaki, 1992).

Table 15.5. Estimated 85% behenic acid production in the United States, Europe, Japan and Southeast Asia, and worldwide for the next four decades.

	1990	2000	2010	2020	2030	2040
			United States			
Volume ($\times 10^6$ kg)	1.8 →	2.9 →	4.4 →	6.5 →	9.6 →	14.2
Annual average increase (%)	7%	6%	5%	4%	3+%	
			Europe			
Volume ($\times 10^6$ kg)	9.1 →	17.8 →	28.8 →	42.6 →	63.1 →	87.4
Annual average increase (%)	7%	5%	4%	4%	3.2%	
			Japan and Southeast Asia			
Volume ($\times 10^6$ kg)	2.7 →	5.9 →	11.3 →	20.0 →	32.5 →	48.0
Annual average increase (%)	8%	7%	6%	5%	4%	
			Worldwide			
Volume ($\times 10^6$ kg)	13.6 →	26.3 →	46.3 →	75.3 →	111.1 →	149.7
Annual average increase (%)	5%	4%	4%	4%	4%	

- Behenamide complex. Used as an anti-tumour agent (Yoshi Pharmaceutical Co., Ltd et al., 1989).
- Pentaerythritol tetrabehenate. Used in long-wear eye cream (Plough, Inc. et al., 1986).
- Behenic triglyceride. Used in a cosmetic emulsion (Nihon Surfactants Industry Co., Ltd, 1980), in a cosmetic stick (Pola Chemical Industries, Inc. et al., 1989) and in a cosmetic make-up base (Kobayashi Kose Co., Ltd et al., 1990).
- Behenyldimethylamine. Used in hair rinses (Nippon Oils and Fats Co., Ltd, 1984).
- Behenylamine hydrochloride. Used in a hair cosmetic (Lion Corp., 1989).
- Dimethyldibehenyl-ammonium chloride. Used in a hair conditioner (Lion Corp., 1981) and hair rinse.
- Dimethyldibehenyl-3-sulphopropyl-ammonium hydroxide salt. Used in skin-cleaning mousses (Procter and Gamble Co. et al., 1986).

Among erucic derivatives, erucamide still finds use as a plastic additive in polyethylene film (Nippon Petrochemicals Co., Ltd et al., 1990), but its use has been extended to polymers other than polyolefins

(Pohrot et al., 1990; Gerhardt and Gribers, 1991; Suzuki et al., 1992). Substituted erucamides like erucyl, palmityl, oleoyl, and stearyl erucamide are finding increasing use in polyolefin films and moulded articles (Park, 1986; Turszyk and Harlen, 1989) and also in other plastics (Mollison, 1984; C.I. Kasei, Co., Ltd, 1985; Efner and Allen, 1992).

Estimates of the markets for 85% behenic acid and 90% erucic acid are summarized for the years 1990 to 2040 (Tables 15.5 and 15.6).

Technical and scientific interest in C_{22} oleochemicals, as measured by the number of patent applications published and patents issued, has shown a marked trend in favour of saturated behenic products over the unsaturated erucic (Table 15.7). Although the relative production of behenic and erucic acid, and their derivatives, is not expected to match this level of interest, production of behenic acid and its derivatives is expected to exceed that of erucic into the 21st century.

Nylon-13,13

The acceptance of nylon-13 and -13,13 as an engineering plastic in the early 21st cen-

Table 15.6. Estimated production of 90–95% erucic acid worldwide for the next four decades.

	1990	1995	2000		2025	2040
Volume ($\times 10^6$ kg)	18.1 $\longrightarrow$	22.7 $\longrightarrow$	34.0[1]	$\longrightarrow$	45.4 $\longrightarrow$	63.5
Annual average increase (%)		4–5%	8.5%	2.5%[2]	2.5–3%	
						$\longrightarrow$ 70.3

[1] Estimated (Sonntag, 1991).
[2] A relative 'slump' is predicted between the years 2000 and 2025 especially if nylon-13 and -13,13 prospects as engineering plastics are dim. Erucyl alcohol development is excluded.

Table 15.7. Patent and publication rate for behenic acid and erucic acid applications between 1965 and 1990.

	Chemical Abstracts	Behenic acid			Erucic acid			Behenic : erucic
Year	Volume	Patents	Publications	Total	Patents	Publications	Total	ratio
1965	62–63	6	11	13	7	1	8	1.63
1970	72–73	24	15	39	20	11	31	1.25
1975	82–83	23	49	72	15	11	36	2.76
1980	92–93	31	38	69	34	29	63	1.09
1981–90	94–113			805			227	3.54

tury may determine, to a large degree, the extent of commercial development of C_{22} oleochemicals. Both nylons can be produced from erucic acid (Fig. 15.1). They are not new, experimental quantities having been prepared for about the past 30 years. They appear to be technically well suited for use in a number of applications as engineering plastics, although there are still serious economic problems. Consideration has been given to nylon-13,13 as a lightweight material for engine automotive parts, exclusive of carburettors. This potential application alone would dwarf all the other existing industrial uses for erucic acid. One estimate holds that the potential market for the automotive engineering plastic use of nylon-13,13 is at 22.7×10^6 kg of equivalent erucic acid in the United States alone (C.P. West, 1986, personal communication).

A major problem involved in the production of nylon-13,13, and similarly the other nylon-13 family members, is the economics of the first step. *Brassica* oils with 48% erucic acid are not economical starting materials because of the declining value of the glycerol by-product and the low value of the unsaturated fatty acid by-product. It is currently held that a break-even point occurs in the raw material at the level of 75% erucic acid, and a 'minimum-profitable programme' results from 80% erucic rapeseed oil, but that $\geqslant 90\%$ erucic rapeseed oil would be desirable (N.O.V. Sonntag, unpublished data).

Recently it has been shown that high erucic acid rapeseed oil contains isomeric C_{22} monounsaturated fatty acids (J. Beare-Rogers, 1990, personal communication). In addition to erucic acid (*cis*-13-docosenoic), the C_{22} monounsaturated fraction of high erucic acid rapeseed oil contains 7–9% of cetoleic acid (*cis*-11-docosenoic) and about 3% of *cis*-9-docosenoic acid. This corresponds to 3.5–4.5% and about 1.5%, respectively, of the total fatty acid composition of the oil.

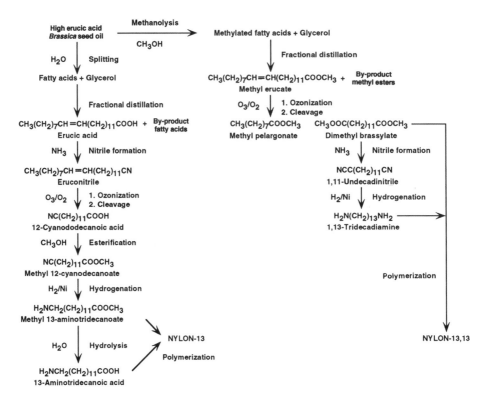

Fig. 15.1. Nylon-13 and -13,13 synthesis from *Brassica* seed oil.

The presence of these C_{22} mono-unsaturated fatty acid isomers can give rise to problems in the quality and performance of nylon-13,13, and to the quality of the pelargonic acid by-product. Methanolysis of the oil will give small amounts of 11- and 9-docosenoic methyl esters, largely insepa-rable from the methyl erucate in the sim-ple fractional distillation (Fig. 15.1). The brassylic acid obtained by distillation fol-lowing ozonolysis will be contaminated with undecanioic acid from the 11-docosenoic acid and will be fractionally distilled with difficulty from it. The by-product pelargonic acid will contain a small amount of tridecanoic acid from the 9-docosenoic acid, some undecanoic acid from the 11-docosenoic acid, which is not entirely separable from the pelargonic acid, and larger amounts of undecanoic acid from the 11-docosenoic acid, also largely inseparable.

The possible effect of brassylic acid contamination on nylon yield, quality and performance is of concern. Moreover, most cost estimates and project evaluations to date have not included relatively expen-sive fractional distillation operations, should these be deemed necessary. On the other hand, experimental quantities of the nylon-13 family probably containing all these contaminants have already been per-formance evaluated and found to be quite satisfactory.

In light of the anticipated large effect of higher concentrated erucic acid raw ma-terials in lowering the costs of producing members of the nylon-13 family research into improvement of production methods is underway. The methyl ester route in-volving alkaline methanolysis of rapeseed oils offers the cost advantage of yielding almost anhydrous glycerol as a by-product, thus eliminating the need to dehydrate the

Table 15.8. Economic comparison of high erucic acid (48%) *Brassica* oil with 80% and 90% erucic acid oils as raw materials for behenic acid production, based on 4.5×10^6 kg of each oil.

	Units	Erucic acid in oil		
		48%	80%	90%
Oil cost	$\times 10^6$ US$	3.5000	5.7000	6.5000
Hydrogenated *Brassica* oil	$\times 10^6$ kg	4.5634	4.5669	4.5669
Yield of distilled hydrogenated *Brassica* seed fatty acids	$\times 10^6$ kg	4.1418	4.1492	4.1516
95% glycerol	$\times 10^3$ kg	381.1	362.4	357.2
at US$1.30 kg^{-1}	$\times 10^3$ US$	491.5	467.4	460.8
By-product fatty acids	$\times 10^6$ kg	1.8029	0.2441	none
at US 40¢ kg^{-1}	$\times 10^6$ US$	0.7154	0.0969	none
Yield of 85% behenic acid	$\times 10^6$ kg	2.3430	3.9052	(90%) 4.1516
Market value 85% behenic acid US$2.53 kg^{-1}	$\times 10^6$ US$	5.9300	9.900	11.4409[1]
Average cost of manufacture all fatty acid kg^{-1}	$\times$ US$	0.49	0.44	0.40
Estimated profit	$\times 10^6$ US$	1.6280	2.9456	3.7542
Overall return on investment	%	29.5	39.1	46.1

[1] 90% behenic acid at US$2.76 kg^{-1}.

glycerol fraction from conventional fat splitting. Improved packed-column fractional distillation offers separation efficiency for both by-product methyl ester fractions and for quality upgrading methyl erucate. In ozonolysis increased yields and lower ozone usage may be achieved with use of specific ozonization catalysts. Primary diamine formation may be maximized by rigid temperature control with the use of secondary amine suppressors. All of these positive process improvements in the methyl ester/diamine route of polymerization remain to be explored.

ECONOMICS

As the 21st century approaches, competition among the various natural sources of behenic and erucic acids is expected to intensify. If *Brassica* oilseeds are to maintain their present position as the major source of these long-chain fatty acids and their derivatives, some significant effort

will need to be directed to increasing the present erucic acid content in these oils. For industrial application, at least, this should be accompanied by a decrease in the oleic, linoleic and linolenic acid contents. The fatty acid profile would also be improved by lowering or removing the isomeric *cis*-5-, *cis*-9- and *cis*-11-monounsaturated C_{22} fatty acids.

The potential overall economics for the production of behenic acid from 48%, 80% and 90% erucic acid rapeseed oil would indicate that using a 90% erucic acid rapeseed oil could sustain a 5-year capital return including interest on a loan, 80% erucic acid rapeseed oil a 10-year capital return including interest on a loan, while with a 48% erucic acid rapeseed oil one would be hard-pressed to sustain a 20-year capital return without including interest on a loan (Table 15.8). This demonstrates the vital need for genetically developed rapeseed oils with increased erucic content for efficient oleochemical production.

REFERENCES

Brown, A.C. (1949) Fat hydrolysis process and apparatus. Issued 11 November 1949 to Emery Industries. *United States Patent* 2,486,630.

C.I. Kasei Co., Ltd (1985) PVC agricultural films. Issued 28 June 1985. *Japanese Patent* 85, 120,740.

Cleary, M.T., Kulprathepanjz, S. and Quezil, R.W. (1985a) Process for separating fatty acids. Issued 18 June 1985 to UOP Inc. *United States Patent* 4,524,030.

Cleary, M.T., Kulprathepanjz, S. and Quezil, R.W. (1985b) Process for separating oleic acid from linoleic acid. Issued 25 March 1985. *United States Patent* 4,578,223.

Davis, P.D. (1988) Imaging materials employing microparticles including a silver initiator. Issued 20 November 1988 to the Mead Corp. *United States Patent* 4,788,125.

Efner, H.F. and Allen, D.E. (1992) Secondary amide-containing poly(ethyleneterephthalate) molding compositions, their preparation, and molding, and glossy molded articles. Issued 22 January 1992 to Phillips Petroleum Co. *European Patent* 467,368.

Eli Lilly and Co., Zimmerman, R.E., Burck, P.J., Jones, C.D. and Thakkar, A.L. (inventors) (1991) Contraceptive method and compositions. Issued 28 April 1981. *United States Patent* 2,264,577.

Fujita, H. and Hayahshi, Y. (1992) Long-lasting hair conditioning emulsions. Issued 18 February 1992 to Mirabon K.K. *Japanese Patent* 92, 49,219.

Fukui, T., Fukomoto, H., Katayama, M., Aralava, K. and Kagami, K. (1989) Issued 30 August 1989 to Canon K.K. *European Patent* 320,504.

Gerhardt, G. and Gribers, J.A. (1991) Stabilizing system for EVA hot-melt compositions containing them. Issued 30 March 1991 to W.R. Grace and Co. *Canadian Patent* 2,025,442.

Gradon, L., Selecki, A., Bolinsky, L.J., Glowacki, J., Uminski, J. and Kulig, A. (1982) Isolation of erucic acid by fractional distillation of rapeseed oil acid in columns with polpak packing. *Przemysl Chemiczny* 61, 98–100.

Grose, R.W. (1977) Crystalline silica. Issued 6 December 1977 to Union Carbide Corp. *United States Patent* 4,061,724.

Grose, R.W. (1978) Crystalline silicates and method for preparing same. Issued 1 August 1978 to Union Carbide Corp. *United States Patent* 4,104,294.

Hashida, J. and Ando, E. (1991) Laminated laser recording medium. Issued 19 February 1991 to Hatsushita Electric Industrial Co. *Japanese Patent* 91, 94,250.

Hull and Co. (1975) *Fatty Acids. North America, A Multiclient Market Survey*. Bronxville, New York, December, p. 383.

Iwata, J., Ogawa, S. and Mayushi, Y. (1991) Preparation of optical recording materials using metallic silver particles. Issued 18 November 1991. *Japanese Patent* 91, 258,589.

Kagami, Y., Mouri, A., Katayama, M., Isaka, K. and Fukui, T. (1990) Image-forming method and image-forming medium. Issued 18 April 1990 to Canon K.K./Oriental Photo Ind. Co., Ltd. *European Patent* 363,790.

Kawaguchi, M. and Hotta, Y. (1991) Process of reversible thermal recording using heat-resistant film. Issued 20 September 1991 to Ricoh Co., Ltd. *Japanese Patent* 91, 215,085.

Kobayashi Kose Co., Ltd, Someya, T. and Yokoyama, H. (inventors) (1990) Cosmetic make-up dispersion. Issued 31 May 1990. *Japanese Patent* 90, 142,716.

Lion Corp. (1981) Hair conditioners containing quaternary ammonium salts. Issued 26 December 1981. *Japanese Patent* 81, 169,615.

Lion Corp. (1982) Particulate softening agents for fabrics. Issued 14 April 1982. *Japanese Patent* 82, 61,769.

Lion Corp. (1989) Hair cosmetics containing cationic and anionic surfactants and protein derivatives. Issued 15 June 1989. *Japanese Patent* 89, 153,611.

Masato, K., Fukui, T., Arahara, K., Takasu, Y. and Kagami, K. (1989) Photosensitive material and image-forming method. Issued 16 August 1989 to Canon K.K. *European Patent* 328,364.

Modern Plastics (1981) *Fatty Amide Plastic Additives* 58, 61.

Mollison, A.N. (1984) Nylon film with improved slip characteristics. Issued 15 December 1984 to DuPont Canada. *United States Patent* 4,490,324.

Naskatsui, H. (1992) Dummy wafer. Issued 14 January 1992 to Canon K.K. *Japanese Patent* 92, 10,409.

Nihon Surfactants Industry Co., Ltd (1980) Novel glycerol fatty acid esters for cosmetic emulsions. Issued 1 November 1980. *Japanese Patent* 80, 139,827.

Nippon Oils and Fats Co., Ltd (1984) Hair rinses containing amine oxides and amino acids. Issued 4 May 1984. *Japanese Patent* 84, 78,114.

Nippon Oils and Fats Co., Ltd, Funada, T., Nakada, M., Uji, Y., Masui, K. and Shibayama, Y. (inventors) (1988) Anti-dandruff shampoo compositions containing saturated higher fatty acids and/or their salts. Issued 4 November 1988. *Japanese Patent* 88, 267,713.

Nippon Petrochemicals Co., Ltd, Harashige, M., Kawamura, T., Kaneko, K., Inoue, T., Todutake, A. and Yochidawa, T. (inventors) (1990) Surface blush- and fibre-resistant polyolefin composition. Issued 8 March 1990. *PCT International Patent* 90, 02,153.

Nishiyama, T. and Tamoaki, S. (1992) Topical cosmetics containing sarkososaponins and surfactants. Issued 31 January 1992 to Shiseido Co., Ltd. *Japanese Patent* 92, 29,917.

Ogawa, S. and Hanashi, Y. (1990) Optical card and information recording and reading using same. Issued 7 December 1990 to Asahi, Chemicals Ind. Co., Ltd. *Japanese Patent* 90, 297,485.

Ogawa, S., Iwata, J. and Hayashi, Y. (1991) Rewritable optical recording material using metallic silver particles. Issued 18 November 1991 to Asahi Chemicals Ind. Co., Ltd. *Japanese Patent* 91, 258,587.

Park, C.P. (1986) Polyolefin foam compositions with improved compositions with improved dimensional stability and their manufacture. Issued 4 December 1986 to Dow Chemical Co. *PCT International Patent* 86, 7076.

Plough, Inc., Suss, H. and Ner, V. (inventors) (1986) Long-wear cosmetics. Issued 30 April 1986, *European Patent* 179,416.

Pohrot, J., Jenne, H., Walte, H.M., Gaesepohl, H. and Benedix, F. (1990) Antiblocking thermoplastic molding compositions. Issued 3 October 1990 to BASF A.-G. *European Patent* 390,019.

Pola Chemical Industries, Inc., Suganuma, J. and Nishimura, K. (inventors) (1989) Cosmetic sticks containing behenic acid triglyceride and ethylene glycol distearate. Issued 17 January 1989. *Japanese Patent* 89, 13,011.

Procter and Gamble Co., Schmidt, R.R., Fortna, R.H. and Seyer, H.H. (inventors) (1986) Mild cleaning mousses. Issued 10 September 1986. *European Patent* 194,097.

Reeves, J.W. (1981) Photothermographic composition and process. Issued 28 April 1981 to Eastman Kodak Co. *United States Patent* 4,264,725.

Shiseido Co., Soyama, Y., Yamaguchi, M., Yamazuki, K. and Kitamura, C. (inventors) (1986) Solvent-based nail lacquer containing soaps. Issued 7 February 1986. *French Patent* 2,568,471.

Shiseido Co., Miyajawa, K., Uchikawa, K., Yamagida, T. and Tamoaki, S. (inventors) (1990) Lightening antidandruff and/or seborrhea, treating compositions containing amine oxides. Issued 20 February 1990. *Japanese Patent* 90, 49,712.

Sonntag, N.O.V. (1988) Surfactants in aqueous emulsification separation of oleic and stearic acids. In: Wasan, T., Ginn, M.E. and Shad, D.O. (eds) *Surfactants in Chemical/Process Engineering*. Marcel Dekker, New York, pp. 169–193.

Sonntag, N.O.V. (1991) Erucic, behenic: new feedstocks for the 21st century. *INFORM, American Oil Chemists' Society Publication* 2, 449–463.

Sonntag, N.O.V. (1992) *21st Century Sources, Technology and Marketing for Erucic and Behenic Acids and Their Derivatives*. Hewin International, Amsterdam, The Netherlands.

Stage, H. (1973) Optimum construction of vacuum distillation plant for the separation of mixtures sensitive to heat and oxidation. *Verfahrenstechnik (Mainz)* 7, 64–70.

Stage, H. (1975) Problems in the distillation of rapeseed fatty acids on a technical scale. *Fette Seifen Anstrichmittel* 77, 174–180.

Stein W. (1968) The hydrophilization process for the separation of fatty materials. *Journal of the American Oil Chemists' Society* 45, 471–474.

Stein, W. and Hartmann, H. (1957) Separation of high molecular organic compound mixtures. Issued 23 July 1957 to Henkel and Cie. *United States Patent* 2,800,493.

Suzuki K., Ito, T. and Nakagawa, H. (1992) Nucleation agents and their use with nucleation accelerators in thermoplastic resin compositions. Issued 11 August 1992 to Mizasawa Industrial Chemicals, Ltd. *Japanese Patent* 92, 220,447.

Takahashi, K. (1991) Silica-containing antiwrinkle cosmetic liquid. Issued 10 September 1991. *French Patent* 2,659,551.

Tsuda, Y., Murashima, R., Yamamouchi, K. and Yokoyama, K. (1990) Emulsified preparations containing perfluorocarbon compound. Issued 28 November 1990 to Green Cross Corp. *European Patent* 399,842.

Turszyk, M.J. and Harlen, G.M. (1989) Plastic compositions for carriers for carbonated beverage cans. Issued 3 January 1989 to Union Carbide Corp. *United States Patent* 4,795,767.

United States Tariff Commission (1970) *Report of Sales of Synthetic Organic Chemicals.*

Yamamoto, H., Konishi, S. and Umezawa H. (1992) Manufacture of high-solids pearlescent luster agent dispersions. Issued 14 February 1992 to Lion K.K. *Japanese Patent* 92, 45,843.

Yoshi Pharmaceutical Industries Co., Ltd/Toyo Jazo Co., Ltd, Ono, M., Arita, M., Okumoto, T. Salto, T., Sakakibara, H. and Fukukawa, S. (inventors) (1989) Preparation of racemic 1-(2,3-dihydroxy-4-hydroxymethyl-4-cyclopenten-1-yl)cytosine derivatives as antitumor agents. Issued 14 February 1989. *Japanese Patent* 89, 42,499.

Zwierzykowski, W. and Ledochowska, E. (1977) Effect of the structure of fatty acids on their wettability by aqueous surfactant solutions and the fractionation of fatty acids. *Tenside Detergent* 144, 257–261.

Zwierzykowski, W. and Ledochowska, E. (1981) The influence of fatty acid structure on their wetting by aqueous surfactant solutions and the fractionation of fatty acids. *Zeszyty Problemowe Postepów Nauk Rolniczych* 211, 279–292.

16 Utilization of Oil as a Biodiesel Fuel

W. Körbitz
Graben 14/3, Vienna, Austria

HISTORICAL PERSPECTIVE

Vegetable oils were used as liquid fuels in the very early days of the combustion engine. At the world exhibition in Paris in 1900, for example, groundnut oil was used to run a small diesel engine. Rudolf Diesel had patented his engine 5 years earlier (Diesel, 1895) and in a subsequent patent application of 1912 observed that 'the use of vegetable oils may be unimportant now but in time may become as important as petroleum or coal-tar products are today' (Syassen, 1991). At that time vegetable oils were cheaper than mineral oils and their different refined forms, but the fossil forms of liquid fuels soon took over.

Oil price increases and supply shocks of the 1970s demonstrated the vulnerability of economies largely dependent on non-renewable and finite fossil energy sources mainly produced in a few, politically un-stable areas of the world. Those shocks stimulated increased efforts in research and development of renewable forms of energy.

Vegetable oils are similar to liquid hydrocarbons in terms of their energy content and physical properties, and when rediscovered were firstly tested as pure unmodified oils in modern engines. There was little success at the outset but the modified methyl ester form that followed proved ideal for diesel engines (Mittelbach *et al.*, 1983; Wörgetter, 1991a). In recent years this methyl ester form of vegetable oil (biodiesel) has become the fuel of choice for the diesel engine, while bioethanol, also a renewable biofuel, was developed for the petrol engine and has been utilized for some years.

Development of modern, intensive farming technologies following the food shortages after World War II resulted in overproduction in some industrialized countries. This led to the introduction of 'setaside' programmes, notably in the United States and the European Union (EU), under which farmers were paid to keep land out of production. In the early 1990s, the EU required larger growers to take 15% (reduced to 12% in 1995) of their arable acreage out of food production (Ferdinand, 1992) but allowed, subject to certain restraints, crops for industrial use to be grown on this land. It is predicted that, by 2000 at the latest, 800,000 ha of oilseed crops will be grown for biodiesel production within the EU (Körbitz,

1994b). The introduction of specifications for biodiesel fuel and the acceptance of biodiesel by engine manufacturers based on engine test results have stimulated further interest in this fuel.

Plant products are increasingly valued for their environmentally friendly properties which can help meet the challenges resulting from air, water and soil pollution. Biodiesel, as well as hydraulic and lubrication oils based on vegetable oils, offers significant advantages in reducing risks to the environment. Key properties of vegetable oils which contribute to their attraction as environmentally friendly alternative fuels and lubricants include:

- high biodegradability,
- low toxicity, both oral and dermal,
- low evaporation, reducing inhalation risk,
- high flashpoint, reducing risk of fire,
- reduced emissions, particularly carbon dioxide, sulphur oxides, soot (particulate carbon matter) and aromatic compounds.

While the original slow-running diesel engine could run on fuel varying in quality, modern diesel engines require high fuel efficiency and a quality standard that ensures low emissions.

BRASSICA OIL AS A SOURCE OF BIODIESEL

Any vegetable oil, including for example used frying oil, can be utilized as a raw material for the production of biodiesel, as long as the fuel produced meets biodiesel fuel specifications. However, preference is given to oils with low saturation and low polyunsaturation.

The degree of saturation of a vegetable oil affects cold-temperature behaviour (as measured by cloud point or cold-filter plugging point). Biodiesel must flow easily at ambient temperatures from the tank, through filters and pumps, to the engine injection nozzles. Oils with a high proportion of saturated fatty acids, such as palm oil, will produce a methyl ester mixture with a high cold-filter plugging point (temperature at which the oil is no longer miscible with an alcohol solvent) and may become solid. While biodiesel produced from palm oil has shown superior engine performance in the hot climate of Indonesia (Schäfer, 1991), oils with a much lower cloud point are required for colder climates.

The degree of polyunsaturation of a vegetable oil affects the tendency for oxidation and polymerization (measured by the iodine number). The higher the unsaturation the greater the chances of oxidation and polymerization during storage. What is more important is the risk of polymerization when traces of fuel seep from the cylinder into the engine oil during operation. This may cause the formation of a thick sludge which can damage the engine.

Double-low *Brassica* seed oil has a favourable fatty acid profile for biodiesel production. New cultivars with increased levels of oleic acid (18 : 1) and reduced levels of linolenic acid (18 : 3) are particularly suited for cooler climates because of their lower saturation and polyunsaturation.

In addition to crude vegetable oil, degummed or refined oil, even used frying oil, may be utilized to produce biodiesel. Degumming adds to the cost of production of the raw starting material but reduces the phosphorus content of the biodiesel produced. Likewise, bleaching and deodorization to produce refined oil reduces impurities but adds to the cost of the initial raw material. In seeking cheaper raw materials technologies have been developed for the esterification of used frying oil (Nye, 1983; Mittelbach and Tritthart, 1988). Technology has even been developed to esterify used hydraulic oils of vegetable origin, which have attracted a market because of their superior biodegradability. Unlike virgin vegetable oils, which usually have a constant fatty acid composition, used oils may vary. However, fatty acid composition may be determined quickly by gas chromatography (see Chapter 11), thus assuring control of product quality. Recycling used frying oil, a waste product, into biodiesel can also contribute to a cleaner environment.

METHYL ESTER (BIODIESEL) PRODUCTION

Vegetable oil methyl esters for use as biodiesel fuel are produced by a reaction known as alcoholysis or transesterification. Methanol is added with a catalyst to extracted oil (see Chapter 12) under alkaline or acidic conditions. While alkaline reactions, produced by the addition of sodium hydroxide, sodium methylate or potassium hydroxide, proceed under mild conditions of ambient pressure and temperatures, acidic reactions require higher pressure and temperature and are technically more elaborate. In each case fatty acids attached to glycerol (glycerine), predominantly triacylglycerols, are transesterified to produce fatty acid methyl esters and glycerol (Mittelbach, 1991).

The transesterification reaction is followed by the separation of the heavier glycerol (density 1.26) from the lighter methyl esters (density 0.88). This can take place as a batch process in settling containers or as a continuous process in tube settlers. After this basic separation both products may require further cleaning. Methyl esters may be freed from traces of soaps (phospholipids) by centrifuging, and from excess methanol by distillation. The recovered methanol may be recycled in the methylation process. The resulting methyl esters are usually suitable for use without further treatment. However, if the original oil is a hydraulic oil containing additives, further distillation may be required.

The crude acidified glycerol contains free fatty acids, which are separable by centrifugation. The free fatty acids can either be used as a heating oil substitute or transesterified to form additional methyl ester product. Following transesterification excess methanol is removed by distillation and recycled.

After cleaning the glycerol product may contain: up to 55 or 65% glycerol, 20–30% soaps, less than 0.5% methanol, 7–12% water and have a calorific value of 22.3 MJ kg^{-1}. This cleaned product may be used directly or further refined to industrial or pharmaceutical grade.

Economic biodiesel production is dependent on the availability of cheap oils (e.g. used frying oil or oil with high chlorophyll content), the degree of conversion of oil to methyl esters and the purity of the final product. Using European production figures, 1 ha of rapeseed can be expected to yield 2900 kg of seed, which may be processed to produce 1050 kg of oil. This in turn will yield 1000 kg biodiesel and 100 kg glycerine.

PROPERTIES OF BIODIESEL

Biodiesel must meet certain fuel specifications to be acceptable to both engine manufacturers and end users. In Austria, for example, comparable specifications exist for biodiesel and diesel fuels and similar specifications have either been drafted or are in the process of being drafted in other countries (Table 16.1). Some parameters distinguish biodiesel from diesel, while others determine the quality of the biodiesel itself.

Density is reduced from 0.915 to 0.88 g cm^{-3} when rapeseed oil is transformed into biodiesel and is virtually the same as diesel.

Viscosity is reduced by transesterification of rapeseed oil from 74 $mm^2 s^{-1}$ to

Table 16.1 Comparison of fossil diesel and biodiesel fuel specifications.

Property	Units	Diesel ÖNORM C1104[1]	Biodiesel ÖNORM C1109[2]	Biodiesel Mercedes[3]
Density at 15°C	g cm^{-3}	0.82–0.86	0.87–0.89	0.875–0.885
Viscosity at 20°C	mm^2 s^{-1}	3.0–8.0	6.5–8.0	6.0–9.0
Flashpoint	°C	>55	>100	>55
CFPP[4] summer	°C	>+5	>0	>−5
CFPP winter	°C	−20	>−15	>−15
CCR[5]	%w	<0.10	−	−
CCRof 100%	%w	−	<0.05	<0.05
Cetan number		>48	>48	>50
Neutralization number	mg KOH g^{-1}	−	<0.80	<0.50
Iodine number	g iodine 100 g^{-1}	−	−	<115
Methanol	%w	−	<0.20	<0.30
Free glycerol	%w	−	<0.02	<0.03
Total glycerol	%w	−	<0.24	<0.20
Total sulphur	%w	<0.15	<0.02	<0.02
Sulphated ash	%w	−	<0.02	<0.02
Oxide ash	%w	<0.01	−	−
Phosphorus	mg kg^{-1}	−	<20	−

−, Not determined; %w = percentage of weight.
[1] Austrian Standards Institute specification C1104 (1990).
[2] Austrian Standards Institute specification C1109 (1994).
[3] Mercedes specification for biodiesel fuel (1991).
[4] CFPP = cold-filter plugging point.
[5] CCR = Conradson carbon residue.

7.0 mm^2 s^{-1}. Thus the resulting biodiesel has virtually the same behaviour as diesel in the injection pump and engine nozzles.

Flashpoint is the temperature at which a liquid becomes flammable. After the removal of methanol it ranges from 100°C to 180°C as compared to 55°C to 60°C for diesel. Thus biodiesel is safer to handle than diesel.

Cold-Filter Plugging Point (CFPP) is the temperature at which the fuel will block fuel filters. Biodiesel has the disadvantage of having a higher cold-filter plugging point than diesel. Commercial cold-filter plugging point depressant additives exist for diesel, which can reduce cold-filter plugging point to −20°C. Similar depressants will be available shortly for

biodiesel fuels from Lubrizol Co. in England.

Conradson Carbon Residue (CCR) is a measure of the tendency for coking and carbon residue formation. Formation of carbon residues on engine nozzles and pistons during combustion can lead to lower engine performance and reduced engine life but the levels obtained with biodiesel are acceptable to engine manufacturers.

Cetan number is a measure of the ignition property of a fuel within the diesel engine. Biodiesel has a higher cetan value than diesel which results in a smoother running engine.

Neutralization number is an index of acidity.

Table 16.2. Physical and chemical properties of selected vegetable oils and tallow.

Property	Units	Canola	Sunflower	Soya	Palm	Tallow
Density at 15°C	g cm^{-3}	0.915	0.925	0.930	0.920	0.937
Viscosity at 20°C	mm^2 s^{-1}	74	66	64	40	solid
Flashpoint	°C	317	316	330	267	–
Cloud point	°C	0	–16	8	31	–
Iodine number		113	131	130	49	45
Cetan number		44	36	39	42	–
Calorific value	MJ kg^{-1}	41	40	40	35	39
Fatty acid composition (%)						
Palmitic (16 : 0)		4	6	8	42	25
Stearic (18 : 0)		1	4	4	5	19
Oleic (18 : 1)		60	28	28	41	40
Linoleic (18 : 2)		20	61	53	10	4
Linolenic (18 : 3)		9	–	6	–	1
Erucic (22 : 1)	2	–	–	–	–	–

Source: Batel (1980), Schütt (1982).
–, Non-existent or negligible.

Iodine number is a measure of the presence of fatty acids with unsaturated double bonds. A level of <115, as achieved with rapeseed oil methyl ester, is desirable as higher values are likely to lead to problems with polymerization in the motor oil.

Methanol content is indicative of processing conditions. Although the engine industry has no objection to higher methanol levels, as it can be viewed as a fuel, levels of less than 0.20% are desirable in order to obtain a high flashpoint and a product of high handling safety.

Free glycerol content is indicative of the efficiency of the settling or separation step in processing. High levels are a sign of poor production control and will cause problems in fuel filters and in the engine.

Total glycerol content is a key quality parameter for biodiesel as it indicates how much of the original oil was transformed into methyl esters. Values above 0.24% will cause engine performance problems in the long run.

Sulphur content is a measure of fuel impurity. Low sulphur content of vegetable oils leads to reduced sulphur oxides in ex-haust emissions of biodiesel, a key advantage of biodiesel fuels over diesel. Low sulphur emissions are also important in prolonging the life of modern catalytic converters, which are used to reduce emissions further.

Sulphated ash content is a measure of inorganic fuel impurities and is a more accurate guide to biodiesel fuel quality than oxide ash, which is more appropriate for fossil diesel.

Phosphorus content is a measure of fuel quality. Phosphorus, like sulphur, can reduce the life of the modern catalytic converter and consequently a limit has been set in recent specifications for biodiesel fuel.

Austrian specifications also state that rapeseed oil methyl ester must be produced from rapeseed oil with a maximum 5% erucic acid, a maximum 15% linolenic acid, free of visible water and of solid particles, and have a maximum 1% of fuel improvement additives (e.g. for low-temperature condition). Comparison of rapeseed oil with other vegetable oils and diesel has shown rapeseed to have a favourable iodine number (Table 16.2). The internal

Table 16.3. Comparison of energy content and energy efficiency of fossil diesel and biodiesel fuel.

Fuel	Density (g cm^{-3})	Calorific value		Efficiency at 1200 rpm[1]
		MJ kg^{-1}	MJ dm^{-3}	
Diesel	0.83	42.9	35.6	38.2
Biodiesel	0.88	37.2	32.9	40.7
Variation		−13.3%	−7.6%	6.5%

Source: Walter (1992).
[1] Efficiency expressed as the ratio of energy input (fuel) to energy output (engine performance).

fuel specifications of Mercedes recommends an iodine number of <115 in order to minimize the potential for polymerization (Schäfer, 1991). Of the vegetable oils available in bulk this recommendation is met only by palm and rapeseed oil. Also of importance is the relative high oxygen content (about 10%) of biodiesel fuels. The absence of oxygen in diesel fuel results in a less efficient combustion process with increased emissions, especially soot (particulate carbon matter) (Syassen, 1991).

ENGINE PERFORMANCE

Engine performance with biodiesel has generally been conducted using unmodified diesel engines. Compared to diesel, biodiesel has about 13% lower calorific value, measured in MJ kg^{-1}, and about 7% lower, measured in MJ dm^{-3}. This would be expected to reduce engine performance by about 8%. However, tests with different engines under a variety of conditions have shown engine performance to be comparable to that with diesel fuels (Schäfer, 1991; Sams and Schindibauer, 1992; Walter, 1992). A more efficient conversion of the energy of biodiesel into engine performance during the combustion process makes up the difference (Table 16.3).

Fuel consumption is greater with bio-diesel compared with diesel. Increases ranging from −4 to +14% and averaging 5% have been reported (Schäfer, 1991; Sams and Schindlbauer, 1992). Some fuel is always transported from the combustion chamber into the engine oil during piston movement. Biodiesel tends to accumulate in the engine oil to a greater extent than diesel because of the more homogeneous nature of biodiesel. Accumulations of 1–10% have been reported in some trials (Schäfer, 1991). However, both Mercedes and Castrol have reported that inclusion of as much as 20% biodiesel in the engine oil did not result in any appreciable adverse lubrication effects (Machold and Dobbins, 1991). Nevertheless, some mineral oil companies have developed engine oils specifically for use with biodiesel, e.g. BP Vanellus FE, Castrol Powermax, Shell Rimula X, Fuchs-Genol Unic 1040 MC, Mobil Delvac 1300 Super, and ÖMV-Elan RME Plus.

There have been many tests on diesel engines ranging from 1500 l Ford cars to 48,300 l Caterpillar stationary drilling engines and including a range of types from prechamber to direct injection. One such series of tests was completed in 1990 at the Institute for Agricultural Engineering at Wieselburg, Austria. The fleet included the leading brands of tractors and engines (Fiat, Ford, J.I. Case, John Deere, KHD, MWM, Perkins, Same, Steyr). Tests

Table 16.4. Manufacturer clearances for use of biodiesel.

Manufacturer	Type of vehicle	Engine	Guarantee	Conditions
Case	Tractor ⎫ Combine ⎭	D-series since 1971 1660 since 1990	1 year	Oil change at 100 h
Deutz-Fahr	Tractor	FL-BFL 912/913	1 year	None
Farymann	Various	All since 1992	Normal	None
Fiatagri	Tractor	All new tractors	Normal	None
Ford	Tractor	All new tractors	2 years	None
John Deere	Tractor	Series 10 to 50	2 years	None
	Combine	All since 1967	2 years	None
Kubota (Japan)	Tractor	All types	Normal	None
Massey-Ferguson	Tractor	All since 1989	1 year	None
Mercedes-Benz	Tractor	All new M-B	1 year	Halve the oil change interval
MVVM (Germany)	Various	266B, 266BL	Normal	None
Same (Italy)	Tractor	All since 1980	4 years	None
Steyr (Austria)	Tractor	All since 1991	Normal	None
Valmet (Finland)	Tractor	All since 1987	Normal	None
Volkswagen	Personal	Golf Ecomatic	Normal	None

Source: Wörgetter (1991b), Körbitz (1994b).

involved 34 tractors running for a total of 25,582 h, during which 170,040 l of fuel were consumed. Motor oil from each engine was analysed at frequent intervals, performance of the engines monitored continuously and every engine was dismantled for internal examination at the end of the test. Bench tests were also conducted to simulate field conditions. Results from these tests together with the establishment of well-defined specifications for biodiesel fuel have led manufacturers of several diesel-powered vehicles to issue clearance for the use of biodiesel in their engines (Table 16.4) (Wörgetter, 1991b). Similar test results were obtained at the FAT-Institute in Switzerland, which provided the bases for engine manufacturers to issue clearances in that country (Wolfensberger, 1993).

INFLUENCE ON THE ENVIRONMENT

Environmental concerns about the use of engine fuels derive predominantly from the amount and types of emissions produced on combustion. Potentially hazardous emissions include carbon monoxide, carbon dioxide, nitrous oxides, sulphur oxides, particulate matter (including soot) and hydrocarbons comprising aldehydes, aromatic compounds and polycyclic aromatic hydrocarbons. Concerns arise partly from possible effects on global warming and partly from effects on local air pollution.

Emissions causing global warming

Whereas great efforts have been made to reduce emission levels by improving engine design, substantial reductions may also be achieved by using alternative fuels. Since plants use carbon dioxide, its production from combustion of biodiesel fuel and

Table 16.5. Classification of the potential hazard of biodiesel emissions to humans, plants and buildings.[1]

Emission	Humans	Plants	Buildings
Carbon monoxide	0	0	0
Carbon dioxide	–	0	+0
Nitrous oxide	0	0	0
Nitrous dioxide	>0	>0	2
Particulate matter	1–2	0	1
Sulphur dioxide	>0	5	3
Particulate matter + sulphur dioxide	4	–	4
Ozone	>0	1–2	0

Source: Elstner and Hippeli (1992).
–, Not determined.
[1] 0 = low, 5 = high hazard level.

reutilization by plants can be viewed from an environmental point of view as a closed cycle. Thus the combustion of biodiesel represents a possible contribution to reducing the greenhouse effect. However, account must be taken of all types and volumes of greenhouse gases released from sowing the oilseed crop to consumption of the biodiesel. In addition to the emissions from biodiesel combustion, consideration must be given to those released during crop production and processing, e.g. field tractors, fertilizer production and oilseed processing. Note should be taken in particular of nitrous oxide, which is produced from inorganic nitrogen compounds under anaerobic conditions by denitrification bacteria.

With intensive rapeseed cultivation in Western Europe it has been estimated that for each kilogram of biodiesel fuel consumed total greenhouse gas emissions may be reduced by as much as 3.2 kg carbon dioxide (Scharmer, 1993). This advantage may be greater with less intensive agricultural systems or with crops requiring lower levels of nitrogen fertilization, e.g. beans. In an attempt to evaluate potential damage a recent study commissioned by the EU calculated that the cost of greenhouse gas emissions in US dollars was $485 tonne^{-1} of carbon dioxide (Hohmeyer, 1993).

Exhaust emission causing local air pollution

Combustion of biodiesel produces emissions which on balance can reduce local air pollution when compared with diesel. In considering emissions which may have a harmful effect on human health, plants and buildings, the Technical University in Munich, Germany, has classified potential hazards (Table 16.5). These results emphasize the synergistic effect of sulphur dioxide in combination with particulate matter in raising the hazard to human beings, while the ozone risk is much lower and is considered to have been overestimated.

While comparisons with diesel have shown carbon monoxide emission increases of 3–10% with biodiesel, one study showed a decrease of up to 30% (Walters, 1992). However, carbon monoxide emissions were reduced by 65% with combustion of biodiesel fuel in engines equipped with a catalytic converter, compared to diesel and no catalytic converter (Walter, 1992).

Generally only minor differences have been observed between biodiesel and diesel fuels for nitrous oxide emissions, although one study in Switzerland (Walter, 1992) and one from Canada (Goetz, 1994) reported a 25% decrease in nitrous oxide emissions, when timing of fuel injection was delayed. This demonstrates the potential for further improvements when unmodified diesel engines are fine-tuned for biodiesel use. While this appears promising in terms of reducing nitrogen oxides, it has to be noted that these are accompanied with slight increases in levels of particulates.

Given the extremely low sulphur content in double-low rapeseed and the biodiesel derived therefrom (specification <0.02% sulphur), there is nearly a 100% reduction in sulphur oxide emissions compared to diesel. Low sulphur content of vegetable oils results in biodiesel fuels with a low sulphur content and much reduced sulphur oxide emissions on combustion. Low sulphur emission also extends the life of catalytic converters (Schäfer, 1992).

Emissions of particulates vary with engine type and test design. Two studies (Sams and Schindlbauer, 1992; Walter, 1992) showed no difference between biodiesel and diesel, while two other studies (Havenith, 1993; May, 1993) gave significant reductions of between 29 and 53% in particulate matter with biodiesel. A study by Goetz (1994) showed a continuous decline in particulate emissions with increasing contents of biodiesel produced from soya oil in a diesel fuel.

Soot emissions have been shown to be reduced by 40–60% compared with diesel (Sams and Schindlbauer, 1992). Reductions were visible as a greyish white smoke from the exhaust pipe rather than the black smoke produced with diesel.

Comparisons of biodiesel with diesel have shown an overall reduction of 13% in aldehyde emissions (Wurst *et al.*, 1991). In the same study cyclic (aromatic) hydrocarbons were shown to have an overall reduction of 30%. There was also evidence of substantial reductions in polycyclic aromatic hydrocarbons, some of which are considered to be carcinogenic or mutagenic, e.g. phenanthrene was reduced by 97%, benzofluoranthene was reduced by 56%, and benzopyrene was reduced by 32% (Wurst *et al.*, 1991).

Although at an early stage with engines designed and set for diesel rather than tuned for biodiesel, the superior fuel characteristics of the latter are revealed in engine performance and emissions. Two trial reports have proposed minor adaptation of the fuel-injection system to improve emission further (Walters, 1992; Goetz, 1994). Moreover, biodiesel and the oxidation catalytic converter are complementary in decreasing carbon monoxide and hydrocarbon emissions further to below levels required internationally.

These comparative studies of biodiesel and diesel, which involved unmodified engines for the most part, gave evidence of improvements in the emissions of sulphur oxides, soot and polycyclic aromatic hydrocarbons. Further research with engines modified and optimized for biodiesel fuel hold promise for added improvements. Combined with beneficial effects of biodiesel on catalytic converters, biodiesel is proving a suitable alternative fuel for meeting the increasingly more stringent national and international emission standards.

Toxicity

The low toxicity of biodiesel fuel has an important environmental advantage. Standard toxicity tests with rats, performed in conformance with the Organization for Economic Cooperation and Development (OECD) guidelines, have shown a median lethal dose (LD_{50}) greater than 2000 mg kg^{-1} with no difference between male and female rats (Körbitz, 1994a). As expected this confirms the extremely low toxicity of vegetable oils, although it is

Table 16.6. Comparative toxicity of biodiesel and fossil diesel to aquatic organisms.

Effective concentration of fuel	Rainbow trout *Onchorhynchus mykiss*	*Daphnia magna*	Cress *Lepidium sativum*	Algae *Selenastrum capricornutum*	Bacteria *Pseudomonas putida*
EC 0[1]					
Biodiesel	100	65	100	100	100
Fossil diesel	1	1	10	0.01	3.2
EC 0[2]					
Biodiesel	–	67	>100	>100	>100
Fossil diesel	1.5	6	25	0.01	>1

Source: Rodinger (1994).
–, Not determined.
[1] Concentration in water at which no effect is observed on any of the populations under test.
[2] Concentration in water at which effects are observed on 10% of the population.

advisable to avoid comparison of vegetable and fuel oils as this might give rise to misunderstanding among laymen and possible misuse of diesel.

Spillage of mineral oils into lakes and waterways can have a damaging effect on water flora and fauna. Standard tests conducted with rainbow trout (*Oncorhynchus mykiss*), *Daphnia magna*, cress (*Lepidium sativum*), algae (*Selenastrum capricornutum*) and bacteria (*Pseudomonas putida*) have confirmed the low toxicity of biodiesel fuels (Rodinger, 1992) (Table 16.6).

An additional advantage of biodiesel is its high biodegradability. Ribarov (1993) showed that within 3 weeks biodiesel fuel was degraded by more than 95% by soil microorganisms, while diesel was only 72% degraded. Low toxicity and rapid biodegradability make the use of biodiesel advantageous in environmentally sensitive areas, areas where groundwater or drinking-water must be protected, and areas where aquatic organisms might be at risk.

High flashpoint and low vapour pressure also make biodiesel a safer fuel to handle, transport and store. The higher flashpoint makes biodiesel less flammable than diesel, while the lower vapour pressure

reduces the risk of inhalation of toxic fumes by workers, and the potential for explosion when mixtures of volatile hydrocarbons and air are exposed to an ignition source (FICHTE, 1994).

ENERGY BALANCE

A cause for general concern, besides the contribution of fossil energy forms to increased carbon dioxide emissions and the greenhouse effect, is the utilization of existing energy resources. Many fossil and finite energy reserves are located in politically unstable areas of the world (Table 16.7). Assuming constant production and consumption levels, present known fossil oil reserves will last for only 42.6 years. Vegetable oil is stored solar power and little extra energy is required to produce this renewable liquid fuel.

The ratio of total energy input to energy output, expressed as the energy balance in a full life cycle analysis, is commonly used to judge alternative energy forms. The key factor in favour of biodiesel on the input side is the free energy from sunlight. Taking into account all other energy inputs, e.g. field tractors, fertilizer

Table 16.7. Geographical supply and distribution of world mineral oil reserves.

Area	Supply (barrels × 10⁻⁹)	Total (%)
Middle East	661.8	66
Latin America	123.8	12
Africa	61.9	6
Former USSR	59.2	6
Asia and Australia	44.6	4
North America	39.7	4
Europe	15.8	2

Source: Walker and Körbitz (1994).
[1] A barrel = 159 l.

production, plant protection, transport and processing, compared with the output energy contained in biodiesel, glycerine and rapemeal, there is a positive balance in favour of output in the range 1 : 4.1 to 1 : 4.5. This allows for slight variation in both input and output energy. This balance will improve further when utilizing cheaper forms of raw material such as used frying oil (Ribarov, 1993).

It takes about 1.18 kg of diesel to produce, distribute and consume 1 kg fossil diesel fuel (ignoring losses due to fires, wars or mismanagement) (Fig. 16.1). Considering all forms of energy needed to produce biodiesel, e.g. in cultivation and processing, and free solar energy, Scharmer (1993) concluded that replacing 1 kg of diesel with biodiesel (rapeseed methyl ester) reduces fossil energy consumption by 0.92–0.79 kg mineral oil equivalent (Fig. 16.2). This does not take into account the superior energy efficiency of biodiesel.

Estimates of the volume of potential biodiesel production have been made for Western and mid-European countries recognizing the limiting factors of total available arable land, arable land available with climatic conditions suitable for rapeseed growing, and allowance for rapeseed production for food. These estimates sug-

gest that biodiesel production may reach 5% market share of fuels sold in the diesel sector, which is in line with long-term EU objectives for biofuel market development. Taking into consideration all known existing production facilities and serious projects in Europe, total production of 800,000 tonnes biodiesel by the year 2000 seems highly probable (Körbitz, 1994b).

ECONOMICS OF PRODUCTION

In view of the relatively short time that biodiesel has been under development, significant achievements have been made in reducing the cost of the end-products. Key factors influencing profitability of biodiesel production are cost of raw material, processing yield and cost, and market prices.

Raw materials costs

The price of rapeseed traded on the world market fluctuates according to demand for food use and this presents a certain risk to economic biodiesel production. For example the world price for rapeseed rose by 49% between December 1992 and April 1994 (*Oil World*, Hamburg). Within the EU, the Common Agricultural Policy regulates that 15% of the arable area should be set aside from food production. This was reduced to 12% for 1995. Non-food crops may be grown on this setaside' and a separate market is developing for industrial oilseeds at a much lower price, which represents an attractive basis for biodiesel production.

Early transesterification technology started with refined rape oil of top grade food quality. This ensured low free fatty acid content and low levels of impurities but increased costs since the raw material represented 80% of production costs. Newer technologies have been developed to utilize crude oil, or oil from one

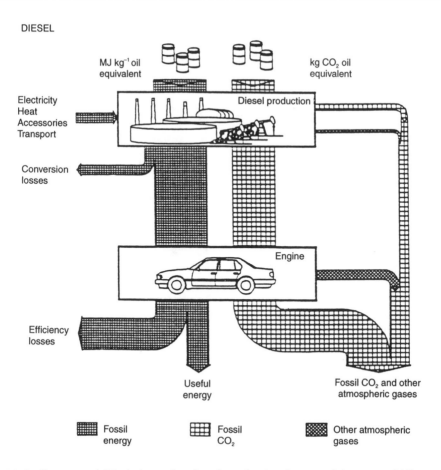

DIESEL

| MJ kg^{-1} oil equivalent | | kg CO_2 oil equivalent |

Electricity
Heat
Accessories
Transport

Diesel production

Conversion losses

Engine

Efficiency losses

Useful energy

Fossil CO_2 and other atmospheric gases

Fossil energy   Fossil CO_2 Other atmospheric gases

Fig. 16.1. Energy and CO_2 balance for diesel production (source: Scharmer, 1993).

degumming operation, which reduces phosphorus content to the biodiesel specification by removing phospholipids.

In the search for even cheaper raw materials technologies have been developed for processing used frying oils and waste fats with high levels of free fatty acids. However, market research has revealed large differences among industrialized countries in awareness of environmental problems with waste vegetable oil disposal and willingness to establish collection systems for recycling.

A recent development involves transesterification of used hydraulic oils derived from rapeseed, which has gained market share because of its biodegradability.

Esterification is more difficult as additives, incorporated to improve properties of the oil for hydraulic purposes, can impede the chemical reaction, thereby increasing cost.

Processing costs

Processing comprises a relatively small portion of the total cost of production, raw material being the major item. Processing costs are low because esterification can be achieved at ambient temperature and pressure, thus avoiding the need for expensive equipment. However, yield of methyl esters is a key factor influencing the economics of biodiesel production. Early technology produced a 95% yield

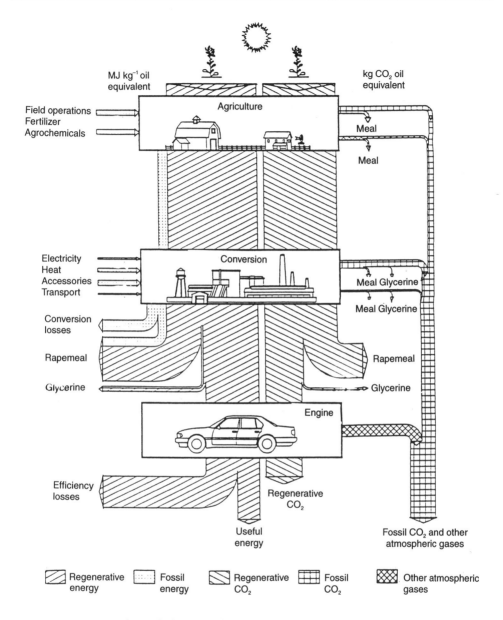

Fig. 16.2. Energy and CO_2 balance for biodiesel production (source: Scharmer, 1993).

of methyl esters but this has increased to 99%, thus reducing waste and improving profitability.

Treatment of the glycerine by-product adds to the overall cost of processing. This cost may be low if the glycerine is to be marketed as a crude unrefined grade but is higher depending on whether the glycerine is refined to 80% or 99.5% purity. The latter is required for the pharmaceutical grade, which may command a market price sufficiently high to cover the additional cost.

Products from a combined oil processing/biodiesel installation will be sold on different markets with prices for each product influenced by the profitability of the relevant specific market. Careful and continuous control of all parameters is essential in order to assure profitability of a biodiesel installation.

Evaluation of the benefits of a biodiesel industry to a national economy will be largely influenced by national considerations and the local economic–political environment. However, an example of a completed feasibility study in the Republic of Ireland demonstrates the potential impact of a biodiesel industry producing 30,000 tonnes year^{-1} (Kinsella, 1994). As a member of the EU, Ireland has to comply with the Common Agricultural Policy, which at the time of the study required 15% of arable land to be set aside from food production. Fere Consultants (1992) showed in a study in France that this can have a negative effect on employment, mainly in the agricultural supply industries. However, the Irish study demonstrated that not only would the 324 jobs, potentially lost through the setaside programme, be retained, but 26 additional jobs would be created to process the 30,000 tonnes year^{-1}. The average value of these jobs would amount to IRP12,850 per job giving a total value of IRP4.5 million year^{-1} (Kinsella, 1994).

Added value in national accounting is a method for calculating gross national product and is defined as:

added value = output (sales) − inputs
(raw material, processing, subsidies)

A comparison in the Irish study of leaving setaside uncropped compared with its use for biodiesel production revealed that 30,000 tonnes year^{-1} of biodiesel production would realize IRP14.20 million year^{-1} while uncropped setaside would yield after subsidy only IRP6.42 million year^{-1}. Thus putting the land into biodiesel production would realize a net increase to the economy of IRP7.77 million year^{-1}.

Balance of trade is another important factor in national accounting. In the Irish study substituting imported fuel for home-produced biodiesel would realize IRP3.34 million year^{-1} for the oil and an additional IRP4.45 million year^{-1} for the protein feed, giving a total saving of IRP7.79 million year^{-1}.

However, when increased employment with its benefits to the economy was balanced against a loss of excise tax resulting from biodiesel being taxed at 10% of that for diesel, a net loss to the economy of IRP4.97 million year^{-1} was predicted.

With current low prices for mineral oils, it is uneconomical to substitute any vegetable oil derivative as a fuel source without some form of support. Given the reduced risk to the environment of biodiesel, environmental concern is one justification for support. Both statutory environmental standards and exemption from excise duties have been used as arguments to support utilization of biodiesel.

Potential markets

It has been estimated that biodiesel could replace 5% of the total diesel market (Körbitz, 1994b). Given the economics of production and the environmental advantages of biodiesel, it is logical that biodiesel should be produced for niche markets. Several niche markets have been identified based on the biodiesel characteristics of biodegradability, low toxicity and low emissions (Table 16.8). Marketing success will depend on local socioeconomic benefits and environmental concerns. Legal enforcement of biodiesel usage may occur in some areas (Körbitz, 1992). There is already a regulation for chainsaw lubricants in Austria (Austrian Standards Institute C2030 specification), which requires not less than 80% biodegradability.

Table 16.8. Niche markets for biodiesel fuel.

Advantage	Niche market
Biodegradability	Operations in drinking-water protected areas Irrigation (pumps) Excavation in groundwater Forestry operations
Low toxicity	Inland waterway shipping Boating in tourism areas Operations in conservation areas and parks
Low emissions	Urban taxi and bus fleets Forestry operations Mining and tunnelling operations Operations in ozone emission restricted areas

Table 16.9. Biodiesel project status: Europe and North America, May 1994.

Austria
Production: 4 units: 500 tonnes capacity each
          2 units: 10,000 tonnes at Aschach (in reconstruction)
                          15,000 tonnes at Bruck, also for used frying oil
Project for used frying oil: 30,000 tonnes (Pischelsdorf)
Standard: established for rapeseed oil methyl ester – Austrian Standards Institute C1190 (1991)
used frying oil methyl ester – by end 1996
Taxation: about 5% of existing mineral oil tax
Marketing: 100% in areas with a high environmental risk

France
Production: 1 unit: 1,000 tonnes at Compiègne by Robbe
          4 units: 20,000 tonnes at Compiègne by Robbe
                          40,000 tonnes at Boussens by Henkel
                          10,000 tonnes at Peronne by Castrol
                          40,000 tonnes at Verdun by ICI and Novamont
Projects: 3 units of 120,000 tonnes each at Rouen, Nogent and in the south-west
Sales contracts: Ministries of Agriculture and Environment, the Elf and Total Petroleum Companies and investors with sales target of 125,000 tonnes by 1995
Standard: established and valid for 5% addition to diesel fuel
Taxation: 0% but government licence obligatory
Marketing: 5% addition to diesel fuel, mixed and distributed through Elf and Total company outlets

Italy
Production: various units: 100,000 tonnes at Livorno by Novamont
          one unit: 10,000 tonnes at Città di Castello (sunflower oil)
Standard: established for motor fuel and heating oil
Taxation: 0% for motor fuel and heating oil but limited to 250,000 tonnes year^{-1}
Marketing: priority for community heating systems

Table 16.9. (contd).

Czech Republic
Production: 3–4 units: 1,000 tonnes each
              1 unit: 30,000 tonnes at Olomouc by MILO
Projects: 2 medium-size units
             10 small-scale units
              1 laboratory-scale trial with used frying oil
Standard: CSN-65 6507, similar to Austrian Standards Institute specification C1190
Taxation: 0% mineral oil tax and value added tax reduced from 23% to 5%
Marketing: 100% in areas where there is a high environmental risk, and also using rapeseed grown on contaminated land unacceptable for food crops.

Slovak Republic
Production: 4 units: 1,000 tonnes each

Germany
Production: 1 pilot unit: 3,000 tonnes at Leer
             1 unit: 60,000 tonnes at Düsseldorf by Henkel
Projects: 1 unit: 10,000 tonnes at Kiel by HaGe
             1 unit: 100,000 tonnes in Bavaria
             1 unit: 60,000 tonnes at Leer starting in 1995
             1 small unit: in Saxonia
Standard: DIN 51.606 preliminary (similar to Austrian Standard C1190 and Mercedes requirement)
Taxation: 0% to begin with
Marketing: Mainly in agriculture, for public transport and taxi fleets and for important tourist areas, e.g. North Sea islands

---

Belgium
Production: 1 unit: 40,000 tonnes at Seneffe by FINA

Denmark
Project: 1 unit: 30,000 tonnes at Otterup
Taxation: 0% when used as heating oil

Sweden
Production: 2 units: 10,000 tonnes at Gothenburg by Swedish Oil AB
                              6,000 tonnes at Skåne by Ecofuel
Project: 1 unit: 5,000 tonnes at Örebro
Marketing: Under trade name 'Scafi 101' (=40% rapeseed methyl ester and from 60% N-paraffin) to reduce soot emissions in areas where workers at risk from inhalation

---

Hungary
Production: 1 unit: 880 tonnes at Visnye
Projects: 1 unit: 18,000 tonnes at Györ for sunflower oil
            15 small units

Ireland
Status: Government study completed
Project: 1 unit: 30,000 tonnes including used frying oil

Table 16.9. (contd).

United States of America
Production: 1 unit with 50,000 tonnes capacity from soya oil by Procter and Gamble
Status: National BioDiesel Development Board (NBB) established to plan marketing strategy, financed by levies and supported by the American Biofuels Association (ABA). Used frying oil and animal fats under consideration as additional raw material.
Project: 1 unit   10,000 tonnes in Nebraska or Iowa
Marketing: mainly for river and canal boats and for public transport where the 'Clean City Programmes' and 'Clean Air Act' Amendments 1990 are relevant

United Kingdom
Status: Study by the Scottish Agricultural College, Körbitz Consulting and the Energy Technology Support Unit published
British Association for Biofuels and Oils formed in November 1993
Engine tests with Reading bus company completed
First biodiesel produced in quantity (East Durham project)

This requirement is easily met by vegetable oils and their derivatives.

To meet market demands and address environmental concerns biodiesel is produced in a number of countries, particularly in Europe, and more production facilities are under construction (Table 16.9). Continued expansion of production may be anticipated as the benefits of biodiesel become more widely accepted.

REFERENCES

Batel, W. (1980) Pflanzenöl für die Kraftstoff- und Energieversorgung. *Grundlagen Landtechnik* 30, No. 2.

Diesel, R. (1895) German patent no. 82168.

Elstner, E.H. and Hippeli, S. (1992) Auswirkungen der Emissionen auf Menschen, Pflanzen und Bauwerke (Impact of emissions on humans, plants and buildings). In: *Proceedings of the 13th International Vienna Engine Symposium*, May 1992. Austrian Society of Automotive Engineers, Vienna, 34 pp.

Ferdinand, Chr. (ed.) (1992) Non-food uses of agricultural products in Europe. In: *Study written by the Club de Bruxelles* (for conference on 'Biofuels in Europe'). Club de Bruxelles, Bruxelles, Belgium.

Fere Consultants (1992) *Economic and Fiscal Effects of the Rapeseed-Methylic-Esters Incorporation into Diesel Oil*. Fere Consultants, Paris, France.

FICHTE (1994) *Handbook of Analytical Methods for Fatty Acid Methyl Esters Used as Diesel Fuel Substitutes*. FICHTE/Technical University, Vienna, Austria, 71 pp.

Goetz, W.A. (1994) Evaluation of biofuels in heavy-duty engines on an engine dynamometer. In: *Proceedings of Bio-Oils Symposium*. Saskatchewan Canola Development Commission, Saskatoon, Canada.

Havenith, C. (1993) Pflanzenölmotoren (Vegetable oil engines). In: *Proceedings of the Symposium Kraftstoffe aus Pflanzenöl für Dieselmotoren* (Fuels from vegetable oils for diesel engines). Technische Akademie Esslingen, Ostfildern, Germany, 14 pp.

Hohmeyer, O. (1993) Evaluation of external costs of CO_2-emissions. *Das Solarzeitalter* 2, 38.

Kinsella, E. (1994) *The Macro-economic Evaluation of the Production of Rapemethylester in Ireland*. Dublin Institute of Technology, Ireland, 34 pp.

Körbitz, W. (1992) Key achievements for biodiesel in Austria and the Czech Republic. In: *Proceedings of the First European Forum on Biofuels*. ADEME, Paris, France, pp. 231–233.

Körbitz, W. (1994a) The technical, energy and environmental properties of biodiesel. In: *Proceedings of Bio-Oils Symposium*. Saskatchewan Canola Development Commission, Saskatoon, Canada.

Körbitz, W. (1994b) Status of biodiesel production in Europe. In: *Proceedings of Bio-Oils Symposium*. Saskatchewan Canola Development Commission, Saskatoon, Canada.

Machold, W. and Dobbins, P.J. (1991). Effect of biodiesel on motor-oils in Diesel engines/Castrol. *Pilotprojekt 'BioDiesel'/Research Reports of the Institute of Agricultural Engineering*, Institute of Agricultural Engineering, Wieselburg, Austria, 25, 34.

May, H. (1993) Nachwachsende Rohstoffe als Kraftstoffe, Schmierstoffe und Funktionsflüssigkeiten für Dieselmotoren im Vergleich zu Dieselkraftstoff (Renewable raw materials as fuel, lubricants and technical liquids for Diesel engines in comparison to Diesel fuel). In: *Proceedings of the Symposium Kraftstoffe aus Pflanzenöl für Dieselmotoren* (Fuels from vegetable oils for diesel engines). Technische Akademie Esslingen, Ostfildern, Germany, 13 pp.

Mittelbach, M. (1991) Methods for the transesterification of vegetable oils. In: *Proceedings of the AVL-list Conference Engine and Environment*. Graz University Graz Institute for Organic Chemistry, Austria, pp. 311–319.

Mittelbach, M. and Tritthart, P. (1988) Diesel fuel derived from vegetable oils. III. Emission tests using methyl esters of used frying oil. *Journal of the American Oil Chemists' Society* 65, 7.

Mittelbach, M., Wörgetter, M., Pernkopf, J. and Junek, H. (1983) Diesel fuel derived from vegetable oils: preparation and use of rape oil methyl-ester. In: *Energy in Agriculture*, Vol. 2. Elsevier Science Publishers, Amsterdam, pp. 369–384.

Nye, M.J. (1983) Methyl ester from used frying oil as a diesel fuel. Thesis, University of Guelph, Canada, 4 pp.

Ribarov, D. (1993) *Handbuch für RME-Anwender* (Handbook for RME users). RME-GmbH, Bruck/Leitha, Austria.

Rodinger, W. (1992) Umweltverträglichkeit von RME im Vergleich zu Diesel-kraftstoff (Environmental parameters of rapeseed methyl-ester in comparison to diesel oil). In: *Proceedings of the Symposium 'RME' – Kraftstoff und Rohstoff*. FICHTE/Technical University, Vienna, Austria, pp. 119–129.

Rodinger, W. (1994) *Neue Daten zur Umweltverträglichkeit von RME im Vergleich zu Dieselkraftstoff* (New data on environmental parameters of rapeseed methyl-ester in comparison to diesel fuel). Bundesanstalt für Wassergüte, Vienna, Austria.

Sams, Th. and Schindlbauer, H. (1992) Untersuchungen über den Betrieb von Dieselmotoren mit RME (Investigations on running a diesel engine with rapeseed methyl-ester). In: *Proceedings of the Symposium 'RME' – Kraftstoff und Rohstoff*. FICHTE/Technical University, Vienna, Austria, pp. 49–66.

Schäfer, A. (1991) Pflanzenölfettsäure-Methyl-Ester als Dieselkraftstoffe/Mercedes Benz. In: *Proceedings of the Symposium Kraftstoffe aus Pflanzenöl für Dieselmotoren* (Fuels from vegetable oils for diesel engines). Technische Akademie Esslingen, Ostfildern, Germany.

Schäfer, A. (1992) Pflanzenölfettsäure-Methyl-Ester als Dieselkraftstoffe/Mercedes Benz. In: *Proceedings of the Symposium Kraftstoffe aus Pflanzenöl für Dieselmotoren* (Fuels from vegetable oils for diesel engines). Technische Akademie Esslingen, Ostfildern, Germany.

Scharmer, K. (1991) Biokraftstoffe für Dieselmotoren (Biofuels for diesel engines). In: *Proceedings of the Symposium Kraftstoffe aus Pflanzenöl für Dieselmotoren*. Technische Akademie Esslingen, Ostfildern, Germany, 24 pp.

Scharmer, K. (1993) Umweltaspekte bei Herstellung und Verwendung von RME (Environmental aspects about RME production and usage). In: *RME Hearing*. Ministry of Agriculture, Vienna, Austria, 4 pp.

Scharmer, K., Golbs, G. and Muschalek, I. (1993) Pflanzenölkraftstoffe und ihre Umweltauswirkungen, Argumente und Zahlen zur Umweltbilanz (Vegetable oil fuels and their influence on the environment – arguments and figures for a full cycle environmental balance). In: *Bericht für die UFOP*. Bonn, Germany, 35 pp.

Schütt, H. (1982) Natürliche Rohstoffe zur Herstellung von Fettalkoholen (Natural raw materials for production of fatty alcohols). In: *Fettalkohole*. Henkel KGAA, Düsseldorf, Germany, pp. 13–49.

Syassen, O. (1991) Situationsanalyse zur Problematik nachwachsender Kraftstoffe (Situation analysis about renewable fuels). In: *Study for the Ministry of Agriculture in Rheinland-Pfalz*. Hemsbach, Germany, 176 pp.

Walker, K.C. and Körbitz, W. (1994) Rationale and economics of a UK biodiesel industry, Vienna,

January 1994. *Report for the British Association of Biofuels and Oils.* Scottish Agricultural College, Aberdeen, UK.

Walters, Th. (1992) Untersuchungen des Emissionsverhaltens von Nutzfahrzeug-motoren am Prüfstand bei Betrieb mit RME (Investigations on emissions of lorry engines in a bench test using rapeseed methyl ester). Swiss Research Institute, Dübendorf. In: *Proceedings of the Symposium 'RME' – Kraftstoff und Rohstoff.* FICHTE/Technical University, Vienna, Austria, pp. 49–66.

Wolfensberger, U. (1993) *List of Diesel Engine Clearances from Manufacturers.* Publication FAT Eidgenössische Forschungsanstalt, Tänikon, Switzerland.

Wörgetter, M. (1991a) Flottenversuch zur Erprobung von Rapsölmethylester als praxistauglicher Kraftstoff (Fleet tests with rapeseed oil methyl-ester as a fuel for daily use). In: *Pilotprojekt 'BioDiesel': Research Reports of the Institute for Agricultural Engineering,* Institute of Agricultural Engineering, Wieselburg, Austria, 25, IV/1–IV/88.

Wörgetter, M. (1991b) *List of Tractors Suitable for Biodiesel.* Bundesanstalt für Landtechnik, Wieselburg, Austria.

Wurst, F., Boos, R., Prey, Th., Scheidt, K. and Wörgetter, M. (1991) Emissionen beim Einsatz von RME an einem Prüfstandmotor (Emissions from a bench test diesel engine using RME). In: *Pilotprojekt 'BioDiesel' Teil II: Research Reports of the Institute for Agricultural Engineering,* Institute of Agricultural Engineering, Wieselburg, Austria, 26, IX/1–IX/64.

17 The Mustard Species: Condiment and Food Ingredient Use and Potential as Oilseed Crops

J.S. Hemingway

18 Postwick Lane, Brundall, Norwich, UK

THE MUSTARD SPECIES

Mustard is principally grown as a source of condiment for the spice trade. Two species are generally grown, each of which derive from different genera of the *Cruciferae* family. *Sinapis alba* (old syns *Brassica hirta* and *Brassica alba*) is commonly known as 'white' or 'yellow' mustard. *Brassica juncea* is known as 'brown' or 'Oriental' mustard, having two forms with dark brown (formerly 'Sarepta') and golden-yellow seed coats, respectively. Apart from being used as a spice, mustards are widely used as leaf and stem vegetables and as a salad crop in the Far East and Southeast Asia, and for green manuring or as a fodder crop mainly in Western Europe (Vaughan and Hemingway, 1959; Rosengarten, 1969). Mustards are also important oilseed crops, particularly brown mustard in the Indian subcontinent, China and the south-western areas of the former Soviet Union, where the crop is grown in a similar manner to spring rapeseed. Some white mustard is grown in Sweden for industrial oil.

Compared to spring rapeseed, mustard is more heat and drought tolerant, more resistant to pod shattering, and has a higher resistance to blackleg disease (*Leptosphaeria maculans*). Although agronomically well adapted to drier climates, the presence of high levels of glucosinolates (approx. $200\,\mu\text{mol g}^{-1}$ of oil-extracted meal) and a significant amount of erucic acid (approximately 25–45%) makes cultivars grown for the spice trade unsuitable as a source of edible oil in many countries where low erucic, low glucosinolate seed is a preferred source. However, Kirk and Oram (1981) in Australia identified, from material which had originated in China, two *B. juncea* strains that were essentially free of erucic acid, and efforts in Canada have succeeded in introducing the low glucosinolate characteristic through interspecific crosses (Love *et al.*, 1990). Thus the crop shows promise for expanding *Brassica* oilseed production into drier climates, such as Western Australia with its short winter growing season (Oram and Kirk, 1992) and the southern prairies of western Canada where rainfall is low (Woods *et al.*, 1991). In cooler climates *B. juncea* mustard matures later than spring rapeseed with higher yields (Woods, 1992).

Sinapis alba and *B. juncea* are cultivated for the spice trade in the same regions of the world with similar husbandry techniques. Morphological differences distinguish each species. *Sinapis alba* foliage and stems are bright green in colour, the leaves deeply veined and bearing surface hairs, and the stems hollow and ribbed. Pods are 2–3 cm in length, with a stiffly hairy and rough surface, and a pronounced long flattened 'beak' at the tip. A pod contains up to eight seeds, and the crop averages 148,000 seeds kg^{-1}. *Brassica juncea* foliage and stems are more glaucous to bluish green, the leaves smoother, and the stems wax coated and pith filled. Pods are up to 6 cm in length, tubular and smooth, with only a short tubular beak. A pod contains up to 20 seeds, and the crop averages 408,000 seeds kg^{-1}. The two species also differ in their glucosinolate content, the source of isothiocyanates (essential oils) which impart the hot 'mustard' taste. *Sinapis alba* seed contain predominantly 4-hydroxybenzyl glucosinolate while *B. juncea* seed contain predominantly 2-propenyl (allyl) glucosinolate.

Black mustard, *Brassica nigra*, was grown along with *S. alba* in the past as a source of seed for the spice trade. However, mainly because of the dehiscent pod, it was impossible to breed cultivars suitable for mechanical harvesting and *B. nigra* was superseded by *B. juncea* in the 1950s. *Brassica juncea* has the added advantage of having a yellow-seeded form which has a thinner seed coat (Hemingway, 1976). The conversion from *B. nigra* to *B. juncea* production occurred remarkably quickly, especially for an industry as traditional as the spice trade. However, the change was made on strong practical and economic grounds. *Brassica juncea* is a closely related species as it is an amphidiploid with *B. nigra* as one of the parents (Hemingway, 1976; Tsunoda *et al.*, 1980). This facilitated the change from *B. nigra* to *B. juncea* as it was possible to isolate cultivars with an identical flavour component.

It should be noted that the *B. juncea* cultivars grown for oil on the Indian subcontinent are unsuitable for condiment purposes as their seeds produce an unpalatable flavour (Vaughan *et al.*, 1963). Nowadays, only minimal tonnages of *B. nigra* are traded on the world market and virtually no seed of another amphidiploid, *Brassica carinata* (Abyssinian mustard), is traded outside north-east Africa.

Seed of *S. alba* and *B. juncea* accounts for about 60% and 40%, respectively, of the world spice trade, with *B. juncea* divided equally between its brown and Oriental forms. Their importance for different products depends on their glucosinolate composition, which in turn determines the isothiocyanate content and thereby the flavour of condiment mustard.

BRIEF HISTORY

The term 'mustard' is believed to be derived from the early use of seeds as a condiment, the crushed seeds being mixed into a paste with the sweet must of old wine and called 'mustum ardens' or hot must. Condiment mustard has been principally used with meat or fish dishes, not least to improve or disguise strong flavours or, indeed, off-flavours, before the development of efficient handling of fresh foods or of modern food preservation techniques.

Mustard seed has been known as a spice since from early times (Man and Weir, 1988). It is described in Sumerian and Sanskrit texts from as early as 3000 BC, in Egyptian texts around 2000 BC, and in Chinese writings before 1000 BC. Seed fragments have been identified from excavations of Indus valley dwellings around 2000 BC and from Egyptian tombs at Thebes of the same period. Many classical scholars mention mustard, including the Greek mathematician Pythagoras and physicians Hippocrates and Dioscorides, and the Roman historians Pliny and Marcus Gabius Apicius.

The Bible, Alexander the Great and Charlemagne all referred to mustard seed, and there are numerous quotes from medieval times onwards which refer to trading in manufactured condiment in the days of the Avignon Popes, Dumas and Shakespeare. Mustard sellers even wore a distinctive uniform in France in the 16th century.

Amongst early spice traders, Arabs traded seed out of Alexandria and Marco Polo is believed to have been involved in the spread of mustard across Asia. Vasco da Gama included the seed in his ships' stores on his first circumnavigation of the Cape of Good Hope, and, as a modern equivalent, the United States Government included mustard seed amongst exploratory botanical material carried in their first space probe (Colmans, 1985).

Two references have passed into everyday usage as figures of speech. The phrase 'as keen as mustard' doubtless derives from the old City of London spice millers, Keen and Company. The phrase 'too old to cut the mustard' is curious as it has almost exclusively North American usage, but must have arisen from the hard physical work when the thick-stemmed *B. nigra* growing in Europe had to be handharvested when still green to prevent the pods from shattering with the resulting loss of seed.

ECONOMIC POSITION

Mustard seed has by far the largest volume among spices traded internationally at the present time, totalling 160,000 tonnes year^{-1}. Mustard is unusual as a spice as the crop requires long cool days with adequate moisture, typical of temperate latitudes. To suit mechanization and economics of advanced farming systems, plant breeders have developed cultivars which give consistent yields of seed of a high quality, suitable for every aspect of its manufacture and utilization as a condiment. Postharvest handling and transportation of the seed in bulk is a further aspect almost unique for items within the spice trade.

The predominant production area is the Canadian prairie provinces of Manitoba, Saskatchewan and Alberta, and the adjacent states of North Dakota and Montana in the United States. Crops of about 170,000 ha annually produce some 120,000 tonnes with an export value of US $35 million. Other important production areas are the Hungarian plains, Poland, and eastern areas of Germany and the United Kingdom. Crops for condiment are grown on a smaller scale in Argentina, Australia, the Balkan countries, France, and Italy.

Condiment manufacturing is concentrated mainly in the United States, France, Germany, Japan, Canada and the United Kingdom. While these countries carry out a considerable international trade in mustard products, there are also many diverse local manufacturers in Europe and in the European-settled areas of the world.

Brassica juncea grown as an oilseed crop is of major importance in the countries of the Indian subcontinent, in China and the southern Ukraine. Indian production statistics for 1990–1991 report a combined rapeseed and mustard production of 4.22 million tonnes (Kothari, 1992). This represented 24.5% of total oilseed product and was second only to groundnut. Rapeseed and mustard production occupied 5 million ha, predominantly (60%) in Pradesh. The oil was used primarily for human consumption and the meal as animal feed. Rapeseed and mustard production is of similar importance in Pakistan and Bangladesh. All of the production is consumed domestically. Additional demand in Bangladesh in recent years has led to the yearly importation of about 70,000 tonnes of seed from Canada.

Brassica juncea is grown widely in China

where it is locally processed and consumed. The southern Ukraine is thought to produce 100,000 tonnes annually for local consumption.

Brassica carinata is grown as an edible oil crop in Ethiopia and Eritrea. Production is consumed locally with virtually none entering international trade.

SEED COMPOSITION AND PROPERTIES

An International Standard (1981) details specifications of cleanliness and composition of mustard seed species and standard analytical methods. The standard states that seed when ground must produce an odour free of mustiness and rancidity. The seed must have no more than 0.7% extraneous material and no more than 2% damaged or shrivelled seed. *Brassica nigra* and *B. juncea* must yield a minimum of 1.0% and 0.7% allyl isothiocyanate, respectively, and *S. alba* a minimum of 2.3% 4-hydroxybenzyl isothiocyanate.

The key difference between *B. juncea* and *S. alba*, which determines their use in condiment products, is the nature of the isothiocyanates (essential oils) derived from glucosinolates when crushed seed comes in contact with moisture. As dry seeds or dry ground products, the mustards are essentially odourless. The pungent isothiocyanates are released when the glucosinolates come in contact with the enzyme myrosinase also present in the seed. *Sinapis alba* contains predominantly 4-hydroxybenzyl glucosinolate (sinalbin), which on hydrolysis gives rise to 4-hydroxybenzyl isothiocyanate known as the 'white principle'. This non-volatile compound produces a 'heat' feeling in the mouth, its instability results in a sensation of sweetness and warmth. The condiment forms of *B. juncea* (and *B. nigra*) contain predominantly 2-propenyl (allyl) glucosinolate (sinigrin) which hydrolyses to yield allyl isothiocyanate. Known colloquially as 'volatile oil', this volatile compound gives a strong olfactory pungency, and is even lachrymatory.

The normal range of isothiocyanate released from the seed is 2.3–3.0% by weight for *S. alba* and 0.8–1.4% by weight for *B. juncea*. Much of the variation is under genetic control, but it is also strongly influenced by climatic effects, which in turn are associated with seed oil content. Oil content varies from 25 to 30% for *S. alba*, 32 to 40% for brown-seeded *B. juncea*, and 33 to 45% for the Oriental *B. juncea*.

As with other *Brassica* seeds, long days, low temperatures and low evaporation rates at higher latitudes result in seed with higher oil content and correspondingly lower glucosinolate content. Not only is the lower glucosinolate content undesirable in a condiment mustard, but the high oil content, especially at the levels found in *B. juncea*, is also undesirable as it makes the seed extremely difficult to mill by the dry-milling processes.

As with other oilseeds, fatty acid composition of the seed is under genetic control and thus manipulable through plant breeding or genetic engineering. However, these possibilities are not presently utilized as altered fatty acid composition has caused distinct deterioration of quality aspects of traditionally formulated mustard products.

A clear inverse relationship exists between oil content and seed protein, which is affected predominantly be environment (see Chapter 10). Protein varies between 27 and 32% for *S. alba* and between 23 and 29% for *B. juncea* (Canadian Grain Commission 1989, 1990, 1991a, 1992, 1993; Hemingway *et al.*, 1993b). Protein content does not have any particular significance in condiment formulations but is of particular importance for *S. alba* when the seed is used as an ingredient in prepared meats.

The seed coat constitutes a significant portion of *Brassica* seed and the proportion is thought to be under genetic control

(Hemingway, 1976). The normal range is 18–24% for *S. alba*, 20–25% for brown-seeded *B. juncea*, and 11–15% for Oriental *B. juncea*. Differences of this order of magnitude are of importance in the economics of the dry-milling process where the seed coat is removed from the product, and in wet milling of *B. juncea* where the seed coat is either left in or a portion screened out of the product. Seed shape and uniformity in size are important characteristics for maximizing seed coat separation. A tendency of the seed coats to break into inseparable small fragments which remain as visible specks can reduce the quality of a product. Variations in seed size of *B. juncea* have been reported together with their effects on yields and processing quality (Hemingway, 1995).

The seed coat of *B. juncea* contains only a small amount of mucilage while the seed coat of *S. alba* contains appreciably more. This is an important characteristic for the spice trade as the mucilage enhances the viscosity of condiment pastes made by the wet milling of whole *S. alba* seed. The mucilage also increases the moisture absorption of whole dry ground seed used as meat ingredients. Both genetics and the environment affect mucilage content and cultivars of *S. alba* have been bred for enhanced mucilage content (Woods and Downey, 1980; Hemingway *et al.*, 1993a, 1994b).

CROP PRODUCTION

Both *S. alba* and *B. juncea* are spring-sown annuals. They are easily and reliably grown on land of reasonable fertility and adequate moisture supply. Both are very responsive to supplements of nitrogenous fertilizer, but neither requires more than normal soil levels of phosphorus or potash. Whereas *B. juncea* has good resistance to periods of drought, *S. alba* flourishes best with higher precipitation and suffers from even modest drought stress. *Sinapis alba* also succumbs quickly to waterlogging of the soil. Both *S. alba* and *B. juncea* prefer neutral to slightly alkaline soil conditions but can tolerate pHs down to 5.6. They are both reasonably tolerant of spring frosts.

Brassica juncea crops can incur damage from the normal range of pests (see Chapter 6), but *S. alba* is immune to those which attack flowers and pods. Neither *S. alba* or *B. juncea* is affected by diseases to any economic extent in areas where they are grown as a source of seed for the spice trade. The necessity for condiment use of production of clean samples of seed requires effective weed control (see Chapter 4). Careful choice of fields is also essential to avoid contamination by seeds of specific problem weeds such as wild mustards (especially *Sinapis arvensis*), cow cockle (*Saponaria vaccaria*), cleavers (*Galium aparine*) and volunteers of other *Brassica* crops, which are all difficult to separate after harvest.

The incidence of internal infection of mustard seeds of both species by the yeast *Nematospora sinecauda* has been recorded over the last 20 years and is now a problem for dry flour processors. High plate counts contravene the specifications for the use of mustard flours as a food ingredient. The internal infection results from 'stigmato-mycosis', the incidental infection from feeding punctures of the false cinch bug, *Nysius niger*. No field control of this vector is possible, as so few insects are involved. However, local control of their alternate host flixweed (*Sisymbrium sophia*) is advantageous. Some millers eliminate the infection by heat-sterilization while others rely on the selection of crops found free of infection, identified in postharvest seed surveys. Peak infection rates were recorded in 1979, 1983 and 1988 in North American producing areas, probably attributable to climatic effects on vector populations.

Mustard provides an ideal break crop in a cereal rotation and utilizes the same equipment as for grain crops. Production may be fully mechanized, eliminating the need for hand-labour.

Mustard crops have prolific reproduction rates and can increase 2000-fold in good conditions. Mean seed yields of 900–1000 kg ha^{-1} are obtained in northern Europe. Expectations on the Canada prairies are lower, mean seed yield being in the order of 350 kg ha^{-1} for *S. alba* and 450 kg ha^{-1} for *B. juncea*, with substantial variation occurring due to differences in available soil moisture. The high northern European yields result from the long season from sowing in March to harvest in early September. Lower yields in Canada are due to a much shorter growing season, typically from mid-May to the end of August. Yields in Hungary are intermediate and vary depending on the presence or absence of summer stresses.

Shallow sowing into a fine, firm seedbed as soon as possible is recommended for optimal production. Seed rates vary from 2.5 kg ha^{-1} in the United Kingdom up to 10 kg ha^{-1} in Canada for *S. alba*, and from 1 kg ha^{-1} in the United Kingdom up to 8 kg ha^{-1} in Canada for *B. juncea*. Row spacings of 17–25 cm are common in Canada and Hungary while in the United Kingdom they range from 34 to 51 cm, which allow for tractor hoeing rather than herbicide weed control. As with rapeseed, seeding rates and row spacing can vary over a wide range without substantially affecting seed yield (see Chapter 3).

Mustard crops advance through four distinct phases of development: (a) seedling, lasting a week to 10 days in which the above-ground plant consists of the hypocotyl and two photosynthetic cotyledons; (b) vegetative, in which the plant develops a basal rosette of leaves over 3 to 6 weeks; (c) flowering, in which the plant bolts rapidly to produce an indeterminate central raceme with branch racemes opening four to five flowers per day over 3 to 4 weeks; and (d) seed filling, which lasts 4 to 8 weeks and is terminated by senescence of the racemes and pods.

Harvesting is invariably by direct combining in Europe, but it is a common practice to swathe the crop in Canada to promote dry-down. Under Canadian conditions care is usually taken to separate later-ripening areas of the field (e.g. around wet sloughs) as the presence of green seed in the sample is a serious factor causing downgrading, and even such small areas can produce enough green seed to downgrade large quantities of seed.

Considerable progress has been made since the 1970s in breeding high-yielding, high-quality cultivars of both *S. alba* and *B. juncea* suitable for fully mechanized production. Quality improvements include seed size, shape and uniformity, colour, and seed coat to kernel ratio, and glucosinolate, oil and seed coat mucilage content. A number of outstanding cultivars of *S. alba* and Oriental *B. juncea* are available, with less progress having been made to improving brown-seeded *B. juncea*. The principal breeding centres have been Norwich, in the United Kingdom, and Saskatoon, Canada, where research has been extended to include agronomic practices. Full descriptions of methods and materials for the husbandry of mustard seed crops for condiments manufacture have been published (Colmans, 1988; Mustard Association, 1989).

Commercial crops are greatly improved by the use of certified seed stocks, particularly in *S. alba*, a cross-pollinating species. Whilst the use of such seed is obligatory in northern Europe it is not in Canada or Hungary and astute processors and contracting companies are wise to specify cultivar.

POSTHARVEST HANDLING

In Canada seed is harvested below 10.5% moisture content, at which level it is safe for long-term storage. Indeed, when mustard seed is thoroughly cleaned and dried there is no risk of deterioration. In countries where weather conditions are variable at harvest seed may need to be gently dried to reduce the moisture to a safe storage level. In such circumstances, it is vital to act quickly as moist seed can deteriorate very rapidly. When drying mustard seed it is important to ensure that the seed temperature never exceeds 52°C or damage to endogenous enzymes may result which on processing will impair hydrolysis of the glucosinolate to the isothiocyante, the hot principle.

Cleaning mustard seed is straightforward, because of the round shape and uniform seed size of well-matured seed lots of good cultivars. Perforated screens, indented cylinders, ducted air and spiral separators are all effective for removing contaminants. Surface dust can be advantageously removed by the use of horizontal or inclined brush conveyors with adequate aspiration. Seed storage is usually in bulk bins of steel, wood or concrete construction. Seed may be conveyed by pneumatic or mechanical means with little damage. Most international trade is in large tonnage lots, but smaller quantities are loaded in standard freight containers, even in 45–50 kg bags if the customer so requires.

As Canada is the principal world source, most world trading is operated under the standard grading systems of the Canadian Grain Commission. There are two main grades and two subgrades, which are established by sampling and inspection. Seed is graded for general appearance (maturity, colour and odour), damaged seed (heating), distinctly green seed, conspicuous admixtures (earth pellets, stones, ergot) and other seeds (particularly inseparable cow cockle, wild mustard and rapeseed/canola) (Canadian Grain Commission, 1991b).

The application of pesticides to mustard crops in Canada is controlled under the provisions of the Pest Control Products Act, 1982, and none may be used which can give rise to dangerous residues. Monitoring of the grading system, residue levels and other aspects of quality are carried out by the Canadian Grain Research Laboratory. Similar systems for avoidance of seed residues apply in the other producing countries.

PROCESSING AND CONDIMENT PRODUCTS

Mustard seeds present three fundamental physical problems to the processor: their small size (*B. juncea* 1.63 mm and *S. alba* 2.22 mm mean diameter), appreciable seed coat content (up to 20% by weight) and substantial oil content (30–45% by weight). Both seed shape and uniformity in size are important characteristics for efficient separation of a seed coat fraction. The seed coat is relatively tough and fibrous in texture, does not contribute to flavour, and may impart a dull greyness or appear as dark specks in the product. Although the oil is confined to oil bodies in the cells of the seed, upon milling it is easily released and can then aggregate to produce a greasy texture.

Brassica juncea with its pungent olfactory flavour and *S. alba* with its buccal hotness are utilized either as distinct single flavours or complementarily combined to manufacture a wide range of condiment products.

Both dry processing and wet processing are carried out. Fine dry-milled flours of both *B. juncea* and *S. alba* are usually produced by reduction milling of the kernels after removal of the seed coat. In Japan, whole seed of *S. alba*, or the press cake of *B. juncea* remaining after partial oil removal, is often ground to a fine powder without prior removal of the seed coat. The wet-milling process varies depending upon the species used. With *S. alba* the seed coat is left in the product whereas with *B. juncea* a portion is removed by screening after wet milling. The amount removed influences the final colour of the resulting paste.

The diversity of mustard flavours and textures and the wide range of components included to enhance flavour or texture have led to many commercial mustard formulations. The largest published compilation is that of Man and Weir (1988). Many of the traditional styles have had national names affixed.

English mustard is a blend of dry-milled flours of both species. The flours may be incorporated into products in paste form with the volatile pungency maintained in the jar by the inclusion of fruit acids.

American 'cream-salad' mustard uses only *S. alba*. Whole seed is soaked to release the mucilage contained in the seed coat, then wet-milled with water in a colloid mill. This greatly increases the viscosity of the product. It is customary to include ground turmeric to produce a bright yellow product as *S. alba* seed imparts only a pale yellow to cream colour. Cream-salad mustards are usually consumed liberally, having a sweet to sharp flavour.

German mustards contain predominantly *S. alba*, but also contain small amounts of brown or Oriental *B. juncea* to add slight pungency. They have variable texture, dependent on the degree of wet-milling applied to the soaked whole seeds. Herbs, spices or vinegar is included according to the individual product.

French mustard, e.g. 'Dijon', contains only brown *B. juncea*. Indeed, in France only this type of seed is permitted under national legislation. The seed is washed and left to soak, then ground with vinegar, grape juice or wine, and fruit acids, salt and spices. Grinding is carried out in two stages. The seed is coarsely ground to break, but not fragment, the seed coats so they can be removed by wet-screening. The

remaining kernels, which wash over the screen, are finely ground to produce a paste. After storing in a vat for about a week, with occasional stirring to complete deaeration, the product is transferred to jars (Man and Weir, 1988).

In addition to the above nationally named mustards, a wide variety of mustards are produced utilizing textures as well as added ingredients. Of these the most traditional is the French 'Meaux', which has been purveyed for over two centuries. Starting from the soaking of brown *B. juncea* seed, vinegars, salt, sugars, honey and spices are variably added, then the mixture is milled through widely set carborundum stones to obtain a particulate or 'crunchy' texture. The mixture is then deaerated and allowed to thicken before placing in jars. Amongst other long-standing traditional styles, 'Bordeaux' is mild and sweet but very dark from the high proportion of seed coats incorporated.

The addition of whole, cracked or partly broken seeds to basic pastes, made either from wet-milling or formulated from dry powder, has become fashionable in recent years. Products range from near-cottage industry or 'local specialities,' of particular appeal to the tourist trade, to purportedly rather modern 'regional styles', and even to national brands marketed as 'whole grains'. Other formulations have become popular which include such ingredients as beer, wines, whisky, honey, lemon peel, garlic, horseradish, peppers, onion, tarragon, chives and other herbs. Indeed, it has been said that, in the United States, a new mustard product is being marketed every week! 'Russian-style', 'Chinese-style' and others may in fact have little national connection. Seasonal specialities include the very pungent paste used with fish dishes at Christmas in Scandinavia.

The highest level of technology is applied in the dry-milling of mustard seeds to produce flours (Duffus and Slaughter,

1980). The process is broadly similar to the roller-milling of wheat, with the main difference necessitated by the small seed sizes and higher oil content. *Sinapis alba* and either brown or Oriental *B. juncea* seed are milled separately, then the flours either used individually as 'straights' or blended in appropriate proportions for the balance of flavour required in the product. Deactivated wheat or rice flour may be added to achieve specified analytical levels of the essential principles (isothiocyanates).

Before dry-milling the seed is thoroughly cleaned, graded and spiralled to obtain uniformity of size and roundness of shape, and dried to less than 3% moisture. Weed seeds which are physically inseparable, such as *Sinapis arvensis* in *B. juncea* and *Galium aparine* and *Saponaria vaccaria* in *S. alba*, can cause problems as their seed coats become brittle during processing and fragment. Sizing enables precise setting of the gap between the mill rollers. The seed is dried to a low water content, then moistened immediately before milling to produce seed coats which are not brittle but slightly elastic. This obviates shattering during milling and consequent inclusion of small specks of seed coat in the product.

The seed is metered on to fluted 'break rollers', between which the seed is nipped to crack open the seed coats, ideally into two hemispheres. These are separated from the coarse particles of cotyledon and radicle by sieving and exposure to turbulent currents of air in 'purifier' machines. This equipment can achieve an extremely efficient separation of kernels and seed coat pieces. A series of passes of the kernel through sets of reduction mills with gradually closer roller settings follows. After each pass, the finer particles are sieved off and the coarser ones returned for further reduction. Sieving is performed using banks of frames covered with bolting cloth of different mesh sizes. Problems with greasiness can occur, particularly with the

higher oil content of *B. juncea*, if the rollers run at too high a temperature or are set too close so as to squeeze rather than pulverize the kernels. Flour rates fall if mills, conveyors and bolting cloth accumulate oil. A failure to control milling temperature by adequate aeration can also result in reduction of myrosinase enzyme activity, which is essential for subsequent release of the hot principle (isothiocyanates).

The resulting flours are fine and free-flowing, and do not normally 'cake' when stored in bulk or packaged form. After blending, the product colour will depend on the proportions of the richer yellow *B. juncea* flour and the light yellow–cream *S. alba*. In some blends the colour is enriched by the addition of curcumine or ground turmeric.

Interest in domestic food preparation and diversity of flavours and textures has led to many recipes to simulate commercial mustard formulations and others with a wide range of additions. The largest compilation is that of Man and Weir (1988).

USE OF MUSTARD AS A FOOD INGREDIENT

Many recipes have been compiled for mustard seed or mustard-containing sauces, pickles and chutneys used with meat and fish dishes, and even in breads, confectionery items and sweets (Man and Weir, 1988).

Of the two species, use of *S. alba* is predominant. However, both species have similar uses in whole, cracked or coarsely ground form, or as dry flours of varying grades of fineness, in commercial brands of sauces, pickles and chutneys, and in some prepared-meat products. They may both be used in whole or ground form or as flours in some spice mixes and blended curry powders.

Flour of *S. alba* with, on occasion, small proportions of *B. juncea*, mainly for colour enhancement, is a principal ingredient in the manufacture of salad dressings and mayonnaise. The flour has important antioxidant and worthwhile bacteriostatic properties in addition to providing some flavour and natural colour. However, its most effective function is as an emulsifying agent which greatly assists in the binding of the water and oil phases of the dressing. Traces of mucilage in the milled flour, resulting from residual seed coat fraction, also help to develop the viscosity of these products.

The seed coat fraction is a principal by-product of the dry milling process. Seed coats of *B. juncea* have very limited application, occasionally being used as a 'filler' after drying and grinding and, if desired, heat inactivation of the myrosinase enzyme. Alternatively, heat inactivation of myrosinase permits incorporation, in low levels, into animal feeds as a source of oil, protein and fibre. The seed coat fraction of *S. alba*, however, is rapidly increasing in value because of its high content of mucilage. The dried, ground or hammer-milled seed coat fraction is in great demand as a thickening agent, contributing viscosity to sauces and other liquid or paste products.

Small, misshapen and damaged mustard seeds removed in the preparation of uniform seed for dry milling are the other by-product. They are utilized in ground form as an ingredient in spice mixes, with a very limited distillation of allyl isothiocyanate from such discarded seeds of *B. juncea* as a flavour ingredient.

Another use made of *S. alba*, which is increasing, is as a minor ingredient in meat products, particularly in North America. The seed used is often of lower harvest grade. It is moistened and heat-treated using a precise temperature : time regime to inactivate myrosinase, then dried and coarsely ground or pin-milled. In the meat industry it is considered an added spice, contributing some background flavour and, from its seed coats, adding mucilage. The mucilage imparts hydrophilic activity which helps to bind more liquid into the product (thus lower cost) and also assists the emulsification of fats present. The presence of mucilage also aids in the clean moulding of compounded meats, e.g. 'skinless sausages'.

Studies are now in progress on the extraction and use of the mucilage of *S. alba* as an ingredient gum of temperate origin. It has unique physical and chemical properties which may find other industrial applications of considerable interest and value (Cui *et al.*, 1993). At an ingredient level of 3%, the 30% protein content of *S. alba* seed increases total protein of the product by almost 1%, and this is advantageous, particularly in countries where animal- and vegetable-derived proteins do not need separate declarations.

REFERENCES

Canadian Grain Commission (1989, 1990, 1991a, 1992, 1993) Annual reports on the quality of western Canadian mustard. Canadian Grain Commission Publications, Winnipeg, Canada.

Canadian Grain Commission (1991b) *Domestic Mustard Seed, Grain Grading Handbook for Western Canada.* Canadian Grain Commission Publications, Winnipeg, pp. 122–132.

Colmans (1985) *The Story of Mustard and How It is Milled.* Colmans of Norwich, Norwich, UK.

Colmans (1988) *Mustard Seed Crop Husbandry*, 5th edn. Colmans of Norwich, Norwich, UK.

Cui, W., Eskin, N.A.M. and Biliaderis, C.G. (1993) Chemical and physical properties of yellow mustard (*Sinapis alba* L.) mucilage. *Food Chemistry* 46, 169–176.

Duffus, C. and Slaughter, C. (1980) *Seeds and their Uses.* John Wiley & Sons, Chichester, pp. 29–31, 132–135.

Hemingway, J.S. (1976) Mustards. In: Simmonds N.W. (ed.) *Evolution of Crops and Plants*. Longman, London, pp. 56–59.

Hemingway, J.S. (1995) *On Seed Size in Mustards*. Report to the Canadian Mustard Association (in press).

Hemingway, J.S., Capcara, J. and Hyra, T. (1993a) *Variations in Seedcoat Mucilage Content of Canadian-grown Yellow Mustard Seed*. Report to the Canadian Mustard Association.

Hemingway, J.S., DeClerq, D.R., Daun, J.K. and Seguin, L. (1993b) *Crop to Crop Variations in Analyzed Quality Aspects of Canadian Mustard Seed*. Report to the Canadian Mustard Association.

Hemingway, J.S., Capcara, J. and Hyra, T. (1994a) *Variations in Seedcoat Mucilage Content of Canadian-grown Yellow Mustard Seed*. Report to the Canadian Mustard Association.

Hemingway, J.S., DeClerq, D.R., Daun, J.K. and Seguin, L. (1994b) *Crop to Crop Variations in Analyzed Quality Aspects of Canadian Mustard Seed*. Report to the Canadian Mustard Association.

International Standard (1981) *Mustard Seed Specification, I.S.O. 1237 - 1981(E)*, 2nd edn. International Organization for Standardization, Geneva, pp. 1–11.

Kirk, J.T.O. and Oram, R.N. (1981) Isolation of erucic acid-free lines of *Brassica juncea*: Indian mustard now a potential oilseed crop in Australia. *Journal of Australian Institute of Agricultural Science* 47, 51–52.

Kothari, S. (1992) Vegetable oils industry profile. *Industrial Directories of India* 18(1)–18(9).

Love, H.K., Rakow, G., Raney, J.P. and Downey, R.K. (1990) Development of low glucosinolate mustard. *Canadian Journal of Plant Science* 70, 419–424.

Man, R. and Weir, R. (1988) *The Compleat Mustard*. Constable, London.

Mustard Association (1989) *Mustard Growers' Manual*. Canadian Mustard Association, Winnipeg, Canada.

Oram, R.N. and Kirk, J.T.O. (1992) Breeding Indian mustard for Australian conditions. In: Hutchinson, K.J. and Vickery, P.J. (eds) *Proceedings of the Sixth Australian Agronomy Conference*. Australian Society of Agronomy, Armidale, New South Wales, pp. 467–470.

Rosengarten, F. (1969) *The Book of Spices*. Livingston Publishing Co., Wynnewood, Pennsylvania, pp. 297–305.

Tsunoda, S., Hinata, K. and Gomez-Campo, C. (eds) (1980) *Brassica Crops and Its Wild Allies. Biology and Breeding*. Japan Scientific Societies Press, Tokyo, Japan, pp. 141–145.

Vaughan, J.G. and Hemingway, J.S. (1959) The utilisation of mustards. *Economic Botany* 13(3), 196–204.

Vaughan, J.G., Hemingway, J.S. and Schofield, H.J. (1963) Contributions to a study of variation in *Brassica juncea* Coss. and Czern. *Journal of Linnean Society (Botany)* 58, 374, 435–447.

Woods, D.L. (1992) Comparative performance of mustard and canola in the Peace River Region. *Canadian Journal of Plant Science* 72, 829–830.

Woods, D.L. and Downey, R.K. (1980) Mucilage from yellow mustard seed. *Canadian Journal of Plant Science* 60, 1031–1033.

Woods, D.L., Capcara, J.J. and Downey, R.K. (1991) The potential of mustard (*Brassica juncea* (L.) Coss) as an edible oil crop on the Canadian prairies. *Canadian Journal of Plant Science* 71, 195–198.

Index

Acidity
    of oil 252–253, 356
    of soil 66, 95, 129
Agrobacterium 182
Agronomic practices 65–81, 94–95, 283,
    377–379
Albugo candida (white rust) 123–126, F124,
    F125, 158–159
Alkali refining 279, 280, 281, 282, 283–285,
    323
Allelopathy 198–199
Allergic reaction 205–209, 324–325
Alternaria (leaf and pod spot) T98, 118–122,
    F119, 131
Aluminium 85
Amino acids
    availability 227, 230, 302–303, 304,
        310–314, 318–319, 325
    composition 227, 228, T228, 303, T303
    degradation 271, 278, 312, 314
    essential 189, 273, 278, 302, 312, 325
    glucosinolate precursor 232
    protein quality 50, 225–227, 302,
        310–314
Amphidiploid 154, 177–178
Analytical methods – organizations 244,
    T245, T246
Animal growth 307, 317, 319, 320, 322, 325
Antioxidant 230, 279, 382
Antisense gene 166, 183, 185, 188, 219
Apetalous character 40, 41, 46, 53, 114, 160

Aphids 69, 145–146, F146, 150
Arabidopsis 181, 182, 185
Arachidic acid (eicosanoic acid) 49, T218,
    219, 239, 340
Argentine rape *see Brassica napus*
Artherosclerosis 292, 294–295
Assimilates (source/sink relations) 34, 41–46,
    47, 70, 72
Aster yellows 132
Axillary buds 33

Backcross breeding 162
Bacterial diseases 132–133
Bees 34, 204–205
Behenic acid T218, 219, 339, 340–348, T343,
    T344, T346, 349, T349
Biodiesel
    development 353–371
    economics of production 363–366
    energy balance 362–363, T358, F364,
        F365
    engine performance 358–359, T359
    environmental concerns 354, 359–362,
        T360, T362
    glycerol byproduct 355, 365
    markets 366–369, T367
    production 355, 363, T367–369
    properties 354, 355–358, T356
    sources 354–355, T357
    specifications 354, T356

Biological control of pests 144, 148, 150–151
Biotechnology 168–169, 177–190
Black leg (*Leptosphaeria maculans*) T98, 114–118, F115, F116, 158, 178, 373
Black rot (*Xanthomonas*) 132
Black spot (*Alternaria*) T98, 118–122, F119, 131
Bleaching 279, 282, 283, 285–286, 287, 354
Bleaching clay 283, 285, 286, 301, 315, 323
Boron 80, T80
Botrytis cinerea (grey mould) T98, 112
Brassica campestris see Brassica rapa
Brassica carinata 2, 3–4, 12, 54, 113, 374
Brassica hirta see Sinapis alba
Brassica juncea
    breeding 182, 219
    characteristics 2, 12, 54–55, 113, 114, 134, 373, 376–377
    condiment use 373, 379–382
    cultivation 2, 12, 66, 373, 375, 377–379
    origin/history 374–375
    processing 379–381
    seed composition 12, 235, 373, 376, 377
    world trade 374
Brassica napus
    breeding 219, 339
    characteristics 1, 11–12, 17, 54, 56, 219, 222
    cultivation 1–2, 12, 65
    origin/history 2–3
    seed composition 219, 227, 229, 231, 235, 254, 301, 306, 339
Brassica nigra 2, 3, 374
Brassica oleracea 2, 3, 4, 154, 156, 166, 178, 182, 221
Brassica rapa
    breeding 182, 219
    cultivation 1–2, 12, 65–66
    morphology/characteristics 1, 11–12, 17, 54
    origin/history 2–3
    sarson/toria 2
    seed composition 219, 227, 229, 231, 235, 254, 301, 306, 325
Brassica species
    characteristics 113, 118, 126
    cultivation 1–2, 4–5, 12, 377–379
    origin/history 2–3, F2, 4–5, 153–154, 374–375
    resynthesis 177
    world trade 5–6, T5, T6, 375–376
Brassica tournefortii 55, 153

Brassylic acid 348
Brown girdling root rot (*Rhizoctonia*) 131–132, 159, 184

Cabbage root fly (*Delia radicum*) 69
Cabbage stem flea beetle (*Psylliodes*) 69, 144, F145
Canola
    biodiesel 354, T357
    breeding 219
    composition 223, 228, 229, 230, 259, 296
    definition 234–235, 257
    erucic acid content 219, 234
    glucosinolate content 235, 237
    nutritional properties 219, 234, 291–297, 301–310
Carbohydrate
    analytical methods 259
    mucilage 259–260
    nutritional effects 308, 311, 312, 318, 319
    seed/meal content 27, 217, 227–230, 304–305
Cardiovascular disease 291, 292, 294–295
Carotenoids 217, 222, 223, 285, 288
Ceuthorhynchus (seed/stem weevil) 144–145, 148–149, F148
Cephalin (phosphatidylethanolamine) 221–222, 279–280
Certified seed 18, 66
Chemical hybridizing agents (CHAs) 167
Chemical mutagenesis 178–179
Chemical weed control *see* Herbicides
Chitin 184
Chlorophyll
    analytical methods 244, 245, 253–254
    content of oil 217, 222–225, 244, T279, T285, 285–288, T289, 355
    content of seed 51, 69, 217, 222, 223–235, T226, 244–245
Cholesterol 223, 291, 292–294, 295, 296, 297
Choline 221, 231, 260, 280, 304, 308–309
Chromosome analysis 2
Club root *see Plasmodiophora brassicae* 128–130, F128
Colour
    mustard paste 380, 381, 382
    oil 223, 279, T279, 283, 285, T285, 286, T286, 288, T289
    seed 230, 318, 378, 379
Combining 86

Conditioning *see* Processing, seed
Cooking *see* Processing, seed, conditioning
Cooking oil 1, 4–5, 185, 325, 373–383
Copper 80, T80
Coronary heart disease 291, 292, 294–295
Crop management
    rotation 67, 81, 94, 98, 100–101, 117, 129, 130, 195
    systems 87
Crop residues 67, 68, 116, 117, 120, 121, 123, 198
Cruciferin 182, 217, 227
Crude oil
    biodiesel, for 354–355, T357, 363
    characteristics 217, 223, T279
Culinary uses 1, 4–5, 185, 325, 373–383
Cultivar
    adaptability 65, 66
    identification by chemical analysis 260
    insect resistance 146
    market requirement 67, 219
    nitrogen response 75
    variety rights/registration 169
    winter hardiness 56
Cybrid 166
Cylindrosporium concentricum (light leaf spot) 122–123, F122
Cystine 227, 303, T303, 319, 325
Cytoplasmic male sterility (CMS) 164–165, 166, 184

Damping off of seedlings 131–132, F133
Dark leaf spot (*Alternaria*) 118–122, 131, F119
Dasineura spp. (pod midge) 148–149, 150
Daylength response 14, 17, 19, 20, 21, 22, 25–29
Deer 199–202, 209
Degree days 14, T18, 23
Degumming 225, 279, 280–283, 323, 354, 364
Dehulling *see* Testa, removal
Delia radicum (cabbage root fly) 69
Dendrograms 168
Denitrification 197
Deodorization 223, 279, 286, 288–289, 354
Dermatitis 223
Deroceras reticulatum (slugs) 68, 97
Desiccants 86, 97
Desolventing-toasting 235, 267, 278
Diacylglycerols 217, 218, 221, 222
Digestibility 229, 230, 258, 269, 273, 278,

301, 303, 304, 305, 308, 309, 310–314, 316, 318–327
Dihaploid breeding 58, 118, 162–163
Direct combining 86
Direct drilling 27, 68
Disease
    breeding for resistance 157–159, 178, 179–180, 183–184
    effect of weeds on 97
DNA finger-printing: RFLP 118, 130, 168–169, 179–182, 189
DNA tagging 182
Dockage 97, 322–323
Docosadienoic acid 252
Downy mildew *see Peronospora parasitica* T98, 126–128, F127, 184
Drilling *see* Sowing
Drought response 12, 19, 53–54, 69, 185, 373, 377

Edible oils – biotechnology 185
Egg taint 317
Eicosedienoic 252
Eicosenoic acid (arachidic acid) 49, T218, 219, 239, 340
Embryo rescue 178
Emergence: temperature requirements 16, 18
Energy
    meal content 269, 301, 302, 305, 308, 310–327
    oil content 291, 292, 293, 296, 301–302, 305, 310–314, 317, 318–319, 322, 323–325
    seed reserve 230
Environment
    biodiesel emissions 354, 359–361, T360, T365, 369
    nitrogen pollution 72, 195–197
    oil composition, effect on 50, 51, 223, 253
    wildlife 199–205, 209
Enzyme inactivation *see* Seed, processing
Epoxy fatty acids 188
Eruca sativa 153, 341–342, 346
Erucamide 341–342, 346
Erucic acid
    accumulation in seeds 49, T218, 219, 221, 235, 373
    analytical methods 251–252
    breeding/biotechnology 155–156, 181, 185, 219–221, 234, 235, 373

Erucic acid (*contd*)
  high oil content 219, 222, 235, 251
  industrial use 187, 219, 339–351, T343,
    T345, T347, T349, 357
  low oil content 219–221, 234, 251, 373
  nutritional effects 219, 291, 296, 319,
    323
  weed, effects on content 97
Erwinia carotovora (soft rot) 132
Erysiphe cruciferarum (powdery mildew) T98
Ethanolamine 221, 280
European brown hare syndrome 202
Extraction of oil *see* Processing, seed
Extrusion *see* Processing, seed 267, 274–275

Fallow 67, 101
Fatty acids
  accumulation in seed 49
  acidulated 283, 301, 315, 323
  analytical methods 251–253
  biodiesel 353, 354, 355, 357, 364, 366
  breeding/biotechnology 179, 180–181,
    183, 184–185, 219, 376
  cultivar identification 260
  essential 219, 291
  free 217, 218, 221, 251, 252–253, 279,
    T279, 280, 283, 285, T286, T289,
    323, 355, 363, 364
  frost, effect on 50
  hydrogenation 219, 286–287
  interesterification 287–288
  long-chain 217, 219, 339–351
  nomenclature 218–219, T218
  nutrition 291–297, F297
  oil composition 49–50, 155–156, 185–188,
    217–222, T220, T221, T222, 279
  pod position, effect on 50
  short-chain 219, 251
  synthesis 181, 185, F186
Fertilization, pollen 34–35
Fertilizer requirements 71–87, 195–197, 209
Fibre
  analytical methods 229–230, 258–259
  meal content 217, 229–230, 305
  nutrition 301–302, 304–305, 309, 312,
    313, 319, 322
  reduction 268–270, 323, 382
  seed content 157, 217, 227–228
  yellow seed 157, 305, 318, 323
Filtration of oil 275–276
Fines 275–276

Fishy egg syndrome 308–309, 319
Flaking 268, 270–271, F270, 323
Flavour 223, 235, 255, 279, 283, 288, 308,
    319, 322, 325, 374, 379, 380, 381, 382
Flea beetle (*Phyllotreta*) 143–144, F145
Flowering
  factors affecting F19, 20, F21, 23, 25–27,
    F25, F33, 53
  flower number 48
  initiation and development 18, 19, F20,
    F21, 22, 23, 27, 33–34, 36, F37, 40,
    44–45, F45
  time of 21, 27, 33–34, 178, 185
  yield potential, in relation to 19
Foots 275
Forage – nutritional value 324
Free fatty acids 217, 218, 221, 251, 252–253,
    279, T279, 280, 283, 285, T286,
    T289, 323, 355, 363, 364
Frost 19, 49, 51, 86, 223, 312, 321, 322, 377
Fungicides 114, 117, 135
Fusarium 131

Gene
  amplification 183
  identification/isolation 180–182
  inhibition 183
  manipulation 177–178
Genetically modified organisms (GMOs)
    159–160, 166, 168–169, 184, 219
Genetic engineering 168–169, 177–190
Genetic fingerprinting: RFLP 118, 130,
    168–169, 179–182, 189
Genotype/environment interaction 66, 219,
    221, 225, 226, 235, 302, 303, 304,
    312, 376, 377
Genotyping 180
Germination 1, 17–18, 66, 227
Globulins 225–227, 303
Glucosinolates
  analytical methods 235, 244, 255–258
  breeding/biotechnology 156–157, 178,
    180, 183, 188–189, 235, 373
  cultivar identification 260
  distribution in plants 50–51, 232
  environment 76, T79, T85, 199, 203,
    235, 376–377
  hydrolysis products 203, 233–237, T233,
    249, 306, 316, 340, 376
  nutrition 217, 234, 279, 301, 304,
    305–308, 312, 314, 318–327, 373

pest/disease reaction 141, 189
processing 235, 271, 273, 278, 379
seed content 50–51, T79, T85, 235,
    T236, 301, 306–307, 373–374, 376,
    378
structure 232–233, T232
Glycerol byproduct 340, 347, 348, 349, 355,
    357
Glycolipids 217, 221–222
Grey mould *see Botrytis cinerea* T98, 112
Growth
    cycle 65, 66
    differentiation 13
    environmental constraints 11, 17, 52, 69
    regulators 44–45, 56–57, T81, 81–82,
        T82
    stages 13, T13, T15–16, 69, 75, 80–82
    temperature response 17, 69
Gums *see* Phospholipids

Half-seed technique 155, 252
Hares 202–204, 209
Harvest 85–87, 95–96, 245–246, 253, 374,
    378
Height, stem 160
Herbicides 67, 68, 69, 98, 99, 101–108,
    159–160, 184, 378
Heritability 154
Hexane 218, 249, 276–278
Hormones 46
Hull *see* Testa
Hybrids 34, 82, 164–167, F165, 178, 184
Hydrogenation 219, 223, 279, 285, 286–287,
    323, 340
Hypersensitivity 206–208

Ideotype 57, 161
Induced variation 178
Industrial oils 185–189, 218, 219, 235, 251,
    339–352, 353–371, 373, 382
Infant formula 296
Inflorescence development 14, 18, 19, F20,
    22, 23, 27, F33, F37, F45
Insecticides 141, 143, 144, 145, 146, 147,
    148, 149, 151, 204–205
Insect pests 141–151, T142, T143, 184
Interesterification 279, 286, 287–288
Intergeneric hybrids 177, 178
Iron 80, T80, 279, 283, 304, 309, 316, 325
Irrigation 32, 49, 53, T84–85, 82–85

Isothiocyanates 198, 233–234, 255–258, 271,
    306, 314, 379, 381, 382
Isozyme marker 179

Lauric acid 156, 187, 219
Leaf
    appearance rate 18, 21–22
    area 19, 23, F26, 27–29, 30, 36, 40, 48,
        55, 184
    growth 18, 19, 20–21, F26, F29, 28–29,
        F44, F45, 53, 55
    initiation rate 18, 20
Lecithin (phosphatidylcholine) 221–222,
    279–280
Lepidoptera 146–147
Leptosphaeria maculans (stem canker, black leg,
    Phoma lingam) T98, 114–118, F115,
    F116, 158, 178, 373
Light leaf spot (*Pyrenopeziza brassicae*) 122–123,
    F122
Lignin 217, 229, 258, 259, 301, 308, 318, 323
Liming 129–130
Linkage and linkage maps 180
Linoleic acid
    accumulation in seed 49, 185
    analytical methods 251–253
    breeding 179, 181, 219
    edible oil composition T218, 219, 222
    industrial oil composition 349
    nutrition 219–299
Linolenic acid
    accumulation in seed 49, 185
    analytical methods 251–252
    breeding 179, 181, 219
    edible oil composition T218, 219, 221,
        222, 292
    industrial oil composition 339, 349, 354,
        357
    nutrition 294–299, 320
Lipoprotein 292–296, 297
Lipoxygenase 260
Liver haemorrhage syndrome 234, 308, 317
Lodging 46, 66, 69, 70, 160
Lubricants 5, 353, 354, 355, 364, 366
Lysine 189, 227, 273, 278, 302, 303, T303,
    304, 313, 314, 316, 319, 322, 324,
    325, 326

Magnesium 79–80, T80, 223, 225, 230, 283,
    304, 309

Male sterility 164–167

Manganese 80, T80, 304, 309

Margarine 185, 223, 286, 287, 288, 289, 293, 294

Mass selection 161

Maturity, plant T17, T18, 25–26, T26, 31–40, 46–48, 160, 312, 379

Mayonnaise 293, 382

Meal *see* Rapeseed meal

Meiosis 19, 80

Meligethes (pollen beetle) 147–148, F147

Methionine 189, 227, 235, 302, 303, T303, 313, 316, 319, 325

Microprojectile bombardment 182

Microspore
    culture 163, 179
    release 19

Midge, pod (*Dasineura* spp.) 148–149, 150

Minor elements 80–81

Miridae 149–150

Moisture content of seed 235, 245–248, 250, 377, 378, 379

Molecular farming 189–190

Molecular marker 178, 179–180

Molybdenum 80, T80

Monoacyglycerols 217, 221

Mucilage 12, 259–260, 377, 378, 380, 382

Mustard
    animal nutrition 322, 325–326
    condiment 235, 375–376, 379–382
    glucosinolates 235, 255, 256, 305, 322, 373, 374, 376, 378, 379–381
    origin/history 374–375
    oil source 235, 301
    seed composition 228–229, 230, 231, 234, 249–253, 254, 257, 259, 260, 376–377
    Sinapis arvensis (charlock/wild mustard) 237, 322
    species 2, 373–374

Mustard oil *see* Glucosinolate, hydrolysis products

Mutagenesis, chemical 178–179

Mycosphaerella capsellae (white leaf spot) 123

Myristic acid: biotechnology 219

Myrosinase 233, 255, 256, 257, 260, 271, 272, 273, 278, 306, 307, 314, 316, 324, 325, 376 379, 382

Napin 182, 183, 217, 227

Near infra-red spectroscopy 244, 248, 250–251, 252, 253–254, 255, 257, 259

Nematicidal activity 199

Net assimilation rate 30, 52

Nitriles 203, T233, 234, 271, 306, 307, 314, 316

Nitrogen
    application 72, 75
    crop response 27, 36, 55, 71–72, T73, T79, 195–197
    environment 72, 195–197
    seed quality 49, 55, 76
    soil 73–74, 195, T196

Nitrous oxide release 197

Nuclear magnetic resonance (NMR) 244, 250

Nutrition
    acidulated fatty acids 301, 315, 323
    energy 310–314
    fatty acid composition 218, 219, 291–300
    fish 318
    gums *see* Phospholipids
    hulls *see* Testa
    meal 301–337
    mustard 325–326
    oil 291–300
    pigs 315–316
    poultry 316–318
    rapeseed as forage 324–325
    ruminants 314–315
    screenings, dockage, damaged seed 322–323
    spent clay 323
    whole seed 320–322

Nylon 346–349, F348

Odour 279, 283, 288–289, 317, 376, 379

Ogura CMS 166

Oil
    acidity 252–253, 356
    analytical methods 249–251
    bodies 190, 225, 227, 271
    breeding 155–156, 157, 179, 181, 185–188
    colour 223, 230, 279, 283, 285, T285, 286, T286, 288, T289
    composition 217–225
    content of seed F31, F35, 36, 53, 49, 65, 76, 77, 80, 83, T84, 121, 178, 181, 217–225, 249, 273, 376
    iodine number 287, 353, 357
    processing 223, 225, 235, 254, 257, 267–290, 302, 307, 308, 312, 314, 316, 319, 321

viscosity 271, 276
  Oleic acid 49, 181, 182, 183, 185, 187,
    T218, 221, 252, T286, 292, 294, 295,
    296, 297, 339, 340, 341, 349, 354
Oleosin 182, 217, T218, 225, 227
Osmotic adjustment 54
Outcrossing 34
Oxalic acid 113, 114
Oxazolidinethiones T233, 234, 256, 271, 306,
    314
Oxidation 223, 285, 288, 295, 296, 297, 353,
    361

Palatability 260, 291, 308-309, 315, 319,
    324
Palmitic acid 49, 185, T218, 219, 287, 294,
    296, 320, 340
Palmitoleic acid T218, 219, 320
Patent law 169
Pectins 217, 228, 229
Pedigree breeding 161-162
Pelargonic acid 348
Pentosans 229, 258, 301
Peronospora parasitica (downy mildew) T98,
    126-128, F127, 184
Peroxide value T286, T289
Pests
    biological control 144, 148, 150-151
    effect of weeds on 97
    resistance to 150, 184
Petroselenic acid 187-188
Phenolics
    analytical methods 260
    nutrition 217, 279, 308-309, 318
    seed/meal content 230-232, T231
    sinapine 231, 260, 304, 308-309, 317,
      318, 319
Pheromones 149
Phoma lingam (Leptosphaeria maculans) T98,
    114-118, F115, F116, 158, 178, 373
Phosphates 55, 76-77
Phosphatidylcholine (cephalin) 221-222,
    279-280
Phosphatidylethanolamine (lecithin) 221-222,
    279-280
Phosphatidylinositol 221-222, 279-280
Phospholipids 217, 218, 221-222, 279-283,
    285, 294, 295, 323, 364
Phosphorus 230, T279, 283, T285, T286,
    354, 357, 364, 377
Photoperiod 19, 21, F21, 23, F24, 25

Photosynthesis 19, 29, 40, 41, 43, F43, F44,
    47, 69
Phyllotreta (flea beetle) 143-144, F145
Phyllochron (leaf appearance rate) 18, 21-22
Physical refining 279, 280, 283
Physiological age 22
Phytate 217, 230, 260, 279, 309-310, 318, 319
Phytophthora megasperma (seedling blight) 131
Phytosphingosine 280
Pigments 223-225, 254, 269, 279, 285, 288,
    322
Plant development
    flowering to maturity F24, F25, 25-26,
      F32, 30-51, 83
    growth stages T13, T15-16, 69, 75, 80-82
    sowing to flowering 14, 16, T18, T19,
      F20, F21, 22-23, F26, F33, 27-30,
      69, 73, T84, 85
Plant population 27, 34, 52, 69-70, 75
Plant growth regulators 44-45, 56-57
Plant variety rights 169
Plasmodiophora brassicae (club root) 128-130,
    F128
Plastochron = leaf, initiation rate 18, 20
Ploidy 154
Ploughing 100
Pod
    abortion 36
    dehiscence 12, 36, F38, 58, 160-161,
      373, 375
    development F32, 34, F37, F38, 36-39,
      F44, 46-48, 51, 83
    loss F35, 40
    number per plant F31, 30-33, 36-37, 41,
      45, 53
    wall photosynthesis 36, 41, 43, 47, 48, 55
Pod midge (Dasineura) 148-149, 150
Polima CMS 166
Polish rape see Brassica rapa
Pollen beetle (Meligethes) 147-148, F147
Pollen culture 163, 179
Pollen
    pesticide contamination 205
    respiratory problems due to 205-208,
      324-325
    type 34
Pollination 34
Polyhydroxybutyrate (PHB) 188
Polymerase chain reaction (PCR) 180
Polysaccharides 217, 227-229, 259, 305, 319
Potassium 77
Prepressing see Processing, seed

Primordia 18, 19
Principal component analysis 168
Processing, seed
    conditioning 271-273, F272
    dehulling 268-270
    desolventizing-toasting 235, 267, 278-279
    extraction 225, 249-250, 273-279
    extrusion 267, 274-275
    flaking 268, 270-271, F270, 323
    mechanical extraction 273-276, F274
    oil settling and filtering 275-276
    prepressing 273-274
    pretreatment 267-273, F269
    solvent extraction 267-279, F277
    solvent recovery 278-279, F277
    tempering 268
Processing, oil
    bleaching 279, 282, 283, 285-286, 287,
        354
    degumming 225, 279, 280-283, F280,
        F281, 323, 354, 364
    deodorization 288-289, T288
    hydrogenation 219, 223, 279, 285,
        286-287, 323, 340
    interesterification 279, 286, 287-288
    refining 279-285, F284
    winterization 286
Production, world 5-6
Protein
    analytical methods 225, 250, 254-255,
        257, 260, 303
    biotechnology 183
    cultural effects 55, 76, 77, T84
    nutrition 230, 231, 235, 301-337, 382
    processing effects 267, 268, 270, 271,
        273, 275, 276, 278
    seed/meal content 49, 50, 217, 225-227,
        235, 254-255, 301-302, 376
Protoplast fusion 178, 182
Pseudocercosporella capsellae (white leaf spot) 123,
    T98
Psylliodes chrysocephala (cabbage stem flea
    beetle) 69, 144, F145
Pyrenopeziza brassicae (light leaf spot) 122-123,
    F122
Pythium spp. (seedling blight) 131

Quality
    breeding for 155-157
    cultural effects 70, 76, 77, 79, 80, 97,
        321, 322

trait loci (QTLs) 180, 185

Radiation response 18, 19, 23, 28-29, 30, 36,
    40-43, F41, F42, 46, 47-48, 52
Raffinose 228-229
Rancidity 223, 376
Random amplified polymorphic DNA marker
    (RAPD) 180
Rapeseed meal
    biotechnology 188-189
    composition 228-229, 302-310, T302,
        T303, T304, T305, T306, T307, 323
    energy 310-314, T311
    utilization as animal feed 225, 314-318,
        324
    utilization as fertilizer 314
Recurrent selection 126, 162
Refining see Processing, oil
Registration of cultivars 169
Restriction fragment length polymorphism
    (RFLP) 118, 130, 168-169, 179-182,
    189
Rhizoctonia solani (root rot) 131-132, 159, 184
Ricinoleic acid 187
Root
    growth F26, 27, 70, F45, 83
    reserves 27, 47
    rot (Rhizoctonia) 131-132, 184
    tuberized 54, 83
Rosette stage of plant development 27, 29
Rotation of crops 67, 81, 94, 98, 100, 117,
    129, 130, 195
Row width 70

Salad dressing/oil 185, 286, 382
Saponins 310
Sarson, brown/yellow 2
Sclerotinia sclerotiorum (stem rot) T98, 111-114,
    F112, 158, 184, 322-323
Screenings 301, 322-323
Seed
    abortion 36, 39, 40, 44, 45
    analytical methods 225, 243-265,
        T247-248, 303
    certification 18, 66
    cleaning, drying and storage 86-87, 97
    composition F31, 217-242
    conditioning see seed processing 271-273,
        F272
    cultural effects on 55, 198

development 31–33, 35, F35, 44, 47, 48–51, 83
dressings 117, 121, 126, 128, 132
number per pod F31, F32, 32–33, 34, F38, 36–39, 40, 41, F42, 46, 48, 53, 83
processing 267–279
quality 18, 28, 48, 66, T66, 117, 120–121, 244–245, 322, 376
seed production F165
size 18, 31–33, 39, 40, 46, 47, 48
Seed coat *see* Testa
Seed drills 71
Seed weevil (*Ceuthorhynchus*) 148–149, F148
Seedling development 131–132, F133
Selection for breeding 126, 161, 162
Self fertility 34, 154
Self incompatibility 166, 178, 184
Shattering/shedding 12, 36, F38, 58, 160–161
Shortening 223, 286, 287, 288, 289
Sinapine 231, 260, 304, 308–309, 317, 318, 319
Sinapis alba (*Brassica hirta*; white/yellow mustard)
animal nutrition 325–326
characteristics 2, 373–374
condiment use 373, 379–382
cultivation 373–374, 377–379
food ingredient 373, 381–382
origin/history 2, 374–375
processing 325, 379–381
seed composition 219, 223, 235, 256, 257, 259, 325, 376–377
world trade 235, 375–376
Sinapis arvensis (charlock/wild mustard) 93, 94, 95, 96, 97, 106, 237, 322, 377, 379, 381
Slugs (*Deroceras reticulatum*) 68, 97
S-methyl cysteine sulphoxide (SMCO) 199, 324
Soapstock 282, 283, 285, 323
Soft rot (*Erwinia carotovora*) 132
Soil
acidity 66, 77, 95, 129
moisture 68, 70, 73, 77, 129
nitrogen 196, T196
preparation/requirements 66–67, 68, 77, 95, 99–100
salinity 12
temperature 16, 17, 68, 69, 70
water 68, 70, 73, 77, 129, 196, F196
weed flora, effect on 95
Solar radiation *see* Radiation response

Solvent extraction F268, 271, 272, 274, 275, 276–279, F277, 282, 306, 324
Somaclonal variation 179, 276–278
Source/sink relations 34, 41–46, 47, 70, 72
Sowing
broadcasting 67, 70–71
drilling 68, 70, 71
seed rate 52, 69, 378
timing T19, F24, F26, F33, 37–39, 40, 41, 46, 52, 68–70, 378
weed control at 99
Stachyose 228, 229, 259
Stagheads (white rust; *Albugo candida*) 123–126, F124, F125, 158–159
Starch 217, 228, 229, 230, 258, 259, 305, 308, 309, 312, 314
Stearic acid 49, 183, 185, T218, 219, 287, 320, 340
Stem
extension 14, 23, F24, 27, 28, 30, 36, 70, 96
height 160
nodes 21–22, F21
photosynthesis F44, F45, 45
Stem canker (*Leptosphaeria maculans*) T98, 114–118, F115, F116, 158, 178, 373
Stem rot (*Sclerotinia sclerotiorum*) 111–114, F112, 158, 184, 322
Stem weevil (*Ceuthorhynchus*) 144–145
Sterols 217, 222–223, T224, 279, 292–297
Stomata frequency T39, 43, 121–122
Stress – nutrient/water 36, 46, 47, 51, 312, 377, 378
Sucrose 228–229, 305
Sulphur
amino acids 227, 302, 323
crop response 77–79, T79, 197–198
deodorized oil 288–289
deposition 198
fuel emissions 353, 357, 359–361
glucosinolate content of seed 51, 76, T79, 235, 256
oil content 271, T279, T286, T289
Swathing 86
Swede rape *see* Brassica napus
Synthetic breeding 161, 163–164

Tannins 231–232, 308, 318, 319
Taxonomy 1–2
Temperature: response 1, 12, 16, T17, 19, 20, 21, F21, 22, 23, F24, 25, F25,

28–30, 39–40, 48, 49, 50, 51, 198
Terpenes 209
Testa
    composition 228, 229, 230, 232, 249,
        258, 279, 301–302, 303, 308, 318,
        323–324, 325, 379
    nutritional value 303–305, 308, 318, 319,
        323–324, 325
    proportion of seed 301, 374, 376–378,
        379
    removal 229–230, 267, 268–270, 271,
        379–382
Thermal development rate 25, F28
Thermal time 14, 15–16, T19
Thiocyanates 198–199, 203, 234, 306, 316,
    320
Thrombosis/thrombogenesis 291, 292,
    294–295
Tissue culture 178, 179, 180, 182
Tocopherols 223, T225
Toria see Brassica rapa
Toxicity 198, 199–204, 271, 306, 307, 314,
    324, 353, 361–362, 368
Transformation, plant 180–182
Transgenics 159–160, 168–169, 184, 185, 187
Triacylglycerol 49, 156, 185, 187, 217,
    218–221, T222, 249
Trypsin inhibitors 184, 310
Tryptophan 235, T303, 319
Turnip rape see Brassica rapa

Variety see Cultivar
Vegetative stage of plant development 23
Vernalization 14, 19, 20, 21, T21, 22, 23,
    T24, 25, 65, 66, T98
Verticillium dahliae (verticillium wilt) 130–31,
    F131, F132
Virus diseases 133–134, 159

Water
    crop response 51, 54, 129
    efficiency of use 54–55

extraction from soil 27
nitrate leaching 195, 196, T196
stress 19, 36, 48, 49, 52, 53, 54, 70, 76,
    83, T84–85
Waxes 188, 279, 286, 374
Wax, leaf surface 105, 123
Weeds
    common flora 54, 93–94, 234
    control 67, 93–109, 378
    determination of admixtures 244–245
    diseases: influence on 117, 130
    glucosinolate content 234, 235–237, 307
    in mustard 377, 378, 381
    in rapeseed 67, 68, 95–98, 234, 235–237,
        244, 307, 322
    seed survival in soil 94, 100, 102, 103
Weevil see Ceuthorhynchus 144–145, 148–149
White leaf spot (Pseudocercosporella capsellae)
    T98, 123
White mustard see Sinapis alba
White rust (Albugo candida) 123–126, F124,
    F125, 158–159
Whole seed, as animal feed 301, 320–322
Wildlife: effect on 199–205, T200, T201,
    T202, T203
Winter cultivars, spring-sown 22
Winter hardiness 56, 69, 159
Winterization see Processing, oil

Xanthomonas (black rot, root rot) 132

Yellow mustard see Sinapis alba
Yellow seeded rapeseed 1, 2, 157, 259, 305,
    308, 318, 319, 323
Yield
    breeding for 161
    components 30–33, F31, 40, T84, 161
    potential 19, 23

Zinc 55, 80, T80, 260, 309, 319